Plant Biotechnology 2002 and Beyond

Plant Biotechnology 2002 and Beyond

Proceedings of the 10th IAPTC&B Congress June 23-28, 2002 Orlando, Florida, U.S.A.

Edited by

Indra K. Vasil

Graduate Research Professor Emeritus,
University of Florida,
President, IAPTC&B

KLUWER ACADEMIC PUBLISHERS

DORDRECHT / BOSTON / LONDON

A C.I.P. Catalogue record for this book is available from the Library of Congress.

ISBN 1-4020-1126-1

Published by Kluwer Academic Publishers,
P.O. Box 17, 3300 AA Dordrecht, The Netherlands.

Sold and distributed in North, Central and South America
by Kluwer Academic Publishers,
101 Philip Drive, Norwell, MA 02061, U.S.A.

In all other countries, sold and distributed
by Kluwer Academic Publishers,
P.O. Box 322, 3300 AH Dordrecht, The Netherlands.

Printed on acid-free paper

Printed in the Netherlands.

Contents

OPENING ADDRESS BY THE PRESIDENT OF THE IAPTC&B

PLENARY LECTURES

---

* Presenting Author

SYMPOSIA

1. Biotic Resistance
Keynote Lectures

Contributed Papers

2. Abiotic Resistance

Keynote Lectures

Contributed Papers

3. Genomics

Keynote Lectures

4. Gene Expression/Silencing/Targeting
Keynote Lectures

Contributed Papers

5. Cell Cycle and Cell Division

Keynote Lectures

Contributed Papers

6. Embryogenesis/Regeneration

Keynote Lectures

Contributed Papers

7. Fruit and Seed Biotechnology
Keynote Lectures

Contributed papers

8. Flower Biotechnology
Keynote Lectures

9. Space Biotechnology
Keynote Lectures

10. Biopharming
Keynote Lectures

Contributed Papers

11. Improvement of Nutritional Quality

Keynote Lectures

Contributed Papers

12. Forest Biotechnology
Keynote Lectures

Contributed Papers

13. Phytoremediation/Phytochemicals
Keynote Lectures

Contributed Papers

14. Biotechnology of Turf and Forage Crops

Keynote Lectures

15. Protoplast/Anther/Embryo Culture

Keynote Lectures

Contributed Papers

16. Biotechnology Regulation, Public Policy and Societal Acceptance

Keynote Lectures

17. Biotechnology in Developing Countries
Keynote Lectures

PREFACE

The 10th IAPTC&B Congress, Plant Biotechnology 2002 and Beyond, was held June 23-28, 2002, at Disney's Coronado Springs Resort, in Orlando, Florida, USA. It was attended by 1,176 scientists from 54 countries. The best and brightest stars of international plant biotechnology headlined the scientific program. It included the opening address by the President of the IAPTC&B, 14 plenary lectures, and 111 keynote lectures and contributed papers presented in 17 symposia covering all aspects of plant biotechnology. More than 500 posters supplemented the formal program. The distinguished speakers described, discussed and debated not only the best of science that has been done or is being done, but also how the power of plant biotechnology can be harnessed to meet future challenges and needs. The program was focused on what is new and what is exciting, what is state of the art, and what is on the cutting edge of science and technology. In keeping with the international mandate of the IAPTC&B, 73 of the 125 speakers were from outside the United States, representing 27 countries from every region of the world. The 10th IAPTC&B Congress was a truly world-class event.

The IAPTC&B, founded in 1963 at the first international conference of plant tissue culture organized by Philip White in the United States, currently has over 1,500 members in 85 countries. It is the largest, oldest, and the most comprehensive international professional organization in the field of plant biotechnology. The IAPTC&B has served the plant biotechnology community well through its many active national chapters throughout the World, by maintaining and disseminating a membership list and a website, by the publication of an official journal (formerly the Newsletter), and by organizing quadrennial international congresses in France (1970), the United Kingdom (1974), Canada (1978), Japan (1982), the United States (1963, 1986, 2002), The Netherlands (1990), Italy (1994), and Israel (1998). In addition, the IAPTC&B has a long tradition of publishing the proceedings of its congresses. Individually, these volumes have provided authoritative quadrennial reports of the status of international plant biotechnology. Collectively, they document the history of plant biotechnology during the 20th century. They are indeed a valuable resource.

We are pleased to continue this tradition by publishing this proceedings volume of the 10th IAPTC&B Congress. I regret that we are not able to publish seven of the lectures in full (only their abstracts are included), as we did not receive manuscripts from their authors in spite of their formal agreement, prior to the Congress and subsequent personal assurances, to provide manuscripts for this volume.

The American and Canadian chapters of the IAPTC&B, the Plant Section of the Society for In Vitro Biology, and the University of Florida hosted the

10th IAPTC&B Congress. The Congress was a true partnership between academia and industry, and was generously supported by both groups (see list of donors/sponsors on back cover). A number of prominent international biotechnology companies and publishers participated in the very successful Science and Technology Exhibit (see accompanying list of exhibitors) The IAPTC&B awarded 84 fellowships to young scientists from 31 countries (see accompanying list of fellowship recipients) to support their participation in the Congress.

The 10th IAPTC&B Congress was indeed a celebration and a showcase of the many achievements of plant biotechnology. It was held in a joyous and celebratory environment, especially during the daily breakfasts, and with accompanying music during lunches, many receptions and the grand banquet.

I thank Tami Spurlin for preparing the camera-ready manuscripts in a uniform style, and Jacco Flipsen of Kluwer Academic Publishers for his assistance and cooperation in the publication of this volume.

Indra K. Vasil

Dedication

With love, to
Vimla, Kavita, Charu and Ryan

Seated (from left): Indra K. Vasil (President), Zhi-Hong Xu (President-Elect), Kavita Vasil. Standing (from left): Kang Chong, Trevor Thorpe (Editor), Dan Brown (Secretary-Treasurer), Mary Brown, Yvonne Thorpe, Vimla Vasil, Yongbiao Xue

Top. IAPTC&B President Indra K. Vasil announcing the election of Zhi-Hong Xu of China as President of the IAPTC&B for 2002-2006, and of Beijing as the host city of the 11th IAPTC&B Congress in 2006. *Bottom.* Zhi-Hong Xu, President of the IAPTC&B (2002-2006).

The logo of the 10th IAPTC&B Congress was designed by Indra K. Vasil and created by Tami Spurlin. The image of the golden spikes of wheat is from the first transgenic wheat plants harvested in 1992 *(V. Vasil et al. 1992. Bio/Technology 10:667-674).*

10th IAPTC&B Congress Donors/Sponsors

The IAPTC&B gratefully acknowledges the support of its many donors and sponsors.

Diamond Level

University of Florida
Genetics Institute
Institute of Food and Agricultural Sciences
Interdisciplinary Centre for Biotechnology Research
Office of the Provost
Office of Research, Technology, and Graduate Education

Platinum Level

Danone Group (France)
Monsanto
Pioneer Hi-Bred
Plant Section-Society for In Vitro Biology
ProfiGen
Syngenta

Gold Level

Canadian IAPTC&B Chapter
Epicyte Pharmaceutical, Inc.
Exelixis Plant Sciences
National Aeronautics & Space Administration (NASA) (USA)
Plant Biotechnology Institute (Canada)
The Scotts Company
USA IAPTC&B Chapter

Silver Level

ArborGen, LLC
Epcot® The Land at Walt Disney World® Resort
RiceTec, Inc.
UNESCO, Paris, France

Bronze Level

Council for Biotechnology Information
Kluwer Academic Publishers
Nestlé (Switzerland)
Paradigm Genetics
Phytomedics
ProdiGene
Weyerhaeuser

The IAPTC&B acknowledges with thanks the publishers of the following international journals for complimentary full page announcements of the 10th IAPTC&B Congress:

- Electronic Joumal of Biotechnology
- Euphytica
- Molecular Breeding
- Nature Biotechnology
- Plant Cell Tissue and Organ Culture
- Plant Molecular Biology
- Society for Experimental Biology
- The Plant Journal

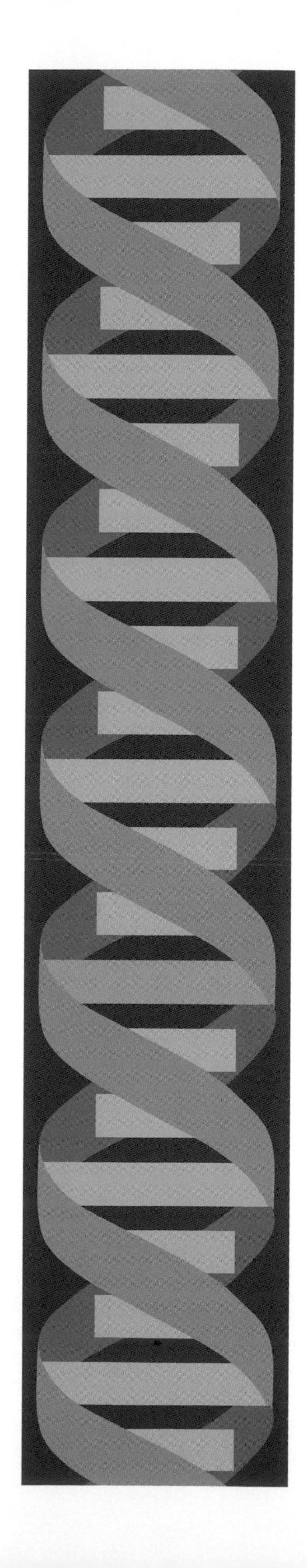

10TH IAPTC&B Congress Fellowship Recipients

Yousef I. Al-Dlaigan, USA
Hanbing An, USA
Michele Auldridge, USA
Ingrid Eloisa Aulinger, Switzerland
Harsh Pal Bais, USA
Ani Lyudmilova Barbulova, Bulgaria
Chhandak Basu, USA
Anna Blaszczyk, Poland
Alvar Carlson, USA
Pick Kuen Chan, Malaysia
Sanjay Velandy Chodaparmbil, Canada
Alfonso Clemente, United Kingdom
Ian S. Curtis, Japan
Daniela Dietrich, Switzerland
Sergey Vladimirovich Dolgov, Russia
Maria Laura Federico, USA
Alicia Fernandez San Millan, Spain
Zulma Isabel Monsalve Fonnegra, Colombia
Celia K. Friedrich, Germany
A. Ganapathi, India
Seedhabadee Ganeshan, Canada
Gokarna Gharti-Chhetri, Sweden
Margy Gilpin, New Zealand
Maram Girgi, Germany
Ramanjini Gowda, India
Wenwu Guo, China
Emma Cecilia Guzman, Israel
Fathi Shukri Hassan, Germany
Lutful Hassan, Bangledesh
Anelia Veneva Iantcheva, Bulgaria
Hani-Al Ibrahim, Israel
Snezana Ivic, Yugoslavia
Pawan K. Jaiwal, India
Sladjana Jevremovic, Yugoslavia
Katherine Kahn, United Kingdom
Kriton Kalantidis, Greece
Fatma Kaplan, USA
Azhakanandam Kasi, India
Patan. Shaik Sha Valli Khan, India
Abha Khandelwal, India
Mariya Vladimirovna Khodakovskaya, Russia
Jacqueline A. Kowalski, Virgin Islands
G. Ravi Kumar, India
Lenka Langhansova, Czech Republic
Lidong Lin, China
Frank Liszewska, Poland
Jihong Liu, China
Aneta Liwosz, Austria
Shengwu Ma, Canada
Pablito M. Magdalita, Philippines
Michael S. Manak, USA
Totik Sri Mariani, Indonesia
Shaheen Banon Mowla, South Africa
Quynh Thi Nguyen, Vietnam
Hector G. Nunez-Palenius, USA
Mohini Nishikant Pathak, India
Simon Poon, Australia
Manoj Prasad, Germany
Volodymr V. Radchuk, Germany
Matthew Frederick Reyes, USA
Gamini Samarasinghe, Sri Lanka
Indra Sandal, India
Valluri Venkata Satyvathi, India
Scott Schaeffer, USA
Liliya Seraletolinova, Germany
Chou Tou Shii, Taiwan
Tee Chong Siang, Malaysia
Nisha Singh, South Africa
Dimuth Siritunga, USA
Claudio Stasolla, Canada
Angelina R. Subotic, Yugoslavia
J. S. Tahardi, Indonesia
Jitendra Kumar Thakur, India
Christopher D. Todd, Canada
Aura Ines Urrea Trujillo, Colombia
Claudia Ines Valencia Delgado, Colombia
K. J. M. Vally, USA
Virendra Mohan Verma, India
Youping Wang, Germany
Tatik Wardiyati, Indonesia
Wei Wei, China
Julia Grace Wilson, Australia
Sonia Wirth, Argentina
Baohong Zhang, USA

10TH IAPTC&B Congress
Science and Technology Exhibitors

Blackwell Publishing
Maiden, MA

BTX, A Division of Genetronics
Sand Diego, CA

CABI Publishing
Wallingford, Oxon, UNITED KINGDOM

Caisson Laboratories, Inc.
Sugar City, ID

COMBINESS
Gent, BELGIUM

Conviron
Hendersonville, NC

Council for Biotechnology Information
Washington, DC

Cytogration
Rockville, MD

Elsevier Science
New York, NY

Enconair
Winnipeg, MB, CANADA

Envrionmental Growth Chambers
Chagrin Falls, OH

Invitrogen Corporation
Grand Island, NY

Kluwer Academic Publishers
Norwell, MA

Monsanto Company
St. Louis, MO

Nature Publishing Company
London, UNITED KINGDOM

Newport Biosystems
Anderson, CA

Olympus America
Melville, NY

Partec GmbH
Munster, GERMANY

Percival Scientific
Perry, IA

PhytoTechnology Laboratories, LLC
Overland Park, KS

Pioneer - A DuPont Company
Johnston, IA

Plant Bioscience Limited
Norwich, Norfolk, UNITED KINGDOM

Sarstedt Inc.
Newton, NC

Science Publishers, Inc.
Enfield, NH

Southern Sun Biosystems
Hodges, SC

Springer-Verlag NY
Secaucus, NJ

Syngenta Crop Protection AG
Basel, SWITZERLAND

Systec GmbH
Wettenberg, GERMANY

Thermo Forma
Marietta, OH

Indra K. Vasil

THE SCIENCE AND POLITICS OF PLANT BIOTECHNOLOGY 2002 AND BEYOND

Indra K. Vasil
1143 Fifield Hall, University of Florida, Box 110690, Gainesville, FL 32611-0690 USA
(email: ikv@mail.ifas.ufl.edu)

The 10th IAPTC&B Congress is the first large gathering of the international plant tissue culture and biotechnology community in the 21st century and the 3rd millennium. As such, it offers us the unique opportunity not only to take a critical look at the contributions of plant biotechnology during the 20th century, but more significantly, to consider how this technology can contribute to human welfare by meeting the many difficult challenges that we face in the 21st century.

The beginnings of plant biotechnology go back to the early 1930's, when Philip White in the United States and Roger Gautheret and others in France established the first continuously growing plant cell cultures. From these modest beginnings, it took nearly six decades - to the early 1990's – to gain a better understanding of the hormonal control of plant development in order to develop efficient techniques for the regeneration of normal and fertile plants from cells and tissues of most of the important plant species. The discovery of the three dimensional architecture of DNA in the 1950's, and of restriction enzymes during the 1970's, led to rapid advances in molecular genetics and the understanding of the structure, function and regulation of plant genes. It was the timely combination of plant cell culture technology with molecular biology during the 1980's that led to the development of transgenic crops and to much of what now constitutes plant biotechnology.

The 10th IAPTC&B Congress is taking place at an important and opportune time. It is a time when we do not have to talk any more only about the potentials and promises of plant biotechnology. Rather, it is a historical milestone when plant biotechnology has come of age. It is a time when we can proudly point to the many transgenic crops that have been grown so far on more than 300 million acres in 15 countries around the world, in addition to the many improved crop varieties developed from various tissue culture techniques. It is a time when transgenic products are becoming common and are being increasingly used by millions of people on a daily basis. In the United States, more than 60% of all processed foods contain transgenic ingredients. The total market for transgenic seed now exceeds $3 billion. It is a time when crop losses caused by pests, pathogens and weeds are being

I. K. Vasil (ed.), Plant Biotechnology 2002 and Beyond, 1-9.

greatly reduced resulting in increased productivity. And, it is a time when plant biotechnology is beginning to have a positive impact on human health, the environment and our shared future by reducing the use of agro-chemicals, and by contributing to the conservation of biodiversity, scarce arable land, and precious water and energy sources. These are, by any standard of measurement, truly remarkable achievements. There is much to be proud of and much to celebrate. This Congress is, therefore, a celebration and a showcase of the contributions of modern plant biotechnology.

The first generation of transgenic plants containing herbicide, insect and virus resistant genes was produced in the late 1980's and early 1990's (Table 1). It took nearly a decade to carry out all of the required field tests and the exhaustive evaluations to satisfy the regulatory requirements, including extensive testing of transgenic foods for nutrient levels, allergens and toxins.

Table 1. First Generation of Transgenic Crops. 1995-2005: The principal benefits of these crops are increased productivity, reduced use of agro-chemicals, conservation of arable land/water/energy, reduced pollution of soil/water/environment, and health benefits owing to reduced pesticide use and seed and other edible plant products with fewer or no toxins.

Resistant to Herbicides
Resistant to Insects
Resistant to Viruses

Transgenic plants were first planted on a large scale in China and the United States in the early 1990s. Since then, the acreage devoted to transgenic crops containing herbicide and insect-resistant genes has increased steadily. In addition, many other useful genes have been introduced into numerous additional crops to improve their quality and nutritional value (including the content of vitamins, essential minerals and healthful antioxidants and fats), and their resistance to a variety of biotic and abiotic stresses. More than 50 of these transgenic crops have been approved for commercial field plantings, and at least 100 more are under trial and regulatory review. A major new development in plant biotechnology research has been the use of plants for the production of vaccines, human therapeutic proteins and pharmaceuticals. This second generation of transgenic plants is now being tested and evaluated for human use and is expected to be released for commercial use before the end of this decade (Table 2).

During the past five years rapid progress has been made in the sequencing of plant genomes, and in understanding the structure, function and regulation of genes. These studies are contributing to our overall understanding of the growth and development of plants, and will surely lead to the third generation of transgenic plants that will be better adapted to a variety of biotic and abiotic stresses, that will provide increased yields and more nutritious and healthful food, and that will produce a variety of drugs and pharmaceuticals for the treatment of human and animal diseases. Such a third generation of transgenic plants is expected to be available for human use after 2015 (Table 3).

Table 2. Second Generation of Transgenic Crops: 2005-2015. Much progress has already been made in the production of these crops. Many of them are now undergoing advanced field or human clinical trials. These crops will offer the highest direct benefit to consumers.

Resistant to Herbicides, Pests and Pathogens
Tolerant to Drought, Salt, Heavy Metal and Temperature
Improved Nutritional Quality (proteins, oils, vitamins, minerals)
Improved Shelf Life of Fruits and Vegetables
Improved Flavors and Fragrances
Elimination of Allergens
Vaccines, Human Therapeutic Proteins, Pharmaceuticals
Phytoremediation

Table 3. Third Generation of Transgenic Crops: 2015 and Beyond. Sequencing of Arabidopsis and rice genomes has been completed; others in various stages of sequencing are Lotus, Brassica, maize, Medicago, poplar, barley, wheat, tomato, potato, soybean and pine. Synteny discovered in cereal genomes will be of significant value in the search for important genes. The discovery and characterization of dwarfing/Green Revolution genes (such as Rht in wheat, sd1 in rice, and GA insensitive in Arabidopsis) will be of much help in manipulating fruit/seed size/number affecting productivity. Recent work has shown that introducing some of the key maize genes involved in C4 photosynthesis into rice significantly increases photosynthetic efficiency and number of grains produced; such plants are also more tolerant to abiotic stress conditions.

Genome Sequencing/Molecular Breeding
Altering Plant Architecture
Manipulation of Flowering Time
Manipulation of Fruit/Seed Quality, Size and Number
Improved Photosynthetic Efficiency
Improved Nutrient Assimilation
Exploiting and Manipulating Heterosis and Apomixis

Various articles included in this volume provide ample documentation for the above advances and developments (Tables 1-3). It is also abundantly clear that the transgenic crops being cultivated today will constitute only a small part of the rapidly expanding plant biotechnology portfolio before the end of this decade, with the availability of many novel plants and plant products. Thus, it would not be an exaggeration at all to conclude that we have just begun, the best is yet to come.

The need to improve crop productivity, to reduce the use of harmful agro-chemicals, and to conserve our valuable natural resources and biodiversity is greater today than ever before. Most of the problems we face in the 21st

century relate, directly or indirectly, to the growing human population. World population stands at 6.2 billion today, and is expected to stabilize at 10-12 billion during the next 50-75 years. Almost all of this increase is projected to take place in the already overpopulated, less developed and poorer regions of Asia, Africa and Latin America. This near doubling of world population, combined with changing dietary habits in China and India as a result of improving economic conditions, will require more than doubling of world food production. The international agricultural community faces this challenge at a time when population is growing faster than increases in food productivity, when the quality and quantity of fresh water supplies are declining, when there is less per capita arable land available for food production (Table 4), when more than 42% of crop productivity is lost owing to various biotic and abiotic factors, and when the use of agro-chemicals which cause soil and water pollution is increasing. Added to this are the enormous post-harvest losses, and the attendant wastage of valuable labor, arable land, energy, fertilizer, and fresh water resources.

Table 4. Roadblocks to increasing food productivity and security. The increasing demand for food must be met primarily by increasing productivity on land already under cultivation with less water and under worsening environmental conditions.

- Population is growing faster than increases in food productivity. Grain production increased 2.1%/year until 1990, slightly ahead of population growth. It has since declined by 0.5%/year. Much of this can be attributed to a progressive decline in the annual rate of yield increase of cereal grains. From a high of 2.5-3% achieved 30 years ago, it has declined to about 1.3%, which is below the rate of population growth.

- From an average of 0.44 hectare in 1961, global per capita arable land declined to 0.26 hectare in 1997, and will likely stabilize at 0.15 hectare by 2050.

- Worldwide drought is one of the biggest problems for food production. Irrigated agriculture requires vast amounts of water (e.g. 17,000 pounds of water is needed to produce one pound of cotton; 4,700 pounds for one pound of rice). Water covers 70% of the earth's surface, yet fresh water comprises only 2.5% of the earth's water. Most of it lies frozen in polar ice caps and glaciers. Less than 1% of the total water is available for human use, including agriculture. Demand for fresh water will exceed supply by 60% by 2025. Genes that help plants to withstand drought have been identified. The same genes allow plants to withstand cold temperatures and high salinity

Clearly, the ultimate solution to each of these problems is reducing population growth, a difficult challenge that is further complicated by social, political, economic and religious considerations. In the hope that national and international efforts will help to stabilize world population in the next few decades, our challenge is to use the power of plant biotechnology toward the solution of the numerous problems caused by population growth by increasing productivity, by reducing crop losses, and by protecting and conserving the environment. Plant biotechnology is not a magic bullet that will solve all of these problems, yet it is becoming abundantly clear that it is the best tool that we have and it can, if used wisely and in a timely fashion, make significant contributions.

It is rather ironic that at a time when international agriculture is under increasing pressure to meet the food needs of the ever increasing population, and when plant biotechnology is beginning to make significant contributions to food productivity and environmental safety, it has become the target of well coordinated and sustained attacks by many environmental and self-appointed watchdog groups, particularly in Western Europe and in some of the developing countries. In the United States, we had faced similar attacks during the 1970's and 1980's. However, after extensive public debate, protest demonstrations, court challenges and congressional hearings, a federal regulatory framework was developed that has served the public and the private interest well. It has allowed the plant biotechnology industry to grow and introduce its products into the market place. The American consumers and farmers have accepted and benefited from transgenic products. Transgenic crops are being grown this year on nearly 100 million acres of American farmland, accounting for 74% of our soybean, 71% of our cotton and 32% of our corn acreages.

It is fortunate and encouraging that China and India, the two most populous countries in the world, with increasing demand for food and worsening environmental problems, have recognized the importance of plant biotechnology in agriculture and have established active and successful research and development programs in plant biotechnology, targeting many regional vegetable and fruit crops, in addition to such staples as wheat, rice, maize, soybean, various pulses, canola, and cotton. It is not too far-fetched to expect that within the next few years these two countries will plant the largest acreages of transgenic crops in the world. Argentina and South Africa are two other developing countries that are increasing their planting of transgenic crops. Indeed, at the present time, nearly 10% of the global acreage of transgenic crops is planted in the developing countries. Thus the argument that plant biotechnology is a tool of the industrialized countries for the exploitation of the developing world is no longer sustainable. It is our hope that the success of plant biotechnology in these countries will encourage similar efforts in other parts of the developing world.

It is true that the first generation of transgenic crops, which contain genes for resistance to herbicides and insects, did not provide any direct benefits to the consumer. Nevertheless, there are numerous indirect benefits, such as the reduced use of pesticides and herbicides, reduced tillage leading to soil conservation, reduced use of natural resources such as petrochemical products and water for the manufacture, transport and application of agro-chemicals, and reduced labor costs. By producing more food on the same amount of land transgenic crops promote conservation and biodiversity by saving wildlife habitats and precious forests from being converted into farmland. Reduced use of pesticides has already shown a marked decrease in illness and death caused by pesticide poisonings in China and South Africa (nearly 500 cotton farmers in China die each year of acute pesticide poisoning).

The vital role of agriculture and food production in human health and nutrition, in poverty alleviation, and in social and political stability, is well known. This was recognized as far back as 1970, when Norman Borlaug was awarded the Nobel Peace Prize for his work that led to the Green Revolution and that helped to save hundreds of millions of lives in the developing countries. Plant biotechnology too can contribute to international peace and security by increasing food production, producing safer and healthier foods, protecting our rather finite natural resources and the environment, and improving human health. It is, therefore, morally and socially irresponsible and indefensible to prevent or delay the applications of plant biotechnology to problems of hunger, health and protection of the environment.

The opponents of plant biotechnology would have us believe that it is an unnatural and unsafe process that produces harmful products, and that it is totally different from plant breeding and selection that account for almost all of our modern crops. The indisputable fact, however, is that humans have engineered crops for nearly 10,000 years. Almost all of our major crops – such as maize, wheat, potato, tomato and others - are man made. Indeed, none of our modern crops are capable of surviving in the wild without human care. The molecular and genetic principles of plant biotechnology and plant breeding and selection are the same. Plant biotechnology is no different from breeding and selection, or for that matter from radiation and chemically induced mutation breeding, except that it is extraordinarily precise and predictable, and is not restricted by taxonomic boundaries. No compelling evidence has ever been presented to show that transgenic crops are innately different the non-transgenic products of breeding and selection.

In retrospect, however, we must share some of the blame for the perception that plant biotechnology is different from plant breeding and selection. During the 1970's and 1980's, when there was a great deal of euphoria over the production of somatic hybrids, doubled-haploid breeding lines and transgenic plants, the plant biotechnology community made a serious error in strategy and judgment when it distanced itself from breeding and selection and established a separate identity for itself. It was us who placed the spotlight on the process and not the product. This has come to haunt us now as it has attracted undue and undeserved attention and opposition. It was also an error not to engage early in the debate on transgenic plants, and to permit the opponents of plant biotechnology to dictate the agenda. As responsible members of the world community, and as scientists, we cannot, and should not, be silent observers of this debate. We must play an active role in the debate on plant biotechnology and make an informed and professional contribution to the public dialogue, emphasizing the many benefits of transgenic crops to human health and the environment.

The opposition to transgenic foods in Europe and elsewhere is based exclusively on political and ideological differences rather than on any credible scientific evidence. On two rare occasions an attempt was made to present scientific arguments against the use of transgenic crops. These involved the allegedly harmful effect of pollen from Bt maize plants on the larvae of the Monarch butterfly, and the alleged transgene contamination of maize in Mexico. More detailed investigations by several research groups have since refuted these claims and showed them to be of dubious scientific value. Indeed, in the Mexican maize story, the journal Nature, in an unprecedented action in its more than 100 year history, was forced to disown the paper published in its own pages.

The consumer, the farmer and the biotechnology industry have been ill served by the sustained campaign of misinformation and unsubstantiated claims of dangers to public health and the environment by transgenic crops and their products. After more than ten years and thousands of field trials in many countries, after nearly a decade of commercial plantings on hundreds of millions of acres, and after transgenic food products having being used by hundreds of millions of humans and farm animals, there is not a single documented instance of illness reported in any human or animal, or of ecological or environmental damage. What then is the basis and rationale for the many restrictions still placed on the field planting and human use of transgenic foods?

The enviable and unblemished record of transgenic crops and their products is the strongest evidence for their safety and wholesomeness. The opponents of plant biotechnology should compare this record with that of the many drugs approved for human use in the United States. In an exhaustive study published recently in the Journal of the American Medical Association, it was reported that 20% of the 548 drugs approved for human use during the past 25 years were later found to have serious or life-threatening side effects. Seven of the drugs possibly contributed to 1002 deaths, and 16 were forced to be withdrawn from the market. In comparison, not a single transgenic food product has ever been shown to have any harmful effects, and none has been withdrawn because of adverse reactions in humans or animals.

The plant biotechnology community has already done more for the environment and the developing countries than the self-proclaimed environmental groups and the so-called friends of the poor. Indeed, the opponents of plant biotechnology have done much harm to their professed cause by slowing down and/or preventing the planting and utilization of transgenic crops around the world. The contributions of the plant biotechnology community, on the other hand, are socially and morally responsible and of considerable humanitarian value. We have every reason to be proud of these contributions.

The rules and regulations adopted for transgenic crops in the 1980's were both prudent and necessary. At that time there were many unknowns about transgenic crops and about their possible effect on humans and the environment. There was a need to establish a database to satisfy the concerns of the general public as well as the scientific community. Three federal agencies, the United States Department of Agriculture, the Environmental Protection Agency and the Food and Drug Administration, were given oversight responsibilities for transgenic crops. The resulting open and transparent system established in the United States has worked well and has served its purpose. It has done much to gain the confidence and support of the American public for plant biotechnology and its products. In light of the demonstrated safety of transgenic crops to humans, animals and the environment, the question must be asked whether it is any more necessary, or even advisable, to continue the expensive, time consuming and burdensome requirements for the public release of transgenic crops and their products (field trials of transgenic crops are 10-20 times more expensive than of similar plants developed by conventional means).

I propose that based on our considerable experience and on the vast amount of information gathered about the safety of transgenic crops over the past decade, it is time for our regulatory agencies to consider whether some or all of the current regulatory requirements can be gradually relaxed and ultimately suspended, except in those rare instances where there is the clear likelihood of risk to human health and the environment. Genuine concerns about gene flow and development of resistance to antibiotics, and pests or pathogens, can be met adequately with currently available and emerging technologies.

The process of deregulation of transgenic crops, controversial and difficult as it may be, needs to begin now because the continuation of the present rules and regulations is entirely unnecessary, unjustified and counterproductive. In order to be effective and acceptable, the process should be open to all points of view. The decisions, however, must be based on science and facts and not on political or ideological considerations. Nearly two decades ago, the United States played a leading and useful role in establishing the rules and regulations for the field planting, evaluation and human use of transgenic crops. It should now play a similar role in having these restrictions relaxed and removed.

In conclusion, it is clear that the challenges we face in the 21st century are greater than those we faced in the last century. Of all the available technologies, plant biotechnology offers the best hope for producing more and better food, fiber and pharmaceuticals, and for protecting, preserving and improving the environment for the benefit of humankind. My own confidence in plant biotechnology comes from knowing that the science behind it is sound, that it is well tested and proven, that it benefits the consumer, the farmer and the industry, and that it protects and conserves the environment. It is for these reasons that I am convinced that plant biotechnology will within the next two decades become an integral part of the international agricultural system. With the United States, China, and lately India, three of the most populous countries in the world serving as examples, we have taken the first steps toward achieving that objective.

WE HAVE JUST BEGUN, THE BEST IS YET TO COME.

Jonathan D.G. Jones

PUTTING PLANT DISEASE RESISTANCE GENES TO WORK

Jonathan D.G. Jones, G. Brigneti and D. Smilde
Sainsbury Laboratory, John Innes Centre, Colney Lane, Norwich NR4 7UH, United Kingdom
(email: jonathan.jones@sainsbury-laboratory.ac.uk)

1. INTRODUCTION

Semi-dominant plant disease resistance (*R*) genes confer recognition of and response to specific races of pathogen that carry a corresponding Avirulence (*Avr*) gene. R proteins are presumed to recognise pathogen Avr gene-encoded products, or compatibility factors, that are likely to be involved in pathogenicity on the host. *R* genes against various important diseases have been used by plant breeders, but when deployed in monocultures, resistance frequently breaks down as races of the pathogen emerge that can overcome the *R* gene through recessive mutations in the corresponding Avr gene. Nevertheless, in nature, *R* genes have been maintained. In Arabidopsis, ~164 homologs of the largest class of *R* genes exist. These *R* genes encode proteins of the nucleotide binding-leucine rich repeat (NB-LRR) class (Dangl and Jones, 2001).

In natural populations and environments, in which nitrogen is limiting, plants have been selected to only turn on the defence response upon attack. Constitutive expression of defence mechanisms is costly and likely to be deleterious, even after application of nitrogen fertiliser. The *R* genes encode the plant "antennae" to ensure defence mechanisms are only expressed when needed. However, if it is so easy to evade detection by mutating an Avr gene, why maintain them? The answer lies in the fact that *R* gene loci are usually extremely polymorphic compared to other loci. *R* gene polymorphism in a population, with each *R* gene allele present at a low frequency, reduces selection in the pathogen population to overcome any particular *R* gene, especially if the corresponding *Avr* gene contributes to pathogenicity on plants that lack that *R* gene. This "balancing polymorphism" is disrupted in conventional agriculture. It is worthwhile to consider possible approaches to deploying *R* gene polymorphism in agriculture, using the example of potato late blight caused by the oomycete *Phytophthora infestans*.

I. K. Vasil (ed.), Plant Biotechnology 2002 and Beyond, 11-17.

2. *R* GENES IN POPULATIONS: THE SIGNIFICANCE OF POLYMORPHISM

In wild populations, natural selection has generated populations that are polymorphic at *R* gene loci, presumably for much longer than plant breeders have been recruiting such diversity for crop improvement (Dixon et al., 2000; Noel et al., 1999). It is clear from many studies that some, though not all, mutations in *Avr* genes lead to reduced pathogen fitness. However, there are also data that suggest *R* genes could lead to reduced plant fitness in the absence of a pathogen, presumably because activating the defence response consumes resources that in the absence of pathogens are best devoted to growth and reproduction. *Rpm1* might confer a fitness cost in the absence of pathogen, since it has been lost so frequently in independent accessions (Stahl et al., 1999). High level expression of *Rps2* results in lethality (Tao et al., 2000). It has been proposed that stacking, or "pyramiding", *R* gene alleles from one species into a single genotype might provide durable resistance (Rausher, 2001). Using transgenic plants, this approach can and should be tested. However, it might select for pathogens that can tolerate the loss of multiple compatibility factors, and may lead to yield penalties. An alternative view is that since polymorphism for *R* genes exists in nature, it may provide a more durable approach to resistance.

The population geneticist Bill Hamilton in an important paper presented the idea that the main purpose of outcrossing was to reduce parasite pressure by sustaining polymorphism at loci that contribute to parasite recognition (Hamilton et al., 1990). According to this model, if a host population is extremely heterogeneous in its recognition capacity, then most isolates of the parasite will not be able to grow on most hosts. In the absence of outcrossing, such polymorphism would be more likely to be lost, even if it is maintained by selection. Furthermore, if sexual recombination between parasites leads to exchange of dominant avirulence genes, then most progeny of most parasites will not be able to find a host. There is still debate about whether such frequency dependent selection, in which rare resistance (recognition) specificities are less likely to be overcome by the parasite, is the main explanation for the enormous diversity of human haplotypes at the major histocompatibility (MHC) locus. An alternative model proposes that this diversity can be explained by overdominance (heterozygote advantage), through which heterozygotes have twice the recognition capacity and resistance of any homozygote (Hughes and Yeager, 1998). Many plant species, such as Arabidopsis, reproduce by self-fertilisation, and overdominance cannot explain the extreme polymorphism of *R* loci compared to other loci in such species. Several commentators have recently highlighted parallels between the way *R* genes and the human MHC act in populations, and suggested that deploying *R* genes in populations might lead to more durable deployment of *R* genes (Bergelson et al., 2001; Dangl and Jones, 2001; Jones, 2001; Pink and Puddephat, 1999).

This idea that *R* gene polymorphism by itself could restrict parasite populations is attractive, but what data are available that test it? It is well known that any individual *R* gene in a crop monoculture is likely to be eventually overcome by parasite mutations, though there is still hope that the *Bs2* gene (Tai et al., 1999), that recognizes *AvrBs2* (Kearney and Staskawicz, 1990) -will prove more durable. *AvrBs2* appears to be indispensable for full virulence in *Xanthomonas campestris*. The analysis of different Avr genes for their contribution to virulence continues to be a valuable and interesting goal (Cruz et al., 2000).

The cost to breeders of continually having to create new varieties that carry new and fleetingly unbroken *R* genes, is substantial. A recent study on rice and rice blast is instructive (Zhu et al., 2000). Two varieties of rice, one low yielding, disease sensitive but high quality, and the other high yielding, more disease resistant and low quality, were grown either individually or as mixtures. In a typical mixture, 4 rows of the resistant variety were grown for every one row of the sensitive variety. Blast incidence in the sensitive variety was reduced 25 fold, and there was even a slight reduction of blast incidence in the resistant variety. A similar study suggesting varietal mixtures can reduce disease incidence has been reported for late blight resistance in potato (Garrett and Mundt, 2000). In the 1970s, the concept of varietal mixtures was championed by Martin Wolfe amongst others (Wolfe, 1985; Wolfe and McDermott, 1994). However, different varieties will not only differ in *R* genes but also in other characteristics such as time to ripening or malting quality, resulting in reduced profitability. Since the *Mla* locus is an allelic series, it is essentially impossible to generate varieties that are genotypically identical at every locus except *Mla*. The isolation of the *Mla* gene has changed that. By transformation it should be possible to introduce several different *Mla* alleles into a variety, and create plant populations that are heterogeneous for pathogen recognition but homogeneous for commercial traits. The idea is illustrated in the Figure, in which plants carrying different *R* genes are given a different stippling. On the right, in a typical monoculture, all plants have the same *R* gene. Any pathogen mutation that enables it to defeat the *R* gene enables explosive population growth of the pathogen, even if the *Avr* mutation results in slightly reduced virulence. On the left, a population that is heterogeneous for 5 different *R* genes has several advantages. Firstly, only 20% of the plants can support growth of any virulent race of the pathogen, so the rate of increase of the epidemic will be correspondingly reduced. Secondly, intense pathogen pressure of avirulent pathogen races may trigger systemic acquired resistance that will reduce

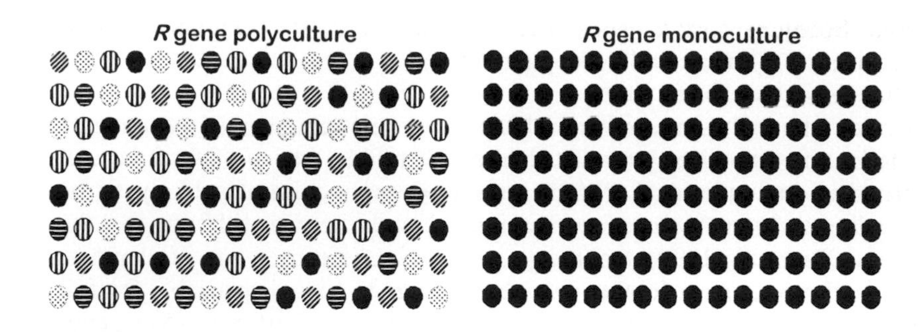

Fig. 1. In an R gene monoculture, any pathogen race can overcome an R gene and cause an epidemic. In an R gene polyculture, for a race with one virulence gene against R gene ●, there are many fewer susceptible plants, so epidemics are slower. Mutations that enable this race to overcome an additional R gene will result in loss of an additional compatibility factor, and be less fit on the first R gene. High pathogen pressure with avirulent races may trigger systemic acquired resistance.

susceptibility to virulent races. And thirdly, any mutation to virulence in a race that already can grow on one *R* gene, enabling it to grow on another *R* gene, may result in slightly reduced parasite fitness on the first *R* gene. If there is a cost to virulence, such mixed pathogen populations will never completely lose the corresponding *Avr* genes. This provides an advantage over the alternative approach, of pyramiding all five *R* genes into the same cultivar; this could easily lead to selection for suppressor mutations that enable a pathogen to tolerate loss of all 5 *Avr* genes without too serious a loss of fitness. With the isolation of *R* gene allelic series, it is time to give this approach serious experimental testing.

Our lab is initiating a program to use this approach to address the problem of potato late blight caused by the oomycete *Phytophthora infestans*. Eleven *R* genes have been identified in the wild potato relative *Solanum demissum*, and bred into cultivated varieties, but in each case, resistance has been short-lived. Currently potato breeders strive to avoid dominant *R* genes, since they interfere with the capacity to detect and manipulate "horizontal resistance" that is conjectured to be more durable. However, a wealth of genetic variation exists in wild *Solanum* species for additional sources of blight resistance. Much of this resistance can only be introgressed into cultivated varieties after somatic hybridization, because of problems of sexual incompatibility with *S tuberosum*.

An alternative approach is to isolate *R* genes from multiple wild species of *Solanum*, and to introduce them into potato by genetic transformation. In principle, the same variety could be transformed with different *R* genes, and multi-line mixtures established that are identical for agronomic traits and polymorphic for blight resistance.

Diploid Solanum species should in principle be easier for map based approaches to gene isolation than tetraploids. This is because diploid genetics is easier than tetraploid genetics, and because through suitable crosses it should be possible to identify lines that are homozygous for a targeted gene, prior to making genomic libraries, avoiding subsequent problems in distinguishing clones of the susceptible and resistant haplotypes in a library made from a heterozygote.

Diploid species fall into two categories, differing in endosperm balance number (EBN). Both the more ancient EBN1 species (including *S. bulbocastanum*), and the more evolved EBN2 species (including *S berthaultii, S. okadae* and *S microdontum*) carry genotypes that show pronounced variation for blight resistance. We are setting out to investigate and "mine" this genetic variation.

To date, a resistance survey has been conducted on 11 diploid EBN1 *Solanum* species and 16 diploid EBN2 *Solanum* species involving 2 to 15 accessions from each species (56 EBN1 accessions and 58 EBN2 accessions). On average, 10 plants were tested per accession in three independent experiments using a detached leaf assay, inoculated with a complex race of *P. infestans*. Genetic diversity for resistance was detected in both EBN1 and EBN2 species. We are extending the survey to include as many diploid species as possible. Frequently, resistant and sensitive individuals were identified within the same accession, suggesting that the accession from which the seed was originally collected was segregating for resistance.

We have generated approximately 180 crosses between susceptible and resistant plants identified in the survey. Most are intra-specific crosses but we have also produced a few inter-specific crosses, especially in cases when susceptible individuals could not be identified within the species. Initial screens of F1 progenies have identified five crosses (2 EBN1, 3 EBN2) that segregate 1:1 for resistance and sensitivity to *P. infestans*. With current advances in the molecular genetics and genomics of *Solanum* and the closely related *Lycopersicon,* we anticipate that this approach will enable us to rapidly identify and isolate multiple genes for blight resistance, and to test whether multiline approaches could provide durable resistance. It will be particularly interesting once map positions start to emerge for these R genes. It could be that many of them map to orthologous loci in the different species and might be considered interspecific alleles of the same R gene. If this were to be the case, isolation and deployment of multiple *R* genes should become even more straightforward.

One last issue has emerged for this strategy. Potato is particular prone to somaclonal variation. Twenty years ago this was perceived as a potentially valuable property; somaclonal variation was spoken of as a powerful new tool in the plant breeder's armory. Sadly, this promise is unfulfilled, and most somaclonal changes are deleterious. To engineer a new and useful trait into a variety such as Russet Burbank, several thousands of transformants need to be produced to find one that carries the trait and does not depart to radically from its parental type. We are exploring whether a shift of potato breeding to the diploid level, coupled with selection for diploid lines that are self compatible and can tolerate inbreeding, might facilitate deployment of transgenes in potato. Somaclonal variation could be "cleaned up" by Mendelian transmission. Once suitable transgenic diploid lines are identified, they can be used for production of defined F1 hybrids, and subsequent re-tetraploidization.

3. ACKNOWLEDGEMENTS

We thank the Gatsby Charitable Foundation and the BBSRC for support for this work, and many colleagues whose discussions with us have helped refine these ideas.

4. REFERENCES

Bergelson, J., Kreitman, M., Stahl, E. A., and Tian, D. C. (2001). Evolutionary dynamics of plant R-genes. Science *292*, 2281-2285.

Cruz, C. M. V., Bai, J. F., Ona, I., Leung, H., Nelson, R. J., Mew, T. W., and Leach, J. E. (2000). Predicting durability of a disease resistance gene based on an assessment of the fitness loss and epidemiological consequences of avirulence gene mutation. Proc Nat Acad Sci USA 97, 13500-13505.

Dangl, J. L., and Jones, J. D. G. (2001). Plant pathogens and integrated defence responses to infection. Nature 411, 826- 833.

Dixon, M. S., Golstein, C., Thomas, C. M., van der Biezen, E. A., and Jones, J. D. G. (2000). Genetic complexity of pathogen perception by plants: The example of Rcr3, a tomato gene required specifically by Cf-2. Proc Nat Acad Sci USA 97, 8807-+.

Garrett, K. A., and Mundt, C. C. (2000). Host diversity can reduce potato late blight severity for focal and general patterns of primary inoculum. Phytopathology *90*, 1307-1312.
Hamilton, W. D., Axelrod, R., and Tanese, R. (1990). Sexual Reproduction As an Adaptation to Resist Parasites (a Review). Proc Nat Acad Sci USA 87, 3566-3573.

Hughes, A. L., and Yeager, M. (1998). Natural selection at major histocompatibility complex loci of vertebrates. Annu Rev Genet 32, 415-435.

Jones, J. D. G. (2001). Putting knowledge of plant disease resistance genes to work. Curr Opin Plant Biol 4, 281-287.

Kearney, B., and Staskawicz, B. J. (1990). Widespread distribution and fitness contribution of *Xanthomonas campestris* avirulence gene *avrBs2*. Nature 346, 385-386.

Noel, L., Moores, T. L., vanderBiezen, E. A., Parniske, M., Daniels, M. J., Parker, J. E., and Jones, J. D. G. (1999). Pronounced intraspecific haplotype divergence at the RPP5 complex disease resistance locus of Arabidopsis. Plant Cell 11, 2099-2111.

Pink, D., and Puddephat, I. (1999). Deployment of disease resistance genes by plant transformation - a 'mix and match' approach. Trends Plant Sci 4, 71-75.

Rausher, M. D. (2001). Co-evolution and plant resistance to natural enemies. Nature 411, 857-864.
Stahl, E. A., Dwyer, G., Mauricio, R., Kreitman, M., and Bergelson, J. (1999). Dynamics of disease resistance polymorphism at the Rpm1 locus of Arabidopsis. Nature 400, 667-671.

Tai, T. H., Dahlbeck, D., Clark, E. T., Gajiwala, P., Pasion, R., Whalen, M. C., Stall, R. E., and Staskawicz, B. J. (1999). Expression of the Bs2 pepper gene confers resistance to bacterial spot disease in tomato. Proc Nat Acad Scien USAmerica 96, 14153-14158.

Tao, Y., Yuan, F. H., Leister, R. T., Ausubel, F. M., and Katagiri, F. (2000). Mutational analysis of the Arabidopsis nucleotide binding site- leucine-rich repeat resistance gene RPS2. Plant Cell 12, 2541-2554.

Wolfe, M. S. (1985). The current status and prospects of multiline cultivars and variety mixtures for disease resistance. Annu Rev Phytopathol 2, 251-273.

Wolfe, M. S., and McDermott, J. M. (1994). Population genetics of plant pathogen interactions: The example of the *Erysiphe graminis-Hordeum vulgare* pathosystem. Annu Rev Phytopathol 32, 89-113.

Zhu, Y. Y., Chen, H. R., Fan, J. H., Wang, Y. Y., Li, Y., Chen, J. B., Fan, J. X., Yang, S. S., Hu, L. P., Leung, H., et al. (2000). Genetic diversity and disease control in rice. Nature 406, 718-722.

Steven Briggs

COMPARATIVE GENOMICS ENABLES A VIRTUAL GENOME OF THE CEREALS

Steven Briggs
Torrey Mesa Research Institute, San Diego, CA 92121 (e-mail: steven.briggs@syngenta.com)

We have completed a 6X draft sequence of the rice genome. The sequence consists of 428 Mb. There are 38 Mb of long repetitive sequence and 390 Mb of sequence that has been assembled into 38,000 contigs. Over 50% of the genome is in 6,000 contigs longer than 20 kb, and more than 90% of the genome is in 20,000 contigs longer than 5 kb. Approximately 10,000 contigs end in long repeat sequences (500 bp), while the other 28,000 contigs are joined together by non-overlapping forward and reverse sequence runs from the same plasmids. Extending the sequence from each end can eliminate the gaps between these contigs. More than 99% of all publicly available rice sequence is found in this dataset. Analysis of the sequence indicates there are about 33,000 genes in rice, comprising 15,000 gene families. Nearly all confirmed or predicted cereal (wheat, maize, barley, rye, oat, sorghum) genes in the public domain were found to have homologs in rice, with a typical sequence identity of 80%-90%. Comparing mapped sequence markers between species confirmed the high degree of synteny that has been inferred from numerous genetic studies. Virtual genomes have been construted for the cereal spccies based upon this synteny. The virtual genomes have been used to predict the gene content and order in quantitative trait loci (QTLs) from species such as maize. More than 2,000 QTLs have been published for the cereals and have been mapped onto our virtual genomes. The virtual genomes immediately provide a set of candidate genes to test as causal agents for each QTL.

Abstract only – no manuscript received.

I. K. Vasil (ed.), Plant Biotechnology 2002 and Beyond, 19.

Martin Yanofsky

THE ROLE OF MADS-BOX GENES IN THE CONTROL OF FLOWER AND FRUIT DEVELOPMENT IN ARABIDOPSIS

Soraya Pelaz, Sarah Liljegren, Adrienne Roeder, Cristina Ferrándiz, Anusak Pinyopich, Lars Ostergaard, Kristina Gremski, Pedro Robles, Gary Ditta, Sherry Kempin and Martin Yanofsky

Section of Cell and Developmental Biology, Division of Biological Sciences, University of California at San Diego, La Jolla, California 92093-0116, USA (email: marty@ucsd.edu)

Keywords Flower development, Fruit development, MADS-box genes

1. INTRODUCTION

Arabidopsis flowers and fruit are typical of the more than three thousand species of Brassicaceae and have been the subject of intensive genetic and molecular studies. Among the many genes that have been identified that control various aspects of flower and fruit development, the MADS-box family has been shown to play central roles. MADS-box genes encode putative transcriptional regulators that play regulatory roles not only in diverse plant species, but also in fungal and animal development. The first Arabidopsis MADS-box gene to be cloned was *AGAMOUS* in 1990, and in the ensuing years, dozens of related genes have been cloned and functionally characterized (Yanofsky et al., 1990; Riechmann and Meyerowitz, 1997; Theissen, 2000). Three general lessons have been learned from these functional studies. The first, is that MADS-box genes play diverse roles in plant development, ranging from the control of flowering time, meristem identity, organ identity, fruit development, and they also appear to play roles during embryo, ovule, seed, root, stem and leaf development. A second lesson that we have learned is that MADS-box genes frequently play multiple roles during development. For example, the *FRUITFULL* gene is involved in leaf development as well as in fruit development. A third lesson is that functional redundancy is prevalent among MADS-box genes. In some cases, single mutants are indistinguishable from the wild type, whereas double mutants carrying mutations in two closely related MADS-box genes display striking phenotypic abnormalities. In this manuscript, we will focus on recent examples from our laboratory that highlight these three basic conclusions of MADS-box gene function.

I. K. Vasil (ed.), Plant Biotechnology 2002 and Beyond, 21-27.

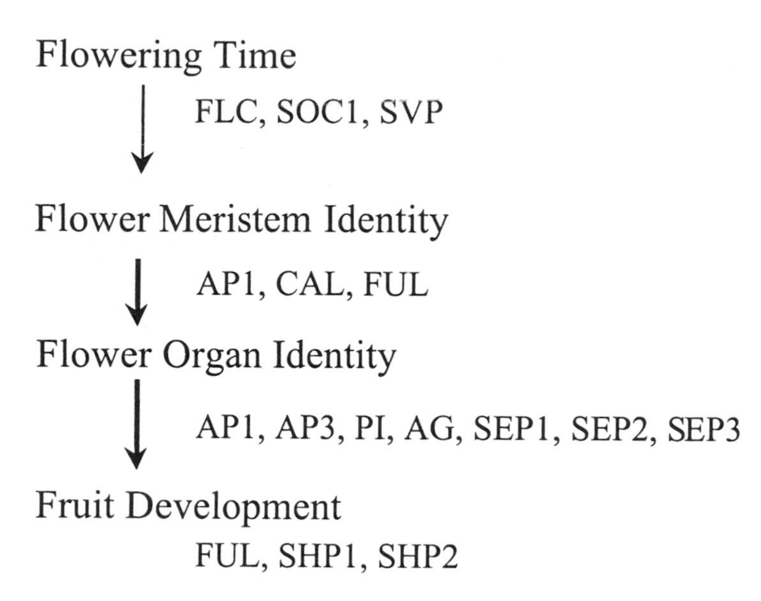

Figure 1. Summary of the roles of MADS-box genes during reproductive development in Arabidopsis.

2. *SEPALLATA* GENES AND THE CONTROL OF FLOWER ORGAN IDENTITY

There is perhaps no more widely recognized area of plant development than the studies that led to the proposal of the ABC model of flower organ development. It is fair to say that this model was conceived at around that same time by both Elliot Meyerowitz and Enrico coen, the former basing his model on studies of Arabidopsis mutants and the latter from his studies in Antirrhinum (Coen and Meyerowitz, 1991; Bowman et al., 1991; Schwarz-Sommer et al. 1990). This model was entirely based on genetic data, but it has since been substantiated through the molecular cloning of the ABC genes, which for the most part turn out to be members of the MADS-box gene family. The basic tenets of this model are that the A function specifies sepals, the combined activities of A and B specifies petals, the combined action of B and C specifies stamens, and C alone specifies carpels. Not only has this model proven useful for studies in the two plant species upon which it based, but it has led to a rash of studies in numerous plant species leading to the notion that this model is generally applicable to most flowering plants.

The ABC model was based on mutants obtained from traditional forward mutant screens, and although these screens have most likely been performed to saturation, they may not have identified functionally redundant genes. Were the ABC genes the only major regulators of flower organ identity or were additional genes required for this process that had been hidden by the redundant function of closely related members of the MADS-box gene family? Among the many MADS-box genes, three closely related genes were good candidates for genes that might play important roles in flower organ specification. These three genes, *SEPALLATA1-3* (*SEP1-3*) (formerly *AGL2*, *AGL4* and *AGL9*, respectively) show extensive sequence similarity and all three show overlapping expression patterns during early flower development. Similar to other organ identity genes, the *SEP* genes are first expressed at late stage 2 or early stage 3 of flower development, and all three show strong expression in petals, stamens and carpels.

To test if the *SEP* genes are involved in flower development, single mutants were obtained through reverse-genetics approaches. Remarkably, none of these single mutants displayed a phenotype that was dramatically different from the wild type. However, triple mutants plants lacking the activities of all three *SEP* genes produced flower in which all of the organs were sepals (Pelaz et al., 2000). Thus, the *SEP* genes encode redundant activities required for petal, stamen and carpel development. Moreover, whereas normal flowers are determinate structures and produce a defined number of organs, these triple mutant flowers were indeterminate, meaning that they continuously elaborated new flower organs, producing a "flower within a flower" or "double flower" phenotype. Because the triple mutant phenotype is strikingly similar, and perhaps indistinguishable, from the bc double mutant plants, we can conclude that the *SEP* genes are required for the activities of the B and C genes. However, it is not the case that the *SEP* genes activate B and C expression, since the onset of B and C gene expression appears unchanged in the *sep* triple mutant flowers. The most likely scenario, which has been proposed by a number of groups is that higher order complexes of MADS-box proteins participate in specifying flower organ identity (Egea-Cortines et al., 1999).

Based on the *sepallata* mutant studies, we have proposed the addition of a D function, encoded by the redundant *SEP* genes, to the previously described model of flower organ identity. In this revised model, A specifies sepals; A, B and D combine to specify petals; B, C and D specify stamens; C and D specify carpels and prevent the indeterminate growth of the flower meristem. [Note: it is worth pointing out that a D activity was previously proposed as a function controlling ovule identity (Angenent et al., 1996; Theissen, 2001). However, because ovule development is occurs much later in flower development, we use the term D-function to relate to its role in flower organ identity.

More than 200 years ago, Goethe made the remarkable suggestion that each of the distinct flower organs represents a modified leaf. In an elegant validation of this speculation, Bowman and Meyerowitz (1991) showed that triple mutants lacking the activities of all three ABC genes produced flowers in which all organs look very much like leaves. Thus, it was fair to conclude that the ABC genes were necessary for flower organ development. However, it was unclear if these genes were sufficient to convert vegetative leaves into each of the distinct flower organs. Indeed, attempts to co-express different combinations of these genes in leaves failed to produce the desired organ conversions, indicating that additional genes must be required to convert leaves into flower organs. Might the recently discovered *SEP* genes represent this missing factor? Indeed, mis-expression of the *SEP* (D function) genes, together with the A and B genes, was sufficient to convert leaves into normal-appearing petals (Pelaz et al., 2000). Now that the major regulators of flower organ identity have been identified, we can begin to unravel the cascade of gene activity that leads to the differentiation of individual cells within organs. In this regard, identifying the direct targets of the organ identity genes will be an essential step in our goal to understand this complex process.

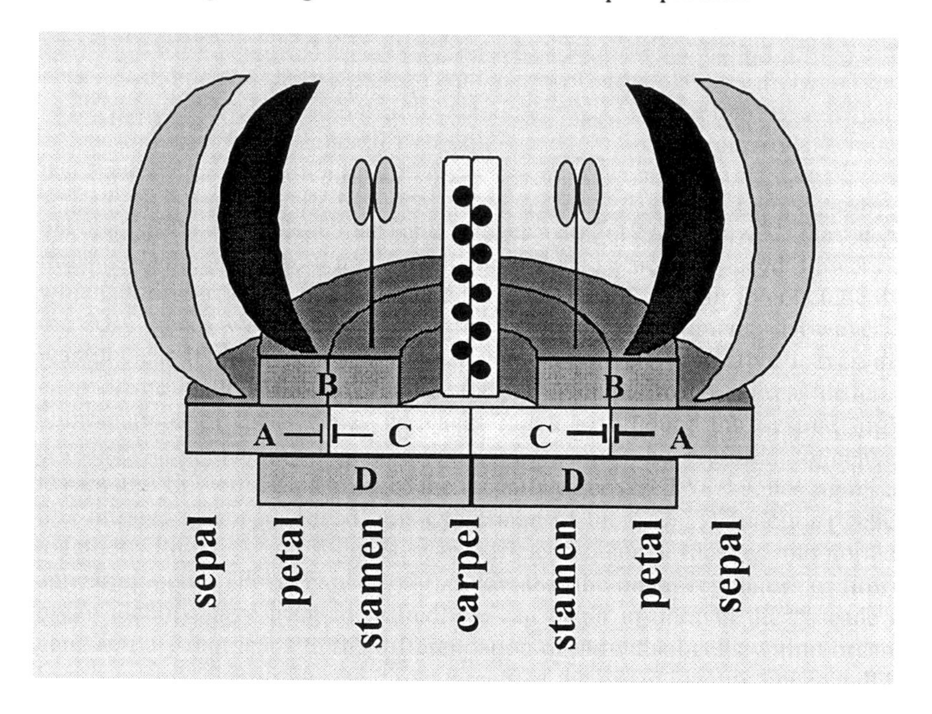

Figure 2: Revised ABCD model of flower organ identity.

2.1. The Role of the *Fruitfull* Gene in Fruit Development

The fruit, which is derived from the fertilized carpels, is the most complex plant organ, and studies of the genes that control fruit development are still in their early stages (Sessions et al., 1995). Among the several genes that have been characterized, the *FRUITFULL* (*FUL*) and *SHATTERPROOF* (*SHP*) MADS-box genes have been shown to play important roles. A cross section of an Arabidopsis fruit (Fig) reveals the valves, which are derived from ovary walls, the replum, and the dehiscence zone.

The *FUL* gene is expressed in valve cells, and *ful*-mutant valve cells fail to elongate and differentiate (Gu et al., 1998; Ferrándiz et al., 1999, 2000). In addition, *ful*-mutant fruit fail to elongate post-fertilization, leading to small and oddly –shaped fruits. Ectopic expression of *FUL* is sufficient to convert dehiscence zone and replum cells into valve cells. These studies indicate that *FUL* is both necessary and sufficient, within this context, for valve cell identity. Whether or not *FUL* directly promotes valve cell identity, or instead, prevents valve cells from adopting an alternative fate, remains to be determined.

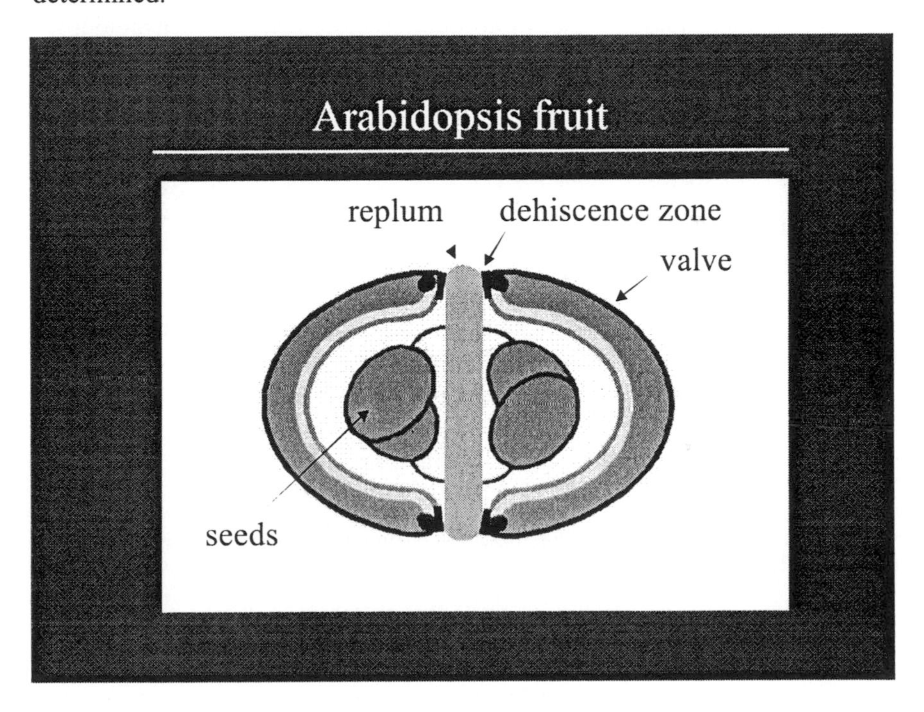

Figure 3: Diagram of a Cross Section of a wild type fruit.

2.2. The Role of *Shatterproof* Genes in Fruit Development

One of the terminal stages of fruit development in Arabidopsis involves separation of cells within the dehiscence zone, allowing the valves to detach from the replum and for the seeds to be dispersed. This process, which is often referred to as fruit dehiscence, or pod shatter, follows a carefully orchestrated series of events within the dehiscence zone cells. The *SHP* genes encode closely related proteins and are expressed in stripes along the valve/replum boundary where dehiscence zone cells differentiate. Although *shp* single mutants cannot be distinguished from the wild type, *shp* double mutants fail to differentiate dehiscence zone cells (Liljegren et al., 2000). One consequence of this failure is that the seeds remain trapped inside and cannot be dispersed through their normal mechanism. Thus, *shp* loss-of-function mutants, as well as *ful* gain-of-function transgenic plants, fail to differentiate a dehiscence zone and thus fail to disperse their seeds.

The genetic control of seed dispersal in Arabidopsis may have important applications in crop plants. For example, significant yield losses in the economically important crop plant canola result from premature seed dispersal. Canola, like Arabidopsis, is a member of the *Brassicaceae*, suggesting that technologies uncovered through basic research in Arabidopsis may be readily transferred into canola. Down regulation of the *SHP* genes, or ectopic expression of *FUL*, offer two distinct genetic approaches for the control of seed dispersal in crop plants.

3. ACKNOWLEDGMENTS

Our Arabidopsis research is supported by grants from the National Science Foundation, the United States Department of agriculture, and from a grant from Bayer CropScience and the UC BioStar program.

4. REFERENCES

Angenent, G. C., and Colombo, L. (1996). Molecular control of ovule development. Trends Plant Sci. 1: 228-232.

Bowman, J.L., Smyth, D.R., and Meyerowitz, E.M. (1991). Genetic interactions among floral homeotic genes of *Arabidopsis*. Development 112: 1-20.

Coen, E.S., and Meyerowitz, E.M. (1991). The war of the whorls: genetic interactions controlling flower development. Nature 353: 31-37.

Egea-Cortines, M., Saedler, H. & Sommer, H. (1999) Ternary complex formation between the MADS-box proteins SQUAMOSA, DEFICIENS and GLOBOSA is involved in the control of floral architecture in *Antirrhinum majus*. *EMBO J.* 18: 5370-5379.

Ferrándiz, C. Pelaz, S., and Yanofsky, M.F. (1999). Control of carpel and fruit development in Arabidopsis. Annu. Rev. Biochem. 68: 321-354.

Ferrándiz, C., Liljegrin, S.J., and Yanofsky, M.F. (2000). Negative regulation of the *SHATTERPROOF* genes by FRUITFULL during Arabidopsis fruit development. Science 289: 436-438.

Gu, Q., Ferrandiz, C., Yanofsky, M.F., and Martienssen, R. (1998) The *FRUITFUL* MADS-box gene mediates cell differentiation during Arabidopsis fruit development. Development 125: 1509-1517.

Liljegren, S.J., Ditta, G.S., Eshed, Y., Savidge, B., Bowman, J., and M.F. Yanofsky. (2000). *SHATTERPROOF* MADS-box genes control seed dispersal in *Arabidopsis*. Nature 404: 766-770.

Pelaz, S., G. Ditta, E. Baumann, E. Wisman and M. F. Yanofsky. (2000) B and C floral organ identity functions require *SEPALLATA* MADS-box genes. Nature 405: 200-203.

Pelaz, S., Tapia-Lopez, R., Alvarez-Buylla, E, and Yanofsky, M. (2000). Conversion of leaves into petals in Arabidopsis. Current Biol. 11: 182-184.

Riechmann, J. L. & Meyerowitz, E. M. (1997) MADS domain proteins in plant development. Biol. Chem. 378: 1079-1101.

Schwarz-Sommer, Z., Huijser, P., Nacken, W., Saedler, H., and Sommer, H. (1990). Genetic control of flower development: homeotic genes in *Antirrhinum majus*. Science 250: 931-936.

Sessions, R.A. and Zambryski, P.C. (1995). *Arabidopsis* gynoecium structure in the wild type and in *ettin* mutant. Development 121:1519-1532.

Theissen, G. (2000) A short history of MADS-box genes in plants. Plant Mol. Biol. 42: 115-149.

Theissen, G. (2001) Development of floral organ identity: Stories from the MADS-house. Current Opp. Plant Biol. 4: 75-85.

Yanofsky, M.F., Ma, H., Bowman, J.L., Drews, G., Feldmann, K. and Meyerowitz, E.M. (1990). The protein encoded by the *Arabidopsis* homeotic gene *AGAMOUS* resembles transcription factors. Nature 346: 35-39.

Dirk G. Inze

THE PLANT CELL CYCLE

Dirk Inze
Department of Plant Genetics, VIB, Ghent, Belgium (email: diinz@gengenp.rug.ac.be)

Cell division has a central role in growth and development of plants and in recent years tremendous progress has been made in the elucidation the molecular mechanisms regulation the plant cell cycle. Progression through the cell cycle is controlled by a class of Ser/Thr protein kinases known as cyclin dependent kinases or CDKs. CDK activity requires association with regulatory cyclins and is furthermore governed by numerous control mechanisms including transcription, translation, proteolysis, phosphorylation, dephosphorylation and interaction with inhibitory proteins (KRPs) and docking factors (CKS). I will provide an overview of the current knowledge on cell cycle regulation in plants and I will give several examples on how this knowledge can be used to inhibit or stimulate cell division in transgenic plants. Furthermore, I will give an update on the identification of the majority of genes (approximately 1300) the expression of which is restricted to a specific cell cycle phase. This genome-wide transcript profiling allowed us to identify a panoply of novel, often plant-specific, genes involved in cell cycle progression.

Abstract only – no manuscript received.

I. K. Vasil (ed.), Plant Biotechnology 2002 and Beyond, 29.

Kazuo Shinozaki

MOLECULAR MECHANISMS OF PLANT RESPONSES AND TOLERANCE OF DROUGHT AND COLD STRESS

Kazuo Shinozaki[1] and Kazuko Yamaguchi-Shinozaki[2]
[1]RIKEN Tsukuba Institute, Tsukuba, Ibaraki 305-0074, Japan (email: sinozaki@rtc.riken.go.jp) [2]Japan International Research Center for Agricultural Sciences (JIRCAS), Tsukuba, Ibaraki 305, Japan (email: kazukoys@jircas.affrac.go.jp)

Keywords Drought, cold stress, gene expression, stress tolerance, signal transduction, microrarray analysis

1. INTRODUCTION

Plants respond and adapt to a variety of environmental stresses including drought, cold and high salinity to survive in severe stress conditions. These stresses induce various physiological and biochemical responses in plants. Moreover, a variety of genes have been described that respond to these stresses at transcriptional level (Shinozaki and Yamaguchi-Shinozaki, 1997, 2000). Their gene products are thought to function in stress tolerance and response. Many stress-inducible genes have been used to improve stress tolerance of plants by gene transfer. It is important to analyze functions of stress-inducible genes not only for further understanding of molecular mechanisms of stress tolerance and response of higher plants but also for improvement of stress tolerance of crops by gene manipulation.

Dehydration triggers the production of abscisic acid (ABA), which, in turn, not only causes stomata closure but also induces various genes. There are at least two ABA-independent as well as two ABA-dependent signal-transduction cascades between the perception of drought-stress signal and the expression of specific genes (Shinozaki and Yamaguchi-Shinozaki, 1997, 2000). *Cis*- and *trans*-acting elements that function in ABA-independent and ABA-responsive gene expression by drought stress have been precisely analyzed (Shinozaki and Yamaguchi-Shinozaki, 2000). In this short article we described recent progress mainly on global analysis of expression profiles of stress responsive gene expression using 7,000 full-length cDNA microarray, and functions of stress-inducible genes. *Cis*- and *trans*-acting factors involved in ABA-independent and ABA–dependent gene expression systems are also described. Signal-transduction pathways in drought stress

I. K. Vasil (ed.), Plant Biotechnology 2002 and Beyond, 31-37.

response, especially roles of two component system and other protein kinsases in stress signaling.

2. EXPRESSION PROFILES OF STRESS-INDUCIBLE GENES USING FULL-LENGTH cDNA MICORARRAY

Full-length cDNAs are essential for correct annotation of genomic sequence, and for functional analysis of genes and their products. We isolated 155,144 RIKEN Arabidopsis Full-Length (RAFL) cDNA clones (Seki et al., 2002). The 3'-end ESTs of 155,144 RAFL cDNAs were clustered into 14,668 non-redundant cDNA groups, about 60% of predicted genes. We also obtained 5'-ESTs from 14,034 non-redundant cDNA groups and constructed promoter database. Furthermore, the full-length cDNAs are useful resources for analyses of expression profiles, functions and structures of plant proteins (Seki et al., 2001). Recently, we prepared a new full-length cDNA microarray containing ca. 7000 independent full-length cDNA groups. We prepared a full-length cDNA microarray containing ca. 7000 independent full-length cDNA groups to analyze the expression profiles of genes under drought, cold and high-salinity stress conditions. The transcripts of 277, 53 and 194 genes increased after drought, cold and high-salinity treatments, respectively, more than 5-fold compared with the control genes. We also identified many drought-, cold- or high-salinity-stress-preferentially-inducible genes. However, we observed strong relationship in the expression of these stress-responsive genes, and found 22 stress-inducible genes that responded to all three stresses. Several gene groups showing different expression profiles were identified by the analysis of their expression patterns during stress-responsive gene induction. Among the drought-, cold- or high-salinity-stress-inducible genes identified, we found 40 (corresponding to ca. 11% of all stress-inducible genes identified) transcription factor genes, suggesting that various transcriptional regulatory mechanisms function in the drought-, cold- or high-salinity-stress signal transduction pathways.

We have analyzed functions of drought- or cold-inducible genes in transgenics by overexpressing their full-length cDNAs. Recently, we showed that overexpression of two stress-inducible genes improves drought-stress tolerance in transgenics. One is a gene for 9-cis-epoxycarotenoid dioxygenase (NCED), a key enzyme in ABA biosynthesis (Iuchi et al., 2001). The other is a gene for galactinol synthase (GolS), a key enzyme involved in raffinose family oligosaccharide biosynthesis (Taji et al., 2002). This indicates that many stress-inducible genes are involved stress tolerance, and that stress-inducible genes are useful for molecular breeding of drought tolerance of transgenic plants.

3. ABA-DEPENDENT AND ABA–INDEPENDENT GENE EXPRESSION IN RESPONSE TO DROUGHT STRESS

As shown in Figure 1, it is now hypothesized that at least four independent signal pathways function in the activation of stress-inducible genes under dehydration conditions: two are ABA-dependent (pathways I and II) and two are ABA-independent (pathways III and IV) (Shinozaki and Yamaguchi-Shinozaki, 1997, 2000). There are at least two ABA-independent signaling pathways in response to cold stress (pathways IV and V). One of the ABA-dependent pathways requires protein biosynthesis (pathway I). Many stress- and ABA-inducible genes encoding various transcription factors have now been reported. These contain conserved DNA binding motifs, such as MYB (ATMYB2) and MYC (rd22BP1) (Abe et al., 1997). These transcription factors are thought to function in the regulation of ABA inducible genes, such as *rd22* in Arabidopsis, which respond to drought stress rather slowly after the production of ABA-inducible transcription factors (pathway I). *Cis*- and *trans*-acting factors involved in ABA-induced gene expression have been analyzed extensively (pathway II). A conserved sequence, PyACGTGGC, has been reported to function as an ABA-responsive element (ABRE) in many ABA-responsive genes (Shinozaki and Yamaguchi-Shinozaki 1997, 2000). Two cDNAs encoding DNA-binding proteins that specifically bind to the ABRE have been cloned by yeast one-hybrid screening and shown to contain the bZIP structure. We named the clones as AREB1 and 2 (Uno et al., 2000). We showed that AREB1 and 2 genes are induced by drought and their gene products are activated by phosphorylation. AREB1 and 2 function downstream from *abi1/abi2* and *era1* in ABA signal transduction pathways.

In the ABA-independent pathways, there are several drought-inducible genes that do not respond to either cold or ABA treatment, which suggests that there is a specific pathway that functions in one of dehydration stress response (pathway III). These genes include *rd19* and *rd21* that encode different cysteine proteases, and *ERD1* that encodes a Clp protease regulatory subunit (Nakashima et al., 1997). Promoter analysis of these genes will give us more information on pathway III.

One of the ABA-independent pathways of drought-stress response overlaps with that of cold-stress response (pathway IV). A cis-acting element including A/GCCGAC, named the Dehydration Responsive Element (DRE) and C-Repeat (CRT), is essential for the regulation of the induction of *rd29A* under drought, low-temperature, and high-salt stress conditions in an ABA-independent pathway. All the DRE/ CRT binding proteins (DREBs and CBFs) contain a conserved DNA binding motif (AP2/ERF motif) (Liu et al., 1998, Stockinger et al., 1997). These five cDNA clones that encode DRE/CRT binding proteins are classified into two groups, CBF1/DREB1 and DREB2. Expression of the *DREB1A (CBF3)* gene and its two homologues

(*DREB1B=CBF1, DREB1C=CBF2*) is induced by low-temperature stress, whereas expression of the *DREB2A* gene and its single homologue (*DREB2B*) was induced by dehydration (Liu et al., 1998). These results indicate that two independent families of DREB proteins, DREB1/CBF and DREB2, function as *trans*-acting factors in two separate signal transduction pathways under low-temperature and dehydration conditions, respectively (Liu et al., 1998, Kasuga et al. 1999). Overexpression of the DREB1A cDNA in transgenic *Arabidopsis* plants not only induced strong expression of the target genes under unstressed conditions but also revealed freezing and dehydration tolerance, which was also shown in the CBF1 transgenics (Liu et al., 1998; Jaglo-Ottosen et al., 1998). Micorarray analysis identified about 30 genes as DREB/CBF target genes (Seki et al. 2001). Genetic analysis of *Arabidopsis* mutants with the *rd29A* promoter::luciferase transgene suggests not only the existence of drought-, salt- and cold- specific signaling pathways in stress-response but also crosstalks between these signaling pathways were also observed (Ishitani et al., 1997, Zhu 2001).

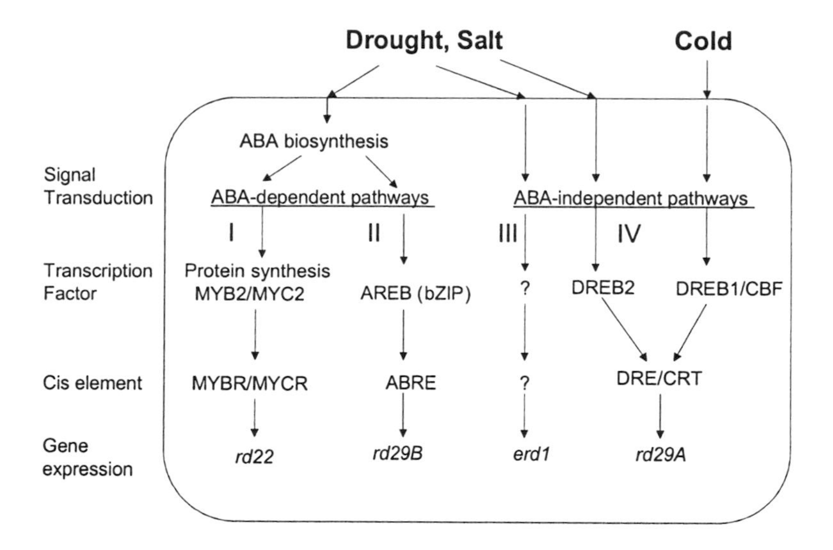

Figure 1. Signal transduction pathways between the perception of drought-stress signal and gene expression. At least four signal transduction pathways exist (I - IV): two are ABA-dependent (I and II) and two are ABA-independent (III and IV). Protein biosynthesis is required in one of the ABA-dependent pathways (I). In another ABA-dependent pathway, ABRE functions as an ABA-responsive element and does not require protein biosynthesis (II). In one of the ABA-independent pathways, DRE is involved in the regulation of genes not only by drought and salt but also by cold stress (IV). Another ABA-independent pathway is controlled by drought and salt, but not by cold (III).

4. SIGNAL PERCEPTION AND SIGNAL TRANSDUCTION IN DROUGHT STRESS RESPONSE

Signal-transduction pathways from the sensing of dehydration signal or osmotic change to the expression of various genes, and the signaling molecules that function in stress signaling have not been extensively studied in plants. Signal transduction pathways in drought stress response have been studied based on the knowledge in yeast and animal systems (Fig. 2). Two component systems function in sensing osmotic stress in bacteria and yeast. In plant, we isolated an *Arabidopsis* cDNA (ATHK1) encoding the two-component histidine kinase, an yeast osmosensor Sln1 homologue, by PCR. ATHK1 has a typical histidine kinase domain and a receiver domain like Sln1 (Urao et al. 1998). The ATHK1 transcript was more abundant in roots than other tissues under normal growth conditions and accumulated in conditions of high salinity and low temperature. Introduction of ATHK1 into a yeast mutant lacking two osmosensors, SLN1 and SHO1, allowed normal growth and activation of the HOG1 MAPK cascade under high osmolarity. These results suggest that ATHK1 can sense and transduce a signal of external osmolarity to downstream targets in yeast. We then generated transgenic Arabidopsis plants that overexpress dominant negative ATHK1 mutant cDNAs. The transgenics showed growth retardation in shoot and roots. Several stress-induced genes were upregulated in the transgenics under unstressed conditions. We think that ATHK1 may function in signal perception during drought stress in *Arabidopsis*. However, there may exist other stress sensing mechanisms, such as mechanical sensors of cytoskeltons and sensors of superoxides produced by stress.

Mitogen-activated protein kinase, or MAP kinase (MAPK), is involved in the signal-transduction pathways associated not only with growth factor-dependent cell proliferation but also with environmental stress responses in yeast and animals. We demonstrated rapid and transient activation of MAP kinase activities of AtMPK4 and AtMPK6 in *Arabidopsis* plants by low temperature, humidity change, wounding and touch (Ichimura et al., 2000). Activation of MPK kinase by osmotic stress is also observed in alfalfa. These observations indicate that certain MAP kinase cascade might function in the signal-transduction pathways in drought stress response.

Many genes for factors involved in the signal-transduction cascades, such as calcium dependent protein kinases and enzymes involved in phospholipid metabolism, such as phospholipase C, phospholipase D and PIP5 kinase, are upregulated by dehydration and cold (Shinozaki and Yamaguchi-Shinozaki, 2000). These signaling factors might be involved in the amplification of the stress signals and adaptation of plant cells to drought-stress conditions (Fig. 2). Transgenic plants that modify the expression of these genes and mutants with disrupted genes will give more information on the function of their gene products.

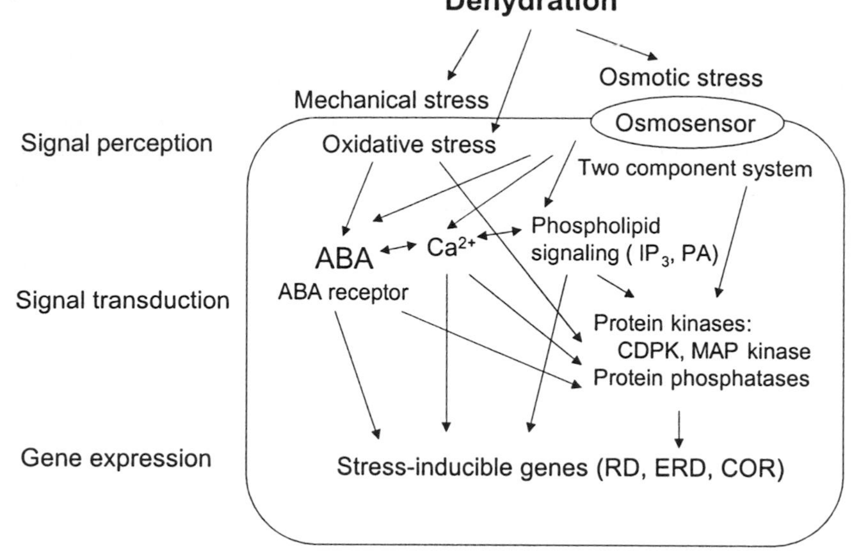

Fig. 2. Second messengers and factors involved in the signal perception and the signal transduction in drought-stress response. Two-component histidine kinase is thought to function as an osmosensor in plants. Ca^{2+} *and* IP_3 *are most probable second messengers of the dehydration signal. Phosphorylation process functions in water-stress and ABA signal-transduction pathways. PI turnover also involved in drought stress response. ABA plays important roles in the regulation of gene expression as well as physiological responses during water stress. Several ABA signal transduction pathways are reported.*

5. REFERENCES

Abe H, Yamaguchi-Shinozaki K, Urao T, Iwasaki T, Hosokawa D and Shinozaki K. 1997 Role of Arabidopsis MYC and MYB homologs in drought- and abscisic acid-regulated gene expression. Plant Cell 9: 1859-1868.

Ichimura K, Mizoguchi T, Yoshida R, Yuasa T, Shinozaki K. 2000 Various abiotic stresses rapidly activate Arabidopsis MAP kinases ATMPK4 and ATMPK6. Plant J. 24: 655-665.

Ishitani M, Xiong L, Stevenson B, Zhu JK. 1997 Genetic analysis of osmotic and cold stress signal transduction in Arabidopsis: Interactions and convergence of abscisic acid-dependent and abscisic acid-independent pathways. Plant Cell 9: 1-16.

Iuchi S, Kobayshi M, Taji T, Naramoto M, Seki M, Kato T, Tabata S, Kakubari Y, Yamaguchi-Shinozaki K, Shinozaki K. 2001 Regulation of drought tolerance by gene manipulation of 9-cis-epoxycarotenoide, a key enzyme in abscisic acid biosynthesis in Arabidopsis. Plant J. 27: 325-333.

Jaglo-Ottosen KR, Gilmour SJ, Zarka DG, Schabenberger O, Thomashow MF. 1998 Arabidopsis CBF1 overexpression induces coe genes and enhances freezing tolerance. Science 280: 104-106.

Kasuga M, Liu Q, Miura S, Yamaguchi-Shinozaki K, Shinozaki K. 1999 Improving plant drought, salt, and freezing tolerance by gene transfer of a single stress-inducible transcription factor. Nature Biotech. 17: 287-291.

Liu Q, Sakuma Y, Abe H, Kasuga M, Miura S, Yamaguchi-Shinozaki K, Shinozaki K. 1998 Two transcription factors, DREB1 and DREB2, with an EREBP/AP2 DNA binding domain, separate two cellular signal transduction pathways in drought- and low temperature-responsive gene expression, respectively, in Arabidopsis. Plant Cell 10: 1491-1406.

Mizoguchi T, Ichimura K, Shinozaki K. 1997 Environmental stress response in plants: the role of mitogen-activated protein kinases (MAPKs). Trends Biotech. 15: 15-19.

Nakashima K, Kiyosue T, Yamaguchi-Shinozaki K, Shinozaki K. 1997 A nuclear gene, *erd1*, encoding a chloroplast-targeted Clp protease regulatory subunit homolog is not only induced by water stress but also developmentally up-regulated during senescence in *Arabidopsis thaliana*. Plant J. 12: 851-861.

Seki M, Narusaka M, Abe H, Kasuga M, Yamaguhci-Shinozaki K, Carninci P, Hayashizaki Y, Shinozaki K. 2001 Monitoring the expression pattern of 1300 Arabidopsis genes under drought and cold stresses by using a full-length cDNA microarray. Plant Cell 13: 61-72.

Seki M, Narusama M, Kamiya A, Ishida J, Satou M, Sakurai T, Nakajima M, Enjyu A, Akiyama K, Oono Y, Muramatsu M, Hayashizaki Y, Kawai J, Carninci P, Itoh M, Ishii Y, Arakawa T, Shibata K, Shinagawa A, Shinozaki K. 2002 Functional annotation of a full-length cDNA collection. Science 296: 141-145.

Shinozaki K, Yamaguchi-Shinozaki K. 1997 Gene Expression and signal transduction in water-stress response. Plant Physiol. 115: 327-334.

Shinozaki K, Yamaguchi-Shinozaki K. 2000 Molecular responses to dehydration and low temperature: difference and cross-talk between two stress signaling pathways. Cur Opin Plant Biol. 3: 217-223.

Stockinger EJ, Glimour SJ, Thomashow MF. 1997 *Arabidopsis thaliana* CBF1 encodes an AP2 domain-containing transcription activator that binds to the C-repeat/DRE, a cis-acting DNA regulatory element that stimulates transcription in response to low temperature and water deficit. Proc. Natl. Acad. Sci. USA 94: 1035-1040.

Taji T, Ohsumi C, Iuchi S, Seki M, Kasuga M, Kobayashi M, Yamaguchi-Shinozaki K, Shinozaki K. 2002 Important roles of drought- and cold-inducible genes for galactinol synthase in stress tolerance in Arabidopsis thaliana. Plant J. 29: 417-426.

Uno Y, Furihata T, Abe H, Yoshida R, Shinozaki K, Yamaguchi-Shinozaki K. 2000 Arabidopsis basic leucine zipper transcriptional transcription factors involved in an abscisic acid-dependent signal transduction pathway under drought and high-salinity conditions. Proc Natl Acad Sci USA 97:11632-11637.

Urao T, Yakubov B, Satoh R, Yamaguchi-Shinozaki K, Shinozaki K. 1999 A transmembrane hybrid-type histidine kinase in Arabidopsis functions as an osmosensor. Plant Cell 11:1743-1754.

Yamaguchi-Shinozaki K, Shinozaki K. 1994 A novel *cis*-acting element in an Arabidopsis gene is involved in responsiveness to drought, low-temperature, or high-salt stress. Plant Cell 6:251-264.

Zhu JK, 2001 Plant salt tolerance. Trends Plant Sci. 6: 66-71.

Anna M.G. Koltunow

ADVANCES IN APOMIXIS RESEARCH: CAN WE FIX HETEROSIS?

Anna M. G. Koltunow[1] and Matthew R. Tucker[2]
[1]CSIRO Plant Industry. PO Box 350, Glen Osmond 5064, South Australia (e-mail: anna.koltunow@csiro.au); [2]Adelaide University, Waite Campus PMB 1, Urrbrae 5064, South Australia (e-mail: matthew.tucker@csiro.au)

Keywords Apomixis, apospory, gametophyte, Hieracium, embryo, endosperm, seed, *Arabidopsis*.

1. INTRODUCTION

1.1. Heterosis

Heterosis or hybrid vigour in plants is defined as the increased size, growth rate or productivity of the offspring resulting from a cross involving parents of different inbred lines of a species or occasionally from two different species. This is exploited in hybrid seed production for agriculture. How heterosis arises is not completely clear; it may result from the fixation of deleterious recessives in the inbreds (or species) or from epistatic interactions among the combined loci (Li et al., 2001; Luo et al., 2001). Heterosis declines in successive generations because of meiotic recombination during gamete formation and genetic segregation. This means that hybrid seed needs to be constantly renewed and parental stocks maintained and stored. Strategies are often required to minimize contamination of hybrid seed with unwanted seed progeny from each parent, further adding to the cost of production. Economies in hybrid seed production could be made if heterosis could be fixed and maintained in successive seed generations.

1.2. Apomixis

The transfer of apomixis, an asexual mode of seed formation, to agricultural crops where it is largely absent may enable perpetuation of hybrid vigour and economies in commercial hybrid seed formation. Apomixis has the potential to fix heterosis because meiosis does not occur during embryo sac formation, embryo formation is fertilization independent with the embryo maintaining a maternal genotype and functional endosperm development occurs with or without fertilization. The fixation of heterosis in crop species by apomixis and the resulting decrease in the costs of seed production should allow the

I. K. Vasil (ed.), Plant Biotechnology 2002 and Beyond, 39-46.

yield advantage of hybrid seed to be passed onto resource-poor farmers (Koltunow et al., 1995).

The modes of apomixis vary in detail amongst the recorded 300 or so species and the capacity for sexual reproduction is retained in most apomicts to varying degrees (Koltunow, 1993). Autonomous apomictic species like dicotyledonous *Hieracium* (Koltunow et al., 1998) and *Taraxacum* (van Baarlen et al., 2000) that avoid meiosis, and initiate fertilization-independent embryo and endosperm development are rare. In the majority of apomicts, fertilization of the central cell is required for endosperm initiation and subsequent seed development. Despite the documented variation, genetic analysis has shown that apomixis is conferred by only a few dominant loci (Bicknell et al., 2000; Martinez et al., 2001; Tas and van Dyk, 1999). Additional modifier loci are important for aspects of the initiation and continuation of apomixis and the amount of viable seed set in *Hieracium* (Koltunow et al., 2000).

Current theories concerning apomixis suggest that common genes are shared in these processes but apomixis enables a short-circuited version of the sexual process, perhaps by inducing altered expression of genes involved in sexual reproduction in both space and time (Koltunow et al., 1993; Grimanelli et al., 2001). The apomixis locus might be comprised of one or many linked genes that might contain rearrangements and structural divergence such as that observed in apomixis associated chromosomes of *Pennisetum* and *Cenchrus* (Ozias-Akins et al., this volume).

1.3. Combining Apomixis And Heterosis - The Status Of Apomixis Research

Apomixis is not prevalent in many crop plants and most crops do not have apomictic relatives to facilitate introgression of the trait by breeding. Success has been obtained towards introgressing apomixis from *Pennisetum* to pearl millet (Dujardin and Hanna, 1989). An introgression program aimed at introducing apomixis to corn from *Tripsacum* is in progress (Savidan, 2000). However, apomixis would only be valuable for breeding and commercial seed production if it could be switched on and off appropriately as required. Greatest value would be obtained from introducing apomixis to cereals.

A high degree of seed sterility is observed in apomictic hybrids in introgression programs and also in genetic studies where crosses are made between apomictic and sexual plants. A balanced ratio of maternal to paternal genomes in the persistent cereal endosperm (Lin, 1984; Haig and Westoby, 1989) and also the ephemeral *Arabidopsis* endosperm (Scott et al., 1998) of 2:1 is critical for seed viability in these sexually reproducing plants.

Autonomous endosperm forming apomicts such as *Taraxacum* and *Hieracium* are insensitive to such requirements and viable seed is produced even though the endosperm is comprised of only maternal DNA. The pseudogamous apomictic grasses *Tripsacum* (Grimanelli et al., 1997) and *Paspalum* (Quarin, 1999) exhibit unbalanced parental genome ratios in their endosperm following fertilization of the central cell and produce viable seeds. However, seed sterility in backcrossed pseudogamous *Pennisetum* correlates with an alternative 4-nucleate embryo sac containing two polar nuclei which does not give rise to a 2:1 maternal, paternal genome ratio in the endosperm following pseudogamy (accompanying paper by Ozias-Akins). An unbalanced genome ratio in the endosperm might pose a barrier to the successful transfer of apomixis to crops (Grossniklaus et al., 2001; Spillane et al., this volume).

Molecular research on apomixis occurs in both sexual and apomictic plants. The biotechnological synthesis of apomixis in sexual plants is being attempted by introducing genes to confer the individual components of apomixis such as embryo sac formation, somatic embryogenesis in the ovule and induction of autonomous endosperm formation. This is being aided by fundamental research identifying genes controlling embryo sac, embryo and endosperm initiation in plants. Successful synthesis of apomixis needs to include control over the expression of the endogenous sexual process with a capacity to induce sexuality and apomixis as required for breeding programs. Characterization of apomixis in functional apomicts includes mapping strategies, mutagenesis, differential screens comparing sexual and apomictic plants and developmental analyses employing markers and molecular tools from apomictic and sexual plants. Attempts to induce apomixis by mutagenesis of sexual male sterile (*Arabidopsis*) plants has resulted in the discovery of the *FERTILIZATION INDEPENDENT SEED* (*FIS*) class of genes that resemble members of the *Drosophila* Polycomb group (Chaudhury et al., 1997; Ohad et al., 1998; Grossniklaus et al., 1998). These genes appear to be required for the appropriate temporal initiation of fertilization-dependent seed development and for early endosperm development in *Arabidopsis* (Chaudhury et al., 2001; Sorenson et al., 2001). The *fis* mutants display fertilization independent endosperm initiation, which is a component of apomixis. However, seed development aborts and the mutants are gametophytic in nature, which contrasts with the dominant sporophytic inheritance of apomixis observed in genetic studies.

2. ANALYSIS OF APOMIXIS IN *HIERACIUM*

An investigation of the role and activity of *FIS* genes in autonomous and pseudogamous apomicts may provide clues towards understanding how apomicts initiate endosperm formation and avoid imprinting barriers to produce viable seeds. Here we provide an update of the current status of the examination of *FIS* class gene function during apomictic and sexual seed formation in *Hieracium*, an autonomous apomict. In addition, we summarize the use of *Arabidopsis* markers to determine the identity of cells initiating apomixis and the relationships between gene expression programs during sexual and apomictic development in *Hieracium*.

2.1. Sexuality And Apomixis Share Gene Expression Programs

Hieracium plants initiate apomixis at different times during ovule development relative to the temporal sequence of sexual reproduction events that are also occurring in the ovules (Koltunow et al., 1998; Tucker et al., 2001). The formation of aposporous initial cells results in a cessation of the sexual process in the majority of the ovules. The mature ovule usually contains a single embryo sac that originates from a single initial cell or the fusion of multiple aposporous structures. The reproductive structures formed in the apomictic pathway are often aberrant compared to those formed in the sexual plant. Embryo and endosperm development is autonomous and some developing seeds do not initiate embryogenesis (Koltunow et al., 1998; 2000).

During *Arabidopsis* ovule development the three *FIS*-class genes are co-expressed late in ovule development near the central cell polar nuclei in the mature embryo sac. After fertilization their expression is also detected near dividing endosperm nuclei during early stages of endosperm formation (Luo et al., 2000). This pattern is consistent with their postulated role in forming a repressive complex that inhibits the expression of endosperm development genes and their later function in regulating endosperm polarity (Sorenson et al., 2001). The chimeric *Arabidopsis FIS::GUS* reporter genes were used as markers in transgenic sexual and apomictic *Hieracium* plants to examine the similarities in gene expression programs in the two different reproductive pathways.

During the late events of ovule maturation and early seed initiation in sexual and apomictic *Hieracium*, the three fusion constructs were expressed in a similar manner and only in reproductive structures. The pattern was comparable with that observed in the late stages of *Arabidopsis* ovule and early seed development indicating the utility of using *Arabidopsis* tools in *Hieracium*. The genes also marked spatially and compositionally altered reproductive structures that are routinely observed in apomictic reproduction such as multiple embryo sacs in a single ovule, seeds forming endosperm but lacking and embryo and aposporous structures positioned in the chalazal portion of the ovule. Collectively these data indicated that common gene expression programs are shared in the later events of both sexual and apomictic pathways in *Hieracium,* supporting the concept that apomixis does not induce a distinct pathway that mimics sexual reproduction.

2.2. Cells Initiating Apomixis In *Hieracium* Do Not Have Megaspore Mother Cell Identity

Another question concerns the identity of the aposporous initial cell. In *Hieracium* plants we have characterised cytologically, differences in timing of aposporous initial cell differentiation during ovule development indicate that this cell could have megaspore mother cell identity, express programs equivalent to a reduced spore or a selected spore. The *Arabidopsis SPOROCYTELESS::GUS* marker (*SPL::GUS*, Yang et al., 1999) was used to examine the identity of the aposporous initial cell in an apomict that usually differentiates its aposporous initial at the time of meiosis in the sexual pathway. Similar to *Arabidopsis*, *SPL::GUS* expression was observed in the megaspore mother cells of sexual and apomictic plants and then faintly and briefly in early megaspore tetrads. Expression was absent from differentiating aposporous initial cells and enlarging selected spores in the sexual pathway. This suggested that in this apomict the initial cell does not have megaspore mother cell or early megaspore identity.

2.3. Cells Initiating Apomixis In *Hieracium* Have Selected Spore Identity

The identity of the aposporous initial was clarified when sexual and apomictic plants expressing the *Arabidopsis FIS::GUS* constructs were examined early in ovule development. Surprisingly *FIS2::GUS* was expressed in both sexual and apomictic *Hieracium* at the time of meiosis but in a differential manner in sexual and apomictic plants. Histochemical *GUS* staining and in situs using *GUS* probes in sexual *Hieracium* and two apomicts distinguished by the number of initials formed and also their mode of embryo sac formation suggested that the identity of the initial when it differentiates close to the time of sexual meiosis is likely to be that of a functional megaspore. This supports the concept that apomixis in *Hieracium* may indeed be a short-circuited version of the sexual process and somatic cells in the plants examined enter the reproductive pathway by differentiating a cell with selected spore identity.

2.4. Apomixis Involves Spatial Alterations In Genes Expressed In The Sexual Pathway

A shift in the pattern of *FIS2::GUS* expression was observed in apomicts relative to sexual plants when the aposporous initial cells formed. A greater zone of cells was marked in apomicts and all of these tissues subsequently degenerated. This suggests that initiation of apomixis may lead to an alteration in gene expression in cells involved in sexual reproduction that results in active displacement of the sexual pathway rather than a passive exclusion resulting from competing structures limited by space restrictions.

2.5. Isolation Of *Hieracium FIS*-Like Genes And Their Down Regulation In Sexual And Apomictic Plants

FIS2-like sequences have been isolated from *Hieracium* using the conserved ACE domain (Chaudhury et al., 2001). As *FIS2*-like genes form a family with a diverse range of functions, silencing strategies will target the ACE domains specifically in the ovule using the *Arabidopsis FIS2* promoter. *FIE* has been isolated from *Hieracium* and the cDNA shows strong conservation with *FIE* sequences from *Arabidopsis* and other species. Isolation of genomic *FIE* sequences and promoters from sexual and apomictic polyploid *Hieracium* plants indicated that the *FIE* sequences from the apomict had diverged from the sexual, containing more non-functional copies and a greater degree of sequence inversions and alterations. Sexual and apomictic plants containing RNAi constructions to down regulate *FIE* in the ovule are being examined and preliminary data in apomicts shows reduced seed set. Conclusions cannot be drawn until data is complete and plants with up-regulated *FIS* activity are also analysed.

3. SUMMARY

Our aim is to understand how the apomictic process is regulated and how it interacts with the sexual pathway during ovule development. An understanding of apomictic processes is crucial to enable the transfer of apomixis to crop species and to regulate sexual and apomictic events to fix hybrid vigour in commercial production. Here we have described that gene expression programs are shared during sexual and apomictic development in *Hieracium*. The aposporous initial cells forming near sexual meiosis do not appear to have the identity of a MMC, supporting the concept that apomixis might be a short-circuited sexual pathway. Spatial changes in gene expression occur in apomictic initiation compared with sexual. Correlation of the expression pattern with cells that will degrade suggests apomixis might displace the sexual process in an active manner. Strategies attempting to engineer apomixis will need to take into account elements common to both sexual and apomictic pathways when developing tools attempting to initiate and control each process for agricultural production. Uncovering methods to produce viable seeds in the absence of parental imprinting is also required.

4. REFERENCES

Bicknell RA, Borst NK, Koltunow AM (2000) Monogenic inheritance of apomixis in two Hieracium species with distinct developmental mechanisms. Heredity 84 (2): 228-237

Chaudhury AM, Ming L, Miller C, Craig S, Dennis ES, Peacock WJ (1997) Fertilisation-independent seed development in *Arabidopsis thaliana*. Proc. Natl. Acad. Sci. USA 94: 4223-4228

Chaudhury AM, Koltunow A, Payne T, Luo M, Tucker MR, Dennis ES, Peacock WJ (2001) Control of early seed development. Ann. Rev. Cell Dev. Biol. 17: 677-699

Dujardin M and Hanna WW (1989) Developing apomictic pearl millet - Characterization of a BC3 plant. J. Gen. Breed. 43: 145-151

Grimanelli D, Leblanc O, Perotti E, Grossniklaus U (2001) Developmental genetics of gametophytic apomixis. Trends Genet. 17 (10): 597-604

Grimanelli D, Hernandez M, Perotti E, Savidan Y (1997) Dosage effects in the endosperm of diplosporous apomictic *Tripsacum* (Poaceae). Sex. Plant Reprod. 10 (5): 279-282

Grossniklaus U, Spillane C, Page DR, Kohler C (2001) Genomic imprinting and seed development: endosperm formation with and without sex. Curr. Opin. Plant Biol. 4: 21-27

Grossniklaus U, Vielle-Calzada JP, Hoeppner MA, Gagliano WB (1998) Maternal control of embryogenesis by medea, a Polycomb group gene in *Arabidopsis*. Science 280: 446-450.

Haig, D. and Westoby, M. (1989). Parent-specific gene expression and the triploid endosperm. Am. Nat. 134: 147-155.

Koltunow AM (1993) Apomixis - embryo sacs and embryos formed without meiosis or fertilization in ovules. Plant Cell 5 (10): 1425-1437

Koltunow AM, Bicknell RA, Chaudhury AM (1995) Apomixis - molecular strategies for the generation of genetically identical seeds without fertilization. Plant Phys. 108 (4): 1345-1352

Koltunow AM, Johnson SD, Bicknell RA (1998) Sexual and apomictic development in *Hieracium*. Sex. Plant Reprod. 11 (4): 213-230

Koltunow AM, Johnson SD, Bicknell RA (2000) Apomixis is not developmentally conserved in related, genetically characterized Hieracium plants of varying ploidy. Sex. Plant Reprod. 12 (5): 253-266

Li ZK, Luo LJ, Mei HW, Wang DL, Shu QY, Tabien R, Zhong DB, Ying CS, Stansel JW, Khush GS, Paterson AH (2001) Overdominant epistatic loci are the primary genetic basis of inbreeding depression and heterosis in rice. I. Biomass and grain yield genetics 158 (4): 1737-1753

Lin B-Y (1984) Ploidy barrier to endosperm development in maize. Genetics 107:103-115

Luo LJ, Li ZK, Mei HW, Shu QY, Tabien R, Zhong DB, Ying CS, Stansel JW, Khush GS, Paterson AH (2001) Overdominant epistatic loci are the primary genetic basis of inbreeding depression and heterosis in rice. II. Grain yield components Genetics 158 (4): 1755-1771

Luo M, Bilodeau P, Dennis ES, Peacock WJ, Chaudhury A (2000) Expression and parent-of-origin effects for *FIS2*, *MEA*, and *FIE* in the endosperm and embryo of developing *Arabidopsis* seeds. Proc. Natl. Acad. Sci. USA 97 (19): 10637-10642

Martinez EJ, Urbani MH, Quarin CL, Ortiz JPA (2001) Inheritance of apospory in babliagrass, *Paspalum notatum*. Hereditas 135 (1): 19-25

Ohad N, Yadegari R, Margossian L, Hannon M, Michaeli D, Harada JJ, Goldberg RB, Fischer RL (1999) Mutations in FIE, a WD polycomb group gene, allow endosperm development without fertilization. Plant Cell 11(3): 407-415

Quarin CL (1999) Effect of pollen source and pollen ploidy on endosperm formation and seed set in pseudogamous apomictic *Paspalum notatum*. Sex. Plant Reprod. 11 (6): 331-335

Savidan Y (2000) Apomixis: Genetics and Breeding. In Plant Breeding Reviews, Vol 18 (ed J Janick) John Wiley and Sons, 13-86

Scott RJ, Spielman M, Bailey J, Dickinson HG (1998) Parent-of-origin effects on seed development in *Arabidopsis thaliana*. Development 125: 3329-3341

Sorenson MB, Chaudhury AM, Robert H, Bancharel E, Berger F (2001) Polycomb group genes control pattern formation in plant seed. Curr. Opin. Plant Biol. 11: 277-281

Spillane C, Steimer A, Grossniklaus U (2001) Apomixis in agriculture: the quest for clonal seeds. Sex. Plant Reprod. 14 (4): 179-187

Tas ICQ, Van Dijk PJ (1999) Crosses between sexual and apomictic dandelions (*Taraxacum*). I. The inheritance of apomixis. Heredity 83: 707-714

Tucker MR, Paech NA, Willemse MTM, Koltunow AMG (2001) Dynamics of callose deposition and β-1,3-glucanase expression during reproductive events in sexual and apomictic *Hieracium*. Planta 212: 487-498

van Baarlen P, van Dijk P, Hoekstra RF, de Jong JH (2000) Meiotic recombination in sexual diploid and apomictic triploid dandelions (*Taraxacum officinale L.*). Genome 43 (5): 827-835

Yang WC, Ye D, Xu J, Sundaresan V (1999) The SPOROCYTELESS gene of *Arabidopsis* is required for initiation of sporogenesis and encodes a novel nuclear protein. Gen. Dev. 13 (16): 2108-211

David C. Baulcombe

OVERCOMING AND EXPLOITING RNA SILENCING

David C. Baulcombe
The Sainsbury Laboratory, John Innes Centre, Colney Lane, Norwich NR4 7UH, UK (e-mail: David.Baulcombe@Sainsbury-Laboratory.ac.uk)

Keywords Cosuppression, functional genomics, protein overexpression

1. INTRODUCTION

RNA silencing is an immune system in plants and animals that allows the cell to detect and eliminate foreign RNA. It plays an important role in defense against mobile DNA and viruses. Here we describe how RNA silencing was discovered and summarize recent progress towards understanding its role and mechanism. We also describe the application of RNA silencing in functional genomics and strategies for protein overexpression that rely on suppression of silencing.

2. COSUPPRESSION AND OTHER SILENCING PHENOMENA

Towards the end of the 1980s a series of unexpected observations with transgenic plants provided the first hints of RNA silencing. These observations indicated that sense orientation transgenes can interfere with the expression of similar endogenous genes (Napoli et al., 1990; van der Krol et al., 1990). The phenomena, called sense- or co-suppression because there is coordinate suppression of both transgenes and endogenous genes, are due to a posttranscriptional process that involves targeted mRNA degradation (Van Blokland et al., 1994).

Some clues to the mechanism of silencing came from experiments with tobacco etch virus (TEV) and potato virus X (PVX). These experiments showed that viral transgenes could confer virus resistance through a mechanism that resembled cosuppression (Lindbo et al., 1993; English et al., 1996) in that it operated at the level of RNA and resulted in coordinate suppression of transgene and viral RNA. Moreover, because TEV and PVX replicate in the cytoplasm, it was concluded that the silencing is cytoplasmic rather than nuclear.

I. K. Vasil (ed.), Plant Biotechnology 2002 and Beyond, 49-58.

Further clues to the mechanism of silencing came from fungi and animals. There is silencing in *Neurospora*, worms, flies and other organisms and, from genetic analysis, it became apparent that a conserved mechanism is involved. For example, the genes encoding an EIF2C/Argonaut-like protein, a putative RNA dependent RNA polymerase and an RNAse D-like protein are a common requirement of silencing in these different organisms (reviewed in (Hammond et al., 2001). The involvement of double stranded (ds) RNA is also feature of RNA silencing in both animals and plants.

A hallmark of silencing is nucleotide sequence-specificity. For example the silencing-mediated virus resistance in transgenic plants is specific for strains of the virus that are very similar to the viral transgene (Mueller et al., 1995). Similarly, in worms, the ds RNA that was injected into the animals determined the specificity of silencing (Fire et al., 1998).

The simplest way to account for the specificity of silencing invoked antisense RNA that would be produced directly or indirectly from sense transgenes or injected dsRNA. However, despite many efforts, it had never been possible to correlate silencing with antisense RNA. The reason, we now know is because the antisense RNA is only 21-25 nucleotides long (Hamilton and Baulcombe, 1999). It had been missed previously because the methods used were not adapted to such small molecules. These molecules are now known as small interfering (si)RNA and are found in both animals and plants. When added to cells and cell free extracts (Elbashir et al., 2001a; Elbashir et al., 2001b) these siRNAs determine the specificity of RNA silencing either *in vivo* or *in vitro*. It is thought that the siRNA guides an RNAse to the target RNA of silencing through a base pairing interaction. The siRNA-RNase complex is known as the RNA induced silencing complex (RISC) (Hammond et al., 2000). (Hamilton et al., 2002; Voinnet et al., 2002).

The siRNA is likely to be derived from a dsRNA precursor because it exists as both sense and antisense of the silencing target (Hamilton and Baulcombe, 1999). Consistent with this idea, there is an RNAseIII with dsRNA binding motifs (Dicer) that processes long dsRNA into siRNA *in vitro* and plays a role in silencing *in vivo* (Bernstein et al., 2001). The combined biochemical, genetic and molecular analyses suggest that the core silencing mechanism, as shown in Figure 1, involves an RNA synthesis step and two RNAse steps. The RNA synthesis step produces dsRNA but can be bypassed if dsRNA is introduced directly (Dalmay et al., 2000). Since this process operates entirely at the RNA rather than DNA level I use the term 'RNA silencing' to replace the previously used 'gene silencing' and 'cosuppression'.

Eventually the understanding of RNA silencing will be considerably more detailed than the scheme shown in Figure 1. There will be other, as yet unknown, proteins to accommodate. In addition it will be necessary to account for features of the RNA silencing that are not yet understood at the molecular level. One of these features is systemic signaling that was discovered in grafting experiments (Palauqui et al., 1997) and when silencing was initiated in localized parts of transgenic plants carrying a GFP transgene (Voinnet and Baulcombe, 1997; Voinnet et al., 1998). The signaling process, like the intracellular phase of the silencing mechanism molecule is highly nucleotide sequence specific. It is likely therefore that the signal specificity determinant will be RNA. However, as yet, the identity of the signal is unknown.

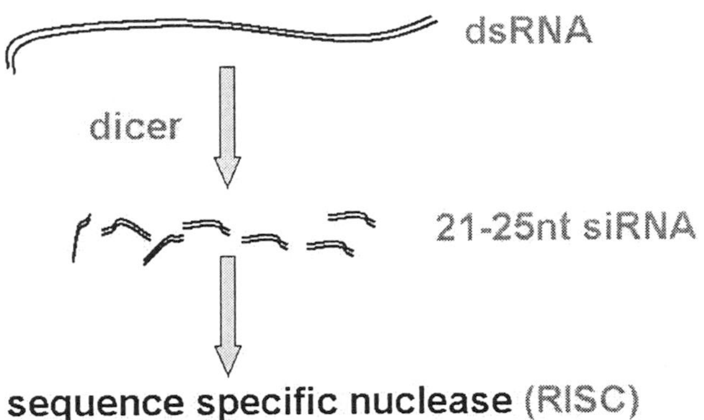

Figure 1.

3. RNA SILENCING AND PROTECTION AGAINST VIRUSES

It was striking that RNA silencing could produce exceptionally strong virus resistance in transgenic plants (Mueller et al., 1995). In many of these plants there was no detectable accumulation of virus in the inoculated leaf and the resistance could not be overcome even by high titre inocula. These observations prompted the speculation that the transgenic resistance is based on a natural resistance mechanism. Consistent with this idea it was found that virus-infected plants contain siRNAs corresponding to the viral genome. In addition it was found that viruses encode suppressor proteins of RNA silencing in transgenic plants (Anandalakshmi et al., 1998; Brigneti et al., 1998; Kasschau and Carrington, 1998). These suppressors of silencing are structurally diverse (Voinnet et al., 1999) and it seems likely that they evolved several times to counteract the antiviral effects of RNA silencing. If the virus encodes a strong suppressor of RNA silencing it would become abundant in the infected cell.

4. VIRUS INDUCED SILENCING

When a plant is infected with a virus vector RNA silencing is targeted against the viral genome including the sequences carried as an insert in the virus vector. Consequently, if the insert corresponds to an endogenous sequence, the corresponding host RNA is targeted by the silencing mechanism and the symptoms on the infected plant resemble the phenotype of a reduced or loss of function mutant (Baulcombe, 1999).

We have tested virus-induced gene silencing (VIGS) with vectors based on tobacco rattle virus (Ratcliff et al., 2001) and PVX (Ruiz et al., 1998) on a range of plants including *Nicotiana benthamiana*, other *Solanaceae* and *Arabidopsis thaliana*. On many of these species there is some evidence for VIGS. However it is normally transient and restricted to regions around the veins. Only on *N. benthamiana* and *N. clevelandii* is the VIGS phenotype extensive although, even on those plants, it is normally transient to some extent.

VIGS is effective with RNA and DNA viruses and has been targeted against the mRNAs of various enzymes (Kjemtrup et al., 1998; Ruiz et al., 1998; Ratcliff et al., 2001). It has also been targeted against a homologue of a *leafy*, a gene required for development of floral meristems so that the infected *N. benthamiana* plants have the appearance of a *leafy* mutant (Ratcliff *et al.*, 2001). In addition we have targeted *Prf* (Salmeron et al., 1996), a gene required for *Pto*-mediated resistance against *Pseudomonas syringae* pv *tabaci* (hereafter *Ps. tabaci*).

From these results we inferred that VIGS would be useful as a general tool of gene identification (Baulcombe, 1999) and initiated a programme to identify proteins involved in disease resistance. We planned a reverse approach in which candidate cofactors of disease resistance could be targeted by VIGS. In addition, because the use of the virus vector is amenable to high throughput applications, it was possible to carry out a forward screen. This forward screen was carried out with PVX vectors carrying *N. benthamiana* RNA inserts. Each PVX clone carried a single insert and 5000 independent clones were tested. Because the RNA inserts were from a normalized cDNA library it is likely that these 5000 clones represent more than 4000 independent genes.

The procedure used for these experiments involves first cloning an insert from a host gene into a TRV or PVX vector. The vector is then inoculated to the host plant and after two or three weeks a second challenge inoculum is applied to test for disease resistance. From this screen we have identified ten viral vector constructs that suppress the various types of disease resistance being tested. In each of these instances the insert in the VIGS vectors is an indicator of target gene identity. However, if there is a multigene family or if conserved domains are present in different genes, the loss of resistance may be due to suppression of genes that are similar to but not the same as that in the vector insert. At present there are no precise guidelines about the similarity requirement in a viral insert and VIGS target. With transgenes the RNA silencing phenotype breaks down if there is more than 20% mismatch with a potential target over several hundred nucleotides (Mueller et al., 1995). However since it is possible to initiate VIGS with only 28 nucleotides of similarity between a virus vector and the target gene (Thomas et al., 2001) the overall measure of nucleotide similarity may not necessarily indicate which genes are the potential targets of VIGS.

To confirm the target of VIGS it is sometimes helpful to target different regions of genes in a family. The 3' untranslated region may be a particularly informative target region because it is often highly variable between different members of a multigene family. Ideally biochemical and genetic approaches will be used to confirm that genes identified by VIGS are required for disease resistance.

A potential development in VIGS is promoter silencing. This technology would be based on observations that VIGS can be targeted to methylate and prevent transcription of a 35S-GFP transgene (Jones et al., 2001). Interestingly the silencing persists in the progeny of the infected plant even though the virus is not transmitted through the seed. To silence endogenous genes it will be necessary to generate virus vector constructs carrying inserts corresponding to the endogenous gene promoters.

There are several potential benefits of promoter silencing over the posttranscriptional effect that takes place when transcribed regions are targeted. It could be, for example, that genes with similar coding sequences could be silenced separately if there is sequence divergence in the promoters. Such similar genes would be difficult to silence differentially if the conserved coding sequence were the target. In addition it might be expected that the promoter silencing would be more persistent than posttranscriptional silencing targeted against transcribed regions.

5. TRANSGENIC SILENCING

VIGS is fast (Baulcombe, 1999). The viral constructs can be assembled within a day or so and the silencing phenotype observed in the infected plants within a further two or three weeks (Ruiz et al., 1998; Ratcliff et al., 2001). However the VIGS persists for only two or three weeks and is then lost.

Moreover VIGS does not affect the growing points and early stages of plant growth and development (Ruiz et al., 1998). To achieve silencing at these stages it is preferable to generate plants carrying a silencing transgene. At present the most effective type of silencing transgene has an inverted repeat configuration so that the targeted gene is expressed as dsRNA (Chuang and Meyerowitz, 2000; Wesley et al., 2001). This dsRNA is processed by Dicer so that corresponding endogenous gene RNAs are targeted by RISC (Figure 1). In principle these inverted repeat constructs could also mediate transcriptional silencing if they were targeted against endogenous promoter sequences.

In future it is likely that developments in silencing technology will exploit the curious phenomenon of transitivity in which the effect of silencing transits along a transgene (Voinnet et al., 1998; Sijen et al., 2001; Vaistij et al., 2002). Transitive silencing will use chimaeric transgenes in which there is sequence of the target endogenous gene is linked in cis to a reporter gene. When transformed into a plant that does not carry any other transgenes the reporter gene will be expressed and there will be no silencing of the endogenous gene. However, when introduced into a plant that was previously transformed to silence the reporter gene there will be two main effects. First the reporter in the chimaeric transgene will be silenced. Second, due to the transitive effect, the endogenous gene will be silenced.

There are likely to be several advantages of transitive silencing over that caused by the inverted repeat constructs (Voinnet et al., 1998; Sijen et al., 2001). First the constructs will be more straightforward because the silencing target will be present only once. In addition, by using different promoters in the silencer construct, it will be possible to tune the silencing effect. For example, if the silencer is expressed only in seed, the silencing will be seed-specific. Similarly, if the silencer is expressed under control of an inducible promoter, the silencing will be conditional on application of the inducer. Other patterns of silencing will be achievable depending on the promoter in the silencer construct.

6. OVERCOMING SILENCING

Transgene expression is often limited by RNA silencing. Presumably if the multiple transgenes are integrated into the genome in a head-to-head or tail to tail configuration the transcripts are produced as inverted repeats with intramolecular dsRNA. Alternatively the transcripts could be a template for an RdRP that generates dsRNA (Dalmay et al., 2000). In situations where high level transgene expression is required, for example in the production of a pharmaceutically active compound, there are now several strategies for overcoming this limiting effect of silencing. For example, if the transgene is introduced into mutant plants that are not competent to carry out silencing, the expression would be higher than in wild type plants. Alternatively this transgene could be introduced together with a second transgene encoding a viral suppressor of silencing.

An example of high level transgene expression involved a viral amplicon that was transformed into plants producing the potyviral HC-pro suppressor of silencing. An amplicon is a viral transgene that encodes the viral replication enzyme as well as the gene to be expressed at a high level. It was designed so that the amplicon RNA includes the cis-acting elements that are required for viral RNA replication. In the absence of the suppressor of silencing the replication enzyme catalyses production of dsRNA that is a potent activator of silencing (Angell and Baulcombe, 1999). However in the presence of the suppressor of silencing the amplicon RNA is replicated freely and consequently there are high levels of the mRNA for the gene to be expressed at a high level. The protein encoded by this RNA can be produced at very high levels.

The suppressors of silencing can also be used to enhance transgenes in a transient assay. Agrobacterium strains carrying the suppressor transgene and the gene to be expressed are mixed and infiltrated into a leaf (Voinnet et al., 2000; Johansen and Carrington, 2001). For the first two days after infiltration the expression is the same irrespective of whether a suppressor is being produced. At later times however there is differential expression so that, in the presence of a suppressor the amount of transgene encoded protein can be up to fifty-fold higher than in its absence. In *N.benthamiana* the strongest suppressor recognized to date is the 19kDa protein of tomato bushy stunt virus. This protein prevents accumulation of the siRNAs that are integral to the silencing mechanism (Hamilton et al., 2002; Voinnet et al., 2002).

7. REFERENCES

Anandalakshmi, R., Pruss, G.J., Ge, X., Marathe, R., Smith, T.H. and Vance, V.B. (1998) A viral suppressor of gene silencing in plants. Proc. Natl. Acad. Sci. USA, 95, 13079-13084.

Angell, S.M. and Baulcombe, D.C. (1999) Potato virus X amplicon-mediated silencing of nuclear genes. Plant J., 20, 357-362.

Baulcombe, D.C. (1999) Fast forward genetics based on virus-induced gene silencing. Curr. Opin. Plant. Biol., 2, 109-113.

Bernstein, E., Caudy, A.A., Hammond, S.M. and Hannon, G.J. (2001) Role for a bidentate ribonuclease in the initiation step of RNA interference. Nature, 409, 363-366.

Brigneti, G., Voinnet, O., Li, W.X., Ji, L.H., Ding, S.W. and Baulcombe, D.C. (1998) Viral pathogenicity determinants are suppressors of transgene silencing in *Nicotiana benthamiana*. EMBO Journal., 17, 6739-6746.

Chuang, C.-H. and Meyerowitz, E.M. (2000) Specific and heritable genetic interference by double-stranded RNA in *Arabidopsis thaliana*. Proc. Natl. Acad. Sci. USA, 97, 4985-4990.

Dalmay, T., Hamilton, A.J., Rudd, S., Angell, S. and Baulcombe, D.C. (2000) An RNA-dependent RNA polymerase gene in *Arabidopsis* is required for posttranscriptional gene silencing mediated by a transgene but not by a virus. Cell, 101, 543-553.

Elbashir, S.M., Harborth, J., Lendeckel, W., Yalcin, A., Weber, K. and Tuschl, T. (2001a) Duplexes of 21-nucleotide RNAs mediate RNA interference in cultured mammalian cells. Nature, 411, 494-498.

Elbashir, S.M., Lendeckel, W. and Tuschl, T. (2001b) RNA interference is mediated by 21-and 22-nucleotide RNAs. Genes Dev., 15, 188-200.

English, J.J., Mueller, E. and Baulcombe, D.C. (1996) Suppression of virus accumulation in transgenic plants exhibiting silencing of nuclear genes. Plant Cell, 8, 179-188.

Fire, A., Xu, S., Montgomery, M.K., Kostas, S.A., Driver, S.E. and Mello, C.C. (1998) Potent and specific genetic interference by double-stranded RNA in *Caenorhabditis elegans*. Nature, 391, 806-811.

Hamilton, A.J. and Baulcombe, D.C. (1999) A novel species of small antisense RNA in post-transcriptional gene silencing. Science, 286, 950-952.

Hamilton, A.J., Voinnet, O., Chappell, L. and Baulcombe, D.C. (2002) Two classes of short interfering RNA in RNA silencing. EMBO J., in press.

Hammond, S.M., Bernstein, E., Beach, D. and Hannon, G. (2000) An RNA-directed nuclease mediates post-transcriptional gene silencing in *Drosophila* cell extracts. Nature, 404, 293-296.

Hammond, S.M., Caudy, A.A. and Hannon, G.J. (2001) Post-transcriptional gene silencing by double-stranded RNA. Nature Reviews Genetics, 2, 110-119.

Johansen, L.K. and Carrington, J.C. (2001) Silencing on the spot. Induction and suppression of RNA silencing in the Agrobacterium-mediated transient expression system. Plant Physiology, 126, 930-938.

Jones, L., Ratcliff, F. and Baulcombe, D.C. (2001) RNA-directed transcriptional gene silencing in plants can be inherited independently of the RNA trigger and requires Met1 for maintenance. Curr. Biol., 11, 747-757.

Kasschau, K.D. and Carrington, J.C. (1998) A counterdefensive strategy of plant viruses: suppression of post-transcriptional gene silencing. Cell, 95, 461-470.

Kjemtrup, S., Sampson, K.S., Peele, C.G., Nguyen, L.V., Conkling, M.A., Thompson, W.F. and Robertson, D. (1998) Gene silencing from plant DNA carried by a geminivirus. Plant J., 14, 91-100.

Lindbo, J.A., Silva-Rosales, L., Proebsting, W.M. and Dougherty, W.G. (1993) Induction of a highly specific antiviral state in transgenic plants: implications for regulation of gene expression and virus resistance. Plant Cell, 5, 1749-1759.

Mueller, E., Gilbert, J.E., Davenport, G., Brigneti, G. and Baulcombe, D.C. (1995) Homology-dependent resistance: transgenic virus resistance in plants related to homology-dependent gene silencing. Plant J., 7, 1001-1013.

Napoli, C., Lemieux, C. and Jorgensen, R.A. (1990) Introduction of a chimeric chalcone synthase gene into Petunia results in reversible co-suppression of homologous genes *in trans*. Plant Cell, 2, 279-289.

Palauqui, J.-C., Elmayan, T., Pollien, J.-M. and Vaucheret, H. (1997) Systemic acquired silencing: transgene-specific post-transcriptional silencing is transmitted by grafting from silenced stocks to non-silenced scions. EMBO J., 16, 4738-4745.

Ratcliff, F., Martin-Hernandez, A.M. and Baulcombe, D.C. (2001) Tobacco rattle virus as a vector for analysis of gene function by silencing. Plant J., 25, 237-245.

Ruiz, M.T., Voinnet, O. and Baulcombe, D.C. (1998) Initiation and maintenance of virus-induced gene silencing. Plant Cell, 10, 937-946.

Salmeron, J.M., Oldroyd, G.E.D., Rommens, C.M.T., Scofield, S.R., Kim, H.-S., Lavelle, D.T., Dahlbeck, D. and Staskawicz, B.J. (1996) Tomato *Prf* is a member of the leucine-rich repeat class of plant disease resistance genes and lies embedded within the *Pto* kinase gene cluster. Cell, 86, 123-133.

Sijen, T., Fleenor, J., Simmer, F., Thijssen, K.L., Parrish, S., Timmons, L., Plasterk, R.H.A. and Fire, A. (2001) On the role of RNA amplification in dsRNA-triggered gene silencing. Cell, 107, 465-476.

Thomas, C.L., Jones, L., Baulcombe, D.C. and Maule, A.J. (2001) Size constraints for targeting post-transcriptional gene silencing and for RNA-directed methylation in *Nicotiana benthamiana* using a potato virus X vector. Plant J., 25, 417-425.

Vaistij, F.E., Jones, L. and Baulcombe, D.C. (2002) Spreading of RNA targeting and DNA methylation in RNA silencing requires transcription of the target gene and a putative RNA-dependent RNA polymerase. Plant Cell, 14, 857-867.

Van Blokland, R., Van der Geest, N., Mol, J.N.M. and Kooter, J.M. (1994) Transgene-mediated suppression of chalcone synthase expression in *Petunia hybrida* results from an increase in RNA turnover. Plant J., 6, 861-877.

van der Krol, A.R., Mur, L.A., Beld, M., Mol, J.N.M. and Stuitji, A.R. (1990) Flavonoid genes in petunia: Addition of a limited number of gene copies may lead to a suppression of gene expression. Plant Cell, 2, 291-299.

Voinnet, O. and Baulcombe, D.C. (1997) Systemic signalling in gene silencing. Nature, 389, 553.

Voinnet, O., Lederer, C. and Baulcombe, D.C. (2000) A viral movement protein prevents systemic spread of the gene silencing signal in *Nicotiana benthamiana*. Cell, 103, 157-167.

Voinnet, O., Mestre, P. and Baulcombe, D.C. (2002) Enhanced transient expression by p19 suppressor of silencing. Submitted for publication.

Voinnet, O., Pinto, Y.M. and Baulcombe, D.C. (1999) Suppression of gene silencing: a general strategy used by diverse DNA and RNA viruses. Proc. Natl. Acad. Sci. USA, 96, 14147-14152.

Voinnet, O., Vain, P., Angell, S. and Baulcombe, D.C. (1998) Systemic spread of sequence-specific transgene RNA degradation is initiated by localised introduction of ectopic promoterless DNA. Cell, 95, 177-187.

Wesley, S.V., Helliwell, C.A., Smith, N.A., Wang, M., Rouse, D.T., Liu, Q., Gooding, P.S., Singh, S.P., Abbott, D., Stoutjesdijk, P.A., Robinson, S.P., Gleave, A.P., Green, A.G. and Waterhouse, P. (2001) Construct design for efficient, effective and high-throughput gene silencing in plants. Plant J., 27, 581-590.

Robert T. Fraley

IMPROVING THE NUTRITIONAL QUALITY OF PLANTS

Robert T. Fraley
Monsanto Company, 800 N. Lindbergh Blvd., St. Louis, MO 63167, USA (e-mail: robert.t.fraley@monsanto.com)

1. INTRODUCTION

In the 18th century, when global population was just under 1 billion, Malthus predicted that unchecked population growth would strain mankind's ability to feed itself. Yet today, when the world's population is over 6 billion, we realize that advances in science and technology have assured that Malthus' prediction did not come true. The Green Revolution of the latter half of the 20th century produced new varieties of grains with improved yields, and new agricultural practices allowed these yields to be produced without greatly increasing the land committed to agriculture. These practices have sustained population growth, but there are indications that we need new agricultural technologies to sustain the growth in population predicted to occur in the next 50 years almost exclusively in the developing world.

Since the introduction in 1996 of the first crops produced by biotechnology, their adoption by farmers has been phenomenal. Global acreage planted to crops developed via biotechnology has increased from 6 million in 1996 to 130 million in 2001, and predictions indicate that plantings will increase in 2002 as well. Farmers realize that herbicide tolerant and insect resistant crops provide numerous advantages, such as reducing reliance on pesticides, increasing yields, and allowing more environmentally sustainable agricultural practices, all which translate into profits. Small scale, poor farmers in the developing world, such as India, China and South Africa, also are adopting biotech crops and realizing profits as well, indicating that biotech crops can be profitable no matter what the scale of the farm.

The biotechnological advances in agriculture of the late 20th century will help us produce enough staple crops to feed a population approaching 9 billion by 2050. New techniques that enable plants to withstand unfavorable growing conditions and resist a variety of pests will be applied to locally familiar crops that will benefit populations of the developing world. This production will occur in developing countries themselves and will boost their agriculture, and thus their economies. We must realize that food redistribution systems, ie, the transport of staples grown in wealthy nations to less fortunate nations, are only short term fixes and do nothing to improve the financial growth and

I. K. Vasil (ed.), Plant Biotechnology 2002 and Beyond, 61-67.

independence of developing countries. This vast increase in food production also will have to occur in spite of static or decreasing arable land and diminishing water supply. But producing enough staple crops will not be enough to meet the nutritional needs of the world's population. Additional advances also are needed to enhance the nutritional qualities of foods such that under-nutrition that still exists in the developing world, and chronic diseases of over-nutrition that affect the populations of the developed world, can be avoided.

2. IMPROVING PROTEIN QUALITY

In the developing world, much of the poor population survives on a limited array of foods — in Asia, rice or wheat; in Latin America and Africa, corn, rice and cassava. The quality of protein in these plant sources is poor in relation to human nutritional needs. Unless animal-derived proteins or bean and legumes are included in the diet along with the staple grain, the diet is likely to be inadequate in protein, a particularly serious problem for growing children. Improving the protein quality of these staple grains is one way to tackle this nutritional problem.

Over the last 30 years there has been an effort to improve the quality of protein in corn using traditional breeding techniques. Now with the techniques of modern biotechnology, it is possible to achieve the same endpoint in a much shorter timeframe. The protein quality, and thus the nutritional value, of corn can be improved by increasing its content of two essential amino acids, tryptophan and lysine. This improvement in corn protein quality is good not only for humans, but also for animals, since much corn is used for animal feed, and feed needs supplementation with essential amino acids to meet the nutritional needs of livestock. The expression of higher levels of tryptophan and lysine will simplify feed preparation and decrease costs for the livestock farmer.

Other efforts are being directed towards improving the protein quality of foods intended for human consumption. For example, the protein quality of rice could be improved by transferring the gene for beta-phaseolin, a lysine-rich protein, from beans into rice, a lysine-poor food. Similarly, the protein quality of soybeans, which are deficient in the essential amino acid, methionine, could be improved by causing the bean to over-express a methionine-rich protein, glycinin.

As world populations increase and become more affluent, the demand for animal sources of dietary proteins also increases. The production of meat and milk, although highly nutritious, requires large inputs of feed grains as well as infrastructures for animal production and processing, resources which are lacking in many areas of the world. Furthermore, lactose intolerance, which is common in much of the world, limits consumption of milk, and the concern about health consequences associated with a high consumption of saturated-fat-rich animal products argues against over-reliance on these products. Thus, there is an opportunity to meet peoples' needs for high quality protein and maintenance of health by improving the palatability of soy-based foods.

In the past the taste of soy protein ingredients was improved by creating relatively expensive and refined soy protein ingredients. Now we can improve soy by modifying its composition. The new approach will result in affordable, nutrient-dense products. The favorable effects of soy on blood lipids also can be improved by modifying soybean protein composition. New products created from these soybeans will help people achieve more heart-healthy diets.

3. OIL SEED COMPOSITION

Vegetable fats serve an essential role in the diet. They not only provide energy and nutrients, but also are essential components in food formulation. In the 1970's nutritional research discovered that too much saturated fat in our diet, primarily from animal sources, could be detrimental to our health. Dietary recommendations were issued, advising that we use more vegetable oils instead of lard and butter. However, for many food formulations, vegetable oils are less than optimal because they oxidize easily or they do not impart the desired functional qualities.

The polyunsaturated fatty acids in vegetable oils need to be chemically hydrogenated to make them more stable and functional, but in the process, *trans* fatty acids are generated. There is now clinical evidence suggesting that *trans* fatty acids are as unhealthy as saturated fatty acids; thus, there is a need for stable, functional vegetable oils but without trans fatty acids.

Modern biotechnology now allows us to manipulate the enzymes involved in fatty acid elongation and desaturation so we can produce vegetable oils with tailor-made characteristics. For example, soybeans, cottonseed, and palm have been modified to express high levels of oleic acid, a monounsaturated fatty acid, at the expense of polyunsaturated or saturated fatty acids. The resulting oils are stable, have the desired functional properties, and since they do not require hydrogenation, they do not contain trans fatty acids. An important additional advantage is that oils rich in oleic acid are considered "heart-healthy."

Another example relates to one of Americans' favorite foods, French fries. The proportion of saturated fatty acids and *trans*-fatty acids in most commercial French fries is about 45% of total fat. However, using modern biotechnology, one can develop a saturated fat-free, mid-oleic vegetable oil in which to cook saturated fat-free, *trans*-free French fries. These fries could be labeled as free of saturated and *trans* fatty acids and could help consumers identify and choose the more heart healthy product.

Our ability to manipulate the fatty acid synthesis pathway in oilseeds also allows us to develop specialty oil products for nutritional purposes. For example, while health experts recommend we consume more omega-3 fatty acids, many people simply do not eat enough fish and seafood, the primary sources, to get enough in their diets. Efforts are now being directed toward developing oilseeds that express these omega-3 fatty acids to provide an alternative to fish and seafood. Vegetable oils high in omega-3 fatty acids could be used in certain food formulations as well as in dietary supplements as an alternative to fish oil capsules.

4. VITAMINS AND MINERALS

Several micronutrients, most significantly vitamin A, iron and iodine, continue to be deficient in the diets of many people, particularly in the developing world. The foods providing these nutrients — fruits and vegetables, animal and fish products — are, unfortunately, unavailable or too expensive for many of these people. While public health strategies, such as supplementation and fortification have been, and continue to be pursued, these are expensive, labor-intensive, and not always successful.

New efforts are now being directed toward "biofortification" a term that describes the increase in levels of essential nutrients in plants, whether via traditional breeding or biotechnology. Researchers at the International Food Policy Research Institute have determined that biofortification of key staple crops can be an efficient and cost-effective approach to solving the world's micronutrient deficiencies. While the up-front investment to breed or bioengineer higher nutrient expression in crops may be high, these are one-time costs. Once the modification is achieved, the costs associated with production and distribution are essentially the same as those required for food production and distribution. In contrast, supplementation or fortification strategies bear repeated, additional costs that over time are greater than those associated with biofortification.

4.1 Vitamin A

Two years ago, a pioneering biotechnologist, Dr Ingo Potrykus, appeared on the cover of Time Magazine, spotlighting his decade-long effort to make rice rich in beta-carotene, the precursor of Vitamin A. Dr Potrykus and coworkers inserted three genes into rice, from the daffodil and from a bacterium, which enabled the rice to synthesize beta-carotene in its endosperm. The resulting "Golden Rice" proved that biotechnology can enhance the nutritional value of food crops.

Dr. Potrykus and others who contributed to the success of this project agree that Golden Rice must be made available free of charge to poor farmers in the developing world. But more work must be done prior to introducing the crop: the traits must be transferred to locally preferred varieties of rice; the level of beta-carotene must be increased and shown to be bioavailable, and the food and environmental safety of the transformed rice must be demonstrated. Golden Rice is still about 5 years away from introduction, but many hope it will be successful in reducing vitamin A deficiency and its devastating consequences of blindness and suppressed immunity.

In addition to rice, other foods also are being genetically enhanced to express high levels of beta-carotene. In India, for example, mustard seed oil, a locally preferred food oil, is being developed to express high levels of beta-carotene. One of the unique advantages of this product is that the beta-carotene is expressed in oil, and since beta-carotene requires fat to be absorbed by the body, it is expected be highly bioavailability. Thus, only a small quantity of the oil, perhaps a few teaspoons, will be needed to meet daily nutritional needs.

4.2. Iron

About one of every three people in the world is deficient in iron; in some cases the deficiency is so severe as to cause anemia and to compromise physical and mental performance. The amount of iron in the diet is only one factor affecting iron nutriture; other factors in the diet, such as amino acids and phytate, impact its absorption in the gastrointestinal tract. The same researchers who developed Golden Rice are leveraging these factors to develop rice genetically improved to deliver not only more iron, but to make the iron more bioavailable. First, they inserted the gene for ferritin, the major iron storage protein, from the common bean into rice. This resulted in a doubling in iron content. Then they over-expressed cysteine-rich metallothionein in rice because cysteine residues in peptides enhance iron absorption. Finally, they inserted a gene to express phytase to reduce the rice's phytate content. Phytate binds iron and reduces its absorption.

Researchers hope that one day Golden Rice will be further enhanced with these traits so not only vitamin A deficiency, but also anemia, can be solved simultaneously.

4.3. Other Nutrients/Functional Components

The diets of Westernized countries are plentiful--perhaps too plentiful for some people, as evidenced by the prevalence of chronic diseases of over-nutrition. Cardiovascular disease is the #1 cause of death in the US and other affluent nations, and scientists recognize that certain aspects of our diets need improving to reduce our risk for this disease.

Although early in the research phase, there are several promising areas where crop biotechnology could enhance the health-promoting properties of foods. For example, it is possible to increase the level of alpha-tocopherol, the most nutritionally important form of Vitamin E, in *Arabidopsis*. If this transformation can be replicated in commercially important oilseeds, the nutritional value of these oils would improve appreciably.

Plant phytosterols are another type of nutritionally significant food component. Phytosterols interfere with absorption of dietary cholesterol, thereby helping to reduce blood cholesterol levels. Margarines containing plant phytosterols recently were introduced on the market in the US and Europe, however, the supply of phytosterols is limited. We have shown that it is possible to over-express phytosterols in soybeans, a measure that could make these food components more widely available.

5. CONCLUSIONS

The first generation of crops produced via biotechnology has been enormously successful, as evidenced by their rapid adoption by farmers all over the world. Analyses of the impact of biotech crops have shown that both large and small scale farmers have realized improved yields, decreased reliance on agricultural inputs, and increased profits. Other benefits include the ability to adopt more environmentally friendly farming practices, such as no-till farming and fewer pesticidal sprayings, and decreased development of mycotoxins in corn.

The next generation of biotech crops is being developed to have improved nutritional qualities that will be recognized and valued by consumers. While these improvements are intended primarily for mature, affluent nations, we also must recognize the critical contribution biotechnology will make to the world's future food security. As the world's population increases by at least

2 billion over the next 50 years, all relevant technologies must be applied to help meet the world's impending food needs. Yields of staple crops must be increased to new highs, and losses to pests and other stresses must be diminished further; these improvements must be available to people of developing countries where virtually all population growth is occurring. Biotechnology is already showing that it can and will make improvements in crops that will have important implications for future food security.

Mich Hein

ANTIBODIES FROM PLANTS: BREAKING THE BARRIERS TO ANTIBODY PRODUCTION

Mich B. Hein and Neil M. Cowen
Epicyte Pharmaceutical, Inc., 5810 Nancy Ridge Drive, San Diego, CA 92121 (email: mbhein@epicyte.com)

Green plants were an important historical source of medicines, and a primary source of humans' first biopharmaceuticals. Because plants are the most efficient producers of protein on the planet, they are now being tapped as a production source for the fastest growing class of new pharmaceuticals: therapeutic and prophylactic proteins. The list of potential protein-based pharmaceuticals includes structural proteins from viral and bacterial pathogens, human antibodies and other immunoglobulins, defensins, enzymes and unique synthetic or chimeric proteins. Because plants are eucaryotes with a well developed endomembrane system, they have proved particularly efficient in producing complex and multimeric proteins. For complex proteins like immunoglobulins, green plants provide breakthroughs in economics, scale of production and the scope of molecules that can be efficiently manufactured. Producing antibodies in large quantities using professional agronomic corps could dramatically increase the use of antibody therapeutics and facilitate new uses for these medicines. While significant improvements are on the horizon, the basic technology for generating transgenic plants that produce a variety of functional antibodies is now well developed. The current challenge is to integrate this robust technology into the existing regulatory framework for developing safe and effective pharmaceuticals. We are witnessing the evolution of crop stewardship, manufacturing and processing technologies that protect health and safety of the public while delivering cost-effective treatments and preventives for human disease.

Abstract only – no manuscript received.

I. K. Vasil (ed.), Plant Biotechnology 2002 and Beyond, 69.

Maurice M. Moloney

PHYTOSYNTHETICS: TRANSGENIC PLANTS AS THE PRIMARY SOURCE OF INDUSTRIAL AND MEDICAL FEEDSTOCKS IN THE 21ST CENTURY

Maurice M. Moloney
SemBioSys Genetics Inc., 2985 23rd Ave NE, Calgary, Alberta, Canada T1Y 7L3 (email: moloneym@sembiosys.ca)

With the completion of the sequencing of the *Arabidopsis* genome in 2000, and the success of numerous sequencing projects, including the human genome, we have entered a new era of plant biotechnology. First, we are beginning to understand at a functional level the roles of whole classes of genes and their protein products. Second, we have at our disposal an enormous resource of genes, including those specifying developmental patterns and those encoding entire metabolic pathways, which can be mobilized into transgenic plants. As a result, we have moved out of the "observational" phase of plant biology into a "synthetic" phase in which we can design plants to perform valuable functions for industry, medicine and the environment. Over the next 25 years, driven by the depletion of fossil fuel supplies and attempts to limit greenhouse gas emission, we will be obliged to develop plants capable of supplying oleochemical feedstocks, biodegradeable structural materials and a wide variety of pharmaceuticals and nutraceuticals. This will require sophisticated approaches to the genetic manipulation of plants and the processing of plant-derived products. In this presentation, the achievements of plant biotechnology and its potential for the sustainable production of oleochemicals, biodegradeable plastics, high-value proteins and complex organic compounds will be evaluated. The necessity to meet the needs for these products using plant resources will be supported using technological, economic and environmental criteria.

Abstract only – no manuscript received.

I. K. Vasil (ed.), Plant Biotechnology 2002 and Beyond, 71.

Ronald Ross Sederoff

QUANTITATIVE INFERENCE IN FUNCTIONAL GENOMICS OF LOBLOLLY PINE (*PINUS TAEDA* L.) USING ESTS AND MICROARRAYS

Matias Kirst[1,2], Arthur Johnson[2], Ernest Retzel[3], Len van Zyl[2], Debby Craig[2], Zhi Jun Li[2], Ross Whetten[2], Christie Baucom[2], Erin Ulrich[2], Kristy Hubbard[2] and Ronald Sederoff[2,4]

[1] Program in Genomics, North Carolina State University
[2] Forest Biotechnology Group, North Carolina State University
[3] Center for Genomic and Computational Biology, University of Minnesota
[4] Corresponding author (e-mail: ron_sederoff@ncsu.edu)

1. INTRODUCTION

Recent advances in high throughput DNA sequencing technology have made possible the identification of large numbers of genes, and the development of cDNA microarrays have made it possible to learn about the timing and level of expression of large numbers of genes. These advances have been particularly important for biological systems that are typically recalcitrant to genetic analysis due to long generations times. Among the most difficult systems are many species of forest trees. Forest trees are important both as dominant species in forest ecosystems and for the commercial value of wood products. Here we report on the assignment of functional categories to six libraries of ESTs from loblolly pine, predominantly from xylem forming tissues. From these libraries, sequences were assigned putative cellular functions based on homologs identified in GenBank or the *Arabidopsis* genome (The Arabidopsis Genome Initiative 2000). We have used the EST data to estimate relative abundance of specific RNA populations within and between libraries. In addition, microarray technology has improved sufficiently so that differences in average signals of 50% or less, can be detected as statistically significant with at least 95% confidence, using improved experimental design and analysis.

2. RESULTS AND DISCUSSION

2.1. Pine ESTs May Be Annotated Against *Arabidopsis*

We have sequenced nearly 46,777 ESTs from tissue specific libraries of loblolly pine and assigned putative cellular functions based on homology to inferred genes of *Arabidopsis*. The libraries represent different types of wood

I. K. Vasil (ed.), Plant Biotechnology 2002 and Beyond, 73-79.

forming tissues, as well as shoot tips and pollen cones (Table 1: Figure 1). It might be expected that wood-forming tissues from immature xylem would be relatively specialized. However, a high level of similarity in the abundance of specific classes is seen when the different libraries are compared.

Table 1. Some statistics for the loblolly pine libraries.

Description of library		Number of ESTs
NXNV	immature xylem vertical springwood	8,490
NXCI	immature xylem compression induced	9,333
NXSI	immature xylem side wood induced	11,812
NXPV	immature xylem deep planings vertical	9,641
ST	shoot tip	5,906
PC	pollen cone	1,595
Total		46,777

NXNV planings represent tissue removed from differentiating xylem using a small block plane to recover lignifying tissue after the soft immature xylem was removed using a vegetable peeler. ESTs selected for this analysis were high quality sequences of at least a hundred base pairs of scores PHRED 20 or higher. The average read lengths for all the libraries was 380bp. Sequences have been assembled into a series of overlapping contigs and singletons using PHRAP. Full descriptions of the contig sequences and images of the contig assemblies are displayed at http://web.ahc.umn.edu/biodata/nsfpine/contig_dir11/.

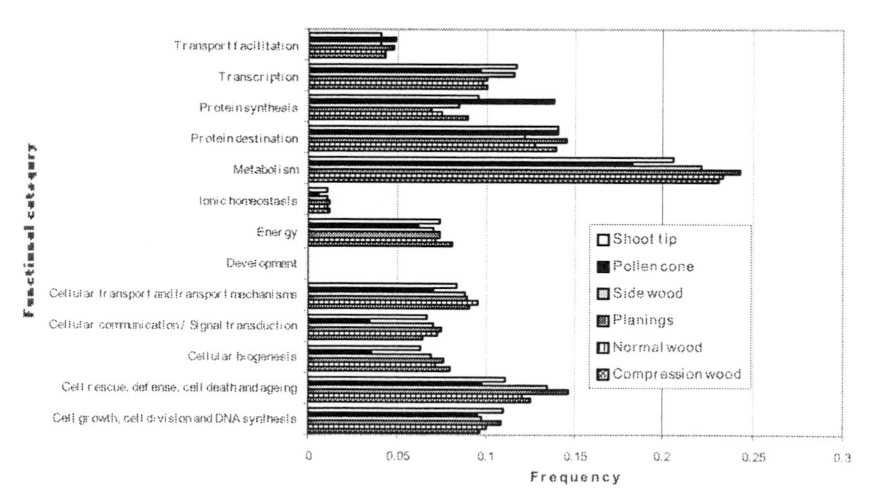

Figure 1. Frequency of individual ESTs from cDNA libraries in functional categories. Vertical order of horizontal bars follows the vertical order in the box.

Table 2. Comparison of EST (transcript) abundance in the pine cDNA libraries. The reference library (left column) is compared to the other libraries (six right columns). Only contigs that are composed by a significantly different number of ESTs in the reference library (P < 0.01) relative to all the others are represented. Six right columns indicate the number of ESTs originated from each library that composes the contig. The number of ESTs from the reference library which is being compared to the others is in bold. Contig annotation based on BLASTX e-value below 1E-5 relative to the A. thaliana gene predicted sequences. Note : contig numbers in the database may be changed when new clones are added to the data set and all contigs are reanalyzed.

NXPV library Contig #	Contig annotation	NXCI 12665	NXNV 12576	NXPV 11550	NXSI 13038	PC 1745	ST 6595
6857	No similarity to A.t.	0	0	**62**	0	0	0
52	Putative protein	0	2	**56**	0	0	0
6844	Putative protein	0	1	**50**	1	0	0
6824	Putative protein	0	5	**38**	1	0	0
6850	Xyloglucan endo-1,4-beta-D-glucanase precursor	0	16	**33**	3	0	0
6791	Putative protein	0	3	**30**	2	0	0
6732	Putative protein	0	1	**27**	0	0	0
6701	Secretory protein - like	0	0	**25**	0	0	0
6691	Secretory protein - like	0	0	**24**	0	0	0
6672	No similarity to A.t.	0	0	**23**	0	0	0
6671	Putative protein	0	0	**23**	0	0	0
6664	Putative protein	0	0	**21**	1	0	0
6811	Putative elicitor-responsive gene	7	3	**20**	4	0	4
6814	Hypothetical protein	3	0	**20**	2	0	0
6620	Xyloglucan endo-1,4-beta-D-glucanase precursor	0	0	**20**	0	0	0
6614	Putative protein	0	0	**20**	0	0	0

2.2 Quantitation Of Gene Expression From EST Abundance: Electronic Northerns And Differential Display

The collection of ESTs, obtained from non-normalized libraries can be used to estimate relative abundance of message for a specific gene or gene family, analogous to the method of estimating transcript abundance on northern blots. The sum of the individual ESTs identified by high sequence similarity provides an estimate of the relative abundance of that particular sequence in the library. It can be compared to estimates for other sequences in the same library, or to the same sequence in different libraries. The statistical significance of these estimates can be calculated (Audic and Claverie (1997).Once the data is organized in this way, it is possible to identify the genes that are differentially expressed. Such calculations represent electronic versions of differential display (Table 2), and highlight the differences between the tissucs. We compared the transcript abundance between all libraries in pair-wise combinations. The results from the comparisons based on the planings library are shown (Table 2). Contigs composed of ESTs that were significantly ($P < 0.01$) more abundant in one library are in bold type. The planings (NXNV) cDNA library has many abundant EST contigs with high similarity to *A. thaliana* genes with unknown function. Contigs 6852, 6844, 6824, 6732, 6671, 6664 and 6614 have weak similarity to *Arabidopsis* protein (At3g53980), a probable lipid transport related protein. Additional data can be seen in the Pine Genome Project web-page http://web.ahc.umn.edu/biodata/nsfpine/contig_ dir1 l/.

2.3. Quantitative Inference And cDNA Microarrays

The technology of microarrays has been heralded as having great promise, but technical problems of quantitation have been significant and limiting (Hedge et al. 2000). In principle, microarrays are a high throughput method for nucleic acid hybridization, subject to the same restrictions regarding specificity, stringency and quantitative inference long used for solution hybridization and southern blots. Experimental design and analysis bears many similarities to field plot design and analysis in agriculture. Recently, improved methods for construction of arrays, hybridization methods, and statistical analysis (Herzberg et al. 2001; Wolfinger et al. 2001) have greatly improved the quality of results. Arrays are still limited with respect to quantitative measurements of RNA abundance. Inferences of significant differences in differential expression may now be made with greatly increased statistical power. However, it is not yet possible to use microarrays to estimate actual mRNA abundance with reasonable confidence. CDNA microarrays show differences in levels of mRNA expression, but the differences in tissues may be far greater than that apparent from microarrays. We suggest that the standard microarray hybridization conditions have not yet been optimized for quantitative inference.

We have recently found that using relatively large numbers of replications, we can resolve differences in expression with strong statistical significance, well below the two fold differences typically used for microarray analysis (Table 3).

Table 3. Example of quantitation of changes in gene expression - Selected results from a microarray experiment show the specific loblolly gene assayed and the change in level of expression found under different conditions of xylem differentiation.

Gene Annotation and clone ID	Standard Error	Fold Change	p-value
No apparent homolog NXCI_133_E11	0.096	0.54	4.56E-10
No apparent homolog NXCI_002_E02	0.063	0.67	4.56E-10
Laccase (Diphenol Oxidase) NXNV_066_B07	0.154	0.74	4.56E-10
No apparent homolog NXCI_083_F01	0.033	0.86	6.77E-11
Laccase (Diphenol Oxidase) NXCI_046_E05	0.040	0.87	6.77E-11
Putative Beta-1,3-Glucanase NXCI_048_G03	0.092	0.89	9.20E-11
Putative T-Complex Protein 1, Theta Sub. NXCI_094_C11	0.094	0.89	4.56E-10
Putative Casein Kinase NXCI_048_E07	0.056	1.76	9.20E-11
Disease Resistance Protein Rpp1-Wsb NXNV_018_H03	0.060	2.04	6.77E-11
Laccase (Ec 1.10.3.2) NXCI_094_C09	0.056	2.78	6.77E-11
Photosystem-I O_2-Evolving Complex P. NXCI_022_G01	0.085	3.77	6.78E-11
Subunit Of Photosystem I NXCI_067_C01	0.078	3.81	4.57E-10
Ethylene-Responsive Small GTP-Binding P. NXCI_022_B10	0.075	3.83	2.74E-10
Putative Glutaredoxin NXCI_026_A11	0.097	4.04	4.57E-10
Putative Basic Blue Protein NXCI_008_C01	0.105	4.10	6.78E-11
Chloroquine Resistance Candidate P. NXCI 018 G05	0.070	4.20	6.78E-11
Putative Basic Blue Protein NXCI_093_F03	0.071	5.24	6.78E-11
Putative Glutaredoxin NXCI_085_E04	0.076	6.43	4.57E-10
Peroxisomal Cu-Containing Amine Oxidase NXCI 137 A01	0.126	8.42	6.78E-11
Heat Shock 70 Kda Mitochondrial Protein NXCI_136_A08	0.119	14.60	6.78E-11

2.4. Experimental Design

Our current experiments use a minimum of 96 and a maximum of 360 replications for a single gene, with an average of 240. Experimental design follows that proposed by Kerr & Churchill (2001) using highly replicated A-Optimal fully balanced complete and incomplete block designs. Methods for PCR amplification, array printing, RNA purification, labeling of cDNAs and hybridization follow methods described by Allona et al. 1998; Whetten et al. 2001; van Zyl et al. in press, and Van Zyl, unpublished results.

2.5. Image Analysis And Data Mining

Scanned images are gridded and intensities are calculated and converted to text files, transformed into log ratios, normalized, and displayed as a work file in SAS. Changes in expression are evaluated using a t test, and corrected for multiple fixed and random effects. Experimental design follows Kerr and

Churchill (2001), and statistical analysis uses the mixed model of Wolfinger et al. (2001).

2.6. The Long-term Goal of Genomics

The power of genomics lies in the integration of large data sets describing molecular, metabolic, cellular, physiological, and morphological changes. The future of plant genomics lies in the use of such integrated data sets, which will create increasing depth of understanding of the complex processes of plant growth, development, adaptation and evolution. The eventual goal would be the ability to predict phenotype from genotype. Will such predictions ever be possible? While we cannot answer this question now, we can begin to define what factors will determine the extent of our ability to construct predictive models. These factors lie in the nature of qualitative and quantitative genetic and molecular interactions.

2.7. Why Is Quantitative Inference Important In Genomics?

If *Arabidopsis* is our guide, then plants have a relatively small number of genes, in the range of 25,000. Our ability to predict the efforts of variation in those genes depends not only on the understanding of specific functions, but on discovery of regulatory networks and the secondary functions of any specific gene product, and on the interactions of products. Many such interactions will be discovered as quantitative effects.

The central question of genomics is then apparent. What is the extent, nature, and number of interactions of the molecular components of living systems, usually characterized as quantitative factors, that contribute to variation, selection, and evolution as the sum of many factors of small effects? The answer to this question will determine the extent to which we will be able to understand or modify a genotype and to predict the phenotype in a well-defined environment.

3. REFERENCES

Allona, I., Quinn, M., Shoop, E., Swope, K., St. Cyr, S., Carlis, J., Riedl, J., Retzel, E., Campbell, M.M., Sederoff, R. and Whetten, R., 1998. Analysis of xylem formation in pine by cDNA sequencing. Proc Natl. Acad. Sci. U S A 95: 9693-9698.

Audic, S. and Claverie, J. M. 1997. The significance of digital gene expression profiles. Genome Research 7: 986-995.

Hedge, P., Qi, R., Abernathy, K., Gay, C., Dharap, S., Gaspard, R., Hughes, J.E., Snesrud, E., Lee, N. and Quackenbush, J. 2000. A concise guide to cDNA microarray analysis. BioTechniques 29:548-562.

Hertzberg, M., Aspeborg, H., Schrader, J., Andersson, A., Erlandson, R., Blomquist, K., Bhalerao, R., Uhlen, M., Teeri, T., Lundeberg, J., Sundberg, B. and Sandberg, G. 2001. A transcriptional roadmap to wood formation. Proc. Natl. Acad. Sci. USA 98:14732-14737.

Kerr, M.K. and Churchill, G. A. 2001. Statistical design and the analysis of gene expression microarray data. Genetical Res. 77: 123-128.

The Arabidopsis Genome Initiative (2000) Analysis of the genome sequence of the flowering plant *Arabidopsis thaliana*. Nature 2000. 408: 791-826.

Van Zyl, L., von Arnold, S., Bozhkov, P., Chen, Y., Egertsdotter, U., MacKay, J., Sederoff, R., Shen, J., Zelena,L. and Clapham, D. Heterologous array analysis in Pinaceae: Hybridization of high density arrays of *Pinus taeda* cDNA with cDNA from needles and embryonic cultures of *P. taeda*, *P. sylvestris*, or *Picea abies*. Comparative and Functional Genomics, in press.

Whetten, R., Sun, Y-H., Zhang, Y. and R. Sederoff. 2001. Functional genomics and cell wall biosynthesis in loblolly pine. Plant Molec. Biol. 47: 275-291.

Wolfinger, R.D., Gibson, G., Wolfinger E.D., Bennett, L., Hamadeh, H., Bushel, P., Afshari, C. and Paules, R.S. 2001. Assessing gene significance from cDNA microarray expression data via mixed models. J. Comp. Biol. 8: 625-637

Lothar Willmitzer

ABOUT STRAIGHT LINES AND COMPLEX CROSSROADS: METABOLISM IS A NETWORK

L.Willmitzer
Max-Planck-Institut für Molekulare Pflanzenphysiologie, D-14476 Golm, Germany (e-mail: willmitzer@mpimp-golm.mpg.de)

Plants are probably the best chemical engineers on this planet. Compounds being made by plants are used not only for food and feed purposes but also serve an important role in delivering material for technical uses. With the advent of plant biotechnology it readily became apparent that these new tools can indeed also be efficiently used to try to metabolically engineer plants. The first generation of compounds which was tackled in this respect comprised the major storage compounds such as carbohydrates (notably starch), oils, proteins and to some extent fibers. As an example data will be presented demonstrating the application of transgenic approaches to engineer novel carbohydrates in plants. In the course of these projects it however also became apparent that whereas for "linear pathways" such as starch biosynthesis essentially simple-minded straight-forward approaches by and large turned out to be successful, more central pathways turned out to be extremely complicated when touched upon. Thus there is a clear need for significantly improving our understanding on metabolism and relating this to gene function. Multiparallel analyses of mRNA and proteins are central to today's functional genomics initiatives. Surprisingly the next level in the realization of genomic information, i.e. the metabolite level, has received very little attention. We therefore set out to develop metabolic profiling as a tool for comprehensive and nonbiased analysis of the metabolic complement of biological systems. Using gas chromatography/mass spectrometry (GC/MS), up to 1000 distinct compounds from *Arabidopsis thaliana* leaf extracts can be detected. Application of this tool to various plant systems and the analysis of the data using various mining tools will be described. It is obvious that the application of metabolic profiling as a new level in system analysis is not limited to plants but has widespread applications ranging from microbiology to pharma.

Abstract only – no manuscript received.

I. K. Vasil (ed.), Plant Biotechnology 2002 and Beyond, 81.

Ilya Raskin

The novel, multicolored, transgenic carnations in front of the podium were produced by Florigene (Australia) and donated by courtesy of Chin-yi Lu and Stephen Chandler

PLANTS AND PHARMACEUTICALS IN THE 21ST CENTURY

Ilya Raskin
Biotech Center, Foran Hall, 59 Dudley Road, Cook College, Rutgers University, New Brunswick, N.J. 08901-8520, USA (e-mail: raskin@aesop.rutgers.edu)

Keywords plant biotechnology, botanical drugs, herbals, botanicals, dietary supplements, nutraceuticals, recombinant proteins, biopharmaceuticals, plant, plants, food supplement, drug discovery, natural products

1. INTRODUCTION

The 20th century became a triumph for the synthetic chemistry-dominated pharmaceutical industry that rapidly proceeded to replace natural extracts with synthetic molecules, which often had no connection to natural products. The spectacular rise of the pharmaceutical industry tremendously benefited disease treatment and prevention, saved countless lives and truly became one of the most outstanding achievements of the 20th century. Human medicines prescribed today still contain phytochemicals valued at $22,608 million in 1997 and projected to reach a value of $30,688.5 million in 2002 with prescription products and OTC herbal remedies each comprising approximately 50% of the market (Anonymous, 1998).

The severed bond between plants and health was not only felt in the area of medicines. By providing a "pill option", the 20th century also diminished the historical connection between food and disease. The universe of plant therapeutic agents presented in this review is summarized and defined in Table 1. Although these definitions may not be ideal, they are helpful in structuring the discussion of the re-emerging connection between seed plants (Superdivision Spermatophyta) and human health.

2. SINGLE INGREDIENT DRUGS

During the 20th century the emphasis gradually shifted from extracting medicinal compounds from plants to making these compounds or their

I. K. Vasil (ed.), Plant Biotechnology 2002 and Beyond, 83-95.

Table 1. Categories of botanical therapeutics

Therapeutic	Description	Example	Availability
Drugs (NCE)	Mostly single active ingredient pharmaceuticals originating from plants	Vinblastine, Taxol or Aspirin	Rx[a] or OTC[b]
Botanical Drugs	Clinically validated and standardized phytochemical mixtures	None in the U.S., several in clinical trials	Rx or OTC
Dietary supplements/ Nutraceuticals	A plant component with health benefits	Garlic or Echinacea extract	OTC
Functional/Medicinal foods	A food engineered or supplemented to provide health benefits	Healthy canola oil, golden rice or edible vaccine	OTC, Rx or Grocers
Recombinant proteins	Pharmaceutical protein expressed and isolated from plants	None commercialized, several in clinical trials	Rx

[a]Rx, prescription drugs; [b]OTC, over the counter

analogues synthetically. Natural products were widely viewed as templates for structure optimization programs designed to make perfect new drugs referred to by industry as a new chemical entity (NCE). In spite of the current preoccupation with synthetic chemistry as a vehicle to discover and manufacture drugs the contribution of plants to disease treatment and prevention is still significant. Today, 11% of the 252 drugs considered as basic and essential by the World Health Organization were exclusively of flowering plant origin with many produced chemosynthetically or made synthetically based on the bioactive discovered in plants (Rates, 2001). The greatest recent impact of plant-derived drugs was probably felt in the anti-tumor area where taxol, vinblastine, vincristine and camptothecin have dramatically improved the effectiveness of chemotherapy against some of the deadliest cancers. The most important pharmaceuticals that are still derived from plants directly or as precursors are listed in Table 2.

The enthusiasm for using plant extracts for the discovery of novel pharmaceutical leads has declined in the last decade, with many pharmaceutical companies closing or downsizing their natural products groups. Throughout human history plants were unchallenged as sources of new drug discovery, but the recent competition from combinatorial chemistry

Table 2. Some of the most economically important pharmaceuticals derived form plants

Name	Structure	Type	Source	Therapeutic use
Alkaloids				
Atropine[a], hyoscysmine, scopolamine	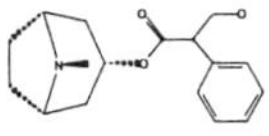	Tropane alkaloids	Solanaceous species	Anticholinergic
Codeine, morphine	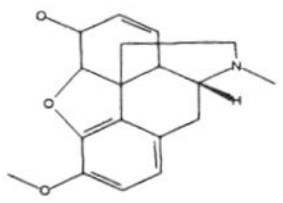	Tropane alkaloids	*Papaver somniferum* L.	Analgesic, antitussive
Cocaine	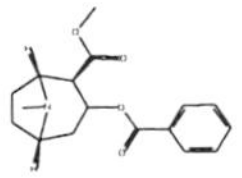	Tropane alkaloid	*Erythroxylum coca* Lamarck	Local anaesthetic
Nicotine	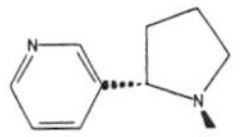	Pyrrolidine alkaloids	*Nicotiana* spp.	Smoking cessation therapy
Reserpine	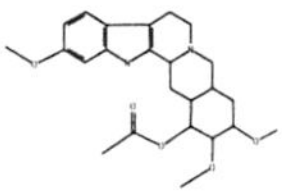	Indole alkaloid	*Rauwolfia serpentina* L.	Antihypertensiv e, psychotropic
Vinblastine, vincristine	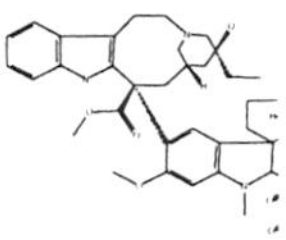	Indole alkaloids	*Catharanthus roseus* L.	Antineoplastic
Physostigmine	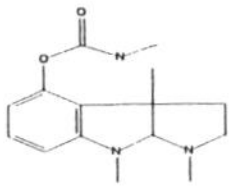	Indole alkaloid	*Physostigma venenosum* Balfor	Cholinergic
Pilocarpine	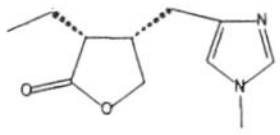	Imidazole alkaloid	*Pilocarpus jaborandi* Holmes	Cholinergic
Quinine	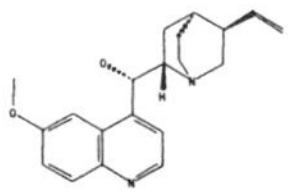	Quiniline alkaloid	*Cinchona* species	Antimalarial
Quinidine	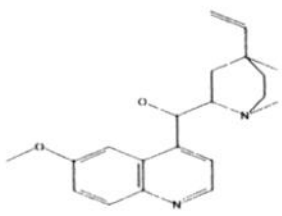	Quiniline alkaloid	*Cinchona* species	Cardiac depressant
Camptothecin[a]	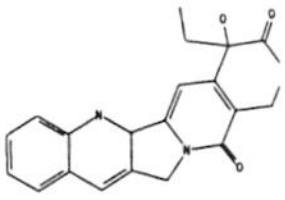	Quinoline alkaloid	*Camptotheca acuminata* Decne.	Antineoplastic
Colchicine	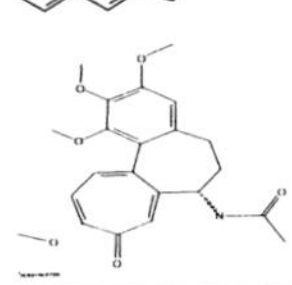	Isoquinoline alkaloid	*Colchicum autumnale* L.	Antigout

Tubocurarine	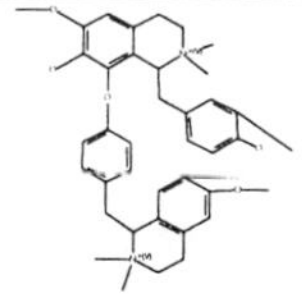	Isoquinoline alkaloid	*Chondodendron tomentosum* Ruiz et Pavon, *Strychnos toxifera* Bentham,	Skeletal muscle relaxant
Galantamine	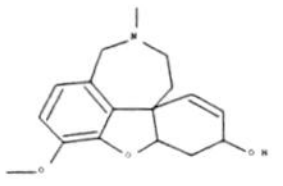	Isoquinoline alkaloid	*Leucojum aestivum L.*	Cholinesterase inhibitor
Emetine	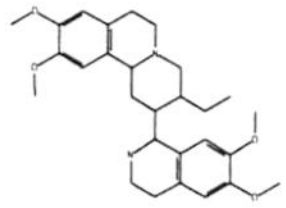	Isoquinoline alkaloid	*Cephaelis ipecacuanha* (Brot.)A. Rich.	Antiamoebic
Yohimbine	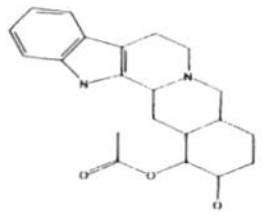	Indolalkylamine alkaloid	*Apocynaceae, Rubiaceae* spp.	Aphrodisiac
Terpenes and steroids				
Taxol and other taxoids[b]	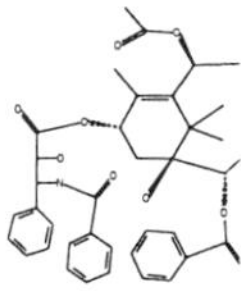	Diterpenes	*Taxus brevifolia* Nutt.	Antineoplastic
Diosgenin[a], hecogenin[a], stigmasterol[a]	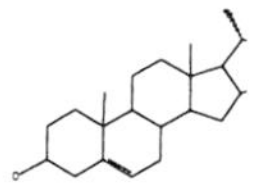	Steroids	*Dioscorea* spp.	Oral contraceptives and hormonal drugs
Glycosides				
Digoxin, digitoxin	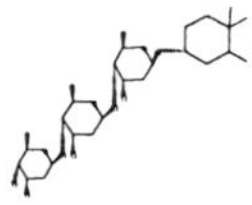	Steroidal glycosides	*Digitalis* spp.	Cardiotonic
Sennosides A and B	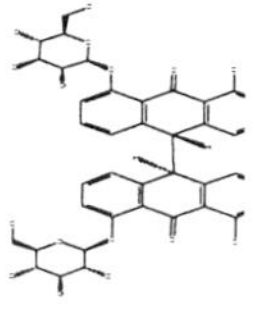	Hydroxyanthracene glycosides	*Cassia angustifolia* Vahl.	Laxative
Others & Mixtures				

Podophyllotoxin [a]	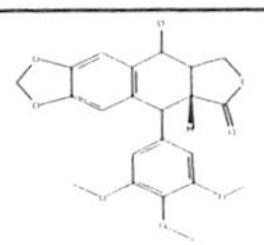	Tetrahydro-naphthalene	*Podophyllum peltatum* L.	Antineoplastic
Artemisinin	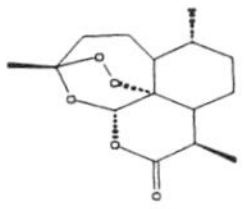	Sesquiterpene lactone	*Artemisia annua* L.	Antimalarial
Ipecac	N/A	Mixture of ipecac alkaloids and other components	*Cephaelis ipecacuanha* (Brot.)A. Rich.	Emetic

[a]Most often used as precursors in chemical synthesis of final products

(Adang and Hermkens, 2001; Schreiber, 2000) and computational drug design (Clark and Pickett, 2000) has put an end to the dominance of natural products in drug discovery. Nevertheless, about 250,000 living plant species contain a much greater diversity of bioactive compounds than any chemical library made by humans including many novel chemotypes.

So why plants being abandoned as sources for NCE while the pharmaceutical industry is having difficulties replacing old products with new and more effective alternatives or developing new products even though many new molecular targets have been recently discovered. A forty percent increase in the R&D spending in pharmaceutical research from 1996 to 2001 did not correct this problem (Bolten and De Gregorio, 2002).

The lack of reproducibility of activity for more than 40% of plant extracts (Cordell, 2000) is one of the major obstacles in using plants in pharmaceutical discovery, despite the great diversity of compounds they synthesize. The activities detected in screens often do not repeat when plants are re-sampled and re-extracted. Moreover, the biochemical profiles of plants harvested at different times and locations vary greatly. In addition, the currently popular high-throughput drug discovery format favors single compounds over mixtures and is not compatible with complex plant extracts in which valuable bioactive molecules are often obscured by pigments and poly-phenols that interfere with screens. Equally important is the lack of efficient, rapid strategies to isolate and characterize NCEs, particularly those present in trace amounts, making phytochemical discovery a complex, laborious task incompatible with short lead discovery times. It often takes six months to isolate and structurally characterize a natural product from a plant extract. This is roughly equivalent to the lifetime of a high-throughput screen for a new target, which is prohibitively long for an ever-accelerating lead discovery race.

Unquestionably, the development of novel technologies that allow rapid isolation and characterization of putative lead molecules and new screening methods more compatible with complex mixtures will be imperative if plants are to return to the mainstream of drug discovery efforts. Another strategy is to exploit the qualitative and quantitative variations in the content of bioactive phytochemicals, which are currently considered major detriments in phytochemical NCE discovery. Different stresses, locations, climates, microenvironments and physical and chemical stimuli, often called elicitors, qualitatively and quantitatively alter the content of bioactive secondary metabolites. Enzymatic pathways leading to the synthesis of these phytochemicals are highly inducible (Ebel and Cosio, 1994). Thus, elicitation-induced, reproducible increases in bioactive molecules, which may otherwise be undetected in screens, should significantly improve reliability and efficiency of plant extracts in drug discovery while preserving wild species and their habitats.

3. BOTANICAL DRUGS

The U.S. Federal Food and Drug Administration (FDA) has recently published a guidance for standardized multi-functional and multi-component plant extracts, referred to as botanical drugs, opening the doors for marketing these products under the New Drug Application (NDA) Approval Process (Anonymous, 2000). In response to the public demand for trustworthy and effective alternatives to NCE pharmaceuticals, the agency proposed abbreviated pre-clinical and clinical testing protocols for botanical drugs derived from plants with a safe history of human use.

In contrast to the Western NCE paradigm, traditional medicinal systems of the East always believed that complex diseases are best treated with complex combinations of botanical and non-botanical remedies that should be further adjusted to the individual patient and a specific stage of the disease. This approach, best articulated and developed in traditional Chinese and Ayurvedic medicinal systems, emphasizes the mutually potentiating effect of different components of complex medicinal mixtures. Ostensibly, plants have adapted a similar strategy in their biochemical warfare with pathogens, which are the main causes of plant disease and death. Relying on a single antibiotic to stop pathogens would probably be evolutionarily suicidal for plants because a resistance would develop.

The future of botanical drugs in the U.S. depends on two factors: sustaining a favorable regulatory environment and developing technologies for the efficient discovery, development and manufacture of botanical drugs. At present, a majority of botanical drugs under development are derived from ethnobotanical sources and traditional medicinal uses. In addition to the creative and innovative technologies needed for new botanical drug discovery, manufacturing botanical drugs presents a challenge not encountered by the modern pharmaceutical industry. To be FDA compliant, the process should involve "seed-to-pill" and "batch-to-batch" standardization of complex phytochemical mixtures, a challenge not encountered by chemical synthesis or single compound extraction processes. As stated above, environmental and genetic factors may dramatically affect the biochemical compositions of plant extracts. Therefore, production of botanical drugs will require genetically uniformed monocultures of source plants grown in fully standardized conditions to assure biochemical consistency and to optimize safety and efficacy in every crop. Fully-controlled greenhouse-based cultivation systems developed for high quality year-round vegetable production are probably more suitable for the future production of botanical drugs.

4. BOTANICAL DIETARY SUPPLEMENTS

Botanical dietary supplements also called botanical nutraceuticals or herbals (Table 3) can be best defined as plant-derived materials with medical benefits aimed at disease prevention or treatment that go beyond satisfying basic nutritional requirements. The use of botanical supplements in the U.S. has increased dramatically after passage of the Dietary Supplement and Health Education Act of 1994 (DSHEA). Under DSHEA, botanical supplements may be marketed with few regulatory impediments providing that disease prevention, curing or detection claims are not made. Instead so-called structure-function claims on products may relate to enhancing or maintaining normal physiological functions of human body. DSHEA does not require the manufacturers of supplements to verify that they are safe and effective.
Although the general public often considers botanical supplements natural and safe alternatives to conventional synthetic pharmaceuticals, there is relatively little scientific evidence behind this belief. In 1999 the global market for herbal supplements exceeded $15 billion with a $7 billion market in Europe, $2.4 billion in Japan, $2.7 in the rest of Asia and $3 billion in North America (Glaser, 1999). The demand for dietary supplements is driven by a variety of factors that include an aging population with substantial disposable income, a growing trend to self-medicate, mistrust in conventional medical establishment, and the perception that natural is healthy and plant products are safe.

Botanical supplements are frequently and rightly criticized for poorly proven efficacy and safety, lack of standardization and quality standards (Osowski et al., 2000; Kressmann et al., 2002) and potential drug-supplement interactions (Izzo and Ernst, 2001). In contrast to ethical drugs, active ingredients of very few botanical supplements have been fully characterized, in spite of the

Table 3. Common botanical dietary supplements sold in the U.S.

1[a]	***Echinacea purpurea*** L. *angustifolia*, DC.and *pallida* Nutt.	*Panax ginseng*, L.A. Mey (Ginseng)	*Serenoa repens* (W.Bartam) Small (Saw Palmetto)	*Ginkgo biloba* L.	*Hypericum perforatum* L. (St. John's Wort)	*Valeriana officinalis* L. (Valerian)	*Allium sativum* L. (Garlic)
2	Shoots, roots	Roots	Fruit	Leaves	Shoots	Roots	Bulb, oil
3	Respiratory infections, immuno-stimulant	Fatigue and stress, high cholesterol, diabetes, gastro-intestinal disorders	Benign prostate hyperplasia (BPH), inflamations, impotence	Dementia, cognitive decline, mental fatigue	Mild and moderate depression, epilepsy	Sleep improvements, anxiety, hypertension	Cancer, high cholesterol, diabetes, arteriosclerosis, hypertension, respiratory infections
4	Polysaccharides, alkylamides, chlorogenic acid, caffeic acid derivatives (echinacosides)	Ginsenosides, panaxans, sequiterpenes	Steroids (beta-sitosterols), flavonoids (isoquercitin, kaempherol, rhoifolin)	Terpene trilactones (ginkgolides), flavonol glycosides	Hyperforin, adhyperforin, hypericin, pseudohypericin, flavonol glycosides	Valeric acids, valepotriates, valtrates	Alliins, allicin, ajoens, oligosulfides
1	***Hydrastis canadensis*** L. (Goldenseal)	*Matricaria chamomilla* L. (German chamomile)	*Silybum marianum* Gaertn. (Milk Thistle)	*Trigonella foenum-graecum* L. (Fenugreek)	*Tanacetum parthenium* Schultz-Bip. (Feverfew)	*Ephedra sinica* Stapf. (Ephedra, Ma Huang)	*Cimicifuga racemosa* Nutt. (Black cohosh)
2	Rhizome, roots	Flowering herb	Ripe seeds	Ripe seeds	Herb	Stems	Root
3	Diarrhea, respiratory and gastro infections, constipation	Intestinal disorders, wound healing, inflammations, anxiety	Liver disorders, lactation problems	Diabetes, loss of appetite, skin inflammation	Migraines, inflammation	Stimulant, obesity, asthma, congestion, fluid retention	Premenstrual symptoms, dysmenorrhea, menopausal symptoms
4	Hydrastine, berberine, canadine	Bisabolols, apigenins, luteolin, coumarins, proazulenes	Silymarins, flavonoids (apigenin, chrysoeriol, quercetin, taxifolin)	Mucilages, steroid saponins (trigofoenosides, dosgenin, foenugraecin), trigonelline	Sesquiterpene lactones (parthenolides, costunolide, reynosin, canin, artecanin)	Protoalkaloids (ephedrine, pseudo-ephedrine), cannins, saponins	Triterpene glycosides (cimifugaside, 27-deoxyactein, actein), isoflavones

[a]1 - Name; 2 – Parts used; 3 – Common indication / use; 4 - Putative active ingredients

significant efforts by many researchers. The difficulty in isolating active ingredients suggests that the therapeutic effects of many botanical supplements are due to the combined effects of many compounds, which are often lost during standard activity guided fractionation, which separates extracts into their individual molecular components. The current botanical food supplement industry, despite its size, is to a large extent marketing driven with almost no sustained R&D efforts directed towards creating credible product pipelines, quality control measures and discovery platforms. Instead, traditional medicinal plants are being repackaged, remixed and remarketed. Creating the environment that rewards better efficacy, quality and safety standards for dietary supplements is a major regulatory challenge that needs to be met in order to sustain the growth of the botanical supplement industry. Allowing more specific claims and some marketing exclusivity in exchange for more thorough pre-clinical and clinical data will be helpful as well as increasing governmental and private research funding of the botanical supplement R&D.

5. Functional/Medicinal foods

This is probably the most known and reviewed area of botanical therapeutics because of its connection to the mainstream of plant biotechnology and molecular biology. As with other botanical therapeutics, the precise definition of functional foods is vague. This definition generally refers to crops engineered or selected to deliver certain health benefits above and beyond those normally present. Botanical functional foods produced by fortification, such as orange juice with calcium, or advertised for their innate health benefits, such as cereals with high fiber, will not be discussed here.

In the area of engineered functional foods, much attention was given to the development of golden rice (Potrykus, 2001; Ye et al., 2000), healthy plant oils from modified oil crops (Bonetta, 2002; Thelen and Ohlrogge, 2002), edible vaccines (Daniell et al., 2001; Walmsley and Arntzen, 2000) and plants with increased levels of essential vitamins and nutrients such as vitamine E (Shintani and DellaPenna, 1998) high lycopene/vitamin C (Frusciante, 2000); metabolic engineering of legumes for high content of bioflavonoids (Forkmann and Martens, 2001), known for their anti-oxidant, anti-cancer and oestrogenic properties.

Presently, a major constraint in engineering secondary metabolites in functional foods is the scarce information about their biosynthetic genes and pathways. Cross-species integration of proteomics and metabolomics with this genetic information will allow a better understanding and utilization of metabolic networks that enhance medicinal properties of foods. In parallel, better technologies for characterizing pharmacologically active compounds in foods have to be developed. Providing that scientific, regulatory and public acceptance issues are solved, the future of plant-based functional foods seems bright and, as a result, grocery and drugs stores may eventually look more alike. Functional foods with clear and direct health benefits for the consumer should lead to greater acceptance of crop genetic engineering, now almost exclusively and controversially used for crop protection.

6. RECOMBINANT PROTEINS

Recombinant proteins such as antibodies, vaccines, regulatory proteins and enzymes represent one of the most rapidly growing segments of the pharmaceutical industry. With over 100 proteins in clinical development today there is a substantial shortage of industrial capacity to manufacture future recombinant drugs (Garber, 2001). During the last decade plants have emerged as promising biopharming systems for commercial production of pharmaceutical proteins. Advantages offered by plants include low cost of cultivation and high biomass production, relatively fast "gene to protein" time, low capital and operating costs, excellent scalability, eucaryotic post-translational modifications (i.e., glycosylation, folding and multimeric assembly), low risk of human pathogens and endotoxins and a relatively high protein yield. These advantages are potentiated by the ease of plant transformation through particle bombardment, electroporation, *Agrobacterium*-mediated transformation, or infection with modified viral vectors (Fischer and Emans, 2000). Plants are generally considered low-cost, safe and relatively fast alternatives to many existing manufacturing systems particularly when large quantities of multimeric recombinant proteins are required (i.e., antibodies). Most major groups of human pharmaceutical proteins have been successfully produced in a diverse variety of crops and model systems such as maize, rice, wheat, soybean, tomato, potato, mustard, oilseed rape, turnip, alfalfa, banana, tobacco and *Arabidopsis* using stable nuclear and plastid transformations as well as transient expression systems such as viruses (Daniell, 2001; Parmenter et al., 1995).

The highest yield of recombinant protein in plants is achieved by chloroplasts expression or possibly by transient viral expression. However, plastid expression suffers from the lack of glycosylation and correct folding of multimeric proteins, while viral expression is problematic for complex proteins and may introduce safety hazards. Seeds (corn) and tubers (potatoes) provide excellent long-term storage compartments for recombinant proteins before purification, but have lower protein yield.

Downstream protein purification is often as expensive as the biomanufacturing and should never be overlooked in the total "cost of goods" equation. At least two approaches have been used successfully to lower the cost of downstream purification of plant-produced proteins: oleosin-fusion technology for heterologous proteins produced in oilseeds (Parmenter et al., 1995), and rhizo- and phyllo- secretion platforms based on continuous, non-destructive recovery of a target protein from plant exudates (Borisjuk et al., 1999; Komarnytsky et al., 2000). The latter also offers the advantage of continuous protein production that integrates the biosynthetic potential of a plant over its lifetime and may lead to higher protein yields compared to single harvest/extraction methods. Nevertheless, the arguments favoring plants are appealing and a growing number of companies are trying to commercialize recombinant protein manufacturing in plants, with most concentrating on pharmacological applications.

7. CONCLUSIONS

Plants are arguably poised for a comeback as sources of human health products. The hopes for this comeback are rooted in the unique and newly appreciated properties of phytochemicals vis-à-vis conventional NCE-based pharmaceuticals and are based on: (i) enormous propensity of plants to synthesize mixtures of structurally diverse bioactive compounds with multiple and mutually potentiating therapeutic effects; (ii) low-cost and highly scalable protein and secondary metabolite biomanufacturing capacity of plants (iii) diminishing return of the single NCE approach to drug discovery and disease treatment and prevention; (iv) cost limitation on the chemical synthesis of complex bioactive molecules; (v) perception that because of the history of human use and co-evolution of plants and humans, phytochemicals provide a safer and a more holistic approach to disease treatment and prevention. While the above properties have been known for a long time, the ability to better exploit the uniqueness of plant therapeutics was acquired only recently due to the dramatic advances in metabolic engineering, biochemical genomics, chemical separation, molecular characterization and pharmaceutical screening. A challenge for phytochemical-based botanical therapeutics is to integrate the growing ability to identify and genetically manipulate complex biosynthetic pathways in plants with better characterization of genetic targets for the prevention and treatment of complex diseases. Similarly, an important challenge is the development of discovery, validation and manufacturing technologies that are compatible with multifunctional phytochemical mixtures. For the recombinant protein manufacturing the main challenge is to commercialize a plant produced therapeutic protein and to demonstrate that plant-based production is, indeed fast, efficient and cost effective.

8. ACKNOWLEDGEMENTS

We thank Barbara Halpern, Ellen Shlossberg and Rainer Becker for their help in preparing this manuscript. The Spanish Secretariat of Education and Universities provided funding for Diego A. Moreno.

9. REFERENCES

Adang, A.E. and Hermkens, P.H. (2001) The contribution of combinatorial approaches to lead generation: an interim analysis. Curr. Med. Chem. 8, 985-998

Anonymous. U.S. Department of Health and Human Services Food and Drug Administration Center for Drug Evaluation and Research (2000) Guidance for Industry Botanical Drug Products. http://www.fda.gov/cder/guidance/1221dft.htm#P131_3293

Anonymous. Business Communications Company, Inc. study, RB-121 (1998) Plant-Derived Drugs: Products, Technologies and Applications

Bolten B.M. and De Gregorio T. (2002) Trends in development cycle. Nature Reviews 1, 335-336

Bonetta, L. (2002) Edible vaccines: not quite ready for prime time. Nature Medicine 8, 95.

Borisjuk, N.V. et al. (1999) Production of recombinant proteins in plant root exudates. Nature Biotechnol. 17, 466-469

Clark, D.E. and Pickett, S.D. (2000) Computational methods for the prediction of drug-likeness. Drug Discov. Today 5, 49-58

Cordell, G.A. (2000) Biodiversity and drug discovery. Phytochemistry 55, 463-480

Daniell, H. (2001) Medical molecular pharming: production of antibodies, biopharmaceuticals and edible vaccines in plants. Trends Plant Biol. 6, 219-226

Ebel, J. and Cosio, E. G. (1994) Elicitors of plant defense responses. Int. Rev. Cytology 148, 1-36

Fischer, R. and Emans, N. (2000) Molecular pharming of pharmaceutical proteins. Transgenic Res. 9, 279-299

Forkmann, G. and Martens, S. (2001) Metabolic engineering and applications of flavonoids. Current Opinion in Biotechnology 12, 155-160

Frusciante, L., et al. (2000) Evaluation and use of plant biodiversity for food and pharmaceuticals. Fitoterapia 71(Suppl. 1), S66-S72

Garber, K. (2001) Biotech industry faces new bottleneck. Nature Biotechnol. 19, 184-185

Giddings, G. et al. (2000) Transgenic plants as factories for biopharmaceuticals. Nature Biotechnol. 18, 1151-1155

Glaser, V. (1999) Billion-dollar market blossoms as botanicals take root, Nature Biotechnology 17, 17-18

Izzo, A A. and Ernst, E. (2001) Interactions between herbal medicines and prescribed drugs: a systematic review, Drugs, Volume 61, 2163-2175

Kressmann, S., et al (2002) Pharmaceutical quality of different Ginkgo biloba brands, The J. Pharmacy and Pharmacology 54, 661-669

Komarnytsky, S. et al. (2000) Production of recombinant proteins in tobacco guttation fluid. Plant Physiol. 124, 927-933

Osowski, S. et al. (2000) Pharmaceutical comparability of different therapeutic Echinacea preperations. Research in Complementary and Natural Classical Medicine 7, 294-300

Parmenter, D.L. et al. (1995) Production of biologically active hirudin in plant seeds using oleosin partitioning. Plant Mol. Biol. 29, 1167-1180

Potrykus, I. (2001) Golden rice and beyond, Plant Physiol. 125, 1157-1161

Rates, S.M.K. (2001) Plants as sources of drugs. Toxicon 39, 603-613

Schreiber, S.L. (2000) Target-oriented and diversity-oriented organic synthesis in drug discovery. Science 287, 1964-1969

Shintani, D. and DellaPenna, D. (1998) Elevating the vitamin E content of plants through metabolic engineering. Science 282, 2098-2100

Thelen, J.J. and Ohlrogge, J.B. (2002) Metabolic Engineering of Fatty Acid Biosynthesis in Plants. Metabolic Engineering 4, 12-21

Walmsley, A.M. and Arntzen, C. J. (2000) Plants for delivery of edible vaccines. Current Opinion in Biotechnology 11, 126-129

Ye, X. et al. (2000) Engineering provitamin A (β-carotene) biosynthetic pathway into (caroteneoid-free) rice endosperm. Science 287, 303-305

Owen R. White

THE REANNOTATION OF THE *ARABIDOPSIS THALIANA* GENOME

Roger K. Smith, Jr., Brian J. Haas, Rama Maiti, Agnes P. Chan, Linda I. Hannick, Catherine M. Ronning, Yong-Li Xiao, Christopher D. Town, Owen R. White
The Institute of Genomics Research (TIGR), 9712 Medical Center Drive, Rockville, MD20850, USA (email: owhite@tigr.org)

Keywords Arabidopsis, genome

1. INTRODUCTION

The small crucifer *Arabidopsis thaliana* emerged as a "model" plant species due to the many experimental advantages the plant has over more economically important plants (reviewed in Meinke, et al., 1998). These economically important plants, mainly crop species, are generally difficult to manipulate in the lab, are large, and have large and complex genomes. Arabidopsis, on the other hand, has a rapid life cycle (6 weeks), a small size, and a small diploid genome (125 Mb, compared to maize, 2,500 Mb, and wheat, 16,000 Mb)(Meyerowitz and Somerville, 1994). These advantages led to an explosion in the number of members of the scientific community using Arabidopsis to investigate biological questions. To date, numerous tools and resource materials exist for those wishing to pose questions using this model plant.

To support the ongoing research efforts in Arabidopsis, the Arabidopsis Genome Initiative (AGI) was formed (Bevan, 1997). This international collaboration facilitated the coordinated sequencing of the genome, and the release of the first completed plant genome was "a scientific event of some importance" (Dennis and Surridge, 2000). Plants evolved along similar paths, and diversification among species is a relatively recent event. Thus, most developmental and physiological processes, and the genes controlling them, are thought to be conserved among various plant species. The elucidation of the genomic sequence of Arabidopsis provides a means for analyzing gene function relevant to a range of plant species.

The multinational effort of AGI produced large amounts of sequence data, but without annotation of genes and other biologically important features, these sequences are of little utility for the research community. AGI groups realized this, and the annotation of the Arabidopsis genome was completed

I. K. Vasil (ed.), Plant Biotechnology 2002 and Beyond, 97-106.

by the individual sequencing groups and released to the public and/or maintained on individual lab's web sites. However, due to the differences between groups, the degree and quality of annotation varied greatly. In addition, annotation methodologies had evolved over the course of the sequencing project and old annotations were not updated based on newly released sequence and annotation. At best, the annotations were heterogeneous and the research community expected the annotation to be uniform and of high quality.

In an effort to improve the annotation, TIGR initiated a multipronged and systematic approach to reannotate the entire Arabidopsis genome. Reannotation permits uniform quality control, systemic updates, easy data parsing, and more comprehensive comparative analysis, providing a valuable resource to the whole research community of plant science. The approach taken at TIGR involves both automated and manual curation, and various computational methods to ensure that the annotation is complete, thorough, and of high quality.

2. Computational Reannotation

To begin the reannotation process of Arabidopsis, the annotation from GenBank and other public databases was captured and the data was imported into a relational annotation database, ATH1. Due to the continuing evolution of the annotation at different centers, it was important to identify genes based on unique identifiers (loci) instead of the protein sequence they encode or any functional name assigned. Comparisons of TIGR annotation to the current contents of genes annotated elsewhere have been made, and loci assignments (BAC-based and chromosome-based) have been linked, providing a stable link between Arabidopsis data repositories.

A series of programs have been written that comprise a functioning automated annotation system, known as Eukaryotic Genome Control (EGC). This set of programs automatically launches numerous analysis programs that were previously manually started, including gene finders, alignment programs, splice site prediction programs, and scripts to load the results of these programs into ATH1. A distributed computer architecture for computationally intensive searches was also employed in the use of these analysis programs. The combined EGC pipeline and parallel computer system have greatly accelerated the acquisition of data required for the detailed examination of genes, making the process more efficient. For example, blast searches and HMM searches for the proteome are performed at least ten times faster than conventional methods of using a single dedicated computer, and can be launched automatically whenever a gene is updated.

There is some redundancy in the current annotation due to the annotation of genes in the overlapping regions of BAC sequences. Using computational methods, TIGR has constructed a contiguous sequence representing each of the 5 Arabidopsis chromosomes based on these overlapping BAC sequences. When constructing the contiguous sequences of the chromosomes, the overlapping, redundant annotation was carefully examined and redundancies were expunged while identifiers for tracking purposes of both genes were retained. We have now completed several passes through the genome, and the current proteome contains approximately 26,000 genes.

3. GENE STRUCTURE ANNOTATION

Assignment of proper intron/exon structures to gene products is a key first step in the reannotation process. TIGR employs computationally derived algorithms and a complex computer interface to verify gene structure manually. Gene structure is validated based on cDNA, EST, and protein matches, as well as gene prediction models. Duplicated gene models are abundant in the genome due to overlapping BAC sequences, but the assembled chromosome sequences and the software used by TIGR in the reannotation eliminates this redundancy. Approximately half of the Arabidopsis genes have curated gene structures, and of those, more than 50% have exon-intron boundaries consistent with cDNA sequence, and another 40% have gene structures similar to those suggested by cDNA sequence.

3.1. Gene Nomenclature and GO Assignments

Once gene structures have been computationally and manually validated, gene products are named according to a hierarchy of specific nomenclature based on database hits to functionally characterized gene products and protein domains. This allows for accurate and consistent naming of the gene products in the genome. If a gene product has a perfect database match to a characterized protein, then the name of the protein is assigned to the gene product in the annotation. During this part of manual evaluation, if there is a discrepancy in the name of a protein, then the name assigned by SWISSPROT (Bairoch and Apweiler, 2000) is frequently used. Thus, gene products with assigned names have a high degree of confidence of function. For gene products that have database matches that are less than perfect and thus the confidence level is not as high, the name of the gene product is given and designated as "putative". This putative naming is given to a gene product in which the annotator believes is functioning as a characterized protein, but has not actually been characterized. If a gene product has a database match to a group of proteins with similar functions but it is not possible to choose a precise name, the name is made more generic and the family name is given.

Finally, if the database matches to a gene product are partial and there are motifs or regions of very good similarity, then the gene product is given the same name as the match but designated as a "related protein."

A nomenclature hierarchy has also been established for gene products that do not have a good database match to characterized proteins or protein domains. In these cases, cDNA, EST, and gene finder predictions are used to name these gene products. If cDNA evidence supports the gene model, then the gene product is designated as an "expressed protein." If only EST data supports the gene model, there is less confidence in the model and the gene product is considered to be an "unknown protein." If no experimental evidence exists for a gene model yet it has been predicted by a gene-finding model, the gene product is designated as a "hypothetical protein."

In addition to adopting and implementing these naming strategies, Gene Ontology (GO) assignments are being given to all genes. The GO Consortium is an international effort to produce a dynamic controlled vocabulary that can be applied to all organisms (www.geneontology.org). GO uses an integrated network to organize and define gene products based on molecular function, biological process, and cellular component. Under the reannotation strategy, gene products in Arabidopsis are defined within this integrated network giving researchers more detailed information about the gene products. Approximately 10% of the Arabidopsis gene products have been defined in this manner and the remaining gene products are expected to be assigned within the next year.

4. GENERATION OF PARALOGOUS PROTEIN FAMILIES IN THE ARABIDOPSIS GENOME

In order to facilitate the reannotation efforts, the approximately 26,000 gene products in the genome have been organized into paralogous families. Paralogous proteins exist due to gene duplications that evolved from an ancestral gene. The grouping of these related proteins allows annotators to better evaluate the function of the predicted genes and to understand the gene duplication. We have identified these families based on Arabidopsis Paralogous Domains (APDs) and Pfam HMM domain organization. APDs were assembled with algorithms developed at TIGR based on Arabidopsis protein sequences annotated in the genome. Pfam is a database of multiple alignments of protein domains or conserved protein regions (Bateman et al., 2002). The alignments represent some evolutionary conserved structure, which has implications for the protein's function. Profile Hidden Markov models (profile HMMs) built from the Pfam alignments can be very useful for automatically recognizing that a new protein belongs to an existing protein family, even if the homology is weak. All of the peptides encoded in the Arabidopsis genome were searched against the Pfam HMM profiles of domains. The search results with scores above the trusted cutoffs of the HMM profiles were kept and alignments of peptides were generated for all the HMM profiles that have more than two protein hits in the genome. The peptide regions that were not covered by the Pfam HMM profiles were gathered into a new set of peptides and an innovative approach was taken to cluster this set of proteins based on their homology. BLASTP searches were performed to establish the relationships among the peptides that were not covered by Pfam profiles. Relatively low stringency criteria were used to link proteins at this stage; if two peptides had an identity above 30% over a 50 amino acid span with BLASTP expected value (E value) lower than 0.001, a link was established between the two peptides.

Because some proteins have multiple domains, non-related proteins can be pulled together into one cluster. These clusters can be rather large and include thousands of proteins. To prevent the problem of single linkage clustering, a parameter called link score was introduced in the clustering process. Two peptides with a link score below the cut-off value were not used as a link to generate a cluster. The link score was defined as the number of common peptide hits shared by two proteins divided by the number of combined peptide hits.

An alignment was generated for each cluster. Alignments generated by HMM profile search and BlastP based clustering were considered to be domains. The paralogous protein families were organized according to the same group of domains they share. A set of 3,844 protein families containing approximately 20,000 proteins was obtained from this paralogous family building process.

5. IDENTIFICATION OF UNANNOTATED GENE MODELS FROM INTERGENIC REGIONS

One of the goals of Arabidopsis genome reannotation was to determine if cryptic and/or overlooked candidate gene models existed that could be included in the current annotation. To identify unannotated gene models, HMM models generated from APDs and gene prediction models obtained from several gene-finding algorithms (Genemark.hmm, Genescan+ and GlimmerA) were used. For each intergenic region that may contain unannotated genes, two types of analyses were performed. In the first analysis, intergenic regions were searched for regions that had multiple gene prediction models and a match with APD HMM models. In the subsequent analysis, intergenic regions that were supported only by multiple gene prediction models were examined.

The HMMMER hmmsearch program was used to identify significant matches to intergenic gene prediction models against the APD HMM models. The intergenic region specified by the coordinates of the predicted gene model was subjected to further analysis. If such predicted gene model co-existed with one or more overlapping gene prediction models, all models were gathered together as one group. More specifically, the grouping was performed if there were two or more gene prediction models present in the same orientation (on the same strand), spanning the same region (at least 100 a.a. overlap), and also having the same domain hit. With this approach, a total of approximately 600 gene prediction models satisfied the criteria of having a APD match and that the protein domain was supported by at least two gene prediction models.

In the second approach, all intergenic regions that contained two or more gene prediction models were identified and examined. The list of gene prediction models generated from the two approaches were further compared to a curated protein sequence database containing known transposon sequences. Any gene prediction models with similarity to transposon sequences were removed, since transposons very often are degenerate. By using these two approaches, it was determined that approximately 1700 gene prediction models satisfied predetermined criteria to be added to the genome annotation.

5.1. Comparative Genomics as a Tool for Gene Discovery and Improving Genome Annotation

Comparisons between genome sequences of evolutionarily related species are emerging as a powerful tool for the identification of functionally important regions that are conserved over evolutionary time. Sequence conservation can be detected within genes, promoters, and regulatory regions. Through the comparison of the Arabidopsis genome to closely related plant genomic sequences, we are identifying regions of nucleotide conservation which may represent conserved functional sequences, and allow us to improve the Arabidopsis genome annotation.

Brassica oleracea was targeted for this comparative study primarily because it is closely related to Arabidopsis and thought to have diverged from a common ancestor approximately 15-20 million years ago (Yang et al., 1999; Koch et al 2001). Percent-identity plots (PIPs), created by PipMaker (Schwartz, et. al, 2000), between previously identified intron-containg *Brassica* genes and the homologous regions of Arabidopsis genomic sequence convincingly demonstrated the conservation of exonic, as well as demonstrated the sequence divergence within introns, supporting the use of

Brassica genomic sequence to detect functionally conserved regions with the Arabidopsis genome.

We performed a pilot study to validate the use of *Brassica* sequence comparisons to identify functionally conserved sequences in Arabidopsis. We generated and examined 16,000 whole-genome shotgun sequence reads from *Brassica oleracea*. Approximately 1/4 of the sequence reads were filtered from the data set due to sequence similarity to the mitochondrial and chloroplast genomes, or to known transposable elements. The remaining sequences were searched against the Arabidopsis genome. The vast majority of the individual *Brassica* sequences (87%) did not have a relevant match to the Arabidopsis genome. Of the sequences matching the Arabidopsis genome, 91% matched genes, and the remaining 9% fell within intergenic regions. Half of the matches to intergenic regions identified missed annotations in the Arabidopsis genome, that is genes which should have been annotated based on our current annotation protocols, but were missing from the annotation data set at that time. The remaining half of the matches to intergenic regions were of greatest interest, consisting of 75 sequences, each sequence representing a candidate novel gene.

To further probe the discovery of novel Arabidopsis genes using this comparative approach, 25 of the conserved sequences were tested for expression using PCR. We extended the conserved sequence matches to identify the longest open reading frames (ORFs) spanning the match. ORFs ranging from 100 to 500 bp were identified and primers were designed based on these sequences for the purpose of amplifying the corresponding region from a set of cDNA populations. Of the 25 selected intergenic regions, 15 were shown to be expressed based on the PCR results.

In light of these encouraging results, we expanded our analysis by generating and examining ~350,000 reads for a total of > 250 Mb, representing 0.3x coverage of the *Brassica* genome. Further exploration of the sequence comparisons using PIP plots yielded a shift in our choice of sequence comparison tool from blastn to blastz, the sequence search and alignment engine underlying the PIPmaker software. The specificity of blastz in identifying conserved functional sequences was demonstrated by our analysis of Arabidopsis gene annotations whose gene structures are completely supported by full-length cDNAs. In this latest set of *Brassica* sequence reads, over 20,000 sequences matched a gene supported by full-length cDNAs, and less than 1% of the relevant sequence match segments fell completely within introns. For the relevant matches that overlapped exons in this quality gene set, greater than 90% of the *Brassica* sequence match regions encompassed exonic sequence. This clearly demonstrated the utility of the *Brassica* sequence comparisons in identifying genic regions, and the existence of exons therein.

After filtering the latest set of *Brassica* sequences for repetitive sequences, the remaining sequences were aligned to the Arabidopsis genome using blastz. In examining this large *Brassica* sequence database, considerable redundancy among the matches was observed. In searches of the individual sequence reads against a non-redundant comprehensive protein database, ~6 *Brassica* sequences matched the same gene on average. To avoid including redundant sequence match data in our analysis, all overlapping matches within the Arabidopsis genome were compressed into single sequence match segments. Upon examining the position of these genome sequence segments that are conserved between Brassica and Arabidopsis, ~90% were anchored to annotated genes. The remaining conserved genome segments lying within intergenic regions suggest the existence of at least 500 previously undiscovered conserved plant genes. Experiments are currently underway to verify the expression of genes within those regions and to obtain the complete cDNA sequence to correctly identify and annotate the newly discovered gene (see below).

6. CLONING AND ANALYSIS OF ARABIDOPSIS HYPOTHETICAL GENES

More than 25% of the genes in the sequenced and annotated Arabidopsis genome have structures that are predicted by computer algorithms or have genomic matches to related species but no supporting expression or protein evidence (see sections above). These "hypothetical" proteins were tested for expression under various growth conditions and in specific tissues. cDNA populations were prepared and used from cold-treated, heat-treated, and pathogen (*Xanthomonas campestris* pv. *campestris*)-infected plants, callus, roots, and young seedlings. 169 of the hypothetical genes have been tested to date, and 138 of them have been found to be expressed in one or more of the cDNA populations. Full-length cDNA sequences from 26 genes were obtained by sequencing and assembling their 5' and 3' end RACE products. Sixteen of the genes have at least one cDNA assembly that precisely supports the predicted intron-exon boundaries, adding only 5' and 3' UTR sequences with the other assemblies arising from alternatively spliced or unspliced introns and/or multiple polyadenylation sites. The cDNA sequences from the remaining 10 genes display differences from their predicted gene structures. Five show major differences from the predicted gene structure. One cDNA assembly displays six additional introns and exons upstream of the predicted ATG start site, which actually falls within an intron, and also shows two polyadenylation sites. The remaining four, either the predicted start codon, stop codon, or both fall in what the cDNA evidence shows to be an intron. Based on these results, it is expected that most of the hypothetical genes are expressed in Arabidopsis and their cDNA sequences can be used to validate or amend their predicted structures.

7. CONCLUSIONS

TIGR is reannotating the entire Arabidopsis genome. Numerous computational methods are being employed to aid in the annotation, and manual curation is ensuring that the annotation is uniform and consistent. Gene structure is validated based on cDNA, EST, and protein matches, as well as gene finder predictions. Specific and descriptive gene nomenclature has been adopted based on database hits to functionally characterized proteins. Related gene products have been grouped together based on paralogy, enabling annotators to better evaluate gene product function. Overlooked candidate gene models have been identified and are being added to the current annotation, and these and other hypothetical genes are being confirmed experimentally. Finally, comparative genomics between Arabidopsis and Brassica is being used to identify functionally important regions and to improve the genome annotation. All of these methods employed in the reannotation efforts will lead to the homogeneous and high quality of annotation that the research community expects.

8. REFERENCES

Altschul, S.F., Gish, W., Miller, W., Myers, E.W., and Lipman, D.J. (1990). Basic local alignment search tool. J. Mol. Biol. 215:403-410.

Bairoch, A., Apweiler, R. (2000) The SWISS-PROT protein sequence database and its supplement TrEMBL in 2000. Nucleic Acids Res. 28:45-48.

Bateman, A., Birney, E., Cerruti, L., Durbin, R., Etwiller, L., Eddy, S.R., Griffiths-Jones, S., Howe, K.L., Marshall, M., and Sonnhammer, E.L. (2002) The Pfam protein families database. Nucleic Acids Res. 30:276-80.

Bevan, M. (1997). Objective: The Complete Sequence of a Plant Genome. Plant Cell 9:476-478.

Dennis, C., and Surridge, C. (2000) *A. thaliana* genome. Nature 408:791.

Koch, M., Haubold, B., and Mitchell-Olds, T. (2001) Molecular systematics of the Brassicaceae: evidence from coding plastidic matK and nuclear Chs sequences. Amer. J. Bot. 88:534-544.

Meinke, D.W., Cherry, M., Dean, C., Rounsley, S.D., and Koornneef, M. (1998) *Arabidopsis thaliana*: A Model Plant for Genome Analysis. Science 282:662-665.

Meyerowitz, E.M., and Somerville, C.R. (1994) Arabidopsis. Plainview, NY: Cold Spring Harbor Laboratory Press.

Schwartz, S., Zhang, Z., Frazer, K.A., Smit, A., Riemer, C., Bouck, J., Gibbs, R., Hardison, R., and Miller W. (2000) PipMaker--a web server for aligning two genomic DNA sequences. Genome Res. 10:577-586.

Yang, Y.W., Lai, K.N., Tai P.Y., and Li, W.H. (1999). Rates of nucleotide substitution in angiosperm mitochondrial DNA sequences and dates of divergence between *Brassica* and other angiosperm lineages. J. Mol. Evol. 48:597-604

BIOSAFE, TRANSGENIC RESISTANCE FOR PLANT NEMATODE CONTROL

Howard J. Atkinson
Centre for Plant Sciences, University of Leeds LS2 9JT, UK (e-mail: h.j.atkinson@leeds.ac.uk)

Keywords proteinase inhibitors, additive resistance, peptides, root-specific promoters, biosafety

1. INTRODUCTION

Nematodes cause about $100 billion of losses to world agriculture annually and their chemical control is essential. Unfortunately, nematicides are among the most environmentally damaging of all crop protection chemicals. Other approaches to control such as cultural procedures are currently inadequate for a variety of reasons centred on cost and effectiveness. The successful cloning of genes for natural resistance may speed development of agronomically acceptable resistant cultivars. My laboratory has taken a distinct approach aimed at developing novel resistance for a wide range of nematode problems on a broad spectrum of crops.

2. EFFICACY OF CYSTATIN MEDIATED RESISTANCE

We have explored the potential of serine and cysteine proteinase inhibitors. In parallel we have studied the proteinases of nematodes to direct rational selection of proteinase inhibitors. Work on the potential of cystatins (cysteine proteinase inhibitors) for plant-parasitic nematode control began with a gene from rice primarily as it was the first phytocystatin to be cloned. Its affinity for papain can be enhanced through proteinase engineering but other plant cystatins with a high affinity for this cysteine proteinase are now known. We have advanced prototype constructs with cystatin genes from hairy roots, through model plants to containment and field trials with a crop plant. A homozygous line expressing the engineered cystatin under the control of the CaMV35S promoter at 0.4 % total soluble protein suppressed *Heterodera schachtii, Meloidogyne incognita* and *Rotylenchulus reniformis* on transgenic *Arabidopsis* plants. This offers the commercial advantage of efficacy against a range of nematodes including crops damaged by several distinct pest species. In collaboration with others, we are studying control of *Meloidogyne*

I. K. Vasil (ed.), Plant Biotechnology 2002 and Beyond, 107-110.

spp on rice as our example monocot crop. We have achieved over 80% resistance using a cystatin under the control of a promoter that favours expression in roots. Field trials to date have been limited to challenge of transgenic potato by *Globodera pallida*. Resistance levels of over 70% have been conferred on the susceptible cultivar Desiree by expressing cystatins under control of the CaMV35S promoter (Urwin et al., 2001) or one that restricts expression to roots.

3. ADDITIVE RESISTANCE TO SUPPORT EFFICACY AND DURABILITY

We have shown that full resistance can be achieved against *G. pallida* by transforming the partially resistant cultivars Sante and Maria Haunca to express a cystatin. We have also shown additive resistance using cowpea trypsin inhibitor and a cystatin expressed as a translational fusion product via a peptide linker that is not cleaved *in planta*. Further targeting of digestive proteinases may not enhance resistance further. Consequently we are studying other enzymes important to digestive processes in nematodes. We have been aided in this by publicly available databases for nematode genes or EST collections. In parallel we have prepared cDNA libraries for nematode intestines dissected from feeding individuals.

3.1. RNA Interference as a Means of Identifying Targets

A key problem is identifying gene expression that is essential for success of the nematode pathogen. We have explored the potential of RNA inference (RNAi) to validate targets. First we used the approach to disrupt growth after suppressing expression of the transcript for a cysteine proteinase. Next we suppressed transcript abundance of the major sperm protein of a cyst nematode. All members of this gene family are expressed in males but only after their feeding has ended. RNAi did suppress transcript abundance in males but, as expected, it did not compromise the growth of either sex. Therefore RNAi can target genes that are first expressed only several days after host invasion. Finally we targeted a gene of unknown function abundantly expressed during invasion and establishment. As a result the total number of parasites at 14 days post invasion was reduced. Therefore we now have a basis for validating potential targets before identifying, cloning and transforming plants with an appropriate anti-nematode gene.

3.2. Peptides Directed at Novel Targets

One limitation in building a highly effective and durable resistance is availability of anti-nematode proteins. Therefore we have explored approaches to overcome this including use of both phage single chain antibody and peptide libraries. As an example, we are designing a line of defence against root invasion. Our lead was the nematicide aldicarb. It is able to disrupt chemoreception of nematodes at low concentration and so prevent root invasion. We found that the anthelmintic levamisole had a similar effect and it lacks mammalian toxicity. Therefore we biopanned bacteriophage peptide display libraries against acetylcholinesterase and a preparation with acetycholine receptors. This was to obtain peptide mimetics of aldicarb and levamisole respectively. Phage clones of interest were obtained and the peptides they displayed were synthesised (Atkinson et al., 2002). Exposure of cyst nematodes to very low concentrations of the peptide mimetics of both aldicarb and levamisole disrupted their chemoreception. The peptides suppressed root invasion and can be made by plant cells. Our aim is to build additive defences based on anti-nematode genes with three distinct modes of action against the parasites. Our expectation is that this will ensure both efficacy and durability of nematode control.

4. BIOSAFETY

We have taken notice of criticism of transgenic crops when prioritising our efforts to assure biosafety. We are studying both the hazards and the risks to non-target invertebrate herbivores arising from expression of a cystatin in potato. A stepped procedure has been developed to identify non-pest associates of UK potato crops and then select the sub-group that use cysteine proteinases as digestive enzymes. We have shown that both the aphid *Myzus persicae* and the leafhopper *Eupteryx aurata* depend on cysteine proteinases. They fail to ingest sufficient cystatin from the phloem or mesophyll cells to be at risk from nematode resistant plants expressing cystatins under control of the CaMV35S promoter. We are looking at indirect effects on other trophic levels but much of our current emphasis is on soil organisms associated with potato crops. Work to date has not identified adverse effects on the soil community but the impact on soil invertebrates resulting from use of current nematicides is avoided.

Nematodes that attack roots do not normally parasite elsewhere in plants. Therefore there is no benefit in directing defences against them via constitutive expression. We have shown that promoters more active in rice or potato roots than green tissues (Green et al., 2002) deliver sufficient cystatin to control nematodes. This adds an additional level of biosafety to the approach.

A specialist toxicology company reported to us that the protein engineered rice cystatin is not toxic to mammals and it is not an allergen. The need to apply the principle of substantial equivalence uniformly has been argued as reason to delay introduction of products from transgenic plants. However food safety issues may not arise for a plant cystatin. They are natural components of the human diet and our saliva contains a cystatin. After cooking, potatoes from some of our nematode resistant transgenic plants probably represent a trivial addition to the normal daily consumption of cystatins.

5. CONCLUSIONS

Durable, effective and biosafe transgene resistance to nematodes can clearly be achieved. We aim to inform policy development and accept the importance of public opinion. Unfortunately the latter has been misinformed and then manipulated by certain pressure groups. There are important benefits to the grower, environment and consumer that would accrue from uptake of transgenic resistance to nematodes by agribusiness. The case for uptake by subsistence growers on a royalty-free basis is also very strong. The approach does not require disturbance to traditional agricultural practices (Atkinson et al., 2001).

6. REFERENCES

Atkinson HJ, Green J, Cowgill S, and Levesley, A. (2001) The case for genetically modified crops with a poverty focus. Trends Biotech. 19:91-96.

Atkinson, H.J., McPherson M.J. and Winter M. (2002) Control of crop pests and animal parasites through direct neuronal uptake patent. Application publication number WO2001EP10004 20010828. International publication number WO 02/17948 A2.

Green J., Vain P., Fearnehough M., Worland B., Snape J. and Atkinson H.J. (2002) Root specific expression analysis of the *Arabidopsis thaliana* tubulin-1 promoter and the constitutive rice ubiquitin promoter in rice plants for nematode resistance. Physiol. Molec. Plant Path., in press.

Urwin PE, Troth KM, Zubko EI, and Atkinson, H.J. (2001) Effective transgenic resistance to *Globodera pallida* in potato field trials. Molec. Breed. 8:95-101.

TRANSGENIC CROPS FOR ENHANCED DISEASE RESISTANCE AND FOOD SAFETY

Jon Duvick
Pioneer Hi-Bred, International, Inc., A DuPont Company, Box 550, Johnston, IA 50131, USA
(email: Jon.duvick@pioneer.com)

Keywords Fumonisin, *Fusarium verticillioides*, ear mold, transgenic, maize, mycotoxin

1. INTRODUCTION

Maize grain is subject to infection by a variety of pre-harvest molds, including *Fusarium verticillioides* (formerly *F. moniliforme*), a major causal agent of *Fusarium* ear mold of maize. *F. verticillioides* produces fumonisins, a family of mycotoxins of concern due to their acute toxicity to certain livestock, and their potential carcinogenicity (Dmello et al., 1999). Fumonisins inhibit sphingolipid biosynthesis in a wide range of organisms, but their mode of action as toxins and carcinogens is not fully understood. Incorporating natural ear mold resistance into elite maize germplasm has been difficult, due to complexity of the trait and the strong environmental component to ear mold symptom development, which complicates screening germplasm.

Transgenes that reduce fungal infection or spread in grain could impact not only grain quality but also safety of the grain for food and feed due to lower mycotoxin levels. For example, *Bt* genes for lepidopteran resistance may indirectly reduce fungal colonization and mycotoxin levels, likely by reducing wounding that promotes infection and spread (Munkvold et al., 1999). Genetic sources of resistance are available, but resistance is polygenic and has been difficult to introgress into elite germplasm. Marker-assisted approaches may be useful here, but early evidence suggests that resistance may be due to small contributions from many loci (Perez-Brito et al., 2001). Transgenes that result in direct inhibition of fungal growth in grain tissues have been proposed as another route to improved resistance, but significantly inhibiting fungal growth with one or a few transgenes presents a significant challenge to biotechnologists.

As an alternative approach, we have identified enzymes that degrade fumonisins to less toxic or nontoxic metabolites, and are testing these

I. K. Vasil (ed.), Plant Biotechnology 2002 and Beyond, 111-114.

enzymes in transgenic maize to determine whether the resulting grain will have reduced toxicity under conditions where ear mold occurs.

2. PROOF OF CONCEPT

We identified microbes resident on or in maize tissue that could metabolize fumonisins as a sole carbon source (Duvick and Rood, 1994; Duvick et al., 1998a). Two species of black yeasts (*Exophiala spinifera* and *Rhinocladiella atrovirens*) as well as a novel bacterium related to *Clavibacter* were isolated and shown to metabolize fumonisins to CO_2. Both fumonisin ester-hydrolyzing and deaminating activities were detected in these cultures. The product of ester hydrolysis, an amino polyol or AP1, has reduced toxicity in sphingolipid biosynthesis assays. The product of N-oxidation of intact or hydrolyzed fumonisin is a polyalcohol with a hemiketal ring in place of the amino nitrogen (Blackwell et al., 1999). Loss of the amine group results is expected to result in substantial loss of biological activity, by analogy with chemical derivatives of fumonisin B1 in which the C-2 amine of fumonisins was altered (Hartl and Humpf, 2000).

A cDNA encoding a fumonisin-specific esterase was cloned from the black yeast, *Exophiala spinifera* (Duvick et al., 1998b) and the gene was engineered into transgenic maize to evaluate the feasibility of fumonisin detoxification *in planta*. Field and greenhouse studies indicate that most of the fumonisin produced in *Fusarium*-infested grain is hydrolyzed when the enzyme is targeted to the cell wall or apoplast (unpublished data). These experiments point to the feasibility of *in planta* detoxification of fumonisin, but since the hydrolysis product of fumonisin retains some toxicity, other genes were sought to be deployed in a similar way in maize.

A gene encoding a fumonisin-specific flavin amine oxidase was subsequently cloned from the same fungus (Duvick et al., 2001). The recombinant enzyme, when expressed in *E. coli,* oxidized the C-2 amine of both FB1 and its hydrolysis product, resulting in a hemiketal polyalcohol, with the release of hydrogen peroxide and ammonia. The enzyme displayed an alkaline pH optimum and was inactive on a wide range of other amine-containing compounds. This amine oxidase cDNA therefore represents an excellent candidate for single-gene detoxification of fumonisins *in planta,* preferably when its product is targeted to the cell wall *via* a secretion signal peptide fusion. Efforts are underway to optimize this protein for extracellular expression and activity at acidic pH, both required for maximum effectiveness in maize.

3. PROSPECTS

Ear molds of maize comprise a complex mixture of fungi (*Fusarium* and other species) that vary according to location and season (Reid et al., 1999). It may be difficult to obtain consistent and complete resistance to all fungi that compromise grain quality using methods currently available to breeders and biotechnologists. Therefore alternative, targeted approaches such as the one outlined above are worth evaluating, since they can potentially remove a toxic substance with minimal effect on the host. As an additional benefit, with certain mycotoxins that also affect plant tissues, their elimination from the host could result in reduced infection or spread on host tissues (Karlovsky, 1999). Whether this will be the case with fumonisins is not clear. Earlier data suggested a link between fumonisin production and aggressiveness in seedling maize (Desjardins et al., 1995), although more recent studies indicate fumonisins are not required for ear mold development following silk inoculation (Desjardins et al., 2000) or stalk disease development (Jardine and Leslie, 1999). Transgenic maize that detoxifies fumonisin can be used to address this question in a different way.

The *in planta* detoxification approach, as with any novel engineered trait, will require careful regulatory scrutiny if it is to be part of a commercial seed product. The toxicological properties of the resulting grain, as well as the metabolites themselves, will need to be carefully evaluated, in addition to standard evaluation of the transgenic protein and DNA insertion event. Experiments to determine the toxicological properties (if any) of purified N-oxidized fumonisin metabolites is currently underway. Ultimately, the successful implementation of a fumonisin detoxification strategy for maize could point thc way towards transgenic approaches to reducing other naturally-occurring toxins in food and feed.

4. REFERENCES

Blackwell, B.A., Gilliam, J. T., Savard, M. E., Miller, J. D., and Duvick, J. P. 1999. Oxidative deamination of hydrolyzed fumonisin B1 (AP(1) by cultures of *Exophiala spinifera*. Natural Toxins 7(1):31-8

Desjardins, A.E., et al., Plattner, R. D., Nelsen, T. C., Leslie, J. F. 1995. Genetic analysis of fumonisin production and virulence of *Gibberella fujikuroi* mating population A (*Fusarium moniliforme*) on maize (*Zea mays*) seedlings. App. Environ. Microbiol. 61(1):79-86.

Desjardins, A.E. and Plattner, R.D. 2000. Fumonisin B-1-nonproducing strains of *Fusarium verticillioides* cause maize (*Zea mays*) ear infection and ear rot. J. Agric. Food Chem. 48(11): 5773-5780.

DMello, J.P.F., Placinta, C.M. , and Macdonald, A.M.C. 1999. *Fusarium* mycotoxins: A review of global implications for animal health, welfare and productivity. Animal Feed Sci. Tech. 80(3-4):183-205.

Duvick, J., and Rood, T. 1994. Isolation of fumonisin-metabolizing fungi from maize. International Mycological Congress, Vancouver, BC, August 14-21, p. 56 (abstract).

Duvick, J., Rood, T., Maddox, J., and Gilliam, J. 1998a. Detoxification of mycotoxins in planta as a strategy for improving grain quality and disease resistance. In Developments in Plant Pathology: Molecular Genetics of Host Specific Toxins in Plant Diseases, Kohmoto, K. Yoder, O.C. (eds.). Kluwer, Dordrecht, The Netherlands, Pp. 369-381.

Duvick, J., Maddox J., Rood, T., and Wang, X. 1998b. Fumonisin Detoxification Compositions and Methods. US Patent 5,716,820.

Duvick, J., Maddox, J., Gilliam, J., Crasta, O., and Folkerts, O. 2001. Amino Polyol Amine Oxidase Polynucleotides and Related Polypeptides. US Patent No. 6,211,435.

Jardine, D.J. and Leslie, J.F. 1999. Aggressiveness to mature maize plants of *Fusarium* strains differing in ability to produce fumonisin. Plant Disease 83(7):690-693.

Hartl, M. and Humpf, H.U. 2000.Toxicity assessment of fumonisins using the brine shrimp (*Artemia salina*) bioassay. Food Chem. Toxic. 38(12):1097-1102.

Karlovsky, P. 1999. Biological detoxification of fungal toxins and its use in plant breeding, feed and food production. Natural Toxins 7(1):1-23.

Munkvold, G.P., R.L. Hellmich, and L.G. Rice. 1999. Comparison of fumonisin concentrations in kernels of transgenic Bt maize hybrids and nontransgenic hybrids. Plant Dis. 83(2):130-138.

Perez-Brito, D., Jeffers, D., Gonzalez-de-Leon, D., Khairallah, M., Cortes-Cruz, M., Velazquez-Cardelas, G., Azpiroz-Rivero, and S, Srinivasan, G. 2001. QTL mapping of *Fusarium moniliforme* ear rot resistance in highland maize, Mexico. Agrociencia 35(2):181-196.

Reid, L.M., Nicol, R.W., Ouellet, T., Savard, M., Miller, J.D., Young, J.C., Stewart, D.W., and Schaafsma, A.W. 1999 Interaction of *Fusarium graminearum* and *F. moniliforme* in maize ears: Disease progress, fungal biomass, and mycotoxin accumulation. Phytopath. 89(11):1028-1037.

TRANSGENIC PAPAYA: A CASE STUDY ON THE THEORETICAL AND PRACTICAL APPLICATION OF VIRUS RESISTANCE

Dennis Gonsalves
Agricultural Research Service, U. S. Pacific Basin Agricultural Research Center, 99 Aupuni Street, Suite 204, Hilo, Hawaii, 96720 USA. (Email: dgonsalves@pbarc.ars.usda.gov)

Keywords Hawaiian papaya industry, papaya ringspot virus, Rainbow, SunUp, pathogen-derived resistance

1. INTRODUCTION

Papaya (*Carica papaya*) is widely grown in the tropics, and has been grown in Hawaii for over a century. Like other crops, papaya is affected by viral diseases, with *Papaya ringspot virus* (PRSV) causing the most damage worldwide. PRSV is a potyvirus that is rapidly transmitted by a number of aphid species in a non-persistent manner. The severity of PRSV world-wide is due to its rapid spread by insects and the lack of resistance in *C. papaya*. Actually, the first report of PRSV was in Hawaii in the 1940s. PRSV severely affected the papaya industry in the 1950s, which was then located on Oahu island. To escape the virus, the papaya industry moved to the Puna district of Hawaii island. Lots of available land, adequate rainfall and sun, an adaptable Hawaiian solo variety 'Kapoho', and the absence of PRSV caused the Hawaiian papaya industry to flourish. It became the Hawaii's second most important fruit crop, behind pineapple, and the Hawaiian solo papaya dominated the mainland US market. Things changed in 1992 when PRSV was discovered Puna. The virus spread rapidly, causing severe damage to Hawaii's papaya industry. This brief article provides a snapshot of the development and successful deployment of genetically engineered transgenic papaya to control PRSV in the Hawaiian papaya industry.

2. RATIONALE FOR DEVELOPING TRANSGENIC PAPAYA FOR HAWAII

The papaya industry expanded rapidly in Puna from about 500 acres in 1957 to about 2,400 acres in 1992. However, the potential threat of PRSV to papaya in Puna was real because the virus was established in the backyards of

I. K. Vasil (ed.), Plant Biotechnology 2002 and Beyond, 115-118

residents of Hilo, which is about 19 miles away from Puna. Hawaii recognized this threat and a small task force from the Hawaii Department of Agriculture was deployed to continually monitor for and rogue infected papaya trees in Hilo and survey for PRSV in Puna. These practices, along with geographical isolation, played a major role in preventing the establishment of PRSV in Puna. However, it was clear that the papaya industry in Puna would be devastated if PRSV became established since Puna had large acres of papaya planted contiguously or in close proximity.

Research was begun in 1978 to develop measures to directly control PRSV in papaya. By 1983 the approach of cross protection was developed and implemented to a limited extent on Oahu where the virus was widespread. Cross protection did not become widely used because the mild strain caused discernible symptoms on papaya, especially with the variety Sunrise. Efforts were made to develop PRSV-resistant transgenic papaya using the coat protein gene of the mild strain of PRSV that had been developed earlier for cross protection. The team consisted of molecular biologist Jerry Slightom of Upjohn Company, papaya horticulturist Richard Manshardt of University of Hawaii at Manoa, tissue culturist Maureen Fitch a graduate student of Richard Manshardt, and myself a plant virologist.

3. 1992: THE COINCIDENTAL DEVELOPMENT OF TRANSGENIC PAPAYA AND DISCOVERY OF PRSV IN PUNA

Transformation of somatic embryo cultures using a biolistic approach was started in 1988. By 1991, one line (55-1) of Sunset showed resistance to PRSV HA in greenhouse inoculations. Since we were interested in quickly testing the transgenic papaya line under field conditions, the line was micropropagated and a field trial of R0 plants of line 55-1 was started in April 1991 on the Oahu island where PRSV was very severe. Coincidentally, PRSV was discovered in Puna in May 1992, setting off major concerns in the papaya industry and putting a more practical meaning to our work on transgenic papaya. Fortunately, by December 1992, our data clearly showed that the transgenic papaya line was resistant to PRSV under field conditions. Unfortunately, efforts to suppress the spread of PRSV in Puna showed signs of failing. By the third quarter of 1994, PRSV was widespread in Puna and efforts to suppress the spread of PRSV was abandoned by Hawaii Department of Agriculture. The papaya industry was in a crisis.

A major benefit of the 1992 R0 field trial was that it gave us a head start in developing cultivars that might be useful for growers in Puna. Indeed, two new transgenic cultivars were developed, SunUp and Rainbow. SunUp is a transgenic Sunset that is homozygous for the coat protein gene, and is red-fleshed. Rainbow is a yellow-flesh F1 hybrid developed by crossing SunUp and nontransgenic yellow-flesh 'Kapoho'. The papaya grown in Puna was nearly all Kapoho. Thus, the 1992 trial confirmed the field resistance of the transgenic line and helped to speed up the development of the two transgenic cultivars. The value of these timely developments became very obvious as PRSV rapidly spread through Puna, and began to destroy Hawaii's papaya industry.

4. COMMERCIALIZE THE TRANSGENIC PAPAYA

To validate our 1992 trial results, a large field trial was established in Puna in October 1995. The results conclusively showed that the transgenic SunUp and Rainbow were resistant, and that these cultivars were of commercial quality. In fact, data showed that Rainbow yielded about 125,000 pounds of marketable fruit per acre per year, while the nontransgenic Sunrise (all of which became infected by PRSV) yielded 5,000 pounds of fruit per acre per year. The papaya growers in Puna were especially impressed by Rainbow because it was yellow-fleshed, bore mature fruit a few months earlier than 'Kapoho', and gave much better yields than Kapoho.

In order for genetically engineered plant to be commercialized, it must be deregulated by various governmental agencies and licenses must be obtained from people or companies that hold the intellectual property rights to the components or processes that were used to create the transgenic plants. The regulatory agencies we dealt with were APHIS (Animal Plant Health Inspection Service), EPA (Environmental Protection Agency), and FDA (Food and Drug Administration). The intellectual property rights were held by several companies, including Monsanto.

Efforts to deregulate the papaya proceeded in a timely manner and APHIS deregulated it in November 1996, EPA in August 1997 and FDA completed its consultation in September 1997. The task of obtaining the licenses for the components of the papaya that were covered by patents were turned over to the PAC (Papaya Administrative Committee), which is composed of papaya growers who have organized themselves under a USDA marketing order, and who pay an assessment fee for each pound of papaya that they sell. Fortunately, these efforts also went well and the necessary licenses were obtained by the PAC by April 1998. Seeds of transgenic papaya were distributed to papaya growers in May of 1998, a decade after the first transformation experiments using somatic embryos.

5. IMPACT OF TRANSGENIC PAPAYA

Farmers quickly planted the transgenic papaya seeds, which was almost all of Rainbow since the farmers in Puna favored this transgenic cultivar. Harvesting of Rainbow was started in 1999, and grower, packer, and consumer acceptance were widespread. The papaya industry had been spared from disaster caused by PRSV. Since 1992 when the virus was discovered in Puna, the yearly amount of fresh papaya sold from Puna had gone from 53 millions pounds in 1992 to 26 million pounds in 1998. In 2001, Puna papaya production rebounded to 40 million pounds of fresh market papaya. We have not found evidence for breakdown in the resistance of the transgenic papaya. The Hawaiian papaya story describes a rather fortunate case where technology development came in a timely manner to help save the papaya industry. The members of the original papaya team (Slightom, Manshardt, Fitch, Gonsalves) were awarded the Alexander Von Humbolt award in 2002 for the papaya work.

6. SELECTED REFERENCES

Ferreira, S. A., Pitz, K. Y., Manshardt, R., Zee, F., Fitch, M., and Gonsalves, D. 2002. Coat protein transgenic papaya provides practical control of papaya ringspot virus in Hawaii. Plant Disease 86: 101-105.

Fitch, M. M. M., Manshardt, R. M., ,Gonsalves D., Slightom, J. L., and Sanford, J. C. 1992. Virus resistant papaya derived from tissues bombarded with the coat protein gene of papaya ringspot virus. Bio/Technology 10: 1466-1472.

Gonsalves, D. 1998. Control of papaya ringspot virus in papaya: A case study. Annual Review of Phytopathology 36: 415-437.

INVESTIGATION AND DEVELOPMENT OF *BACILLUS THURINGIENSIS* INSECTICIDAL PROTEINS FOR EXPRESSION IN TRANSGENIC PLANTS

David J. Ellar
Biochemistry Department, Cambridge University, Tennis Court Road, Cambridge, CB2 1GA, UK (email: dje1@mole.biol..cam.ac.uk)

In 1901 Ishiwata reported that the spore-forming bacterium *Bacillus thuringiensis* (Bt), caused death of the silkworm (*Bombyx mori*) (Dulmage and Aizawa, 1982). Subsequently Bt has been found to produce three groups of novel protein toxins that are specific for a large number insect crop pests and some nematodes (de Maagd et al., 2001). The Cry and Cyt toxins (δ-endotoxins) are produced as cytoplasmic inclusions during sporulation and the Vip toxins (Vegetative Insecticidal Proteins) are secreted from the cells during vegetative growth. In vivo the Cry and Cyt toxins bind to insect-specific receptors on the surface of gut epithelial cells and in a second, irreversible step insert into the cell membrane to form leakage channels that result in cell death by colloid osmotic lysis (Knowles and Ellar, 1987). The receptors for some Cry toxins have been shown to be aminopeptidase-N and cadherin-like proteins exposed on the gut epithelial cell surface. Much less is known about the mechanism of the secreted Vip proteins and their receptors, but a recent report (Estruch et al US patent 5,877,012) claims that Vip3a (89kDa) binds to the epithelial cell surface protein Tenascin-X and causes cell lysis by triggering apoptosis. Vip1 and Vip2 are binary toxins.

Because Bt Cry toxins and Vip toxins target insect midgut epithelium, they must first be eaten by susceptible larvae. In bacteria the Cry toxins are expressed at a very high level (20% of the bacterial protein) and deposited as insoluble crystalline protoxins. In transgenic plants these proteins are produced in much smaller quantities. Regardless of the source, the Cry toxins must be solubilised to be active and this is achieved by a combination of the pH and proteases in the insect gut. Cry toxins are inactive protoxins and are converted to the active form by insect gut proteases. Lepidopteran and dipteran midguts have a high pH (>9.5) and contain a range of proteases that are optimally active at this pH. Many important coleopteran pests have much lower gut pH's (pH4.5-6.8). Proteolytic processing of the Cry toxins occurs at the C-terminus, the N-terminus, or both, depending on the toxin. The group of Cry toxins with a molecular mass of ~130kDa (e.g. Cry1-type toxins) are processed to toxic fragments with molecular weights of 60-70kDa). This active moiety is generated by removal of 500-600 amino acids from the C-

I. K. Vasil (ed.), Plant Biotechnology 2002 and Beyond, 119-129.

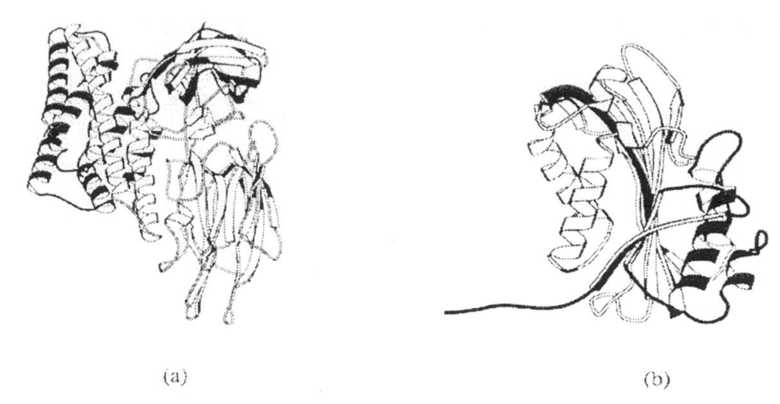

Figure 1. (a) Three-dimensional structure of Cry3Aa (Li et al., 1991) (b) Three-dimensional structure of Cyt2Aa1 (Li et al., 1996)

terminus and approximately 30 amino acids from the N-terminus. Protein sequence alignments show that those Cry toxins with a molecular mass of ~70-kDa (e.g. Cry2-type, Cry3A-type, Cry10Aa1 and Cry11Aa1) can be considered as naturally truncated versions of the 130-kDa δ-endotoxins (Höfte and Whitely, 1989; Hodgman and Ellar, 1990). These protoxins undergo N-terminal processing but little or no C-terminal cleavage.

The protein nature of these toxins coupled with genetic engineering offers great potential for pesticide improvement and has allowed them to be expressed in plants as systemic biopesticides. An increasing number of transgenic crop plants expressing Cry toxins have now been produced including tobacco, tomato, cotton potatoes, maize, rice, eggplant, broccoli, chickpeas and peanuts. The X-ray structure of the first Cry toxin (Figure 1a. Li et al., 1991) revealed putative membrane insertion and receptor binding domains whose functions are being explored by intensive mutagenesis and domain swapping. The first Cyt toxin structure (Figure 1b; Li et al., 1996) showed it to be entirely different from the Cry toxins - despite their similar toxic mechanism. The structure of the coleopteran-active ViP2 resembles that of an ADP ribosylase (Rydel, T personal communication). The 42kDa complex is comprised of a 29kDa protein (Vip1A) and a 13kDa protein (Vip2A). It is suggested that Vip1A binds the membrane and creates an entry pathway for Vip2, which is typical of ADP ribosylating toxins using NAD as substrate.

At the time of writing 109 different Cry toxins and 7 different cyt toxins have been identified (http://www.biols.susx.ac.uk/home/Neil_Crickmore/Bt/holo2.html). Among the Cry toxins are single examples that have been shown to kill major lepidopteran and coleopteran pests. Some Cry toxins are also effective against dipteran insects. There are also some broad spectrum Cry toxins that are effective against two (Cry2Aa) or even three (Cry1Ba) orders of insect.

This toxin portfolio is a rich resource from which to select genes for plant expression and hence biological pest control. To exploit this resource to the full however certain difficulties have to be overcome. Firstly the potency of some of the most attractive toxins for the target pest is low by comparison with chemical pesticides. Consequently effective control in transgenics would place an undesirable expression burden on the plant. Notice that this is not an intrinsic problem deriving from the protein nature of the pesticide, since some of the natural Bt toxins and toxin combinations are as potent as DDT. Four main factors appear to govern the specificity and potency of the Cry toxins in target insect midguts: crystal solubilization, protoxin activation, receptor binding and pore formation. Therefore research needs to be directed to understanding the molecular determinants of each of these steps if we are to devise genetic protocols to drive up the potency of all the toxins. This would be an impossible task if Cry toxins with different insect targets showed major differences in structure. However as more Cry toxin structures are solved, their striking structural similarity despite wide target diversity (Fig 2), tells us that the overall toxic mechanism is the same for all pest targets and that subtle changes in molecular structure are responsible for major differences in specificity and potency. Furthermore the modular (three-domain) structure of these similar Cry toxins immediately suggests the possibility of creating novel toxins by domain swapping.

Interestingly a number of reports have shown that in addition to N and C-terminal processing by insect gut enzymes, both ~130kDa and ~70kDa Cry toxins may be cleaved internally, with the resulting fragments remaining tightly associated through non-covalent interactions. Cry3A for example, was found to be converted into three distinct polypeptides (49, 11 and 6 kDa) by cleavage at the beginning of α-helix 4 in Domain I and at the end of β-sheet 19 in Domain III (Carroll et al., 1997). We have also found that a Cry1B protein is proteolyzed between putative helices 3 and 4 when treated with gut extract from *P. brassicae*, a susceptible insect species (Carroll and Ellar, unpublished). Cry1Ba is reported to exhibit dual lepidopteran and coleopteran toxicity, with the coleopteran activity being enhanced by prior solubilization and trypsin treatment (Bradley et al., 1995). Conceivably inter-helical processing in domain I is important for its coleopteran activity. In addition to Cry3A and Cry1Ba, proteolysis within putative domain I regions with retention of toxic activity has been reported for both Cry4B (Angsuthanasombat et al., 1993) and Cry2Aa (Nicholls et al., 1989). In seeking to engineer increased potency of these toxins for transgenic expression it is important to know whether these internal cleavages are essential for activity or merely the first step towards destruction of the protein by the insect. For Cry 3A the internal cleavages are clearly part of the solubilization process. Thus Cry3A is only soluble under strongly alkaline or acidic conditions *in vitro* (Koller et al., 1992) and yet the pH in susceptible beetle larvae is neutral. This paradox was resolved when Carroll et al., (1997)

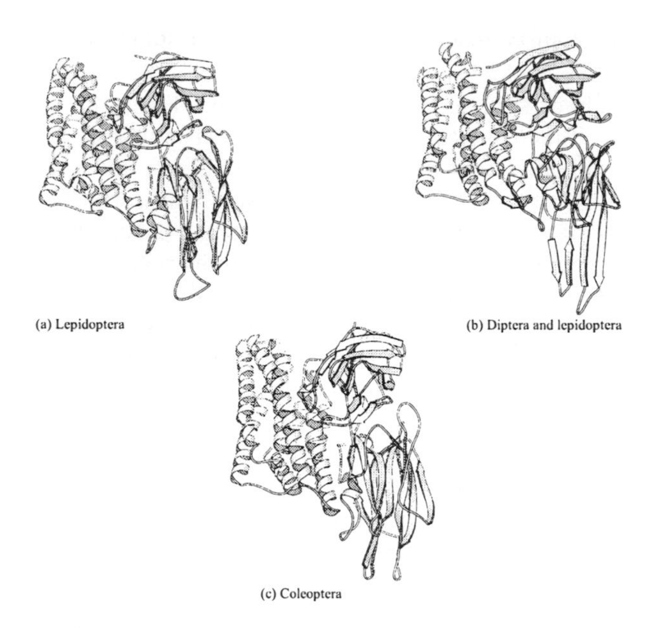

Figure 2. B. thuringiensis Cry toxins toxic to 3 different insect orders; (a) Cry1Aa; (b) Cry2Aa; (c) Cry3Aa

showed that the extra charges on the protein resulting from the two internal cleavages were sufficient to render Cry3A soluble at neutral pH at 3 mg/ml. Most importantly, the tripartite Cry3A product retained full activity against susceptible beetle larvae and exhibited specific binding to *Leptinotarsa decemlineata* midgut membranes. Thus toxin solubilization and activation occur simultaneously.

This inter-helical proteolytic nicking in domain I of Cry toxins may also introduce the flexibility into the Cry toxin structure needed to allow unfolding and penetration of all or part of the structure into target insect membranes (Li et al., 1991). If internal Cry toxin proteolysis is important in membrane insertion then it may be occurring on the membrane surface catalyzed by membrane proteases. We have looked further at the biochemistry of a number of toxins in the presence and absence of the target membrane and also in insects that differ markedly in their susceptibility to a given toxin. Cry1Ac is at least 2000 times more toxic towards *P. brassicae* larvae than towards *M. brassicae* larvae. This difference provides a useful model with which to investigate the contribution of internal proteolysis to toxin potency and insect resistance. Our experiments with this sytem (Lightwood et al., 2000) have shown that Cry1Ac binds specifically to BBMV from both insects, but the forms of the toxin associated with the midguts of the two insects differed. In the resistant *M. brassicae*, Cry1Ac binds as a doublet of 60 and 58-59 kDa with the higher molecular weight form representing approximately 90% of the bound toxin. In the highly sensitive *P. brassicae* the toxin also binds as a doublet but with products of 60 and ~56 kDa being present in similar amounts. We conclude that the two insects have different Cry1Ac-binding and/or post-binding mechanisms. The generation of a ~56-kDa form of Cry1Ac may be a prerequisite for high activity and differential processing of membrane-bound Cry1Ac may account for the large difference in toxin potency that exists between these two insect species. We believe that the sites at which processing occurs are recognized by membrane-associated proteases to generate the nicked forms of membrane-bound Cry1Ac. Conceivably the interaction of Cry1Ac with the *P. brassicae* receptor(s) and/or membrane may induce a specific conformational change, that does not occur on binding to *M. brassicae* BBMV, that consequently exposes amino acids to an increased level of proteolytic attack. In this way potency/sensitivity would be determined *both* by receptor binding and protease specificity. Using a chemiluminescent detection method, (Lightwood et al., 2000) Cry1Ac was found to be present as a doublet in midguts from intoxicated *P. brassicae* larvae and as a single band in the *M. brassicae* membrane. These findings support the view that differential toxin proteolysis may play a role in membrane binding, and thus toxicity, *in vivo*. In other studies with Cry1Ac and *M. sexta* (Jones, Carroll and Ellar, unpublished) we observed a doublet of 55kDa and 53kDa after binding to BBMV. Sequence analysis showed that the 53kDa product resulted from proteolysis at two positions between α-helices 1 and 2a and between α-helices 2a and 2b in Domain 1. By membrane rescue experiments we were able to show that the 53kDa product had inserted into the membrane. The same doublet was also observed after activation by gut enzymes in the absence of BBMV, but significantly the 53kDa band was *insoluble*. Two Cry1Ac mutant toxins that had lost the ability to bind irreversibly to the membrane (N135Q and E129D) did not undergo this proteolysis when incubated with the BBMV. We interpret this tentatively to show that activation of the protoxin by gut enzymes results in exposure of hydrophobic toxin segments. If this event occurs without simultaneous membrane contact, the insertion-competent hydrophobic product precipitates. If however the proteolytic activation occurs in proximity to the membrane, the toxin can insert and further steps leading to pore formation can take place.

Bacillus thuringiensis insecticidal toxins have been variously shown to bind specifically to aminopeptidase-N and cadherin-like proteins exposed on the gut epithelial cell surface. We previously identified aminopeptidase-N (APN) as a putative receptor for the *Bacillus thuringiensis* Cry1Ac1 δ-endotoxin in a *Manduca sexta* brush border membrane vesicle preparation by immunoblot analysis (Knight et al., 1994) and subsequently cloned it from a *M. sexta* midgut epithelium cDNA library (Knight et al., 1995). This is a 120kDa major transmembrane glycoprotein in the brush border epithelial cell membrane attached to the membrane surface via a glycosyl phosphatidyl inositol (GPI) anchor. To provide unequivocal evidence for the functioning of the APN as a receptor *in vivo* we decided to express this lepidopteran protein in the midgut of the Dipteran host, *Drosophila melanogaster*, an insect that is not susceptible to this bacterial toxin. In order to achieve this targeted expression we utilized the GAL4 enhancer trap technique (Brand and Perrimon, 1993; Phelps and Brand, 1998), a system that allows the directed, ectopic expression of any cloned gene in a cell or tissue specific manner.

The GAL4 enhancer trap technique generates two transgenic Drosophila lines. The first contains a genomic insertion of the GAL4 gene under the control of an enhancerless promoter, whose expression is directed in relation to its relative position of integration within the genome. This random insertion enables the generation of a library of GAL4 expression lines. The second line contains a genomic insertion of the UAS-*X*, GAL4 responsive plasmid, were *X* denotes a gene of interest. Crossing the two transgenic lines enables the targeted expression of the desired gene *X*, in a pattern reflecting that of GAL4 expression. In order to utilize this system the *M. sexta* APN cDNA was subcloned into the GAL4 responsive plasmid, pWRUAS-pA, generating the *in vivo* expression construct pUAS *APN*. (Gill and Ellar, unpublished; Gill, 1999) Preceding plasmid construction, standard fly genetics were employed in order to establish a homozygous UAS *APN* transgenic line that could be used to target the APN expression within a desired tissue type e.g. the midgut tissue. We used the GAL4 enhancer line, *48Y* to drive the expression of UAS *APN* in the midgut tissue of the developing larvae.

The larvae resulting from the UAS *APN/48Y* cross were collected immediately after hatching and placed onto media containing varying concentrations of activated Cry1Ac1 toxin. When using 50ng/μl of activated Cry1Ac1 toxin all the Drosophila larvae expressing the APN ceased feeding and eventually died within two/three days. This assay was repeated in triplicate, using over 100 larvae. All assays resulted in 100% mortality of the APN-expressing larvae. Control larvae resulting from the cross incubated in media containing no toxin developed normally through all larval stages and pupation and emerged as viable adults. Therefore the observed larval mortality resulted from the presence of both the APN and the Cry1Ac1 toxin. In triplicate assays control progeny, representing only UAS *APN* and *48Y* larvae fed in media containing the same concentration of activated Cry1Ac1 toxin (50ng/μl) and also in media containing toxin at a concentration of up to 1μg/μl all developed as expected and produced viable adults. These results from the midgut APN expression studies demonstrate *in vivo* functioning of the *M. sexta* APN as the receptor for the Cry1Ac1 toxin. Although the findings are conclusive in terms of mortality and morphological damage in the presence of the APN, the level of APN expression in the Drosophila midgut was substantially less than that observed in *M. sexta* midgut BBMV preparations. This reduced level of expression may in part explain the higher concentration of activated Cry1Ac1 toxin required to cause 100% larval mortality to the transgenic APN expressing lines.

For the lepidopteran specific Cry1Ac toxin one specificity-determining group on the APN receptor has been identified as N-Acetyl Galactosamine (GalNac) (Knowles and Ellar, 1986; Knowles et al., 1991) Surface-plasmon-resonance measurements showed that GalNac competitively displaced Cry1Ac from APN. Analysis of Cry1Ac binding to *M. sexta* APN in a supported lipid monolayer distinguished an initial rapid, reversible (low-affinity) phase from a slower, irreversible (high-affinity) phase (Cooper et al., 1998) Only the low-affinity phase was GalNAc-sensitive. The higher-affinity phase, which followed first-order kinetics, was sensitive to neither GalNAc nor reagents that disrupt protein-protein interactions. This suggests that the GalNAc-mediated binding to APN is rapidly followed by a rate-limiting step in the mechanism, that promotes toxin insertion into the lipid. In order to positively identify the receptor-binding site and investigate the structural repercussions of initial binding, we have determined the crystal structures of the Lepidoptera-specific Cry1Ac both in the free state and in a complex with its specificity-determining ligand GalNAc (Li et al., 2001; D. J. Derbyshire, J. Li and D. J. Ellar, unpublished work).

Unlike Cry1Ac, Cry1Aa does not bind GalNac. A striking difference between the two toxins is in domain III, where, relative to Cry1Aa, Cry1Ac displays a six-residue insertion (Li et al., 2001). The large insertion is accompanied by a twist and rotation of the outer sheet (which faces the solvent) of the β-sandwich. This inserted sequence in Domain III is unique to Cry1Ac in the toxin family and forms a projecting loop that bends back towards the outer sheet, as if forming a lid over a shallow ligand-binding cavity (Li et al. 2001). Mutagenic analysis of this region confirmed that residues here are important to GalNAc-mediated receptor binding by Cry1Ac (Burton et al., 1999). The structure of the Cry1Ac-GalNAc complex (Li et al., 2001) identified seven key residues forming the high-affinity receptor-binding site, which are located both in the insertion loop and on the strands opposite it. GalNAc specificity is determined by hydrogen bonding from side-chain atoms of two arginines and a glutamine residue, as well as main chain atoms of the insertion loop, to the ring oxygen of the sugar: the O-4 atom (distinguishing galactose from glucose) and the acetyl group (distinguishing GalNAc from galactose). The arginines are stabilized in their ligand binding conformation by stacking with aromatic residues located on the sheet. Interestingly, comparison of the Cry1Ac structures with and without bound GalNAc showed an increase of temperature factors in large segments of the helical domain I of the bound form relative to the molecular average. This suggests that when the high-affinity site involving the insertion in domain III is occupied by GalNac an increase of mobility in the pore-forming domain is facilitated. This may indicate how binding could trigger the initiation of the major conformational change required for the toxin to enter the target membrane. Further work is necessary to uncover the mechanism by which the signal is carried from the binding site to the pore-forming elements.

Cry toxin membrane insertion is proposed to begin with penetration of an α-4/α5 hairpin from one molecule (Masson et al., 1999; Gazit et al., 1998) and to proceed to a final oligomeric pore complex of four monomers (Knowles and Ellar, 1987; Vie et al., 2001). We are attempting to identify those regions in Cry toxins that are necessary for the oligomerization step. As part of this study (Cooper et al., 1998; Tigue et al., 2001) we replaced Asn135 in α-helix 4 of the lepidopteran-specific toxins Cry1Ac1 and Cry1Ab5, with glutamine (N135Q). Both mutants were non-toxic to *Manduca sexta.* An interesting feature of the mutants was that they both bound to *M. sexta* brush border membrane vesicles (BBMV) but did not form pores in the membrane as measured by a BBMV permeability assay. This suggested a deficiency in pore formation for these domain I mutants. Oligomerisation studies were performed in order to investigate the effect of the N135Q mutation on the oligomerisation properties of the two Cry1A toxins. In solution, both mutants and the corresponding wild type toxins were all present as monomers of approximately 65kDa. In contrast, others (Aronson et al., 1999; Kumar and Aronson, 1999) have found that Cry1Ac1, but not Cry1Ab5 can be present as oligomers in solution. Oligomers of Cry1Ac1 in solution were not observed in this study and this may be due to differences in toxin preparation and buffer composition. When incubated with BBMV, the mutant toxins bound to the membrane and appeared as monomers suggesting that the lack of toxicity may be caused by a deficiency in oligomerisation. In contrast the majority of the wild-type toxins appeared in the membrane pellet as monomers and higher molecular weight species (dimers and trimers). The mutation of Asn135 to Gln, therefore, appears to affect oligomerisation. According to helical wheel models (Masson et al., 1999; Kumar and Aronson, 1999) Asn135 may face the lumen of the pore suggesting a role in channel function. Evidence suggests, however, that charged residues are important in this respect (Masson et al., 1999). For example, mutation of the charged residue Asp136 in Cry1Ac, which faces the lumen, abolished ion channel activity in *M. sexta* vesicles. The mutation of Asn135 to Gln, however, results in an increase in side-chain length but does not alter the charge. Thus far only residues in α-5 have demonstrated a role in oligomerisation (Aronson et al., 1999; Kumar and Aronson, 1999) but our results suggest that residues in α-helix 4 residue may also be important in this step. As both mutant toxins are also pore formation deficient this suggests a critical role for oligomerisation in the functioning of the Cry toxins.

The number of monomers comprising the pore is still unclear. Knowles and Ellar (1987) predicted a pore size of 1-2nm, which would be able to accommodate four to six monomers. In their model, Masson et al., (1999) have also predicted a pore consisting of at least four molecules. More recently, Vie et al., (2001) used atomic force microscopy to show that the pore formed by Cry1Aa1 in a lipid environment may consist of four subunits.

REFERENCES

Angsuthanasombat, C., Crickmore, N. and Ellar, D.J. (1993) Effects on toxicity of eliminating a cleavage site in a predicted interhelical loop in *Bacillus thuringiensis* CryIVB δ-endotoxin; FEMS Microbiol. Letts. 111:255-262.

Aronson, A. I., C. Geng, and L. Wu. (1999) Aggregation of *Bacillus thuringiensis* Cry1A toxins upon binding to target insect larval midgut vesicles. Appl. Environ. Microbiol. 65:2503-2507.

Bradley, D., Harkey, M. A., Kim, M. K., Biever, K. D. and Bauer, L. S. (1995) The insecticidal CryIB crystal protein of *Bacillus thuringiensis* subsp. *thuringiensis* has dual specificity to coleopteran and lepidopteran larvae. J. Invertebr. Pathol. 65:162-173.

Brand, A. H. & Perrimon, N. (1993) Targeted gene expression as a means of altering cell fate and generating dominant phenotypes. Development 118:401-415.

Burton, S.L., Ellar, D.J., Li, J. and Derbyshire, D.J. (1999) N-acetylgalactosamine on the putative insect receptor aminopeptidase-N is recognized by a site on the domain III lectin-like fold of a *Bacillus thuringiensis* insecticidal toxin. J. Mol. Biol. 287:1011-1022.

Carroll, J, Convents, D., Van Damme, J., Boets, A., Van Rie, J. and Ellar, D.J. (1997) Intramolecular Proteolytic Cleavage of *Bacillus thuringiensis* Cry IIIA δ-Endotoxin may facilitate its coleopteran toxicity. J. Invert. Pathol. 70:41-49.

Cooper, M.A., Carroll, J., Travis, E.R., Williams, D.H. and Ellar, D.J. (1998) *Bacillus thuringiensis* Cry1Ac toxin interaction with *Manduca sexta* aminopeptidase N in a model membrane environment. Biochem. J. 333:677-683.

de Maagd, R.A., Bravo, A. and Crickmore, N. (2001) How *Bacillus thuringiensis* has evolved specific toxins to colonize the insect world. Trends Genet. 17:193-199.

Dulmage, H.T. and Aizawa, K. (1982) Distribution of *Bacillus thuringiensis* in nature; In "Microbial and viral pesticides" (Kurstak E., Ed.), pp. 209-237; Marcel Dekker, Inc., New York.

Gazit, E., P. La Rocca, M. S. Sansom, and Y. Shai. (1998) The structure and organisation within the membrane of helices composing the pore-forming domain of *Bacillus thuringiensis* δ-endotoxin are consistent with an "umbrella-like" structure of the pore. Proc. Natl. Acad. Sci. USA 95:12289-12294.

Gill, M. (1999) PhD Thesis, University of Cambridge

Hodgman, T.C. and Ellar, D.J. (1990). Models for the structure and function of the *Bacillus thuringiensis* δ-endotoxins determined by compilation analysis. J. DNA Sequencing & Mapping 1:97-106.

Höfte, H. and Whiteley, H.R. (1989) Insecticidal crystal protein of *Bacillus thuringiensis*. Microbiol. Rev. 53:242-255.

Knight, P.J.K., Crickmore, N. and Ellar, D.J. (1994) The receptor for *Bacillus thuringiensis* CryIA(c) delta-endotoxin in the brush border membrane of the lepidopteran *Manduca sexta* is aminopeptidase N. Mol. Microbiol. 11:429-436.

Knight, P. J. K., Knowles, B. H. and Ellar, D. J. (1995) Molecular cloning of an insect aminopeptidase-N that serves as a receptor for *Bacillus thuringiensis* CryIA(c) toxin. J. Biol. Chem. 30:17765-17770.

Knowles, B.H. and Ellar, D.J. (1986) Characterisation and partial purification of a plasma membrane receptor for *Bacillus thuringiensis* var. *kurstaki* lepidopteran-specific δ-endotoxins J. Cell Sci. 83:89-101.

Knowles, B.H. and Ellar, D.J. (1987) Colloid-osmotic lysis is a general feature of the mechanism of action of *Bacillus thuringiensis* δ-endotoxins with different specificity; Biochim. Biophys. Acta 924:509-518.

Knowles, B.H., Knight, P.J.K. and Ellar, D.J. (1991) N-acetyl galactosamine is part of the receptor in insect gut epithelia that recognizes an insecticidal protein from *Bacillus thuringiensis*; Proc. R. Soc. Lond. B 245: 31-35.

Koller, C.N., Bauer, L.S. and Hollingworth, R.M. (1992) Characterization of the pH-mediated solubility of *Bacillus thuringiensis* var. *san diego* native δ-endotoxin crystals. Biochem. Biophys. Res. Comm. 184: 692-699.

Kumar, A. S. M., and A. I. Aronson. (1999). Analysis of mutations in the pore-forming region essential for insecticidal activity of a *Bacillus thuringiensis* δ-endotoxin. J. Bacteriol. 181:6103-6107.

Li, J., Carroll, J. and Ellar, D.J. (1991) Crystal structure of insecticidal δ-endotoxin from *Bacillus thuringiensis* at 2.5Å resolution; Nature 353:815-821.

Li, J., Koni, P.A. and Ellar, D.J. (1996) Structure of the mosquitocidal δ-endotoxin Cyt2Aa1 from *Bacillus thuringiensis* sp. *kyushuensis* and implications for membrane pore formation; J. Mol. Biol. 257:129-152.

Li , J., Derbyshire D. J., Promdonkoy, B. & Ellar D. J. (2001) Structural implications for the transformation of the *Bacillus thuringiensis* δ-endotoxins from water-soluble to membrane-inserted forms. Biochem. Soc. Trans. 29:571-577.

Lightwood, D.J., Ellar, D.J. & Jarrett, P. (2000) The Role of Proteolysis in Determining the Potency of the *Bacillus thuringiensis* Cry1Ac δ-endotoxin. Applied Environm. Microbiol. 66:5174-5181.

Masson, L., B. E. Tabashnik, Y.-B. Liu, R. Brousseau, and J.-L. Schwartz. (1999) Helix 4 of the *Bacillus thuringiensis* Cry1Aa toxin lines the lumen of the ion channel. J. Biol. Chem. 274:31996-32000.

Nicholls, C.N., Ahmad, W. and Ellar, D.J. (1989); Evidence for two different types of insecticidal P2 toxins with dual specificity in *Bacillus thuringiensis* subspecies; J. Bacteriol. 171, 5141-5147.

Phelps, C. B. and Brand. A. H. (1998) Ectopic gene expression in Drosophila using the GAL4 system. Methods-A Companion to Methods in Enzymology 14: 367-379.

Tigue, N.J., Jacoby, J. and Ellar, D.J. (2001) The α-helix 4 residue, Asn135, is involved in the oligomerisation of the Cry1Ac1 and Cry1Ab5 *Bacillus thuringiensis* toxins. Applied Environ. Microbiol. 67: 5715-5720.

Vie, V., N. Van Mau, P. Pomarède, C. Dance, J. L. Schwartz, R. Laprade, R. Frutos, C. Rang, L. Masson, F. Heitz and C. Le Grimellec. (2001). Lipid-induced pore formation of the *Bacillus thuringiensis* Cry1Aa1 insecticidal toxin. J. Membrane Biol. 180:195-203.

TRANSFORMATION OF COTTON PRODUCTION THROUGH THE USE OF GENETICALLY IMPROVED COTTON

Frederick J. Perlak, Mark Oppenhuizen, Karen Gustafson, Richard Voth, Saku Sivasupramaniam, David Heering, Boyd Carey, Robert A. Ihrig and James K. Roberts
Monsanto Company, Chesterfield, MO 63198 USA (email: frederick.j.perlak@monsanto.com)

Keywords Pest control, Biotechnology, cotton, transgenic crops

1. INTRODUCTION

Over time, the production of cotton has dramatically improved. Better cotton varieties are now available. Improvements in agricultural practices with the introduction of chemical insecticides, herbicides and mechanization have increased productivity and efficiency. The introductions of cotton varieties, which contain genetically engineered traits, have transformed cotton production for the better. One such trait, Bollgard® cotton, confers resistance to lepidopterous insect pests that attack cotton such as tobacco budworm, *Heliothis virescens*; cotton bollworm, *Helicoverpa zea*; and pink bollworm, *Pectinophora gossypiella* (Perlak ct al., 2001). Introduced in 1996, Bollgard cotton has changed the way farmers approach insect control in their cotton fields. It allows growers to reduce their insecticide use (Carpenter, 2001) while improving their productivity and insect control. It is the only cotton bio-engineered trait for insect control approved in the United States.

Roundup Ready® cotton, genetically engineered to withstand glyphosate herbicide by the introduction of a bacterial gene (Nida et al., 1996a) was approved and commercialized in 1997. This product has simplified weed control in cotton, reducing the number of herbicides used and the number of trips required across fields to control weeds. Roundup Ready® cotton has been widely accepted and is grown on over 70% of the US cotton acres today as Roundup Ready®/Bollgard® cotton with both traits stacked in a single variety (35% of the acres) or as Roundup Ready® cotton alone.

I. K. Vasil (ed.), Plant Biotechnology 2002 and Beyond, 131-134.

2. DESCRIPTIONS OF BOLLGARD® COTTON AND ROUNDUP READY® COTTON

The adoption of transgenic cotton in the United States has been dramatic and has increased every year since the introduction of Bollgard® cotton in 1996 (Figure 1).

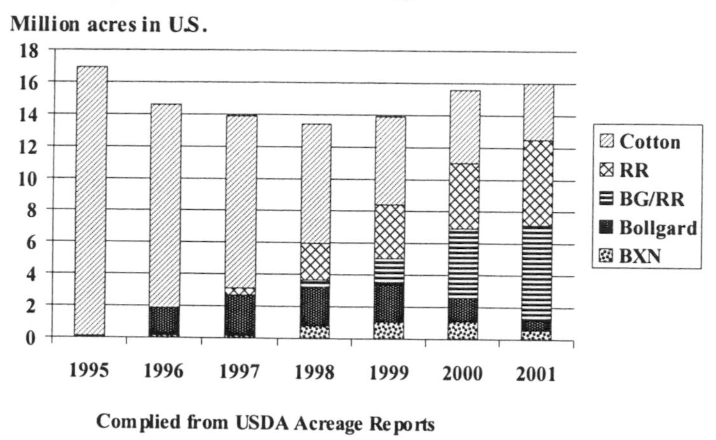

Figure 1. The adoption of transgenic cotton in the United States. The numbers are compiled from USDA acreage reports.

Bollgard® cotton contains a synthetic version of essentially the full-length cryIA(c) gene. The gene was introduced into cotton via Agrobacterium – mediated transformation. The event that is commercial has excellent activity versus tobacco budworm and pink bollworm and good activity against cotton bollworm. Under conditions of very high insect infestations of cotton bollworm, sprays of chemical insecticides over Bollgard® cotton have been shown to provide additional value to the grower.

Roundup Ready® cotton has been commercially available in the United States since 1997. It contains a gene encoding 5-enolpyruvylshikimate -3 – phosphate synthase (EPSPS) isolated from a bacterial source (Agrobacterium sp. CP4; Nida et al., 1996a). Glyphosate, the active ingredient in Roundup herbicide, binds to the plants EPSPS and blocks aromatic amino acid synthesis. The bacterial EPSPS gene (CP4) introduced into cotton is tolerant to glyphosate. The expression of the CP4 EPSPS is targeted to the chloroplasts and confers glyphosate tolerance to cotton. Vegetative tolerance to glyphosate is excellent in Roundup Ready® cotton. However, reproductive tolerance in Roundup Ready cotton can be compromised if glyphosate is applied "over the top" of the cotton in the field after cotton has reached the 5th leaf growth stage.

3. SAFETY OF BOLLGARD® AND ROUNDUP READY® COTTON

The specificity and the safety profile of the Bt protein produced in Bollgard® cotton (Cry1Ac) has been maintained. The Cry1Ac protein is known to be extremely selective against lepidopterous insects (McClintock et al., 1995). Studies have confirmed the spectrum of insecticidal activity of the plant produced protein to target and non-target species (Sims and Martin, 1997). After six years of commercial use, Bollgard® cotton has provided the specificity and activity anticipated when it was first commercialized in the United States in 1996. A detailed discussion of the safety of the Cry1Ac protein has been covered in several references (Betz et al., 2000; Perlak et al., 2001).

Environmental, food and feed safety studies have demonstrated that Roundup Ready® cotton is safe and substantially equivalent to currently available commercial varieties (Nida et al., 1996b)

4. CONCLUSIONS

Bollgard® cotton and Roundup Ready® cotton have had a dramatic effect on cotton production. Both of these new technologies offer significant benefits for the grower, the cotton industry and the environment. Bollgard® cotton can reduce a grower's use of chemical insecticides, improve his insect management program with better control of lepidopterous insects while reducing his insect control costs, and improving yield. Roundup Ready® cotton can simplify weed management for the cotton grower providing improved weed control and reduced herbicide use. It can provide increased flexibility to the cotton grower to adopt "reduced tillage" weed management systems, which can promote soil conservation and reduced production costs. Both technologies have been widely adopted by cotton growers in the United States and both have demonstrated the potential to improve productivity and reduce costs without affecting fiber quality.

5. REFERENCES

Betz, F.S., B.G. Hammond, and R.L. Fuchs. 2001. Safety and advantages of *Bacillus thuringiensis*-protected plants to control insect pests. Regulat. Toxicol. Pharmacol. 32:156-173.

Carpenter, J. 2001. GM crops and patterns of pesticide use. Science 292:637.

McClintock, J.T., C.R. Schaffer, and R.D. Sjoblad. 1995. A comparative review of the mammalian toxicity of *Bacillus thuringiensis*-based pesticides. Pestic. Sci. 45:95-105.

Nida, D.L., K.H. Kolacz, R.E. Buehler, W.R. Deaton, W.R. Schuler, T.A. Armstrong, M.L. Taylor, C.C. Ebert, G.J. Rogan, S.R. Padgette and R..L. Fuchs. 1996a. Glyphosate-tolerant cotton: Genetic characterization and protein expression. Journal of Agricultural and Food Chemistry 44:1960-1966.

Nida, D.L., S. Patzer, R. Stpanovic, P. Harvey, R. Wood, and R.L. Fuchs. 1996b. Glyphosate-tolerant cotton: The composition of the cottonseed is equivalent to that of conventional cottonseed. J. Agric. Food Chem. 44: 1967-1974.

Perlak, F.J., M. Oppenhuizen, K. Gustafson, R. Voth, S. Sivasupramaniam, D. Heering, B. Carey, R.A. Ihrig, and J.K. Roberts. 2001. Development and commercial use of Bollgard® cotton in the USA – early promises versus today's reality. The Plant Journal 27: 489-501.

Sims, S.R. and J.W. Martin 1997.n Effect of *Bacillus thuringiensis* insecticidal proteins CryIA(b), CryIA(c), CryIIA, CryIIIA on *Folsomia candida* and *Xenylla grisea* (Insecta: Collembola). Pedobiologia 41:412-416.

BIOTECHNOLOGICAL SOLUTIONS FOR WEED PROBLEMS – THE NEXT GENERATIONS

Jonathan Gressel
Weizmann Institute of Science, Rehovot, Israel (email: jonathan.gressel@weizmann.ac.il)

Keywords Herbicides, transgenic herbicide resistance, herbicides, introgression, self biocontrol, allelochemicals, weeds, biocontrol

1. INTRODUCTION

There has been a considerable, and much-justified hagiography of the use of transgenics in weed science (Duke, 1996). Indeed, the vast majority of transgenic crops cultivated are herbicide resistant crops. They have been exceedingly useful in the selective control of weeds in crops where the same weeds could not be controlled. They have had extreme eco-environmental success in South America, allowing minimum tillage of fragile, erodable soils. This review will not continue the hagiography, as that will not indicate what could be done to develop better crops.

The present transgenic herbicide resistant crops will eventually be viewed in a similar manner as the Model T Ford, a major breakthrough that changed transportation forever, which was inexpensive, and fulfilled a need that possibly was not easily envisaged by most futurist gurus of the time. The Model T was slow, inefficient, not that reliable, and grossly unsafe, a breakthrough, yet a clunker. The present herbicide resistant crops should be viewed in the same manner. The Model T evolved into the diverse products of high quality we now have, none of which would have been possible if the Model T and following generations, each better than the previous, had not been built and widely sold. Similarly, we cannot have the next generations of transgenic crops without having had the present ones to learn from. We must realize that transgene technology can do more than make herbicide resistant crops. Herbicides must be augmented with other transgenic and conventional technologies to remain sustainable. All these issues have been discussed at great length in Gressel (2002).

I. K. Vasil (ed.), Plant Biotechnology 2002 and Beyond, 135-138.

2. NEEDS NOT BEING MET

2.1. Weed Needs

There are still many unmet needs for weed control, especially in the developing world, where industry, through lack of foresight, does not envisage markets (while continuing to develop products for the over-saturated developed world markets). There are still no good herbicides to control many perennial weeds such as *Cyperus* spp., both in field crops, pasture, and forestry. New herbicides are needed to control grass weeds, due to evolution of herbicide resistance, as well as due to the genetic and metabolic proximities of the grass weeds to graminaceous crops (Gressel, 2002).

The over reliance on *Arabidopsis* genomics for herbicide discovery will preclude finding specific herbicides or herbicide synergists based on the specific properties of such intransigent weeds having very different genetic structures from *Arabidopsis*. The present herbicide resistant constructs are being used to generate resistant crops, despite some having closely related weeds. Indeed, red-rice, a feral form of domestic rice is becoming the worst weed in that crop, due to the global move towards direct seeding of rice because of manpower constraints for hand transplanting.

2.2. Preventing/Mitigating Gene Introgression from Crops to Weeds

In the latter case of rice, and in other major crops such sorghum, barley, oilseed rape (and in some areas wheat and sunflowers) transgenic herbicide resistant crops should not be released unless they contain failsafe mechanisms to preclude introgression into closely-related weeds, or are engineered to mitigate the effects of introgression. Apomixis, plastome transformation, and "terminator" technologies can delay introgression, but as they are not absolute, cannot fully preclude it. Typically, >0.2% pollen transmission of plastome traits is revealed when semi-dominant nuclear markers are used to detect large numbers of hybrids. This amount is sufficient to rapidly allow gene flow of traits with high selective value, such as herbicide resistance. Tandem constructs bearing a herbicide resistance gene flanked by weed-deleterious genes (dwarfing, anti-shattering, anti secondary dormancy genes) can mitigate the effects of introgression by rendering offspring of rare hybrids unfit to compete with cohorts and crops (Gressel, 1999). Such anti-introgression failsafe mechanisms should be stacked with each other where the risks of generating resistant weeds are great, to obtain maximal protection from gene flow.

2.3. Novel Application Technologies

Biotechnology possibly has more to offer the developing world than the marginal increases in utility/economics derived from them in the developed world. For example, treating seeds of herbicide-resistant maize with minuscule amounts of herbicide triples crop yield where parasitic *Striga* spp. typically flourish and decimate maize yields, i.e. much of sub-Sahara Africa (Kanampiu et al., 2002).

2.4. Better Herbicide Resistant Crops

Most resistant crops are not as resistant as those bearing mutations in the same genes, because the crops are functionally heterozygotes bearing the native gene and the transgene. Better varieties may be achieved by first deleting the native gene or by chimera plastic conversion of the native gene. Better promoters and more efficient transgenes will decrease fitness drag.

3. WHY JUST HERBICIDE-RESISTANT CROPS?

Profits from herbicide and seed sales were the dynamo behind the present generation of crops. Other paradigms will need to be found for other possible ways to biotechnologically rid crops of competing weeds. Examples of alternative biotechnologies are:

3.1. Hypervirulent Biocontrol Agents

Arthropods, fungi, and bacteria could be engineered to sufficient virulence to specifically decimate weeds and other pests at an agronomically adequate level (Vurro et al., 2001). This too would require failsafe mechanisms to preclude transgene introgression into crop pathogens (Gressel, 2001).

3.2. Transgenic Allelochemical Emitting Crops

Efforts are being made to discover specific allelochemical genes and to obtain their overexpression to control weeds (Duke et al., 2001). They would have to be without too much autotoxicity and without genetic drag on yield to render them competitive with herbicides.

3.3. Self Biocontrol

Some outcrossing weed species could be controlled by engineering chemically-induced deleterious genes into multicopy transposons, having them quickly disseminate through the population and then turning on the inducible promoter (Gressel and Levy, 2001), as had been proposed and demonstrated for arthropod pests (Grigliatti et al., 2001).

4. GETTING OUT NOVEL TECHNOLOGIES

Much basic research on weed biochemistry, genetics and molecular biology will be needed to apply some of the novel concepts above, and others that will evolve. While public good may derive from some, no short term profit can be foreseen with some technologies, necessitating public R & D.

5. REFERENCES

Duke, S.O. (ed.). 1996. Herbicide Resistant Crops: Agricultural, Environmental, Economic, Regulatory, and Technical Aspects. CRC Press, Boca Raton, Pp. 420.

Duke, S.O., S.R. Baerson, F.E. Dayan, I.A. Kagan, A. Michel, and B.E. Scheffler. 2001. Biocontrol of weeds without biocontrol agents. In: M .Vurro, J. Gressel, T. Butts, G. Harman, A. Pilgeram, R. St.-Leger, and D. Nuss (eds.) Enhancing Biocontrol Agents and Handling Risks, IOS Press, Amsterdam. Pp. 96-105

Gressel, J. 1999. Tandem constructs: preventing the rise of superweeds. Trends in Biotechnology 17: 361-366.

Gressel, J. 2001. Potential failsafe mechanisms against the spread and introgression of transgenic hypervirulent biocontrol fungi. Trends in Biotechnology 19: 149-154.

Gressel, J. 2002. Molecular Biology of Weed Control. Taylor and Francis, London

Gressel, J., and A. Levy. 2000. Giving *Striga hermonthica* the DT's. In: B.I.G. Haussmann, D.E. Hess, M.L. Koyama, L. Grivet, H.F.W. Rattunde, and H.H. Geiger (eds.) Breeding for Striga Resistance in Cereals., Margraf Verlag, Weikersheim. Pp. 207-224

Grigliatti, T.A., T.A. Pfeifer, and G.A. Meister. 2001. TAC-TICS: transposon-based insect control systems. In: M .Vurro, J. Gressel, T. Butts, G. Harman, A. Pilgeram, R. St.-Leger, and D. Nuss (eds.) Enhancing Biocontrol Agents and Handling Risks, IOS Press, Amsterdam. Pp. 201-216.

Kanampiu, F.K., J.K. Ransom, J. Gressel, D. Jewell, D. Friesen, D. Grimanelli, and D. Hoisington. 2002. Appropriateness of biotechnology to African agriculture: *Striga* and maize as paradigms. Plant Cell Tissue and Organ Culture 69: 105-110.

Vurro, M., J. Gressel, T. Butts, G. Harman, A. Pilgeram, R. St.-Leger, and D. Nuss (eds.). 2001 Enhancing Biocontrol Agents and Handling Risks. IOS Press, Amsterdam.

DISCOVERY, DEVELOPMENT, AND COMMERCIALIZATION OF ROUNDUP READY® CROPS

Gregory R. Heck, Claire A. CaJacob and Stephen R. Padgette
Monsanto Company, Chesterfield, MO 63198, USA (email: gregory.r.heck@monsanto.com)

Keywords: Glyphosate, herbicide, EPSPS, Roundup Ready, resistance

1. INTRODUCTION

The herbicide Roundup® whose active ingredient is glyphosate (N-[phosphonomethyl]-glycine) was introduced as a broad spectrum, post emergent herbicide in 1974. Extensive research by a number of groups showed that glyphosate's herbicidal activity resulted from inhibition of the enzyme 5-enolpyruvyl-shikimate-3-phosphate synthase (EPSPS) a key enzyme in the shikimate pathway that is responsible for the biosynthesis of aromatic amino acids in plants and microbes (reviewed, Franz et al., 1997).

Since even small amounts of Roundup® can cause severe injury to plants, it could not be applied "over-the-top" on crops. This changed in 1996 when the first crops tolerant to Roundup® were introduced. However, work to develop this technology began in the early 1980's. Three approaches were investigated in an effort to generate Roundup® tolerance in plants: (1) overexpression of EPSPS, (2) introduction of an EPSPS with decreased affinity for glyphosate, and (3) introduction of a glyphosate degradation gene. Two key discoveries from the latter two approaches enabled the current Roundup Ready® (RR™) technologies. A glyphosate resistant EPSPS was identified and isolated from *Agrobacterium sp.* strain CP4 (CP4 EPSPS) and a glyphosate metabolizing enzyme, glyphosate oxidase (GOX) which converts glyphosate to glyoxylate and aminomethylphosphonic acid was isolated (reviewed, Padgette et al., 1996).

2. CURRENT AND DEVELOPING RR PRODUCTS

An appropriate combination of tolerance enabling transgene(s) and expression elements (e.g. promoters) is needed to deliver commercial-grade glyphosate tolerance in a crop species. The first product to achieve this goal

I. K. Vasil (ed.), Plant Biotechnology 2002 and Beyond, 139-142.

was RR soybean launched in 1996 (Padgette et al., 1996) using the enhanced cauliflower mosaic virus promoter (e35S) and CP4 EPSPS gene. Tolerance greater than 0.84 kg acid equivalents of glyphosate per hectare (label rate) was achieved, which effectively manages weeds while providing a large margin of crop safety, facilitating adoption on approximately 20 million US hectares by 2001 (approximately 68% of total US hectares planted in soy, NASS, 2001).

Similarly, RR canola was launched in 1996 with a double cassette combination that combined figwort mosaic virus (FMV) promoter driven expression of CP4 EPSPS to maintain flux through the pathway and GOX to metabolize glyphosate. RR cotton was commercialized in 1997 with constitutive expression from the FMV promoter driving a codon-optimized synthetic nucleotide version of CP4 EPSPS. Finally, the latest RR corn event, NK603, was launched in 2001, bearing tandem transgenes: the rice actin 1 promoter and e35S promoter, respectively, driving complementary CP4 EPSPS expression patterns for full tolerance.

Since the introduction of the current commercial RR crops, work has continued to improve and expand utility of this technology. The current RR cotton product is limited to an early (4-leaf stage) over-the-top spray since later applications may cause a reduction in the number of set bolls (primarily due to reduced fertilization). Studies have shown that a late, non-registered Roundup® application damages the male reproductive tissues in cotton plants, in particular, tapetum and developing microspores (Chen and Hubmeier, 2001). This leads to male sterility. A second generation RR cotton product is in development that uses different expression elements to enhance male tissue expression and create improved glyphosate tolerance beyond the 4-leaf stage. This product will permit greater flexibility to the applicator and improved weed control options.

Extending the benefits of the RR system to other crops also continues. A RR wheat product is currently under development. This product will provide wheat farmers all the benefits of this management system and enable them to manage currently hard to control weeds. Collaborative efforts will bring forward other products: RR creeping bentgrass to the professional golf course market (with The Scotts Co.) and RR alfalfa to forage (in conjunction with Forage Genetics International).

3. PRODUCT STEWARDSHIP

Herbicide resistant weeds are now known for each major herbicide class (Heap, 2002). Resistance to glyphosate is rare despite over 28 years of use and treatment of millions of hectares. Currently 3 resistant species *Lolium rigidum* (Australia 1996), *Eleusine indica* (Malyasia 1997) and *Conyza canadensis* (USA 2000) have been investigated and shown to be resistant through demonstrated genetic differences between biotypes (putatively resistant *Lolium multiflorum* in Chile has recently been reported). Altered glyphosate uptake, metabolism, translocation, EPSPS gene amplification, or level of EPSPS expression have not been implicated as major contributors to the observed resistances. Rather the mechanisms remain undefined (*Lolium*, Lorrainne-Colwill et al., 1999) or are at least partially attributable to EPSPS active site mutation (*Eleusine*, Tran et al., 2000), or are under investigation (*Conyza*, Heck et al., 2001).

To ensure long term utility of glyphosate, programs have been developed to manage occurrences of glyphosate resistant weeds. For instance, control of resistant *Conyza* can be achieved by a mix of alternative herbicide treatments and agricultural practices. This includes rotation to RR corn plus use of atrazine, cultivation, burn-down using tank mixes of glyphosate plus 2,4-D, or in-crop tank mixes of glyphosate and cloransulan-methyl when treating RR soybean fields. These management strategies have effectively dealt with resistant *Conyza* in Delaware, USA, where resistance was first reported. To help mitigate future development of resistance, use of prescribed glyphosate label rates and timing of application are advised. Where agronomically appropriate, herbicides with other modes of action are recommended.

4. BENEFITS OF RR CROPPING SYSTEMS

The use of glyphosate tolerant crops produces benefits for the farmer and the environment. The economic benefit to the farmer is realized in a number of factors: (1) farm efficiency is improved as yields are optimized and foreign matter in grain is reduced, (2) production costs are reduced as input costs are reduced ($216 million savings in weed control in soybean in 1999) (Carpenter and Gianessi, 2001), (3) an alternate weed control strategy utilizing a key herbicide is enabled, and (4) crop injuries are reduced.

This weed control system has several environmental benefits over traditional weed management systems. It encourages the adoption of conservation tillage which reduces soil erosion, increases soil moisture, reduces fuel use, and reduces CO_2 release thus increasing carbon retention in the soil. It has also resulted in reduced herbicide use. Nineteen million fewer pesticide applications were applied in RR soybeans in 1999 (Carpenter and Gianessi, 2001). Finally this system enables the use of Roundup® instead of older herbicides which are soil persistent, mobile and have higher toxicity. Roundup® is highly unlikely to move to groundwater and is degraded by soil microorganisms improving water quality.

5. REFERENCES

Bourque, J., Y-C.S. Chen, G.R. Heck., C. Hubmeier, T. Reynolds, M. Tran, and D. Sammons. 2001. Investigations into glyohosate resistant horseweed. Abstract 73. Proceedings 2001 North Central Weed Science Society Conference. December 11-13, 2001, Milwaukee, WI.

Carpenter, J.E., and L.P. Gianessi. 2001. Agricultural Biotechnology: Updated benefits estimates. National Center for Food and Agricultural Policy, Washington, DC.

Chen, Y.S., and Hubmeier, C. 2001. Histochemical analysis of male reproductive development in glyphosate-tolerant cotton *(Gossypium hirsutum)*. Proceedings of the 18th Asian-Pacific Weed Science Society Conference, Beiging, China. Pp. 437-441.

Franz, J.E., Mao, M.K., and Sikorski, J.A. 1997. Glyphosate: A Unique Global Herbicide. Am. Chem. Soc., Washington, DC.

Heap, I. 2002. The international survey of herbicide resistant weeds. Online. Internet. May 20, 2002. www.weedscience.com.

Lorraine Colwill, D. F., T. Hawkes R., P. Williams H., S. Warner A.J., P. Sutton B., S. Powles B., and C. Preston. 1999. Resistance to glyphosate in Lolium rigidum. *Pesticide Science.* 55: 489-491.

National Agricultural Statistics Service (NASS) 2001. Agricultural Statistics Board. US Department of Agriculture.

Padgette, S.R., Re, D.B., Barry, G.F., Eichholtz, D.E., Delannay, X., Fuchs, R.L., Kishore, G.M., and Fraley, R.T. 1996. New Weed Control Opportunities: Development of Soybeans with a Roundup Ready Gene. Herbicide-Resistant Crops: Agricultural, Environmental, Economic, Regulatory, and Technical Aspects. S.O. Duke (ed.). Boca Raton, Florida. pp. 53-84

Tran, M., S. Baerson, R. Brinker, L. Casagrande, M. Faletti, Y. Feng, M. Nemeth, T. Reynolds, D. Rodriguez, D. Schafer, D. Stalker, N. Taylor, Y. Teng and G. Dill. 1999. Characterization of glyphosate resistant Eleusine indica biotypes from Malaysia. Proceedings 1(B) Asian-Pacific Weed Society Conference, pp 527-536.

ENGINEERING VIRUS-INDUCED AFRICAN CASSAVA MOSAIC VIRUS RESISTANCE BY MIMICKING A HYPERSENSITIVE REACTION IN TRANSGENIC CASSAVA

Peng Zhang[1,*], Johannes Fütterer[1], Petra Frey[1], Ingo Potrykus[1], Johanna Puonti-Kaerlas[2] and Wilhelm Gruissem[1]
[1] Institute of Plant Sciences, ETH-Zentrum / LFW E 17, CH-8092 Zürich, Switzerland
[2] European Patent Office, D-80298 Munich, Germany
* Author for correspondence, e-mail: zhang.peng@ipw.biol.ethz.ch

Keywords *Manihot esculenta*, genetic transformation, ACMV resistance, barnase and barstar, virus induced cell death

1. INTRODUCTION

African cassava mosaic disease (ACMD) is caused by African cassava mosaic virus (ACMV), a whitefly-transmitted *geminivirus* with bipartite genome denoted DNA-A and -B. It has been rated one of the most important diseases of cassava in Africa. Due to high heterozygosity and the low fertility of many local varieties, it is difficult to produce ACMV resistant varieties by traditional breeding such as the inter- and intra-specific crosses. Biotechnology provides an alternative approach to complementing the efforts in traditional breeding. Current progress in cassava genetic transformation makes it possible to produce ACMV-resistant cassava from different virus resistance approaches.

Here we report the development of a new ACMV resistance strategy by mimicking a hypersensitive reaction using barnase and barstar genes in *planta*. It is known that a basal low level of transcription takes place from both sides of ACMV promoters (Hong et al., 1996). In our earlier regulation study of the bidirectional promoter by a dual luciferase assay, we further found that in ACMV DNA-A the virion-sense (AV) promoter was up-regulated while the complementary-sense (AC) promoter was down-regulated after a transient viral infection (Frey et al., 2001). We took advantage of the unique regulation properties of the bidirectional ACMV promoter during virus infection and the barnase/barstar based block/anti-block system to mimic a hypersensitive reaction in plant cells. Therefore, the new strategy was designed for ACMV resistance by introducing a virus-induced cell death

I. K. Vasil (ed.), Plant Biotechnology 2002 and Beyond, 143-145.

system based on barnase, the ribonuclease produced by *Bacillus amyloliquefaciens,* and its specific inhibitor, barstar, into cassava cells.

The barnase gene was cloned under the control of AV promoter (*AVp*), which is *trans*-activated by the TrAP protein of ACMV. Synchronously, the barstar gene was driven by the AC promoter (*ACp*) to counteract the basal expression of barnase. Upon viral infection the ratio of barnase/barstar would be expected to shift in favor of the barnase due to the up-regulation of *AVp* by viral protein TrAP, resulting in local cell death before the virus can spread to adjacent cells. In order to adjust the expression level of barnase, constructs with different additional short open reading frames in front of the barnase gene were designed and used for transforming cassava via particle bombardment-mediated transformation.

Among 24 transgenic plant lines which have a normal phenotype, only three of them were confirmed the presence of ACMV DNA-A promoter, barnase gene and barstar gene (Fig. 1A). Southern analysis of these three plant lines showed the integration of the barnase and bastar genes, but transgene rearrangements were also observed (Fig. 1B). The basic expression level of *barnase* and *barstar* was detected at RNA level by RT-PCR in 2 plant lines (Fig. 2). An *in vitro* viral replication assay using leaves of transgenic plants compared with wildtype plants could show the reduction of viral replication in transgenic leaves by 86% to 99%. Virus infection test will be conducted to verify the ACMV resistance in these plants.

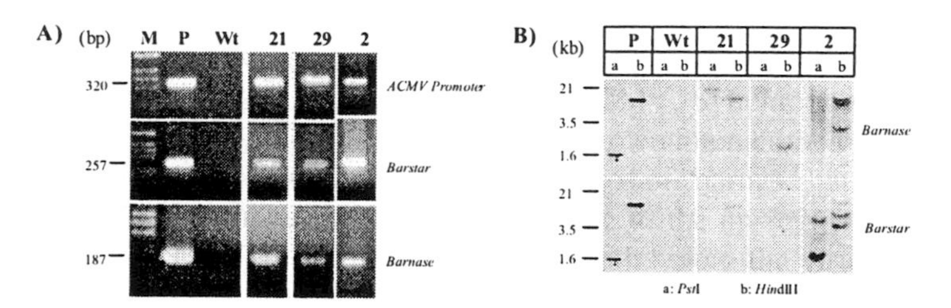

Figure 1. PCR (A) and Southern (B) analyses of transgenic cassava plant lines transformed with barnase-ACMV promoter-barstar construct. M, Molecular marker; P, Plasmid control; Wt, wildtype control; 21, 29 and 2, independent transgenic lines.

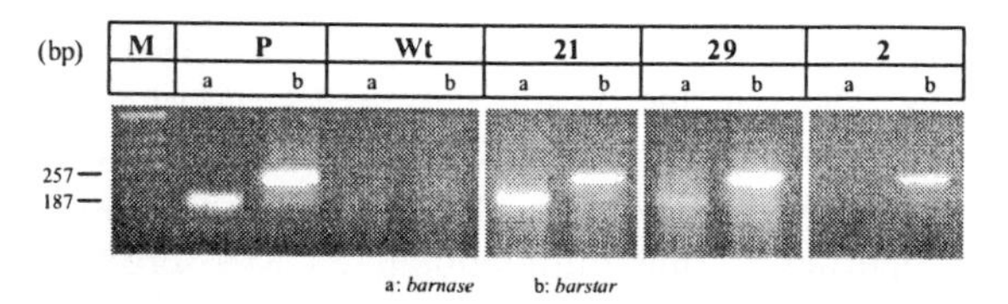

Figure 2. RT-PCR analysis of transgenic cassava plant lines transformed with barnase-ACMV promoter-barstar construct. M, Molecular marker; P, Plasmid control; Wt, wildtype control; 21, 29 and 2, independent transgenic lines.

2. REFERENCES

Hong Y, Saunders K, Hartley M and Stanley J (1996) Resistance to geminivirus infection by virus-induced expression of dianthin in transgenic plants. *Virology* 220:119-127.

Frey PM, Scharer-Hernandez N, Futterer J, Potrykus I, Puonti-Kaerlas J (2001) Simultaneous analysis of the bidirectional African cassava mosaic virus promoter activity using two different luciferase genes. *Virus Genes* 22: 231-242.

DISEASE RESISTANT TRANSGENIC COTTON TO PREVENT PREHARVEST AFLATOXIN CONTAMINATION

K. Rajasekaran, T.J. Jacks, J.W. Cary, and T.E. Cleveland
USDA-Agricultural Research Service, Southern Regional Research Center, New Orleans, LA 70124 (email: *krajah@srrc.ars.usda.gov*)

Keywords biotechnology, chloroperoxidase, food safety, *Gossypium*, haloperoxidase, mycotoxin, synthetic peptide

1. INTRODUCTION

Mycotoxin contamination is a major problem in food and feed crops such as corn, cotton, peanut and tree nuts. We are developing transgenic cottons that resist the aflatoxigenic fungus, *Aspergillus flavus*. Crop losses due to pathogens result in economic plus food and feed safety concerns. Current biotechnology-related approaches to improve disease resistance in plants have been reviewed recently (Ouchi, 2001; Punja, 2001; Rajasekaran et al., 2002). We provide below two examples from our laboratory, which utilize novel gene constructs that impart disease resistance to transgenic tobacco and cotton.

2. HALOPEROXIDASE

When plants and animals are attacked by microorganisms, an early response is the generation of microbicidal hydrogen peroxide (H_2O_2). This occurs in a two-step reaction catalyzed by NAD(P)H oxidase and superoxide dismutase, as summarized in the following equation:

$$NADPH + H_2O + O_2 \rightarrow NADP^+ + {}^-OH + H_2O_2$$

Animals carry this defensive response a step further by converting H_2O_2 to a deadlier microbicide, a hypochlorite (${}^-OX$) such as hypochlorite ((${}^-OCl$), via haloperoxidases:

$$H_2O_2 + {}^-Cl \rightarrow H_2O + {}^-OCl$$

Plants lack haloperoxidases that catalyze this conversion and therefore miss an effective antimicrobial system that exists in animals.

To rectify this, we proposed plant transformation with genes for nonplant haloperoxidases. The effectiveness of nonplant haloperoxidases (combined with H_2O_2) in killing the plant pathogen *Aspergillus flavus in vitro* (Jacks et

I. K. Vasil (ed.), Plant Biotechnology 2002 and Beyond, 147-150.

al., 1991, 1999) supported the prospect that the defense capacity of plants could be greatly improved by haloperoxidases. These enzymes, however, contain heme that would require expression of a plethora of genes. Not only does incorporation of a heme into an enzyme molecule entail post-translational processing, but also the particular heme derivative required by the enzyme is not available in the plant targeted for transformation. Consequently, a transformation scheme would require the addition of a profusion of genes coding for the multi-component synthesis of each heme derivative. To circumvent these problems, we decided to transform plants with a gene for nonheme chloroperoxidase from *Pseudomonas pyrrocinia*. The enzyme CPO-P, which contains neither heme nor metal cofactor, catalyzes the oxidation of alkyl acids with H_2O_2 to form peracids (van Pée, 1996):

$$AcOH + H_2O_2 \rightarrow H_2O + AcOOH$$

where Ac is an acyl group. AcOOH spontaneously oxidizes halides (^-X) to hypohalites (^-OX):

$$AcOOH + {}^-X \rightarrow AcOH + {}^-OX$$

Hypohalites and alkyl peracids exhibit potent lethality against *A. flavus* (Jacks et al., 1999, 2000).

The utility of the CPO-P gene was first demonstrated using a transgenic tobacco model system (Rajasekaran et al., 2000a). For instance, leaf extracts of transgenic plants not only exhibited increased resistance in vitro to *A. flavus*, the amount of resistance was proportional to the amount of CPO-P enzyme activity expressed in the tissue (Jacks et al., 2000). *In planta* resistance to *Colletotrichum destructivum*, causal agent of anthracnose, also increased in transgenic tobacco (Rajasekaran et al., 2000a). Similar results were obtained with transgenic cotton plants, which showed significant resistance in vitro to *A. flavus, Verticillium dahliae* and a vascular pathogen *in planta* (e.g., *Thielaviopsis basicola*) (unpublished observations). The enzymically catalyzed reaction and mechanism responsible for imparting resistance to the transgenic tissues, however, are yet to be determined (Jacks et al., 2000, 2002).

The utility of nonplant haloperoxidase in protecting transgenic plants and their progenies against microbial infection is readily apparent from the results of our studies. Other plant species are also currently being transformed with the CPO-P gene and being tested for resistance to pathogens.

3. SYNTHETIC ANTIMICROBIAL PEPTIDES

Naturally occurring antimicrobial peptides of plant or non-plant origins have been successfully evaluated in several transgenic plants (Punja, 2001; Rajasekaran et al. 2002). These natural hydrophobic or amphipathic peptides, however, are readily degraded by fungal and plant proteinases. Synthetic linear peptides with broader antimicrobial activities capable of resisting such degradation by plant proteinases (De Lucca et al., 1998; DeGray et al., 2001) are gaining importance, thanks to automated, combinatorial chemical synthesis. We demonstrated for the first time in our laboratory the effective use of one such synthetic peptide (D4E1; with a sequence of FKLRAKIKVRLRAKIKL) in transgenic tobacco first (Cary et al., 2000) and in cotton (Rajasekaran et al., 1999, 2000b) for broad-spectrum control of fungal and bacterial phytopathogens. *In vitro* results using crude leaf extracts from transformed cotton plants (R_0 and R_1) indicated significant control of *Verticillium dahliae*, a pathogen very sensitive to the antifungal peptide (Rajasekaran et al., 2001). In anti-*A. flavus* assays *in vitro*, we observed reduced number of colonies although the results were not highly significant compared to controls. This might due to either reduced level of gene expression or rapid degradation of the peptide in ground extracts by plant proteinases. For example, we observed rapid degradation of D4E1 added to cotton leaf extracts within 30 minutes of incubation. An improved *in situ* assay using immature cottonseeds inoculated with a Green Fluorescent Protein-expressing *A. flavus* strain showed that the transgenic plants are capable of delaying and reducing the fungal advance in both seed coat and cotyledons. In addition, results from preliminary *in planta* assays with R_1 progeny for seedling diseases caused by the vascular pathogen, *Thielaviopsis basicola*, showed improved resistance (unpublished observations).

In summary, we have demonstrated the novel utility of a bacterial non-heme chloroperoxidase gene and a linear synthetic peptide for control of disease causing microbes, including saprophytic mycotoxigenic fungi. Continued effectiveness of these gene constructs in transgenic plants under field condition is yet to be evaluated.

4. REFERENCES

Cary J.W., K. Rajasekaran, J.M. Jaynes, and T.E. Cleveland. 2000. Transgenic expression of a gene encoding a synthetic antimicrobial peptide results in inhibition of fungal growth *in vitro* and *in planta*. Plant Sci. 154: 171-181.

DeGray G, K. Rajasekaran, F. Smith, J. Sanford, and H. Daniell. 2001. Expression of an antimicrobial peptide via the chloroplast genome to control phytopathogenic bacteria and fungi. Plant Physiol. 127: 852-862.

De Lucca A.J., J.M. Bland, C. Grimm, T.J. Jacks, J.W. Cary, J.M. Jaynes, T.E. Cleveland, and T.J. Walsh. 1998. Fungicidal properties, sterol binding, and proteolytic resistance of the synthetic peptide D4E1. Canadian J. Microbiol. 44: 514-520.

Jacks T.J., P.J. Cotty, and O. Hinojosa. 1991. Potential of animal myeloperoxidase to protect plants from pathogens. Biochem. Biophys. Res. Comm. 178: 1202-1204.

Jacks T.J., A.J. De Lucca, N.M. Morris. 1999. Effects of chloroperoxidase and hydrogen peroxide on the viabilities of *Aspergillus flavus* conidiospores. Mol. Cell. Biochem 195 169-172.

Jacks T.J., De A.J. Lucca, K. Rajasekaran, K. Stromberg, and van K.-H Pée. 2000.. Antifungal and peroxidative activities of nonheme chloroperoxidase in relation to transgenic plant protection. J. Agric. Food. Chem. 48: 4561-4564.

Jacks T.J., K. Rajasekaran, K.D. Stromberg, A.J. De Lucca, and K.-H. van Pèe. 2002. Evaluation of peracid formation as the basis for resistance to infection in plants transformed with haloperoxidase. J. Agric. Food. Chem. 50: 706-709.

Ouchi S. 2001. Biotechnology as an approach to improving disease resistance in plants, p. 251-264. In: N. T. Keen, S. Mayama, J. E. Leach, and S. Tsuyuma (eds.), Delivery and Perception of Pathogen Signals in Plants. APS Press, St. Paul, Minnesota.

Punja Z.K. 2001. Genetic engineering of plants to enhance resistance to fungal pathogens - a review of progress and future prospects. Canad. J. Plant Pathol. 23: 216-235.

Rajasekaran K., J.W. Cary, T.J. Jacks, K.D. Stromberg, and T.E. Cleveland. 1999. Inhibition of fungal growth by putative transgenic cotton plants. In J.F. Robens (ed.), Proc. Aflatoxin Elimination Workshop, USDA, ARS, Beltsville, MD, p 64.

Rajasekaran K., J.W. Cary, T.J. Jacks, K.D. Stromberg, and T.E. Cleveland. 2000a. Inhibition of fungal growth in planta and in vitro by transgenic tobacco expressing a bacterial nonheme chloroperoxidase gene. Plant Cell Rep. 19: 333-338.

Rajasekaran K., J.W. Cary, T.J. Jacks, C.A. Chlan, and T.E. Cleveland. 2000b. Transgenic cottons to combat preharvest aflatoxin contamination: An update. In J.F. Robens (ed.), Proc. Aflatoxin Elimination Workshop, USDA, ARS, Beltsville, MD, p 104.

Rajasekaran K., K.D. Stromberg, J.W. Cary, T.E. Cleveland. 2001. Broad-spectrum antimicrobial activity in vitro of the synthetic peptide D4E1. J. Agric. Food. Chem. 49: 2799-2803.

Rajasekaran K., J.W. Cary, T.J. Jacks, and T.E. Cleveland. 2002. Genetic Engineering for Resistance to Phytopathogens. In K Rajasekaran; T.J. Jacks; J.W. Finley (eds.), Crop Biotechnology, American Chemical Society, Oxford Press, Washington, DC.

van Pèe K.-H. 1996. Biosynthesis of halogenated metabolites by bacteria. Ann. Rev. Microbiol. 50: 375-399.

EVALUATION OF TRANSGENIC HERBICIDE (GLUFOSINATE AMMONIUM) RESISTANT SUGARCANE (*SACCHARUM* SPP. HYBRIDS) UNDER FIELD CONDITIONS

Sandra J. Snyman[1] and Noel B. Leibbrandt[1,2]
[1]South African Sugar Association Experiment Station, Private Bag X02, Mount Edgecombe, 4300, KwaZulu Natal, South Africa (email: snyman@sugar.org.za)
[2]current address: Coastal Farmers Co-operative, PO Box 1003, Umhlanga Rocks, 4320, KwaZulu Natal, South Africa (email: noel@coastals.co.za)

Keywords agronomic performance, transgene stability, sucrose yield

1. RATIONALE AND APPROACH

There is little published information available on the agronomic performance of herbicide resistant sugarcane. In addition, transgene stability in vegetatively propagated monocotyledonous plants has not been well documented. Acceptance of transgenic crops for commercialisation will only be possible if it can be demonstrated that the introduced foreign gene is expressed in a plant that retains its agronomic characteristics.

In this study, a well characterised but non-commercial, transgenic cultivar (NCo310), line 22.2, transformed with the *pat* gene, conferring resistance to the herbicide Buster® (active ingredient glufosinate ammonium; 200 g/l; Aventis) was used as a model plant for field assessment. The aims were: 1) to establish transgene stability over several ratoons and to compare herbicide sensitivity at different application rates in a preliminary trial, and 2) to compare morphological and agronomic characteristics of transformed and untransformed plants in a large-scale field trial.

2. SUMMARY OF FINDINGS

Results from the preliminary field trial indicated that Buster applied at a rate of 5l/ha was lethal to untransformed plants. Transformed plants showed no phytotoxic symptoms at rates of up to 7 l/ha, and the *pat* gene was stably expressed over two ratoon crops.

A large-scale field trial was carried out to determine the agronomic performance of transgenic sugarcane under four weed control programmes.

I. K. Vasil (ed.), Plant Biotechnology 2002 and Beyond, 151-152.

Sugarcane stalk morphology (height, diameter, population density and fibre content) and the incidence of common sugarcane diseases such as smut (causal agent *Ustilago scitaminea*) and sugarcane mosaic virus, as well as susceptibility to the sugarcane borer, *Eldana saccharina*, were compared in transformed and untransformed sugarcane at harvest of the first ratoon. No significant differences were observed.

A comparison of yield was also made in the four weed control treatments, which were T1: repeated Buster (5 l/ha) application; T2: conventional pre-emergence herbicide cocktail, followed by Buster application; T3: conventional pre- and post-emergence herbicides; T4: hand weeding only. In the Buster treatments, T1 and T2, transformed cane performed better than untransformed cane. In T1, untransformed cane died. Treatments T2 and T3 are successful weed control regimes, with T2 transformed and T3 untransformed cane containing significantly higher fresh mass and sucrose content at harvest, when compared to other weed treatments (Table 1). This suggests that *in vitro* culture- and transformation-derived plants are not compromised in terms of yield potential.

When a cost comparison was carried out, the most economical treatments were T1 transgenic cane treated with Buster, and T2 transgenic cane treated with pre-emergent herbicides plus Buster. However, the pricing of the herbicide to which resistance has been engineered is crucial if herbicide resistant sugarcane is to be commercially viable.

Table 1. A comparison of sugarcane biomass (cane Mg /ha), sucrose yields (estimated recoverable crystal, ERC Mg/ha) and cane quality (ERC % cane) for the four weed control treatments in transformed and untransformed cane in the ratoon crop. Means are of eight replicates and are followed by different alphabetical letters if differences are significant (ANOVA).

Weed control treatment for first ratoon	Category	Cane mass (Mg/ha)	Sucrose yield (ERC Mg/ha)	Cane quality (ERC % cane)
T1 (Buster)	Untransformed	0	0	0
	Transformed	59.8 ab	8.2 a	13.8
T2 (cocktail plus Buster)	Untransformed	56.3 a	7.7 a	13.6
	Transformed	68 b	9.5 bc	13.9
T3 (conventional)	Untransformed	67.5 b	9.7 c	14.3
	Transformed	60.1 ab	8.3 ab	13.9
T4 (hand-hoeing)	Untransformed	60.5 ab	8.2 a	13.6
	Transformed	53.9 a	7.4 a	13.7
ANOVA comparison of category means				
SED		1.8	0.31	0.21
LSD (0.05)		3.8	0.65	0.44

TRANSGENIC APPLE PLANTS EXPRESSING VIRAL EPS-DEPOLYMERASE: EVALUATION OF RESISTANCE TO THE PHYTOPATHOGENIC BACTERIUM *ERWINIA AMYLOVORA*

Viola Hanke[1], Klaus Geider[2] and Klaus Richter[3]
[1]BAZ, Institut für Obstzüchtung, 01326 Dresden, Germany (e-mail: v.hanke@bafz.de)
[2]MPI für Zellbiologie Rosenhof, 68526 Ladenburg, Germany
[3]BAZ, Institut für Epidemiologie und Resistenz, 06449 Aschersleben, Germany

Keywords *Agrobacterium tumefaciens*, apple, amylovoran lyase, *Erwinia amylovora*, transformation

1. INTRODUCTION

The cultivated apple is the most important fruit crop in Europe. In various countries, large breeding programmes were established for apple to select new stable cropping, disease resistant varieties with an excellent fruit quality. Apple has been also one of the prime targets for genetic manipulation in fruit tree species since in 1989 James et al. reported on transgenic apple plants of the non-commercial cultivar 'Greensleeves'. Subsequently, transgenic plants were produced using *Agrobacterium* infection of leaf explants obtained from proliferating shoot cultures (Maheswaran et al., 1992; Sriskandarajah et al., 1994, Yao et al., 1995; De Bondt et al., 1996; Puite and Schaart, 1996). Most studies on apple transformation have focused on the transmission of agronomically important genes to utilize improved genotypes of cultivars commercially established in fruit production. The improvement is mainly focused on apple storage ability (Yao et al., 2000) and on resistance to main diseases (Ko et al., 1999) and insects (James et al., 1993).

This paper reports on the transformation of apple aimed on an improvement of resistance to fire blight. Fire blight is a serious disease, caused by the bacterium *Erwinia amylovora*, affecting apple, pear and other members of the *Rosaceous* family. The utilization of genes encoding lytic proteins for fire blight resistance was reported previously (Ko et al., 1999; Norelli et al., 1999). Beside others, we used a gene encoding an EPS-depolymerase (amylovoran lyase) in transformation studies which was derived from an *Erwinia amylovora* bacteriophage (Kim and Geider, 2000; Hanke et al., 2002)

I. K. Vasil (ed.), Plant Biotechnology 2002 and Beyond, 153-157.

2. MATERIAL AND METHODS

2.1. Bacterial Strains, Plasmids, Plant Transformation And Regeneration

For transformation experiments proliferating in vitro cultures of 8 different apple cultivars (*Malus domestica* Borkh.: 'Elstar', 'Jonagold', 'Pilot', 'Pinova', 'Pirol', Reka', 'Remo', 'Retina') and one clonal rootstock AU 56-83 were used. The *Agrobacterium tumefaciens* strains LBA44404, EHA 105 and KYRT1 with plasmid pBinAR containing the ΦEa1-depolymerase gene (Kim and Geider, 2000), separated from the phage promoter and expressed via the CaMV35S promoter, were chosen (Hanke et al., 2002). A previously described protocol for leaf disk transformation and plant regeneration was applied (Norelli et al., 1996; Hanke et al., 1999 and 2000). Selection was performed on 100mg/l kanamycin based on the *nptII*- marker gene which was used in transformation experiments.

2.2. Molecular And Greenhouse Evaluation Of The Plants

Assessment of NPT II protein expression, of the integration and expression of the *dpo* gene were performed as described by Hanke et al. (2002). All regenerated shoots producing higher levels of NPTII protein compared to the non-transformed controls and showing specific bands for the marker and the foreign gene after PCR were designated as transgenic plants and clonally propagated in vitro. The transgenic plants were rooted and transferred to greenhouse conditions according to Bolar et al. (1999) for further evaluation of resistance.

Besides, the transgenic lines were also evaluated for fire blight resistance using an in vitro assay as described by Hanke et al. (2002).

3. RESULTS AND DISCUSSION

3.1. Efficiency Of Transformation

Production of genetically transformed plants depends both on the ability to integrate foreign genes into target cells and the efficiency with which plants are regenerated from genetically transformed cells.

The regeneration frequency (number of putative transgenic shoots regenerated per 100 explants) in experiments using different apple genotypes and various *Agrobacterium tumefaciens* strains was between 0 and 12.1 %

depending on the genotype (Table 1). However, with delayed selection on kanamycin the number of putative transformed shoots decreased due to a non-stable gene integration. On average 85.3% of the regenerated shoots were escapes not surviving further selection on medium containing antibiotics.

Table 1. Regeneration frequency (%) in apple using an EPS-depolymerase gene for transformation experiments.

Genotype	Number of leaf segments	Number of regenerated shoots	Frequency of putative transformed shoots (%)
AU 56-83	2848	187	6.57
Elstar	712	49	6.88
Jonagold	1824	135	7.40
Pilot	1360	24	1.76
Pinova	5416	375	6.92
Pirol	1104	22	1.99
Reka	1488	42	2.82
Remo	1224	148	12.09
Retina	1048	0	0.00

The efficiency of regeneration reported here is comparable with the regeneration efficiency reported for other cultivars (James et al., 1989; Maheswaran et al., 1992; De Bondt et al., 1996; Puite and Schaart, 1996).

3.2. Evaluation Of Resistance To The Fire Blight Bacterium

Resistance or susceptibility to infection by *Erwinia amylovora* was estimated first by inoculation of leaves from transgenic in vitro shoot cultures with *gfp*-labeled bacteria. The zones of fluorescence were scored by rating the strength and extent of fluorescence, and by considering bacterial movement from the inoculation site into leaf veins, their outbreak into parenchyma and also the development of necrotic tissue. Out of 83 transgenic lines carrying the EPS-depolymerase gene, 61 of the lines were significantly less susceptible than the parental cultivar 'Pinova'.

In 2000 and 2001, 53 individual grafted shoots from transgenic lines of the cv. 'Pinova' and the appropriate control plants (on average 14 shoots/line) were evaluated by artificial inoculation. Initially the ex vitro plants consisted of a single own-rooted shoot which was pruned at a height of 10 cm above soil level to adjust the growing conditions of all plants to be inoculated. The vigorously growing shoot regrowth was inoculated by bisecting the two youngest unfolded leaves with scissors dipped in 1×10^9 cfu of *Erwinia amylovora* high virulent stains per ml. The lesion was measured 4 weeks after inoculation and the severity of infection was calculated as % of the total shoot length necrotized after lesion extension had ceased. The level of resistance varied among transgenic lines. Compared to the mother genotype 'Pinova' the transgenic lines derived from, there were lines more and less resistant than 'Pinova'. The highest level of resistance was detected in the wild species *Malus robusta* with nearly no infection. 4 weeks after inoculation, three transgenic lines were significant more resistant than 'Pinova'. Nevertheless, after two months these differences compensated. A high level of variation was detected among plants of a single line. These results confirm the problems which occur in assessment of resistance to the fire blight pathogen (Brisset et al., 1988; Donovan, 1991; Donovan et al., 1994). Using different *E. amylovora* strains and combining the results from the greenhouse evaluation and the in vitro evaluation a correlation of r = 0.5 was found. However, further experiments are necessary to minimize the variation among plants of a line, to adjust the conditions of the in vitro assay and the greenhouse assessment.

4. REFERENCES

Bolar J.P., J.L. Norelli, H.S. Aldwinckle, V. Hanke.1999. Hort Science 33: 1251-1252

Brisset M.-N., J.-P. Paulin, M. Duron. 1988. Agronomie 8 : 707-710

De Bondt A., K. Eggermont, I. Penninck, I. Goderis, W.F. Broekaert. 1996. Plant Cell Rep. 13: 587-593

Donovan A.1991. Ann.appl. Biol.119: 59-68

Donovan A.M., R. Morgan, C. Valobra-Piagnani, M.S. Ridout, D.J. James, C.M.E. Garrett. 1994. J. Hort. Science 69: 105-113

Hanke V., K. Düring, J.L. Norelli, H.S. Aldwinckle.1999. Acta Hort. 489: 253-256

Hanke V., J.L. Norelli, H.S. Aldwinckle, I. Hiller, G. Klotzsche, K. Winkler, J. Egerer, K. Richter. 2000. Acta Hort. 538, 611-616

Hanke V., W.-S. Kim, and K. Geider. 2002: Acta Hort. In press.

James D.L., A.J. Passey, D.J. Barbara, M. Bevan. 1989. Plant Cell Rep. 7: 658-661

James D.J., A.J. Passey, A.D. Webster, D.J. Barbara, A. M. Dandekar, S.L. Uratsu, P. Viss. 1993. Acta Hort. 336, 179-

Kim, W.-S. and K. Geider. 2000. Phytopathology 90: 1263-1268.

Ko K., J.L. Norelli, S.K. Brown, E. Borejsza-Wysocka, K. Düring, and H.S. Aldwinckle. 1999. Acta Hort. 489: 257-258

Maheswaran G., M. Welander, J.F. Hutchinson, M.W. Graham, D. Richards. 1992. J. Plant Physiol. 139: 560-568

Norelli J.L., J.Z. Mills, M.T. Momol, and H.S. Aldwinckle.1999. Acta Hort. 489: 273-278

Puite K.J. and J.G. Schaart. 1996. Plant Science 119: 125-133

Sriskandarajah S., P.B. Goodwin, J. Speirs. 1994. Plant Cell Tissue Organ Cult. 36: 317-329

Yao J.L., D. Cohen, R. Atkinson, K. Richardson, B. Morris. 1995. Plant Cell Rep. 14: 407-412

Yao J.L., D. Cohen, R. Atkinson, B. Morris. 2000. In: Y.P.S. Bajaj (ed). Transgenic Trees. Biotechnology in Agriculture and Forestry 44, Springer Verlag, 153-170

EXPRESSION OF AN ALTERED ANTIMICROBIAL HORDOTHIONIN GENE IN BARLEY AND OAT

Jianming Fu[1,2], Ronald W. Skadsen[2] and Heidi F. Kaeppler[1]
[1]Department of Agronomy, University of Wisconsin, Madison, WI 53706;
[2]USDA/ARS/Cereal Crops Research Unit, 501 Walnut St., Madison, WI 53705 (email: jianmingfu@facstaff.wisc.edu)

Keywords antifungal proteins, cereal crops, hordothionin

1. INTRODUCTION

Alpha-hordothionin (HTH) is specifically produced in barley endosperm. Purified HTH has antimicrobial activity against a wide range of pathogenic microbes. Native HTH does not protect barley from infection because HTH is confined to endosperm, and many pathogens infect seeds through outer tissues. Therefore, it would be desirable to express HTH gene in floral tissues. We describe transformation and expression analyses of an altered HTH gene driven by a constitutive promoter in transgenic barley and oat.

2. MATERIALS AND METHODS

A HTH cDNA (Hth1) of nearly full length was cloned from a cDNA library constructed from barley (cv. Morex) developing endosperm. A truncated cDNA version (Hth2) was developed by deleting 3' and 5'UTRs and the 18 nts encoding the 6 amino acids between the first methionine and the second methionine in the coding sequence. The Hth2 was cloned in pAHC25, replacing gus. The resulting plasmid Hth2/pAHC was used for transformation of immature embryos of barley cv. Golden Promise and calli derived from apical meristems of an elite oat cultivar Belle according to Wan and Lemaux (1994) and Torbert et al. (1998), respectively. Extraction of genomic DNA and total RNA from leaves of greenhouse-grown plants and northern blot analyses were conducted (Skadsen et al., 2000). PCR was conducted using an upstream primer UBI and a downstream primer NOS to amplify both bar gene (730 bp) and Hth2 gene (520 bp) simultaneously.

3. RESULTS AND DISCUSSION

Approximately 170 barley plants were grown in the greenhouse. Integration of the Hth2 into barley genomes was confirmed by PCR analyses (Figure 1,

I. K. Vasil (ed.), Plant Biotechnology 2002 and Beyond, 159-160.

Lanes 2- 8, Upper panel). The transgenic barley plants were derived from at least 6 independent events, as demonstrated by Southern blot analysis (data not shown), including 1 bar-only line (Lane 3) which probably resulted from plasmid fragmentation. Northern blot analysis showed that all lines had mRNAs transcribed from the transgene Hth2 (Lanes 4-8, Middle panel) except the bar-only line (Lane 3). Seventy oat plants were grown in the greenhouse. Stable transformation was confirmed by PCR analysis (Lanes 11 to 15). Southern blot analysis showed that the plants were derived from at least 15 independent events, including 3 bar-only lines (data not shown). Similar to barley, HTH mRNA was detected in transgenic oat (Lanes 11, 12, 14 and 15). Efforts to detect HTH protein derived from the transgene in both transgenic barley and oat are currently underway in our laboratory.

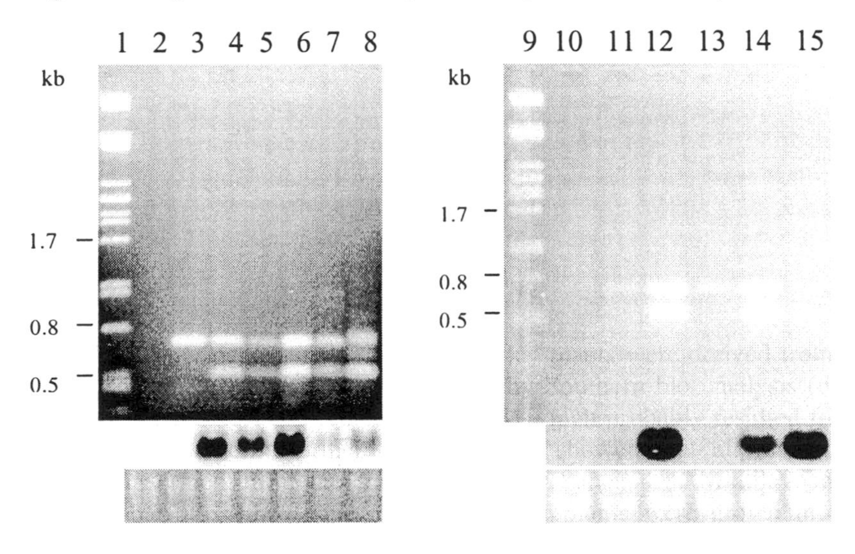

Figure 1. PCR and northern blot analyses. Lanes 1 and 9, λ/pstI DNA markers. Lane 2, non-trans barley. Lanes 3 to 8, transgenic barley. Lane 10, non-trans oat control. Lanes 11 to 15, transgenic oat. Upper panel, PCR. Middle panel, northern. Lower panel, Total RNA gel.

4. REFERENCES

Skadsen R.W., P. Sathish and H.F. Kaeppler. 2000. Expression of thaumatin-like permatin PR-5 genes switches from the ovary to the aleurone in developing barley and oat seeds. Plant Science 156: 11-22.

Torbert K.A., H.F. Kaeppler, H.W. Rines, G.K. Menon and D.A. Somers. 1998. Genetically engineering elite oat genotypes. Crop Sci. 38: 1685-1687.

Wan Y., P. M. Lemaux. 1994. Generation of large numbers of independently transformed fertile barley plants. Plant Physiol. 104:37-48.

THE DUAL FUNCTION OF CHITINASES IN DEFENCE RESPONSES AS WELL AS DURING NODULATION IN LEGUMES

Carin I. Jarl-Sunesson, Ulrika Troedsson and Felicia Andersson
Cell and Organism Biology, Lund University, S-221 00 Lund, SWEDEN (e-mail: carin.jarl@cob.lu.se)

Keywords *Galega orientalis*, *Trifolium*, *Hordeum*, nodulation, pathogenesis, PR-proteins

1. INTRODUCTION

How can the plant distinguish between, on one hand a beneficial colonization by Rhizobium or mycorrhiza and on the other hand a harmful attack by a pathogenic microorganism? It is intriguing that those processes are combined by several common steps. One such class of enzymes with a dual function are the chitinases. Chitinases forms one of the different classes of Pathogenesis-Related proteins expressed by most plants as a response to pathogenic attack. Chitinases are also expressed during colonization by *Rhizobium* as well as mycorrhiza. The signal molecules produced by the *Rhizobium*, the nod factors, are chemically lipo-chito-oligo saccharides, making them a theoretical substrate for the chitinases. One theory is that chitinases are also involved in the signalling pathway in the recognition process. One way of studying those questions is by transforming plants with sense- and antisense vectors of the genes involved. A prerequisite for this approach is the availability of protocols for plant regeneration and transformation.

2. PLANT REGENERATION AND TRANSFORMATION

We have chosen to work with barley (*Hordeum vulgaris*)and two different legumes, goat's rue (*Galega orientalis)* and red clover (*Trifolium pratense)*. The protocol for the regeneration and transformation of Swedish cultivars of barley has been published elsewhere (Jarl, 1999). For the legumes, three different protocols for regeneration starting from different explants

I. K. Vasil (ed.), Plant Biotechnology 2002 and Beyond, 161-162.

have been developed. Different basal media and hormone combinations were compared. A protocol for regeneration by shoot induction on immature embryos from *Galega* have been published (Collén and Jarl, 1999). By including TDZ in the regeneration media for *Galega*, embryogenesis in high frequency was obtained from mature seeds. The regenerated plants from the embryos could evetually be transferred to the greenhouse.

Five Swedish commercial cultivars of red clover have been tested. In addition to the different genotypes, basal media and hormone combinations for calli induction, shoot/embryo induction and rooting were investigated. Regeneration capacity differed among the different cultivars. For mature seeds of red clover, a shoot induction containing TDZ gave the highest frequency of shoot induction. After root induction, regenerated plants could be transferred to the greenhouse.Transformation using *Agrobacterium tumefaciens* on mature seeds after wounding by particle bombardment gave the best results.

3. CHITINASES

Barley plants transformed with vectors containing a chitinase gene (courtesy of T Bryngelsson and D Collinge) in antisense and sense have been regenerated. Transformation was done by particle bombardment of immature embryos. The transformation was confirmed by PCR analyses. A lowered production of the chitinase in antisensed plants was indicated by Western. Activity studies are in progress. Using a cDNA library isolated from nodulated root from *G. orientalis* (Kaijalainen et al., 2002), we have been able to isolate and sequence 4 different chitinases, as well as other genes possibly involved in the nodulation progress. Imunolocalization of chitinases in legume roots after induction by *Rhizobium*, mycorrhiza and pathogenic fungi are in progress.

4. REFERENCES

Collén A.M.C. and Jarl C.I. 1999. Comparison of different methods for plant regeneration and transformation of the legume *Galega orientalis* Lam. (goat's rue). Plant Cell Rep. 19: 13-19.

Jarl C.I. 1999. Stable transformation of regenerated plants of Swedish breeding lines of barley (*Hordeum vulgare* L.) by particle bombardment. Hereditas 130:83-87.

Kaijalainen S, Schroda, M. and Lindström, K. 2002. Cloning of nodule-specific cDNAs of *Galega orientalis*. Physiol Plant. 114: In press.

TRANSGENIC OIL PALM WITH STABLY INTEGRATED CpTI GENE CONFERS RESISTANCE TO BAGWORM LARVAE

Ruslan Abdullah, Christine Chari, Winnie Yap Soo Ping and Yeun Li Huey
School of Bioscience & Biotechnology, Faculty of Science & Technology, Universiti Kebangsaan Malaysia, 43600 UKM-Bangi, MALAYSIA (email: russzn@pkrisc.cc.ukm.my)

Keywords Transgenic, oil palm, CpTI, insect resistance, bioassay

1. INTRODUCTION

Insect predation of plants contributes towards reduction of crop yield worldwide. In oil palm alone, loss of yield due to insect attack is obvious. The most important insect pests of oil palm are the bagworms, particularly *Metisa plana* Walker. Bagworms caused severe defoliation of oil palm leaves as they fed on them. Synthetic insecticides have been applied to address losses due bagworm, but with unsatisfactory results. In addition most of the insecticide applied were wasted and it could also lead to toxicity to non-target insects. Thus efforts are in progress to produce transgenic plants resistant to insect. The use of transgenic plants offers several advantages over conventional insecticides. Earlier work on cowpea protease trypsin inhibitor (CpTI) gene on other crops showed that it has the potential to confer resistance against common insect pests especially members of the Lepidoptera, Coleoptera and Orthoptera. And most importantly, it exhibits low or no toxicity to mammals (Pusztai et al., 1992). When ingested, the larval gut protease was unable to digest proteins consumed. This eventually led to larvae death due to starvation (Gatehouse and Boulter, 1983). The CpTI gene was successfully transferred into oil palm giving rise to transgenic plants with stably intergrated gene. This report further highlights the effectiveness of CpTI gene against the common insect pest of oil palm, *Metisa plana* Walker.

2. MATERIALS AND METHODS

A 5-year old transgenic oil palm plant (*Elaeis guineensis* Jacq. var. Tenera, CpTI P_8) carrying CpTI gene was used. The plant was obtained from immature embryos bombarded with cowpea trypsin inhibitor (CpTI) gene.

I. K. Vasil (ed.), Plant Biotechnology 2002 and Beyond, 163-165.

PCR, Southern Blot and nucleotide sequence analyses were carried out as per standard protocols described elsewhere.

Newly hatched first instar larvae of *Metisa plana* Walker (Lepidoptera:Psychidae) were used in the bioassay study. Leaves from CpTI P_8 were washed and exposed for larvae feeding. Observations were carried out over a period of 35 days.

3. SUMMARY OF FINDINGS

The presence and integration of the CpTI gene in the transgenic CpTI P_8 plant was confirmed using PCR, Southern Blot and nucleotide sequence analyses (Figures 1, 2). Analyses were carried out on leaf periodically sampled from each frond as the plant matures into flowering. The expected 314 bp fragment corresponding to the CpTI coding region was present in all 46 fronds produced throughout the 5-year period. The nucleotide sequence of the PCR product was 100% homologous to the corresponding fragment from the initial plasmid used.

Leaves from both Control and Transgenic CpTI P_8 plants exposed to *Metisa plana* for a period of 35 days showed significant difference in the degree of leaf defoliation recorded (Figure 3). Leaves from Control plant were severely damaged compared to only marginal defoliation on leaves from the CpTI P_8 plant carrying the CpTI gene.

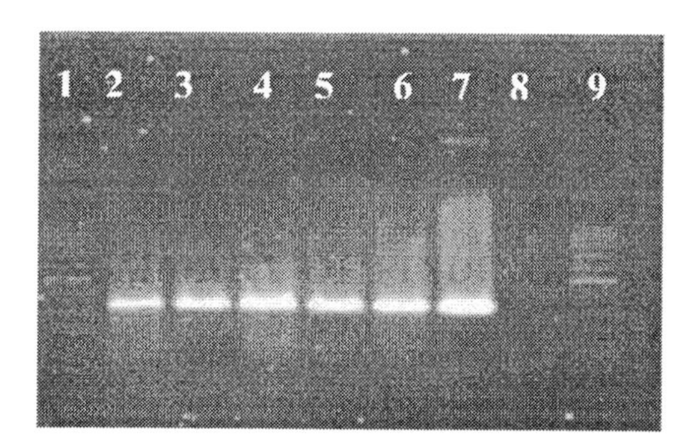

Figure 1. PCR amplification of CpTI gene in fronds of 5-year old CpTI P_8 Plant. Lane 1&9:100bp ladder; Lane 2-6: DNA from frond of CpTI P_8 Plant; Lane 7: Positive Control; Lane 8: Negative Control

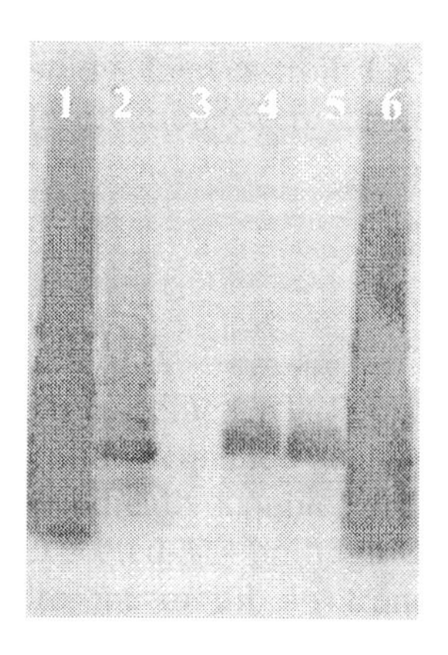

Figure 2. Southern blot analysis of CpTI P_8 plant. Lane 1&6: 100bp ladder; Lane 2: Positive Control; Lane 3: Negative Control; Lane 4&5: DNA from frond of CpTI P_8 plant.

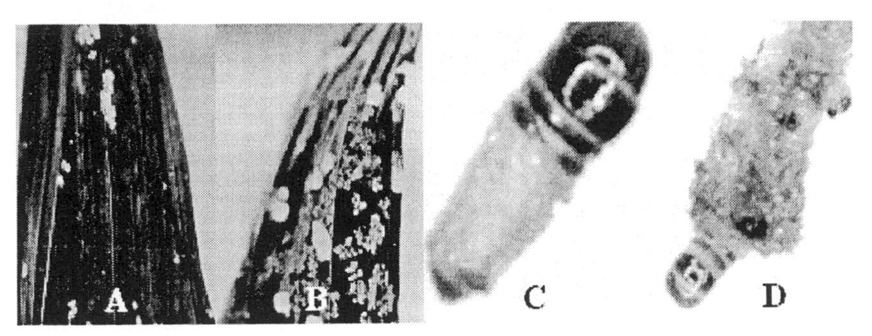

Figure 3: Leaf defoliation caused by Metisa plana larvae feeding on leaves from transgenic CpTI P_8 Plant (A) compared to leaves from Control plant (B) after 35 days. Newly hatched Metisa plana larvae (C) begin to construct its bag only after 2 days feeding on the oil palm leaves (D).

It was also observed that all larvae feeding on leaves from Control plant survived the bioassay period, by which they were already at the second and third instar stage. However, all larvae feeding on leaves from CpTI P_8 failed to survive but died whilst still at the same larval instar stage. The death is possibly due to the expression of the CpTI gene that encodes for the production of trypsin inhibitor. The inhibitor was postulated to bind to the digestive proteases and in turn inhibit proteolysis in the mid-gut of the larvae, resulting in the inability to digest consumed protein. This later led to death as in the case of larvae feeding on CpTI P_8 plant. However, this was not the case for larvae feeding on Control plant. These observations suggested that the stably integrated CpTI gene resulted in increased resistance the larvae of *Metisa plana,* a common insect pest for oil palm.

4. REFERENCES

Pusztai, A., Grant, G., Brown, D.J., Stewart, J.C. and Bardocz, S. 1992. British Journal of Nutrition. 68:783-791.

Gatehouse, A.M.R. and Boulter, D. 1983. Journal of the Science of Food and Agriculture 34:345-350.

PRODUCTION OF SALT TOLERANT RICE BY INTRODUCTION OF A GENE ENCODING CATALASE, *kat* E

Kenji Nagamiya[1,3], Kimiko Nakao[1], Shamsul H. Prodhan[1], Tsuyoshi Motohashi[1], Keiko Morishima[1], Sakiko Hirose[2], Kenjiro Ozawa[2], Yasunobu Ohkawa[2], Tetsuko Takabe[3], Teruhiro Takabe[4], and Atsushi Komamine[5]

[1]Tokyo University of Agriculture, Atsugi, Kanagawa, Japan; [2]National Institute of Agrobiological Sciences, Tsukuba, Ibaraki, Japan; [3]Faculty of Agriculture, Nagoya University, Nagoya, Japan; [4]Research Institute, Meijo University, Nagoya, Japan; [5]The Research Institute of Evolutionary Biology, 2-4-28, Kamiyoga, Setagaya-ku, Tokyo, 158-0098, Japan (e-mail:khf10654@nifty.ne.jp)

Keywords Catalase, *Oryza sative*, rice, salt tolerance, transgenic plant

Salt loading is one of the most serious environmental stresses that cause adverse affects on the growth of plants, leading to decrease in the productivity of crops. At present, 10% of available land for cultivation of crops is polluted with salt stress in75 countries in the world.

In order to expand available land for crop production by reducing effects of salt stress, salt tolerant plants should be produced by gene engineering. When plants are subjected to salt stress, radical oxygen species such as superoxide anion, hydrogen peroxide, hydroxyl radical and others occur and cause damage to most plant functions.

Hydrogen peroxide (H_2O_2) is the most stable radical oxygen species. Therefore, the decomposition of H_2O_2 is pivotal in the protection of cells from oxidative damage. Kaku et al. (2000) reported that overexpressing of catalase gene enhances tolerance for salt stress in a freshwater Cyanobacterium because catalase acts as a quencher of damage by H_2O_2.

In the represent report, we transformed a Japonica rice (*Oryza sativa* cv. Nipponbare) with a gene encoding catalase, *kat* E, derived from *E.coli* which decomposes H_2O_2. Salt tolerance of the T_0 transgenic plants and their progenies was investigated.

Seeds and callus of Japonica rice (*Oryza sativa* cv. Nipponbare) were used throughout the present experiments. The seeds kept at 4°C were sterilized with sodium hypochrolite solution and cultured on N_6 medium supplemented with 2mg/l 2,4-D at 25°C. Two weeks after callus induction, callus was

I. K. Vasil (ed.), Plant Biotechnology 2002 and Beyond, 167-170.

infected with *Agrobacterium tumefaciens* EHA101 carrying pIG121/Hm/ *kat* E, the construction of which is shown in Fig.1.

Nos-P	NPTII	Nos3'-Nos-T	P	katE	T	P	HPT	T

Figure 1. Construction of pIG121HmkatE Nos-P, Nos promotor; NPTII, Kanamycin resistant gene; Nos3'-Nos-T, Nos3'-Nos-terminator; P, CaMV35Spromotor; katE Catalase gene; T, Nos terminator; HPT, Hygromycin resistant gene

After three days of co-cultivation with EHA101 in the medium containing acetosyringone (10mg/l), callus was sterilized with 500mg/l carvenicillin for two weeks, and transferred to N_6 selection medium containing 50mg/l hygromycin and 100mg/l carvenicillin for 2 weeks. After selection callus was transferred to MS medium containing 50mg/l hygromycin, 100mg/l carvenicillin, 1mg/l NAA and 2mg/l BAP for shoot regeneration. Regenerated shoots were transferred to hormone free MS medium for root formation. Regenerated plantlets were acclimated with soil and submerged with water in a growth incubator at 30°C (light phase), 25°C (dark phase) with 12 hours light at 72 μmol m^{-2} s^{-1} under 62% relative humidity. Progenies were obtained from 15 transgenic T_0 plants (T_0-1 to T_0-15 strain).

The salt tolerance of T_0 transgenic rice plants was evaluated. Growth of T_0 transgenic plants was not suppressed and they formed seeds by selfing in 100 mM sodium chloride solution. On the contrary, the non transformed rice plants could not survive even in 50mM NaCl solution for more than 10 days. Seeds were harvested from T_0-1 to T_0-15 strains of T_0 transgenic plants and salt tolerance was examined at the germination stage. Seeds of progenies of T_0-1 to T_0-15, designated as T_0-1 to T_0-15 respectively, were germinated in the petridishes filled with OmM and 100mM NaCl solution. In 100mM NaCl solution none of seeds of non transgenic plants germinated. On the contrary, 30 – 70% of T_1 seeds of *kat* E transgenic plants germinated.

Salt tolerance of T_1 plants was investigated using a T_1 plants strain (strain 13) where the presence of *kat* E gene was confirmed by PCR southern hybridization. Seeds of the T_1 plants and non transgenic ones were sowed and germinated plants were cultivated in the absence of NaCl. When the plants began to form seeds and seeds were maturing, the plants were transferred to 250mM sodium chloride solution with fertilizer. Non transgenic plants could not grow and died within 7 days after transfer to NaCl solution while the T_1 transgenic plant could continue to grow even 14 days after transfer and produce mature seeds.

Growth rate, periods required for flowering, seed formation and maturation of T_1 plants were similar to those of non transgenic plants when T_1 plants were cultivated in the absence of NaCl. Seeds fertility was lower in T_1 plants (57.9% in average of 76 T_1 plants from 10 strains) than that of non transgenic plants (86.7%). However, seed fertility of T_1 plants derived from T_0 plant strain, which showed relatively high seed fertility (17.1%), was 76%, a similar value to that of non transgenic plants. Percentages of productive tiller in T_1 plants were similar to those of non transgenic plants.

The presence of *kat* E gene in T_0 and T_1 transgenic plants was confirmed by PCR, PCR Southern or Southern blot hybridization. Since only a single band of *kat* E gene appeared in Southern blot analysis of a T_1 plant, it is suggested that a single copy of *kat* E gene was introduced in the genome of transgenic plant. In T_1 plants, *kat* E bands were observed in 56 T_1 plants from 76 T_1 plants. The segregation ratio of the presence to the absence of *kat* E gene in T_1 plants (56:20 = 2.8:1) was approximately 3:1, suggesting that the transferred gene, *kat* E, was inherited stably.

Catalase activities in 4 strains of T_1 plants were measured. As shown in Table 1, catalase activities of *kat* E transgenic T_1 plants were approximately 1.5 to 2.5 fold higher level of catalase activities than those of non transgenic plants as shown in Table 1.

Table 1. Catalase activities of T_1 plants and non transgenic plants

No. of strain of T_1 plants	2	6	9	14	15	Non Transgenic Plants
Relative Activity of catalase	173	249	147	257	142	100

In the present study, we transformed Japonica rice plant (*Oryza sativa* cv. Nipponbare) with *kat* E gene encoding catalase derived from *E. coli* and examined salt tolerance of the obtained transgenic rice plants. T_0 plants obtained could grow and form seeds in the presence of 100mM NaCl. T_1 plants could grow in the presence of 250mM NaCl for 14 days during maturation of seeds, while non transgenic plants died within 7 days even in the presence of 50mM NaCl. The catalase activity of T_1 plants increased to approximately 1.5 to 2.5 fold higher level than that of the non transgenic plants. These results indicate that overexpression of catalase is due to enhancement of salt tolerance in rice plants by acting as a scavenger for H_2O_2, a radical oxygen species.

Many efforts have been made to try to produce salt tolerant rice plants by introduction of genes encoding proteins involved in rescuing plants from stresses, such as scavengers of radical oxygen species (ascorbate peroxidase, superoxide dismutase), osmo regulators (glycinebetain synthase, proline synthase, and Na^+/H^+ antiporter), promotive agent for photorespiration (chloroplast glutamine synthase). However, salt tolerance in transgenic rice plants, which have been reported so far, is to be able to grow in 150mM NaCl solution at maximum for 1 to 2 weeks (Sakamoto et al., 1998; Tanaka et al., 1999; Saijo et al., 2000; Hoshida et al., 2000). Therefore, the transgenic rice plants reported here, which overexpress catalase by introduction of *kat* E derived from *E. coli*, is one of the most tolerant salt stress rice plants for salt stress.

In particular, this is the first report that seeds were obtained from progenies of salt tolerant transgenic T_0 rice plants and plants of progenies showed inheritable high salt tolerance.

Attempts are under way to apply the procedure reported here to Indica rice plants such as Kasalath, BR-5 and other varieties to produce salt tolerant Indica rice plants, and to produce marker-free salt tolerant rice plants using MAT Vector (Ebinuma and Komamine, 2001). Furthermore, investigation of drought and cold tolerance of transgenic rice plants obtained here is intended. In conclusion, overexpression of catalase is an efficient approach to produce salt tolerant rice plants and findings reported here indicate the feasibility of producing salt tolerant transgenic plants to expand available lands for cultivation of crops on the earth.

REFERENCES

Ebinuma H., A. Komamine, 2001. MAT (Multi-Auto-Transformation) Vector system. The oncogene of Agrobacterium as positive markers for regeneration and selection of marker-free transgenic plants. In Vitro Cell. Dev. Biol. Plant 37:103-113

Hoshida H., Y. Tanaka, T. Hibino, Y. Hayashi, A. Tanaka, T. Takabe and T. Takabe. 2000. Enhanced tolerance to salt stress in transgenic rice that overexpresses chloroplast glutamine synthase. Plant Mol. Biol. 43:103-111

Kaku N., T. Hibino, Y-L. Meng, Y. Tanaka, E. Araki, T. Takabe and T. Takabe. 2000. Effects of overexpression of *Escherichia coli kat* E and *bet* genes on the tolerance for salt stress in a fresh water Cyanobacterium *Synechococcus* sp. PCC7942. Plant Sci. 159:281-288

Saijo Y., S.Hata, J. Kyozuka, K. Shimamoto, K. Izui. 2000 Over-expression of a single Ca^{2+}-dependent protein kinase confers both cold and salt/drought tolerance on rice plants. Plant J. 23:319-327

Sakamoto A. A., N. Murata, A. Murata. 1998. Metabolic engineering of rice leading to biosynthesis of glycinebetaine and tolerance to salt and cold. Plant Mol. Biol. 38:1011-1018

Tanaka Y., H. Hibino, Y. Hayashi, A. Tanaka, T. Takabe, T. Takabe. 1999. Salt tolerance of transgenic rice overexpressing yeast mitochondrial Mn SOD in chloroplast. Plant Sci. 148:131-138

TRANSGENIC TOBACCO OVEREXPRESSING GLYOXALASE I AND II SHOW ENHANCED TOLERANCE TO SALINITY AND HEAVY METAL STRESS

Sneh Lata Singla-Pareek, Veena, M. K. Reddy, Brad W. Porter[1], Frank F. White[1] and Sudhir K. Sopory
Plant Molecular Biology Lab, International Centre for Genetic Engineering and Biotechnology, Aruna Asaf Ali Marg, New Delhi 110 067, India (e-mail: sopory@hotmail.com)
[1]Department of Plant Pathology, Kansas State University, Manhattan, KS 66506, USA

Keywords Genetic engineering, glyoxalase I, glyoxalase II, heavy metal, methylglyoxal, salinity, transgenic tobacco

1. INTRODUCTION

Genetic manipulation of crop plants for enhanced abiotic stress tolerance holds a great promise for sustainable agriculture. Various genes are being tested for their potential to confer stress tolerance (see Singla-Pareek et al., 2001). In this respect, we have genetically manipulated the glyoxalase pathway in which two enzymes, glyoxalase I (EC 4.4.1.5) and glyoxalase II (EC 3.1.2.6) act coordinately to convert 2-oxoaldehydes into 2-hydroxyacids using reduced glutathione as a cofactor. The reaction catalysed by glyoxalase I and glyoxalase II is as follows:

$$\text{METHYL GLYOXAL} \xrightarrow{\text{GSH}} \text{HEMITHIO ACETAL} \xrightarrow{\textit{Glyoxalase I}} \text{S-D LACTOYL GLUTATHIONE} \xrightarrow{\textit{Glyoxalase II}} \text{D-LACTIC ACID + GSH}$$

One of the main role of glyoxalase system is the glutathione-based detoxification of methylglyoxal (MG), which is a potent mutagenic and cytotoxic compound known to arrest growth and reacts with DNA and protein and increase sister chromatid exchanges (Thornalley, 1990).

I. K. Vasil (ed.), Plant Biotechnology 2002 and Beyond, 171-174.

The precise physiological significance of the glyoxalase system is still not clearly defined but it has been proposed to be involved in various functions like regulation of cell division and proliferation, microtubule assembly, and in protection against oxoaldehyde toxicity. In fact, this system has been often regarded as a ``marker for cell growth and division" (Paulus et al., 1993). In response to abiotic stresses, glyoxalase I has been shown to be upregulated in tomato in response to salt stress, osmotic and phytohormonal stimuli (Espartero et al., 1995) while the glyoxalase II from plants is yet to be characterized in detail.

In the present study, we have overexpressed glyoxalase I from *Brassica juncea* and glyoxalase II from *Oryza sativa* in transgenic tobacco. The transgenic plants have been found to tolerate higher levels of salinity as well as heavy metal stress clearly suggesting the importance of glyoxalase pathway in stress tolerance.

2. MATERIALS AND METHODS

Full length glyoxalase I (glyI) cDNA has been previously cloned from *Brassica juncea* (Genbank accession no. Y13239) and glyoxalase II (glyII) cDNA has now been cloned from *Oryza sativa* cDNA library (Genbank accession no. AY054407) as well as from *Pennisetum glaucum* (Genbank accession no. AF508863). For overexpression in transgenic tobacco, glyI was cloned in pBI121 binary vector (Clonetech) with uidA and nptII as the screenable and selectable markers respectively. The glyII gene has been cloned in pCAMBIA1304 with gfp:uidA as the screenable and hptII as the selectable marker. Tobacco (*Nicotiana tabaccum* cv petit Havana) leaf discs were transformed with either glyI or glyII genes alone or glyII gene in glyI transgenic plants so as to get the double transformants following a leaf disc transformation procedure (Horsch et al., 1985) with *Agrobacterium* containing glyI or glyII. The individual transformants were selected on appropriate antibiotic while the double transformants were selected on the mixture of kanamycin and hygromycin as different selectable markers were chosen for each glyI and glyII gene constructs.

Putative transgenic plants were screened using the histochemical GUS assay. The transgenic lines were further confirmed by first carrying out PCR for the presence of transgene using tobacco genomic DNA as the template. Finally, Southern hybridization has been carried out to confirm transgenic lines as described in Veena et al. (1999).

Leaf discs of 1 cm diameter were cut from wild type and transgenic plants and floated in 6 ml solution of methylglyoxal (5 to 20 mM, 48 h), sodium chloride (200 mM to 1 M, 72 h) or zinc chloride (5 to 20 mM, 96 h) or sterile distilled water which served as experimental control. The chlorophyll content was measured as described in Veena et al. (1999).

3. RESULTS AND DISCUSSION

We have previously shown that overexpression of glyI in transgenic plants results in improved detoxification of MG and also confers tolerance to high levels of salinity (Veena et al., 1999). This trait has now been found to be functionally and genetically stable which prompted us to genetically manipulate the entire glyoxalase pathway. For this, we have now cloned the full length gene for glyII from *Oryza sativa* as well as from *Pennisetum glaucum* and in the present study, *Oryza sativa* glyII clone has been used.

The glyII cDNA was engineered into tobacco plants either independently (i.e. under wild type situation) or in concert with glyI (i.e. in the glyI transgenic line) to produce double transgenics. Significantly higher levels of both enzymes accumulated in various transgenic plants thus ensuring that the overexpressed protein is functionally active in the cellular milieu of the transgenic system. Following this, the survival of transgenic plants under stressful conditions was analysed. The double transgenic plants showed tolerance to higher levels of salinity. Importantly, in the leaf disc assay, the explants could tolerate even up to 1 M NaCl (which is although a non-physiological level of salinity) for 3 days and the capacity of the leaf discs from double transformants to tolerate such a high degree of salinity was more than the independently transformed glyI and glyII plants. There was a total loss in chlorophyll content in the wild type plants while the glyoxalase transformants protected this degradation (Figure 1).

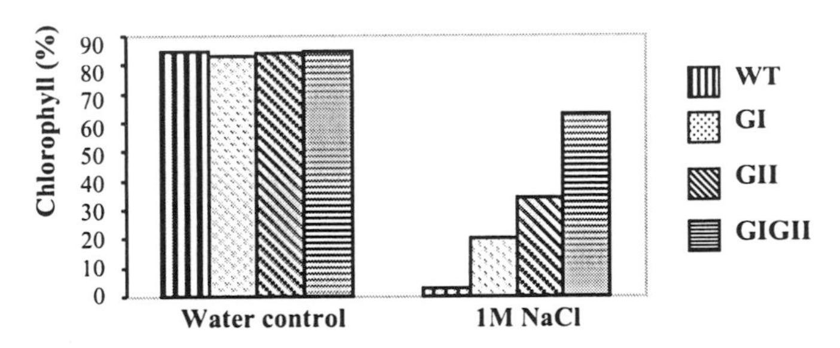

Figure 1: *Salinity tolerance of various transgenic plants. Leaf discs of wild type (WT), glyI (GI), glyII (GII) and glyI+II double transformants (GIGII) were exposed to 1M sodium chloride for 3d and total chlorophyll was measured. Note the difference in retention of chlorophyll in WT and transgenic plants.*

The T1-generation seedlings also could complete their life cycle under 200 mM NaCl. In fact, upon measuring the endogenous levels of NaCl in these transgenic plants, it was found that Na^+ ions were maximally sequestered out in the older leaves. The double transformants were also found to tolerate upto 20 mM zinc chloride for 4 days reflecting their heavy metal tolerance (data not shown). The loss of chlorophyll was minimum in double transformants as compared to either of the gene alone. Upon analysing the methylglyoxal sensitivity of these transgenic plants, it was found that the double transgenics could detoxify MG many folds higher than the independently transformed glyI and glyII plants (data not shown). Taken together, these findings clearly indicate the potential of glyoxalase system in engineering salinity and heavy metal tolerance in higher plants.

4. ACKNOWLEDGEMENTS

The research was supported from the internal grants of ICGEB.

5. REFERENCES

Espartero J., I. Sanchez-Aguayo and J. Pardo 1995. Molecular characterization of glyoxalase I from a higher plant: upregulation by stress. Plant Mol. Biol. 29: 1223-1233.

Horsch R.B., J.E. Fry, N.L. Hoffmann, D. Eichholtz, S.G. Rogers and R.T. Fraley. 1985. A simple and general method for transferring genes into plants. Science 227: 1229-1231.

Paulus C, B. Knollner and H. Jacobsen 1993. Physiological and biochemical characterization of glyoxalase I, a general marker for cell proliferation, from a soybean cell suspension. Planta 189: 561-566.

Singla-Pareek S.L., M.K. Reddy and S.K. Sopory 2001.Transgenic approach towards developing abiotic stress tolerance in plants. Proc. Indian natn. Sci. Acad. (PINSA) B67: 265-284.

Thornalley P.J. 1990. The glyoxalase system: new developments towards functional characterization of metabolic pathways fundamental to biological life. Biochem. J. 269: 1-11.

Veena, V.S. Reddy and S.K. Sopory 1999. Glyoxalase I from *Brassica juncea*: molecular cloning, regulation and its over-expression confer tolerance in transgenic tobacco under stress. Plant J. 17: 385-395.

PLANT MOLECULAR RESPONSES TO PHOSPHATE-STARVATION

Derek W.R. White, Roy J. Meeking, Erika Varkonyi-Gasic, Ruth J. Baldwin and Andrew Robinson
Plant Molecular Genetics Laboratory, Grasslands Research Centre, AgResearch, Private Bag 11008, Palmerston North, New Zealand (e-mail: Derek.white@agresearch.co.nz)

Keywords Phosphate, nutrient starvation, molecular response, Pi-transporters, gene expression patterns.

1. INTRODUCTION

The availability of phosphorus, an essential macronutrient for plant growth, is often limiting, even in soils with a high P content. To overcome limitations in P supply plants have evolved a variety of highly specialized physiological, biochemical and molecular response mechanisms to acquire inorganic phosphate ions (Pi) from the soil (see Raghothama, 1999). These mechanisms include; changes in plant morphology, the secretion of protons, organic acids, RNases and acid phosphatases, from roots to solubilize P-containing compounds in the soil, and enhanced Pi uptake through the activation of Pi transporters.

The cloning of plant high affinity Pi transporters (Muchhal et al., 1996; Leggewie et al., 1997; Smith et al., 1997; Daram et al., 1998; Lui et al., 1998a; Lui et al., 1998b) has done much to advance our understanding of the molecular regulation of plant Pi acquisition. Localization of phosphate-starvation inducible Pi transporter gene expression in root epidermal and root hair cells indicates a role in uptake of inorganic P from the soil/root interface (Daram et al., 1998; Lui et al., 1998a; Chiou et al., 2001). There is also evidence that expression of Pi transporter genes is regulated by the internal Pi concentration in plant tissues (Lui et al., 1998a; Dong et al., 1999).

In order to gain new insights about the molecular events underlying plant phosphate-starvation responses we have examined the spatial pattern of expression of the tomato *LePT2* Pi transporter gene. Transgenic tobacco plants, expressing a *LePT2* promoter-*gus* transgene, were also used to establish a model system for the molecular analysis of Pi-starvation response and the identification of genes activated in the roots of seedlings grown under Pi-deficient conditions.

I. K. Vasil (ed.), Plant Biotechnology 2002 and Beyond, 175-178.

2. *LePT2/NtPT2* EXPRESSION PATTERNS

Chromosome walking techniques were used to isolate a 1.5kb 5′ upstream promoter DNA fragment from the tomato *LePT2* high affinity Pi transporter gene. A *LePT2* promoter-*gus* reporter gene cassette was constructed, incorporated into tobacco, and the progeny of transgenic plants were assayed for the spatial pattern of Pi-starvation induced GUS histochemical localization (Figure 1).

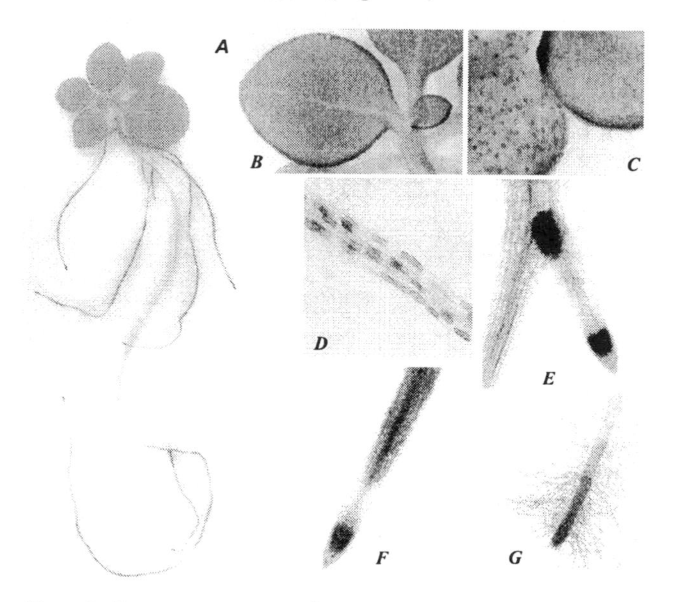

Figure1. Expression pattern of a LePT2 promoter-gus transgene in transgenic tobacco seedlings grown in Pi deficient nutrient solution. Seeds were germinated on Pi-agarose medium, grown for 2 weeks and transferred to hydroponic solution containing 15μM Pi for a further 3 weeks. Expression in (A) whole seedling, (B) expanding leaf, (C) stomata guard cells, (D) cortex of old roots, (E) lateral roots, (F) primary root tips, and (G) the root hairs of older roots.

Besides the expected root epidermal/root hair expression pattern (Lui et al., 1998a) GUS activity was also localized in; (1) the cortical and vascular initials immediately adjacent to the root apical meristem, (2) at the site of lateral root formation, (3) in individual cortical cells of older roots, (4) in the vascular bundles of roots, (5) on the margins of rapidly expanding immature leaves, and (6) in the stomata guard cells of hypocotyls and expanding leaves. This expression pattern occurred only under conditions of Pi-starvation. The results of whole mount *in situ* hybridization and RT-PCR analysis of *NtPT2* (tobacco) and *LePT2* (tomato) gene expression in Pi starved seedlings were consistent with the results from *LePT2* promoter-*gus* transgene expression in transgenic tobacco plants.

The complex pattern of *LePT2* expression indicates a role for Pi transporters in the internal reallocation of inorganic phosphate between tissues and cells in seedlings grown under conditions of Pi starvation. *LePT2* may have a role in moving Pi from older primary roots to young lateral roots that establish new root/soil interfaces. Expression on the margin of expanding leaves and in the stomata guard cells may be due to a higher Pi requirement for the physiological functioning of these cells.

2.1. Identification of Pi-starvation induced genes

Lines of *LePT2* promoter-*gus* transgenic tobacco plants were used to analyze the relationship between Pi transporter expression and nutrient solution Pi concentration. The transgene is highly expressed in seedlings grown in 0-15μM Pi, but not significantly expressed in seedlings grown at concentrations above 50 μM Pi. These results were used to establish a model tobacco seedling system that can be used to detect Pi-starvation induced gene expression. Suppressive subtractive hybridisation was used to prepare an enriched cDNA library from transcripts isolated from the roots of seedlings grown under Pi-starvation conditions. Clones from this cDNA library were then screened, using replicate 384 spot nylon microarrays, to identify root specific, Pi starvation-induced, genes. An example of one of these genes is *Pit1* (Figure 2), a gene previously identified as Al and Pi-starvation induced in tobacco cell cultures (Ezaki et al., 1995). *Pit1* expression is root specific, up regulated by Pi-starvation treatment, and unaffected by K-starvation.

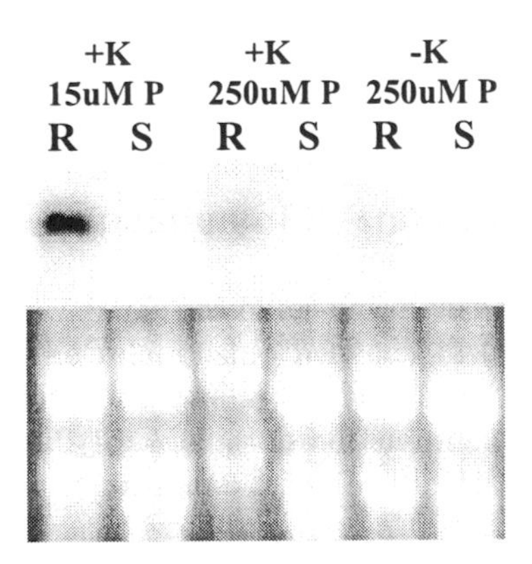

Figure 2. Root-specific, Pi-starvation induced expression of Pit1. Northern blot analysis of total RNA isolated from roots (R), or shoots (S), of tobacco seedlings growth in nutrient solution with 15 μM P or 250 μM P, with (+K) or without (-K) potassium.

3. REFERENCES

Ezaki B., Yamamoto Y., and H. Matsumoto. 1995. Cloning and sequencing the cDNAs induced by aluminium treatment and Pi starvation in cultured tobacco cells. Physiol. Plantarum 93:11-18.

Dong B., Ryan P.R., Rengel Z., and E. Delhaize. 1999. Phosphate uptake in *Arabidopsis thaliana*: dependence of uptake on the expression of transporter genes and internal phosphate concentrations. Plant Cell & Environment 22: 1455-1461.

Raghothama K.G. 1999. Phosphate acquisition. Annu. Rev. Plant Physiol. Plant Mol. Biol. 50: 665-693.

Liu C., Muchhal U.S., Uthappa M., Kononowicz A.K., and K.G. Ragothama. 1998a. Tomato phosphate transporter genes are differentially regulated in plant tissues by phosphorus. Plant Physiol. 116: 91-99.

Daram P., Brunner S., Persson B.L., Amrheim N., and M. Bucher. 1998. Functional analysis and cell-specific expression of a phosphate transporter from tomato. Planta. 206: 225-233.

Smith S.E., Ealing P.M., Dong B., and E. Delhaize. 1997. The cloning of two *Arabidopsis* genes belonging to a phosphate transporter family. Plant J. 11: 83-92.

Muchhal U.S., Pardo J.M., and K.G. Raghathama. 1996. Phosphate transporters from the higher plant *Arabidopsis thaliana*. Proc. Natl. Acad. Sci. USA. 93: 101519-110523.

Leggewie G., Willmitzer L., and J.W. Riesmeier. 1997. Two cDNAs from potato are able to complement a phosphate uptake-deficient yeast mutant: identification of phosphate transporters from higher plants. Plant Cell. 9: 381-392.

Lui H., Trieu A.T., Blaylock L.A., and M.J. Harrison 1998b. Cloning and characterization of two phosphate transporters from *Medicago truncatula* roots: Regulation in response to phosphate and colonisation by arbuscular mycorrhizal (AM) fungi. Mol. Plant-Microbe Interact. 11: 14-22.

Chiou T-J., Lui H., and M.J. Harrison. 2001. The spatial expression patterns of a phosphate transporter (MtPT1) from *Medicago truncatula* indicate a role in phosphate transport at the root/soil interface. Plant J. 25: 281-293.

ENGINEERING ENHANCED NUTRIENT UPTAKE IN TRANSGENIC PLANTS

José López-Bucio, Verenice Ramírez, Fernanda Nieto, Aileen O`Connor and Luis Herrera-Estrella
Departamento de Ingeniería Genética de Plantas, Centro de Investigación y Estudios Avanzados, Km 9.6 carretera Irapuato-León. Irapuato, Gto. México (email:jolopez@ira.cinvestav.mx)

Keywords Nutrient uptake, transgenic plants, citrate overproduction, root architecture

1. INTRODUCTION

Poor soil fertility is a critical yield-limiting factor in most cultivated lands. In acid soils, which represent nearly 40% of the worlds arable land, P fixation and Al toxicity limit agricultural practices, whereas in alkaline calcareous soils, which represent more than 25% of the earth surface, P and Fe deficiency constrain crop production. In these soils, farmers need large amounts of fertilizer to meet crop requirements. Fertilizer use as a mean to correct nutrient deficiencies may involve high monetary and ecological costs, especially when soil chemical conditions lead to the fixation in the soil of the applied fertilizer (Raghothama, 1999). In many countries, the development of cultivars with increased capacity to extract nutrients from the soil is an urgent requirement. A primary strategy in overcoming P-deficiency is to identify plant species or genotypes that can best solubilize P from sparingly soluble native sources, such as rock phosphate. Recently, genetic engineering has become an important tool for plant breeding. Molecular techniques are available for the analysis and manipulation of traits that confer an efficient use of phosphorus, which include organic acid synthesis/excretion and root branching patterns. The use of molecular techniques in combination with basic and applied research in the field of plant nutrition will play a decisive role in developing crops more efficient in the use of soil nutrients such as P.

I. K. Vasil (ed.), Plant Biotechnology 2002 and Beyond, 179-182..

2. ADAPTATIONS OF PLANTS TO ENHANCE NUTRIENT CAPTURE

Plants adapted to nutrient poor soils have particular responses to increase nutrient acquisition from the environment. These include an expanded root surface area through increased formation of lateral roots and root hairs, synthesis and exudation of organic acids and enhanced expression of NO^{3-}, NH^{4+}, and PO^{4-} transporters. In addition, they commonly live in symbiosis with soil microorganisms such as mycorrhizal fungi and N_2-fixing bacteria (Raghothama, 1999; López-Bucio et al., 2000b). An especially successful adaptation of plants to low fertility soils is the formation of cluster roots, also termed proteoid roots, which occur in a wide range of plant families including the Leguminosae, Betulaceae, Casuarinaceae and Proteaceae. The cluster roots combine adaptations to root branching, rhizosphere acidification and nutrient uptake in a fascinating way. Each cluster is composed of groups of lateral roots arising from the pericycle. Initiation of lateral roots is linked to a number of factors including phosphate and iron deficiency. The proteoid root zone is responsible of exudation of large amounts of organic acids, including citrate and malate, as well as phosphatases and protons. These exudates function in phosphorus acquisition and may also play a role in mobility and uptake of other minerals such as nitrate and iron (Dinkelaker et al., 1995).

3. IMPROVED PHOSPHORUS UPTAKE IN TRANSGENIC PLANTS THAT OVERPRODUCE CITRATE

When plants grow under low P conditions, an increase in the exudation of organic acids into the soil is a common response. It has been proposed that release of citrate, malate and oxalate into the rhizosphere improves P mobilisation and uptake. We have been studying the effect of organic acid overproduction on plant adaptation to low fertility soils by using transgenic plants that express a bacterial citrate synthase gene. Current work with tobacco, Arabidopsis, papaya and maize transgenic lines, indicate that aluminum tolerance and the P uptake ability are significantly enhanced in citrate overproducing plants. Under greenhouse conditions, the transgenic tobacco plants yielded more fruit and leaf biomass than controls when growing in alkaline soils with low P content. Very recently, we found that Fe uptake can be increased in Arabidopsis and sorghum plants by overproducing citrate.Taken together, these results suggest that organic acid overproduction in transgenic plants represent a promising alternative to produce novel plant varieties better adapted to grow under marginal, low fertility soil conditions (De la Fuente-Martínez et al., 1997, López-Bucio et al., 2000a).

4. MOLECULAR MECHANISMS CONTROLING ROOT ARCHITECTURAL RESPONSES TO PHOSPHORUS AVAILABILITY

In Arabidopsis (Brassicaceae), increased formation of lateral roots and root hair elongation occurs in response to nutrient starvation, particularly to low phosphate availability (Figure 1). We have used this response to study the molecular mechanisms that control the root architecture in plants. We found that at P-limiting conditions (<50 μM), the Arabidopsis root system undergoes major architectural changes in terms of lateral root number, lateral root density, and primary root length. Treatment with auxins and auxin antagonists indicate that these changes are related to an increase in auxin sensitivity in the roots of P deprived Arabidopsis seedlings (López-Bucio et al., 2002). The root response of Arabidopsis plants to a low P supply, has allowed the isolation of mutants affected in this process and the identification of genes that participate in the regulation of changes in the root architecture of the Arabidopsis root system. It is expected that some of these genes could be used to manipulate root architecture in transgenic plants.

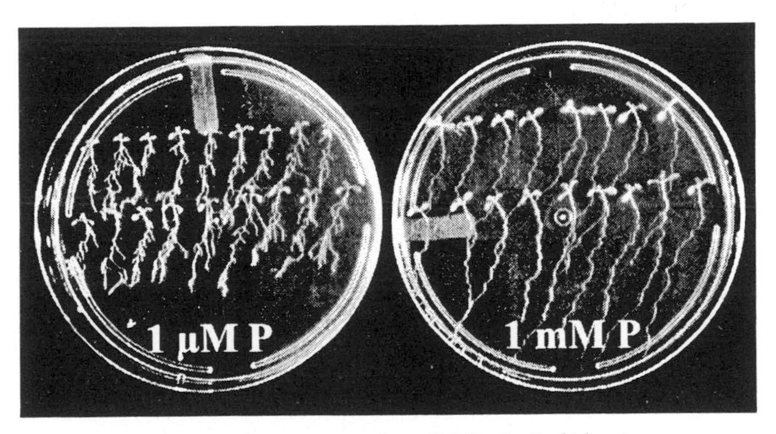

Figure 1. The root architectural responses to P availability in Arabidopsis

5. CONCLUSIONS

Increased physiological efficiency in P acquisition may be an attractive breeding target for crop plants. Mechanisms that would enhance P acquisition include better P mobilization through root exudates, such as organic acids and superior root architecture. We have shown that by overproducing citrate in transgenic plants, it is possible to increase their P uptake capacity. Through a better understanding of the mechanisms controlling the root architectural responses to P availability, it would be possible to identify genetic determinants useful for manipulating root systems. The major objective of this work is to develop P-efficient plants for cultivation in marginal soils of the world.

6. REFERENCES

De la Fuente J.M., Ramírez-Rodríguez V., Cabrera-Ponce J.L. and L. Herrera-Estrella. 1997. Aluminum tolerance in transgenic plants by alteration of citrate synthesis. Science. 276:1566-1568.

Dinkelaker B., Hengeler C. and H. Marshner. 1995. Distribution and function of proteoid roots and other root clusters. Bot. Acta. 108:183-200.

López-Bucio J., Martínez de la Vega O., Guevara-García A., and L. Herrera-Estrella. 2000a. Enhanced phosphorus uptake in transgenic tobacco plants that overproduce citrate. Nat. Biotechnol. 18:450-453.

López-Bucio J., Nieto-Jacobo M.F., Ramírez-Rodríguez V., and L. Herrera-Estrella. 2000b. Organic acid metabolism in plants: From adaptive physiology to transgenic varieties for cultivation in extreme soils. Plant Sci. 160:1-13.

López-Bucio J., Hernández-Abreu E., Sánchez-Calderón L., Nieto-Jacobo M.F., Simpson J., and L. Herrera-Estrella. 2002. Phosphate availability alters architecture and causes changes in hormone sensitivity in the Arabidopsis root system. Plant Physiol. 129:244-256.

Raghothama K.G. 1999. Phosphate acquisition. Annu. Rev. Plant Physiol. Plant Mol. Biol. 50:665-693.

PRELIMINARY CHARACTERIZATION OF THE DREB GENES IN TRANSGENIC WHEAT

Alessandro Pellegrineschi[1], Jean Mecel Ribaut[1], Richard Thretowan[2], Kazuco Yamaguchi-Shinozaki[3], and David Hoisington[1]
[1]Applied Biotechnology Center and [2]Wheat Program, CIMMYT, Apdo Postal 6-641, 06600 Mexico, D.F. [3]Biological Resources Division, Japan International Research Center for Agricultural Sciences (JIRCAS), Ministry of Agriculture, Forestry, and Fisheries, 2-1 Ohwashi, Tsukuba, Ibaraki 305-8686, Japan. (e-mail: A.Pellegrineschi@cgiar.org)

Keywords Wheat, transgenic plants, DREB gene, drought tolerance

1. INTRODUCTION

Drought is one of the major causes of yield loss in crops, significantly affecting world food production. Considerable efforts have been devoted during the past decade to improving the level of drought tolerance in wheat and extensive research has been conducted in the areas of breeding, physiology and agronomy. To date, reasonable progress has been achieved through conventional breeding (Trethowan, 2001). However this approach is slow, time consuming, and due to the genetic complexity of the wheat plant's response to water-limited conditions, potential for further progress is uncertain. Recently, several candidate genes have been identified that could improve a plant's response under water-limited conditions (Soderman et al., 2002). These genes can be divided in three classes; firstly, genes where very strong evidence already has been published demonstrating their significant role under abiotic stresses conditions; secondly, genes that have been identified as of interest but still require further evaluation (Liu et al., 1998); and thirdly, those genes that have not been evaluated in a plant, and therefore still need to be confirmed under experimental conditions. Of particular interest are the DREB genes, identified in *Arabidopsis* and rice, which have been reported to increase the level of tolerance to abiotic stresses in *Arabidopsis*.

In this study we have introduced the DREB1A gene identified from *Arabidopsis*, and involved in the plant response to several abiotic stresses. Our preliminary experiments conducted with the DREB1A gene demonstrated that when introduced in wheat, the transformed T_1 plants survived a short and intensive water stress and the plantlet stage, while the controls were completely dry.

I. K. Vasil (ed.), Plant Biotechnology 2002 and Beyond, 183-187.

2. MATERIALS AND METHODS

Two plasmids were used in the co-transformation experiments. The plasmid pACH25 contained the *gus*A reporter gene and the *bar* selectable marker (conferring resistance to PPT), with each gene under the control of the maize ubiquitin promoter. The construct containing the DREB1A cDNA, was cloned into the *Eco*R1 site of the plasmid pBIG. The rd29 promoter was cloned in the *Hin*d III site of the plasmid. This plasmid was predicted to give the complete DREB1A transcript only under stress conditions.

Microprojectile-mediated transformation of wheat was carried out according to the procedure described in Pellegrineschi et al. (2002a). Immature embryos 0.8–1 mm long were selected and placed on an E3 medium supplemented with 15% (w/v) mannitol for about 8 h before bombardment. Bombardment was carried out with the Biolistic PDC-1000/He instrument (Bio-Rad). Fifty microliters of gold particles (60 $mg \cdot ml^{-1}$ at 1:1 ratio of 1.0 and 0.6 μm diameter gold particles) were coated with plasmid DNA carrying the DREB1A plasmid, and the selection marker plasmid DNA (pACH25) in a ratio of 3:1 (wt/wt) and accelerated with a helium pressure of 900 psi. After the bombardment, the embryos were transferred to E3 medium, supplemented with 2.5 mg 2,4 D for somatic embryo induction. Twenty days later, the regenerating embryos were transferred to E3 selection media containing 5 $mg \cdot l^{-1}$ glufosinate ammonium and allowed to grow in a growth chamber at 26°C for 30 days. Plantlets resistant to glufosinate ammonium were transferred again to E3 selection medium and allowed to grow as described by Pellegrineschi et al. (2002b) After shoots had reached a height of 1–3 cm, the plantlets were transferred to rooting media containing MS medium (Murashige and Skoog, 1962) plus 5 $mg \cdot l^{-1}$ glufosinate ammonium, (Sigma). After 2 weeks, the plantlets were transferred to soil and grown in a greenhouse to maturity.

Total genomic DNA was isolated from 1 g fresh weight leaf material using the Nucleon Phytopure Plant DNA Extraction Kit, according to the manufacturer's protocol (Amersham Life Sciences). All the Basta™ resistant plants were analyzed by PCR for the presence of the *bar* and *DREB1A* genes. The primers used to detect the *bar* gene were: forward primer 5'-GTCTGCACCATCGTCAACC-3' and reverse primer 5'-GAAGTCCAGCTGCCAGAAAC-3', and for the *rd29A* promoter: forward primer 5'-AAGCTTGCCATAGATGCAATTAATC-3' and reverse primer 5'-AGCTTTTGGAAAGATTTTTTCTTTCCAA-3' as described in Kasuga et al. (1999). The PCR reactions were carried out in a total volume of 25 µl, comprising 10 ng of wheat genomic DNA, 50 mM KCl, 10 mM Tris-HCl buffer, pH 8.8, 3 mM $MgCl_2$, 0.1% Triton X-100, 0.24 mM each dNTP, 0.04 units Taq DNA polymerase and 0.16 µmol of each primer. For PCR analyses of both genes, DNA was denatured at 94°C for 1 min, followed by 30 amplification cycles (94°C for 30 sec., 64°C for 2 min., 72°C for 2 min).A 50 μg aliquot of DNA was digested overnight at 37°C with an appropriate restriction enzyme. The digested DNA was fractionated in a 1.0% agarose gel, transferred to a positively-charged nylon membrane (Boehringer Mannheim), and hybridized to digoxigenin (DIG)-dUTP labeled probes, according to the manufacturer's instructions (Boehringer Mannheim). The entire plasmid for transformation, DIG-labeled by nick-translation, was used as a probe. Detection was achieved using the DIG Luminescent Detection Kit (Boehringer Mannheim) and the hybridization signals were observed following exposure to Fuji X-ray film at 37°C for 40 min. Filters were probed first with the *bar* probe, stripped for rehybridization by washing twice for 15 min in 0.2 N NaOH, 0.1% (w/v) SDS at 37°C, and probed with *rd29A*.

Surviving in vitro plantlets were selected for Basta resistance by spraying the plantlets with 0.3% Basta solution at the 5-6 leaf stage. The surviving plantlets were then selected for the presence of the DREB1a gene by PCR, and finally phenotypically selected for tolerance to drought. Drought conditions were generated in the biosafety greenhouse to ensure that the phenotypic evaluations were conducted under controlled conditions. The drought evaluation test was first calibrated by subjecting control plants to various levels of water deficit. It was found that withholding water for 15 days gave the best discrimination. Subsequent drought evaluations using transgenic and nontransgenic control plants were conducted by subjecting 20 plants, selected at the same developmental stage, to this drought regime. Plants were selected from several lines identified from T_1 transgenic plants that showed the presence of the DREB transgene in PCR and Southern blots. The photoperiod was set at 16 h of light and 8 h darkness; temperature between 18 and 30°C; and humidity at 75%.

3. RESULTS AND DISCUSSION

More than 20 independent transformation experiments (bombarded rings) were conducted using Bobwhite SH 98 26 wheat cultivars (Pellegrineschi et al., 2002a). For the preliminary screening of transformed lines, a solution of 0.3% Basta herbicide was sprayed at the 5-6 leaf stage. Molecular analyses for *bar* presence were carried out independently, and 100% of the Basta-resistant plants exhibited *bar*, while only 113/447 were positive to *rd29DREB1A* (data not shown). Southern blot analysis of DNA from a T_1 progeny using the two probes, consisting of coding regions of the DREB1A and *ba*r genes, indicated that the two transgenes were present in their entirety and integrated into the genome of the transgenic plants. Among the 113 lines positive to *DREB*1A, 13 were selected based on their simple segregation patterns of the transgenes: 3:1 in progenies derived from selfed plants and 1:1 in progenies derived from transgenic plants crossed with the controls plants

(Table 1). The expression of the two genes differed when determined by RT-PCR. The presence of mRNA for the *bar* genes, whose expression was driven by the maize ubiquitin promoter, was easily detected in the transgenic plants; transcripts of the DREB1a gene, driven by the RD29a promoter, were detected after 2 days of water stress.

Table 1: Segregation patterns of the 12 lines chosen for the stress tolerance experiments. T = tolerant to drought stress; S = susceptible to drought stress.

Plant number	Segregation ratio of the drought tolerant phenotype in transgenic line crossed with controls (T:S)	Segregation ratio of drought tolerant phenotype in transgenic line crossed with controls (T:S)	Segregation ratio of the DREB1a gene transgenic line crossed with controls (PCR)	Segregation ratio of the DREB1a gene of transgenic line selfed (PCR)
8174	1:1	3:1	1:1	3:1
8195	1:1	3:1	1:1	3:1
8207	3:1	15:1	3:1	15:1
8208	1:1	3:1	1:1	3:1
8209	1:1	3:1	1:1	3:1
8218	1:1	3:1	1:1	3:1
8223	3:1	15:1	3:1	15:1
8230	1:1	3:1	1:1	3:1
8245	1:1	3:1	1:1	3:1
8402	1:1	3:1	1:1	3:1
8406	1:1	3:1	1:1	3:1
8424	1:1	3:1	1:1	3:1

After sowing, a delay in germination was observed in the transgenic lines. Control plants started to germinate after 2-3 days, while the DREB plants showed nonuniform germination under both stressed and fully irrigated conditions. However, no differences related to the presence of the transgene were observed within the transgenic lines during the first days of water stress conditions.

At this stage, no significant differences were observed between the development of the transgenic and control lines. At the fourth-fifth leaf stage, the plants were subjected to water stress by simply withholding water. The control plants and transgenic plants were randomly distributed on the trays. After 10 days without water, the transgenic wheat lines started to show differences in drought tolerance, determined by the visible wilting of the leaves. The control plants began to show drought symptoms (loss of turgor and bleaching of the leaves) after 10 days of stress. Severe symptoms (death of all leaf tissue) were evident in the control samples after 15 days without water. The selected transgenic plants either did not show any symptoms or showed reduced leaf turgor after 10-12 days without water.

4. REFERENCES

Liu Q, Kasuga M, Sakuma Y, , Abe H, Miura S, Yamaguchi-Shinozaki K, and Shinozaki K. (1998) Two transcription factors, DREB1 and DREB2, with an EREBP/AP2 DNA binding domain separate two cellular signal transduction pathways in drought- and low-temperature-responsive gene expression, respectively, in *Arabidopsis*. Plant Cell 10:1391-1406.

Murashige T. and Skoog F (1962) A revised medium for rapid growth and bioassays with tobacco tissue cultures. Physiol. Plant. 15:473-497.

Pellegrineschi A, Noguera L.M., Skovmand S., Brito R.M., Velazquez L., Hernandez R., Warburton M., and Hoisington D. (2002a) Identification of highly transformable wheat genotypes for mass production of fertile transgenic plants. Genome 45:421-430.

Pellegrineschi A., Brito R.M., Velazquez L., Noguera L.M., Pfeiffer W., McLean S., and Hoisington D. (2002b) The effect of pretreatment with mild heat and drought stresses on the explant and biolistic transformation frequency of three durum wheat cultivars. Plant Cell Rep. 20:955-960.

Soderman EM, Brocard IM, Lynch TJ, and Finkelstein RR (2002) Regulation and function of the Arabidopsis ABA-insensitive4 gene in seed and abscisic acid response signaling networks. Plant Physiology 124:1752-1765.

Trethowan, R.M., Crossa, J., Ginkel, M. van and Rajaram., S. 2001.
Relationships among Bread Wheat International Yield Testing Locations in Dry Areas. Crop Sci. 41:1461 - 1469.

ARABIDOPSIS β-AMYLASE INDUCTION DURING TEMPERATURE STRESS

Fatma Kaplan and Charles L. Guy
Plant Molecular and Cellular Biology Program and Department of Environmental Horticulture, University of Florida, Gainesville, Florida 32611. (e-mail: FKaplan@mail.ifas.ufl.edu)

Keywords beta-amylase, temperature shock, starch, maltose

1. INTRODUCTION

Beta-amylase activity is found everywhere starch is present in plants such as in seeds, tubers, and leaves (Avigad and Dey, 1997). It is an exoamylase that hydrolyses α 1,4 glycosidic linkages of polyglucan chains at the non-reducing end to produce maltose (4-O-α-D-Glucopyranosyl-β-D-glucose). The reducing glucose of the maltose disaccharide product of the reaction is in the β-form, hence, the name β-amylase. The primary physiological role of β-amylase is considered to be in starch breakdown. This is somewhat controversial because this enzyme is considered by some to be unable to attack native starch without prior digestion by other amylolytic enzymes like α-amylase (Beck and Ziegler, 1989).

Compatible solutes (osmoprotectants, osmolytes) are low molecular weight organic molecules that accumulate under stress conditions. They are thought to stabilize proteins and membranes and contribute to cell osmotic pressure under stress conditions (Yancey et al., 1982). There are three general types of osmoprotectants: methylamines (betaines), amino acids (proline), and polyols (glycerol, sucrose) (Yancey et al., 1982). The protective effect of polyols and sugars depend on their concentration and molecular weight. Among carbohydrates, trisaccharides are more effective than disaccharides and disaccharides are generally more effective than monosaccharides as protective agents (Santarius, 1973).

We are interested in why some β-amylases are transcriptionally induced in response to acute temperature shock, and what benefits the product of the reaction may provide to the plant during stress.

I. K. Vasil (ed.), Plant Biotechnology 2002 and Beyond, 189-191.

2. RESULTS

Arabidopsis thailana cultivar Columbia seedlings were grown in controlled environment at 21°C $\pm$ 1 under a light irradiance of 25-30 μmol m^{-2} sec^{-1} photosynthetically active radiation with a 15/9 hr light/dark cycle. To better understand how β-amylase transcription and function is influenced by temperature, a step up/step down temperature response profile was conducted where sixteen day-old Arabidopsis plants were treated with the following temperatures: 20, 25, 35, and 45°C for 1 hr for heat shock and 20, 15, 10, and 0°C for 12 hr for cold shock. Plants kept at 20°C served as the control for both treatments. Three replications were done for each treatment. Leaf samples were taken for RT-PCR and sugar analysis.

RT-PCR analysis: β-amylase7 (β-amy7; AT3G23920) and β-amylase8 (β-amy8; AT4G17090), both encode plastid-localized enzymes, were found to be induced under heat and cold shock (Figure 1A and B). Beta-amy7 and β-amy8 showed the highest expression at 45 and 0°C, respectively. Other enzymes involved in starch degradation like α-amylase, phosphorylase b, and isoamylase, were repressed or unchanged under heat and cold shock temperatures except for a modest α-amylase induction at 10°C (not shown).

Carbohydrate Analysis: Soluble sugars were extracted with hot 80% aqueous ethanol from 10 mg (dry-weight) leaf tissue, and passed through an Amberlite® ion exchange column to remove charged molecules. Neutral sugars and sugar alcohols were separated in a diluted sodium hydroxide gradient (10, 80, and 140 mM) using a Dionex HPLC PA10® column at the University of Florida Glycobiology Core lab. Maltose, maltitol, sucrose, glucose and fructose content of samples were determined.

Maltose accumulation (Table 1) showed the same profile as the steady state β-amy8 transcript accumulation under cold stress with the highest accumulation at 0°C for 12 hr. Under heat stress, while β-amy7 transcript increased, maltose content was not changed from control levels (Table 1). No maltitol, a sugar alcohol of maltose, was detected under any condition. Sucrose, a compatible solute of the cytosol, was accumulated gradually under heat and cold shock, as was glucose and fructose (Table 1); both components of sucrose.

In conclusion, induction of beta-amylase expression under a variety of heat (from 20 to 45°C for 1 hr) and cold (from 20 to 0°C for 12 hr) shock temperatures appears to be a specific temperature stress response. Other genes involved in the starch degradation pathway were either unchanged or repressed under heat and cold shock. At least for cold shock, RT-PCR and carbohydrate analysis suggest a connection between the induction of beta-amylase expression and maltose accumulation. The purpose of this maltose accumulation is under investigation.

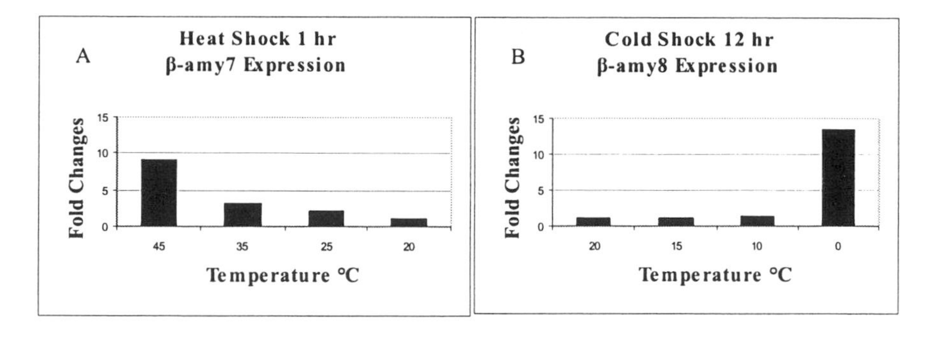

Figure1: Fold changes in (A) β-amy7 (AT3G23920) and (B) β-amy8 (AT4G17090) gene expression using RT-PCR under heat and cold shock. Three replications are done for each treatment. Control 20°C is 1-fold.

Table 1. Changes in soluble sugar content under heat and cold shock

	Heat Shock 1 hr Temperature at °C				Cold Shock 12 hr Temperature at °C		
	45	35	25	20	15	10	0
Sugars	nmol/mg DW				nmol/mg DW		
Fructose	0.11±0.01	0.25±0.06	0.14±0.03	0.27±0.22	0.09±0.01	0.92±0.80	1.83±0.70
Glucose	0.17±0.05	0.39±0.09	0.18±0.12	0.25±0.14	0.23±0.03	0.91±0.91	1.66±0.17
Sucrose	0.83±0.20	0.78+0.25	0.40*	0.47±0.33	0.43±0.12	0.67±0.29	1.80±0.68
Maltose	0.02±0.01	0.06±0.04	NA	0.03±0.02	0.02*	0.03±0.01	0.69±0.63
Maltitol	ND	ND	ND	ND	ND	ND	ND

ND, Not Detected; NA, Not Available; DW, Dry Weight

Three replications were done for each treatment. Control 20 °C included 6 replications.

*Only one replication was available.

3. REFERENCES

Avigad G., and P. M. Dey. 1997. Carbohydrate metabolism: storage carbohydrates. In Plant Biochemistry. P. M. Dey and J. B. Harborne (eds). Academic Press. Pp. 143-204.

Beck E., and P. Ziegler. 1989. Biosynthesis and degradation of starch in higher plants. Annu. Rev. Plant Physiol. Plant Mol. Biol. 40: 95-117.

Santarius K. A. 1973. The protective effects of sugars on chloroplast membranes during temperature and water stress and its relationship to frost, desiccation and heat resistance. Planta 113: 105-114.

Yancey, P. H., Clark, M. E., Hand, S. C., Bowlus, R. D., and Somero, G. N. 1982. Living with water stress: evolution of osmolyte systems. Science 217: 1214-1222.

EXPRESSION OF CEREAL PEROXIDASE AND OXALATE OXIDASE GENES IN TOBACCO RESULTS IN ALTERATIONS IN PLANT DEVELOPMENT AND PROGRAMMED CELL DEATH IN CELL CULTURES

Philip J. Dix[1], Emma Burbridge[1], Søren K. Rassmussen[2] and Paul F. McCabe[3]
[1]Biology Department, National University of Ireland Maynooth, Co. Kildare, Ireland (e-mail: phil.dix@may.ie);
[2]Environmental Science and Technology Department, Mil-301, Risø National Laboratory, DK-4000 Roskilde, Denmark (e-mail: soren.rasmussen@risoe.dk);
[3]Botany Department, University College Dublin, Ireland (e-mail: paul.mccabe@ucd.ie)

Keywords Peroxidase, oxalate oxidase, hydrogen peroxide, lignin, xylem, cell death

1. INTRODUCTION

Peroxidases and oxalate oxidases are enzymes associated with the production and scavenging of hydrogen peroxide in plants. Peroxidases can influence developmental processes, including auxin metabolism (Normanly, 1997) and lignification (Mader and Amberg-Fisher, 1982), as well as responses to biotic (Thordal-Christensen et al., 1992) and abiotic stresses (Jansen et al., 2001). Oxalate oxidase (germin) is although believed to have several roles in plant development and response to pathogens (Bernier and Berna, 2001), and has recently been shown to afford resistance to insect predation (Ramputh et al., 2002). Hydrogen peroxide (H_2O_2), levels might reasonably be expected to be altered by modifications in activities of peroxidase (H_2O_2-utilising) and oxalate oxidase (H_2O_2-generating) enzymes. H_2O_2 itself may act as a signaling molecule, triggering programmed cell death and the production of lignified cells. This may be defence related, as in the "oxidative burst" associated with hypersensitive response to pathogens (Levine et al., 1994), or connected to normal developmental processes, such as xylogenesis. The latter involves both programmed cell death, and the action of peroxidase in the cross-linking of lignin precursors.

The current investigation was initiated to further dissect the roles of oxidative enzymes and hydrogen peroxide in plant developmental and stress responses involving programmed cell death. The study involves characterisation of

I. K. Vasil (ed.), Plant Biotechnology 2002 and Beyond, 193-196.

transgenic tobacco plants expressing elevated levels of either a barley peroxidase, or a wheat oxalate oxidase or both. Plants expressing the oxalate oxidase gene (nc5 plants) were obtained from F. Bernier and have been described previously (Berna and Bernier, 1997). Plants expressing thc barley peroxidase gene *prx*8, were obtained by *Agrobacterium*-mediated transformation of wild type, or nc5 tobacco plants. Details of the production and analysis of these plants are to be described elsewhere. The native genes used for the investigations include targeting signals believed to lead to the preferential secretion of the recombinant protein into the cell wall. Both genes are under control of a constitutive, CaMV-35S, promoter.

2. LIGNIN, XYLEM AND PLANT GROWTH

Apoplastic fluid was assayed for ability to oxidase the lignin precursor syringaldazine. nc5 (oxalate oxidase) plants showed no difference when compared to wild-type tobacco, whereas there was a greater than doubling of syringaldazine oxidation in prx8 (peroxidase) plants. Prx8/nc5 plants gave intermediate values. These results suggest that apoplastic barley peroxidase in transgenic tobacco is can increase oxidation of lignin precursors, and therefore might influence lignification and xylem production. This finding is supported by measurements of extractable lignin in the stem. Prx8 plants produce significantly more lignin than wild type plants, while the expression of oxalate oxidase in nc5 plants has no effect. However, in contrast to the syringaldazine oxidation results, there was no reduction in the peroxidase-induced increase in lignin caused by the simultaneous expression of oxalate oxidase.

Peroxidase expression on its own (prx8 plants) leads to an increase in the number of xylem vessels in stem sections, but a decrease in their mean diameter. The overall width of the band of xylem tissue did not change significantly. Prx8/nc5 plants give a similar increase in vessel number, without the reduction in diameter. Therefore there is a significant increase in the width of the xylem band.

Alterations in the vascular network of plants could influence plant nutrition and therefore growth. Neither peroxidase nor oxalate oxidase expression individually influenced the growth of tobacco plants, monitored over a 35 day period. However, they appeared to exhibit a concerted effect. Prx8/nc5 plants showed a 20% reduction in plant height, compared to wild-type plants, or single transformants.

3. PROGRAMMED CELL DEATH

Elevated levels of H_2O_2 in oxalate oxidase expressing plants were confirmed by incubation of leaf discs with tetramethylbenzidine (TMB) and peroxidase. Blue coloration is an indication of the presence of H_2O_2 and was found in nc5, but not wild-type discs. The effect of this H_2O_2 on cell death is supported by Evan's blue staining which reveals a much higher incidence of dead cells at the periphery of nc5 leaf discs. Very low efficiency of *Agrobacterium*-mediated transformation of nc5 leaf discs may also be linked to this high cell mortality at potential infection sites. Other observations supporting the role of oxalate oxidase in plant stress include the elevated levels of ascorbate oxidase activity found in the leaves of nc5 plants, which is further increased by the simultaneous expression of barley peroxidase.

Programmed cell death was investigated in cell suspension cultures initiated from the different tobacco lines. Cells succumbing to programmed cell death (PCD) are readily identified, and distinguished from necrotic cells, by their distinctive morphology. Cell cultures were subjected to a range of temperatures and PCD determined. The critical temperature, at which differences between the lines could be clearly detected, was 45°C. At this temperature almost 60% of the prx8 cells died though PCD, almost double the value for wild type cells. Expression of oxalate oxidase (nc5) plants also resulted in a small increase in PCD. However, when combined with peroxidase, (prx/nc5) it almost completely negated the effect of peroxidase expression alone.

4. CONCLUSION

The results show that modification of levels of enzymes which generate or utilise/detoxify H_2O_2 can influence a raft of developmental and defence responses in plant cells, particularly in induction of PCD, and the production of lignified cells involving PCD. Further analysis of the structural and physiological consequences of expression of these enzymes should facilitate a more thorough dissection of the multiple biological activities of H_2O_2, as a toxic metabolite, as a signaling molecule, and as a reactant in the synthesis of key biopolymers.

5. ACKNOWLEDGEMENTS

This work was carried out through a basic research award from Enterprise Ireland.

6. REFERENCES

Berna A. and Bernier F. 1997. Regulated expression of a wheat germin gene in tobacco: oxalate oxidase activity and apoplastic localization of the heterologous protein. Plant Mol. Biol. 33: 417-429.

Bernier F. and A. Berna. (2001). Germins and germin-like proteins. Plant do-all proteins. But what do they do exactly? Plant Physiol. Biochem. 39: 545-554.

Jansen M.A.K., van den Noort R.E. Adillah M.Y., Prinsen E., Lagrimini L.M. and R.N.F. Thorneley. 2001. Phenol-oxidising peroxidases contribute to the protection of plants from ultraviolet radiation stress. Plant Physiol. 126: 1012-1023.

Levine A., Tenhaken R., Dixon R. and C. Lamb. 1994. H_2O_2 from the oxidative burst orchestrates the plant hypersensitive disease resistance response. Cell 79: 583-593.

Mader M. and V. Amberg-Fisher. 1982. Role of peroxidase in lignification of tobacco cells. Plant Physiol. 70: 1128-1131.

Normanly J. 1997. Auxin Metabolism. Physiol. Plant. 100: 431-442.

Ramputh A.I., Arnason J.T., Cass L. and J.A. Simmonds. 2002. Reduced herbivory of the European corn borer (*Ostrinia nubilalis*) on corn transformed with germin, a wheat oxalate oxidase gene. Plant Science 162: 431-440.

Thordal-Christensen H., Brandt J., Cho B.H., Rasmussen S.K., Gregersen, P.L., Smedegaard-Petersen, V. and D.B. Collinge. 1992. cDNA cloning and characterization of two barley peroxidase transcripts induced preferentially by the powdery mildew fungus *Erysiphe graminis.* Physiol. Mol. Plant Pathol. 40: 395-409.

FUNCTIONAL GENOMICS OF RICE BY T-DNA TAGGING

Gynheung An
Pohang University of Science and Technology, Pohang 790-784, Korea (e-mail: genean@postech.ac.kr)

Keywords Gene tagging, GUS, insertional mutants, reverse genetics, rice, T-DNA

1. INTRODUCTION

Rice has become a model for monocot plants because of the accumulation of molecular information for this species, its efficiency in transformation, small (430-Mb) genome, and economic importance. Its gene content is comparable to that of other grasses, such as wheat, maize, barley, rye, and sorghum. Because of the conservation of gene sequences and order among cereals, the structural and functional analyses of rice have broad practical implications for these other economically important crops (Gale and Devos, 1998). Thus, genomic information for rice should be widely applicable when developing products and technologies from other cereal crops.

As the rice genome sequence nears completion, a large effort is in progress for determining the functions of its genes. One of the most popular techniques for deducing gene function is the use of knockout plants (Jeon and An, 2000). Among the various methods, Ac/Ds, Tos17, and T-DNA are most frequently used for generating pools of insertional mutant lines.

2. T-DNA INSERTIONAL MUTAGENESIS

In the past four years, we have produced approximately 100,000 transgenic rice lines that carry a T-DNA insertion (Jeon et al., 2000). This T-DNA is inserted into rice chromosomes at an average of 1.4 genetic loci per plant. Assuming that these insertions are random, the probability of finding one within a given gene from our T-DNA insertional lines is over 60%. Rice has an endogenous retrotransposon, Tos17, that becomes active during tissue culture (Hirochika et al., 1996). Our T-DNA insertional mutants have been raised via *Agrobacterium*-mediated transformation of embryo cultures, and

I. K. Vasil (ed.), Plant Biotechnology 2002 and Beyond, 197-200.

carry an average of four new copies of Tos17. Therefore, this element should also generate insertional mutations.

3. GENE TRAPS WITH THE *GUS* AND *GFP* REPORTERS

The binary T-DNA vectors used for generating insertions contain the promoterless glucuronidase (*gus*) reporter gene next to the right border. These gene-trap vectors are designed to detect gene fusion between *gus* and the endogenous gene that is tagged by T-DNA. To increase gene-trap efficiency, a synthetic intron, carrying triple splice donors and acceptors, is placed between the T-DNA right border and the *gus* reporter. Therefore, an insertion of T-DNA, in the proper orientation, into either exons or introns can generate translational fusions. GUS assays of the developing seedlings and flowers have shown that 2 to 5% of the transgenic plants are GUS-positive in the tested organs. Some of them also show tissue-specific expression patterns.

We have also used the green fluorescence protein (*gfp*) gene as a reporter. Because chlorophyll interferes with the GFP assay, we have, so far, examined only mature seeds and etiolated seedlings. Gene-trap efficiency has been about 1 to 2% for our tested organs. Because the GFP assay can be conducted without destroying the tissues, the *gfp*-trap system is useful in identifying genes inducible by environmental or chemical stimuli. It will also work well for examining the cellular locations of fusion proteins.

Analysis of the T-DNA flanking sequences of trapped genes has resulted in identification of knockout genes, which are disrupted by T-DNA insertion. Some of these lines have shown co-segregation of T-DNA and a mutant phenotype. Some of the most frequently found mutants are those affecting color, including albino, chlorina, and pale-green appearances. For example, T-DNA insertion into a chlorophyll synthase gene causes a pale-green phenotype. Likewise, disruption of the magnesium chelatase gene by T-DNA or Tos17 results in seedling lethal phenotypes due to a lack of chlorophyll and underdeveloped chloroplasts. Fertility is often affected by mutations as well. Dwarf mutants also are frequently found, such as those caused by T-DNA insertion in OsKn3, a homeobox gene, which results in semi-dwarf phenotypes similar to those observed from d6 mutants. In addition, one mutation that was due to T-DNA insertion has been found to be highly homologous to *Arabidopsis* CER, a gene necessary for biosynthesis of very long-chain fatty acids.

4. REVERSE GENETICS APPROACHES

DNA pools have been prepared from 20,000 lines, and additional pools are being developed. Each pool contains DNAs from 50 to 100 lines; each super pool comprises 500 lines. These pools can be screened for T-DNA insertion into the gene of interest, using primers located within the gene and the T-DNA, and can also be used for screening Tos17 insertions. When sufficient homology is evident, insertions of T-DNA or Tos17 into a group of genes can be screened simultaneously by designing degenerated primers. Because PCR efficiency decreases with larger amplified DNA fragments, more than one primer is necessary when the gene is longer than 3 kb. From our T-DNA pools, we have identified T-DNA tags within several MADS box genes. Tags in other genes also are being screened, and a DNA pool-screening service is available.

To more efficiently use our lines, we are systematically determining the flanking sequences of each T-DNA insertion. We have employed the inverse PCR method, which uses the PstI site and other restriction enzymes located within the T-DNA. After enzyme-digestion and self-ligation, nested inverse PCR is performed to amplify the genomic sequence flanking the T-DNA. It is important to select a restriction enzyme that cuts once in the T-DNA but not in the vector backbone. If a restriction site is found within the vector backbone, amplification frequency of that backbone is high, due to the frequent insertion of the entire vector into the plant chromosome. Tandem T-DNA repeats, which cause a problem when multiple T-DNAs are inserted into a single locus, can be identified by the size of their amplified bands and, thus, avoided. Flanking-sequence information has been obtained from about 70% of our lines, using three different restriction enzymes. Analysis of the tagged sequences (via public databases) has shown that more than 40% of the T-DNA is inserted within a gene. Assuming that rice has 30,000 genes (with an average length of 3 kb) and that the insertions are random, about 20% of the T-DNA should be located within a gene. However, our observations have suggested that either the number of genes in the rice chromosomes is >30,000 or that T-DNA prefers gene-rich areas. A consortium has been organized to determine the T-DNA flanking sequence, which data will be made available to the public.

5. ACTIVATION TAGGING

T-DNA insertion does not always result in an obvious mutant phenotype, largely because of gene redundancy. Plant chromosomes are known to contain a large number of duplicated genes. Therefore, knocking out only one may have little effect. In these cases, increased gene expression often causes a change in phenotype. This can be achieved through gene activation, i.e., by placing the gene under an enhancer element. We have found that the 4 x 35S enhancer sequence is effective with rice genes. Random insertions of this enhancer element via *Agrobacterium*-mediated transformation generates activation of the surrounding genes. It can also activate the *gus* reporter gene located within the same T-DNA, with GUS-trap efficiency increasing at least two-fold when the 4 x 35S enhancer is used. We have generated 50,000 independent rice plants using an activation-tagging vector that carries this enhancer and the *gus* reporter.

6. REFERENCES

Gale M.D., and Devos K.M. 1998. Comparative genetics in the grasses. Proc. Natl. Acad. Sci. USA 95: 1971-1974.

Hirochika H, Sugimoto K, Otsuki Y, Tsugawa H, Kanda M. 1996 Retrotransposons of rice involved in mutations induced by tissue culture. Proc. Natl. Acad. Sci. USA 93: 7783-7788.

Jeon J.S. and An G. 2001. Gene tagging in rice: A high throughput system for functional genomics. Plant Science 161: 211-219.

Jeon J.S., Lee S., Jung K.H., et al. 2000. T-DNA insertional mutagenesis for functional genomics in rice. Plant J. 22: 561-570.

A PLANT APPROACH TO SYSTEMS BIOLOGY

Keith R. Davis
Paradigm Genetics, Inc., 108 Alexander Drive, Research Triangle Park, NC 27709, USA
(e-mail: kdavis@paragen.com)

Keywords Functional genomics, Arabidopsis, phenomics, metabolomics, data integration

1. INTRODUCTION

Large-scale sequencing of complete genomes and the establishment of large compilations of sequence information derived from Expressed Sequence Tags (ESTs) and full-length cDNAs have provided an overwhelming amount of information concerning the physical structure for thousands of specific genes. This initial success in structural genomics has required development of new approaches to efficiently determine the function of the large number of genes for which little or now functional information is available. In the field of plant research, the completion of the genome sequence for the model plant *Arabidopsis thaliana* (Bevan et al., 2001) triggered a major transition from structural genomics to functional genomics in higher plants. The efficient determination of gene function requires the development of robust functional analysis platforms that allow the reproducible analysis of large numbers of genes in a rapid and high throughput manner.

Paradigm Genetics, Inc. has established a plant functional analysis platform that is based on integrated platforms that can be used to determine the effects of specific perturbations in gene expression on morphological, biochemical, and molecular phenotypes. These large unified data sets allow correlations to be made between specific changes in gene expression and specific metabolites with changes in plant development. This GeneFunction Factory™ provides a powerful tool for identifying genes that can be utilized for improving plants via genetic engineering.

2. ARABIDOPSIS FUNCTIONAL GENOMICS

A key feature of the Arabidopsis functional genomics program at Paradigm is the focus on high throughput methods for measuring large numbers of plant characteristics, including phenotypic traits, gene expression profiles, and

I. K. Vasil (ed.), Plant Biotechnology 2002 and Beyond, 201-204.

metabolite profiles. A major emphasis has been placed on developing rigorous and reproducible systems for growing and analyzing plants as well as informatics tools for evaluating the large amount of data generated by these studies. The current Paradigm functional analysis platform is suitable for analyzing any type of transgenic plant or mutant with the appropriate wild-type control as well as plants treated with specific chemicals. Gene function can be investigated under standard growth conditions or under specific environmental or nutritional stress conditions.

The foundation of the Paradigm Arabidopsis functional analysis platform is a detailed phenotypic analysis (phenomics) that uses a data collection strategy based on a series of distinct growth stages that describe the entire Arabidopsis life cycle. This growth-stage-dependent methodology is based on the existing BCCH scale previously developed by crop breeders (Lancashire et al., 1991). Phenomics data is collected using two distinct platforms; an early developmental analysis of plants grown for two weeks on a defined media in petri plates and a soil-based system that allows development to monitored for the entire life cycle. Data collection in these two phenomics platforms is a two-step process that involves a set of core measurements taken throughout the entire analysis period that allows a growth stage to be determined. At specific growth stages representing key developmental landmarks, e.g., the transition to flowering, more detailed measurements are then triggered. These measurements include a number of quantitative measurements that are either taken manually or are derived from computerized images. In addition, destructive harvests are made to provide tissues for biomass determinations and molecular analyses using biochemical profiling (BCP) of metabolites and gene expression profiling (GEP) using high-density DNA microarrays . A key component of this functional analysis platform is that all aspects of the process are integrated using a laboratory information management system (LIMS) that has been designed specifically for this platform. This LIMS system greatly increases the throughput of the platform while insuring that high quality data are easily deposited into a comprehensive database that allows efficient mining of these complex data sets.

2.1. Arabidopsis Phenomics

The Paradigm Arabidopsis phenomics platforms allow the characterization of over 75 different traits in a highly reproducible manner. A detailed description of the Arabidopsis phenomics platform and its utility for uncovering subtle phenotypes in mutant plants has been published (Boyes et al., 2001). In addition to quantitative measurements of specific organs, such as flower size, number of siliques, etc., the phenomics platforms allows the evaluation of developmental progression throughout the life cycle. This allows the facile detection of changes in development that does not result in any detectable changes in organ size or shape. Examples of growth-stage changes observed in a group of Arabidopsis mutants are shown in Figure 1.

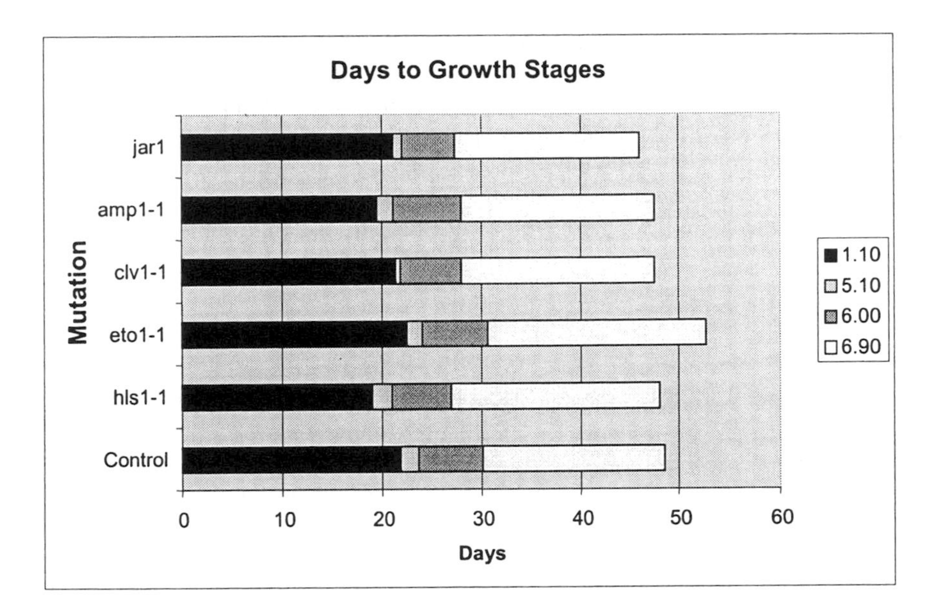

Figure 1. Growth stage progression of Arabidopsis mutants analyzed using the soil phenomics platform. Growth stage 1.10, ten rosette leaf stage; growth stage 5.10, first flower bud visible; growth stage 6.00, first flower open; growth stage 6.90, flower production complete.

A good example of the utility of the phenomics platform and analysis of the growth stage progression data in Figure 1 is that of *jar1*, a jasmonic acid resistant mutant. *jar1* plants are not significantly different from wild-type plants with respect to organ morphology, however, there are indeed subtle changes in growth stage progression throughout the transition to flowering. This phenotype was not previously observed by the more casual inspection that typically is associated with mutant evaluation.

2.2. Biochemical And Gene Expression Profiling

In addition to the detailed phenotypic characterization, the gene function analysis platform includes comprehensive BCP and GEP analyses. The BCP platform is designed to detect a wide variety of metabolites. The core analysis platforms for BCP include GC/MS, LC/MS and CPI/MS, all of which utilize time of flight (TOF) mass spectrometers. Our current platform can detect at least 300 different low molecular weight compounds in a standard Arabidopsis extract, including polar and non-polar organic species as well as metal ions.

The GEP platform is the newest addition to Paradigm's GeneFunction Factory™. This platform is based on the use of custom DNA microarrays developed in cooperation with Agilent Technologies. Current efforts are focussed on the use of a 16,000-feature array. This GEP platform provides broad coverage of the Arabidopsis genome and adds significantly to the functional analysis platform. Assuming that the same tissues that are subjected to BCP are also analyzed by GEP, over 60,000 additional data points per construct are collected. Thus, a standard analysis using all the plant gene function analysis platforms results in the collection of over 65,000 individual data points for each gene analyzed.

3. CONCLUSIONS

The genomics field has clearly entered a phase where the focus is on converting the large amount of available structural genomics information into useful functional information. Model plant systems such as Arabidopsis and rice will play a major role in the elucidation of gene function in higher eukaryotes. The results obtained using these model systems will have a major impact on understanding the molecular basis for plant processes and have direct application to the improvement of crop plants and the development of novel plant products.

4. REFERENCES

Bevan M., K. Mayer, O. White, J. A. Eisen, D. Preuss, T. Bureau, S. L. Salzberg, and H. W. Mewes. 2001. Sequence and analysis of the Arabidopsis genome. Curr. Opin. Plant Biol. 4(2):105-110.

Boyes, D.C., A. M. Zayed, R. Ascenzi, A. J. McCaskill, N.E. Hoffman, K. R. Davis, and J. Gorlach. 2001. Growth stage-based phenotypic analysis of Arabidopsis: A model for high throughput functional genomics in plants. Plant Cell 13:1499-1510.

Lancashire, P. D., H. Bleiholder, T. van der Boom, P. Langeluddeke, R. Stauss, E. Weber, and A. Witzenberger. 1991. A uniform decimal code for growth stages of crops and weeds. Ann. Appl. Biol. 119:561-601.

PHYSCOMITRELLA PATENS AS A NOVEL TOOL FOR PLANT FUNCTIONAL GENOMICS

Ralf Reski
Freiburg University, Plant Biotechnology, Sonnenstr. 5, D-79104 Freiburg, Germany (email: ralf.reski@biologie.uni-freiburg.de)

Keywords EST, gene targeting, homologous recombination, moss

1. INTRODUCTION

The post-genomic era relies on a variety of high-throughput approaches for the establishment of stringent gene/function correlations (e.g. Holtorf et al., 2002a). One of the most powerful tools is reverse genetics, where the targeted disruption of a specific ORF leads to a loss-of-function mutation, which in turn establishes the biological role of this specific gene. Unlike all other land plants analyzed so far, the moss *Physcomitrella patens* exhibits a very high rate of homologous recombination in its nuclear DNA, making gene targeting approaches in this plant as convenient as in yeast or in ES cells of mice (Reski, 1998, 1999). We, therefore, have developed Physcomitrella as a novel tool in plant functional genomics.

2. DEVELOPMENT AND CELL BIOLOGY

In mosses the dominating generation is the haploid gametophyte, which facilitates analysis of loss-of-function mutations without complex back-crosses. Futhermore, several tissues are only one cell-layer thick, making them unique targets for advanced cell biological studies. Thus, we have identified the first organelle dividing protein in any eukaryote (Strepp et al., 1998), provided evidence for a plastoskeleton (Kiessling et al., 2000; Reski, 2002), and identified a novel mechanism for dual, but distinct targeting of nuclear-encoded proteins into mitochondria and plastids (Richter et al., 2002). In order to better understand the role of cell walls during plant growth, we expansins were analyzed by targeted knock-out (Schipper et al., 2002). In order to study embryo development in *Physcomitrella* we developed a highly standadized protocol for the induction of sexual reproduction in this species (Hohc ct al., 2002a).

I. K. Vasil (ed.), Plant Biotechnology 2002 and Beyond, 205-209.

3. PHYSIOLOGY, TISSUE CULTURE AND TRANSFORMATION

Physcomitrella is far more robust against abiotic stresses like drought and salt than many seed plants including *Arabidopsis*, making the moss a valuable source for the identification of novel resistance genes (Frank and Reski, 2002). Moreover, Physcomitrella is rich in secondary metabolites and in very long and unsaturated fatty acids promoting human health. Subsequently, a variety of novel genes like elongases and desaturases were identified (Girke et al., 1998; Zank et al., 2002; Sperling et al., 2002). Likewise, functional knockout of the *apr* gene revived an old route of sulfate assimilation in plants (Koprivova et al., 2002).

Most transformation protocols of Physcomitrella rely on direct DNA transfer to protoplasts. For large-scale plant production a bioreactor regime was established that allows a tight control of growth and differentiation (Hohe et al., 2002b) which was subsequently optimized for mass production of protoplasts (Hohe and Reski, 2002). Protoplast regeneration was studied in detail (Bhatla et al., 2002) and protocols were optimized for efficient regeneration (Schween et al., 2002a). Furthermore, the effects of different parameters on transformation as well as on gene targeting were elucidated (Hohe et al., 2002c), leading to very efficient gene targeting (Egener et al., 2002a).

4. TRANSCRIPTOME AND SATURATED MUTANT COLLECTION

The Physcomitrella genome is about three times as big as the Arabidopsis genome and is distributed on 27 chromosomes (Reski et al., 1994). Small EST-sequencing efforts (Reski et al., 1998; Machuka et al., 1999) revealed that Physcomitrella is a rich source of novel genes although the majority of genes are significantly conserved between moss and seed plants. In a more thorough approach normalized and subtracted cDNA collections covering all steps of the life cycle have been sequenced. About 110,000 ESTs were generated and from that a clustered database was constructed covering more than 95% of the transcriptome (~ 25,000 genes) matching ~ 50% of the ~ 26,000 Arabidopsis genes (Rensing et al., 2002).

As a prerequisite for a large-scale gene/function correlation study, we are establishing a collection of transgenics with insertion mutations in most expressed genes. Low-redundancy moss cDNA libraries are mutagenized in *E. coli* and used for plant transformation. These disruption constructs are expected to target preferentially expressed genes. Among the first over 20,000 transgenics ~ 16% deviated from wild-type in a variety of developmental, morphological and physiological characteristics (Egener et al., 2002b), exceeding by far the respective figures from similar *Arabidopsis* approaches. Our constant production capacity is 160 transgenics per day, 800 per week. For optimization different promoters were tested (Holtorf et al., 2002b), and efficient protocols for DNA isolation (Schlink and Reski, 2002) and high-throughput PCR were developed (Schween et al., 2002b).

Thus, *Physcomitrella* has been established as a novel tool in plant functional genomics, utilizing the unique moss features like homologous recombination and a variety of genes so far not identified in higher plants. Details can be found under www.plant-biotech.net and we are open for the establishment of novel co-operations.

5. ACKNOWLEDGEMENTS

This work was made possible by a variety of dedicated colleagues and co-workers, all cited in the references. Financial support came from BPS, DFG, A.-v.-H.-Stiftung, FCI, MWK and Freiburg University.

6. REFERENCES

Bhatla S.C., J. Kiessling, and R. Reski. 2002. Observation of polarity induction by cytochemical localization of phenylalkylamine-binding receptors in regenerating protoplasts of the moss *Physcomitrella patens*. Protoplasma 219:99-105.

Egener T., A. Hohe, J.M. Lucht, and R. Reski. 2002a. High throughput gene targeting by protoplast transformation of *Physcomitrella patens*. Submitted.

Egener T., J. Granado, M.-C. Guitton, A. Hohe, H. Holtorf, J.M. Lucht, S. Rensing, K. Schlink, J. Schulte, G. Schween, S. Zimmermann, E. Duwenig, B. Rak, and R. Reski. 2002b. High frequency of phenotypic deviations in *Physcomitrella patens* plants transformed with a gene-disruption library. Submitted.

Frank W., and R. Reski. 2002. *Physcomitrella patens* is highly tolerant against drought, salt and osmotic stress. Submitted.

Girke T., H. Schmidt, U. Zähringer, R. Reski, and E. Heinz. 1998. Identification of a novel Δ6-acyl-group desaturase by targeted gene disruption in *Physcomitrella patens*. Plant J. 15:39-48.

Hohe A., and R. Reski. 2002. Optimisation of a bioreactor culture of the moss *Physcomitrella patens* for mass production of protoplasts. Plant Sci., in press.

Hohe A., S.A. Rensing, M. Mildner, D. Lang, and R. Reski. 2002a. Day length and temperature strongly influence sexual reproduction and expression of a novel MADS-box gene in the moss *Physcomitrella patens*. Plant Biology, in press.
Hohe A., E.L. Decker, G. Gorr, G. Schween, and R. Reski. 2002b. Tight control of growth and cell differentiation in photoautotrophically growing moss (*Physcomitrella patens*) bioreactor cultures. Plant Cell Rep., in press.

Hohe A., T. Egener, J. Lucht, H. Holtorf, C. Reinhard, G. Schween, and R. Reski. 2002c. DNA conformation and plant pre-culture influence both, transformation efficiency and gene targeting efficiency in Physcomitrella. Submitted.

Holtorf H., M.-C. Guitton, and R. Reski. 2002a. Plant functional genomics. Naturwissenschaften, online-publication available.

Holtorf H., A. Hohe, H.-L. Wang, M. Jugold, T. Rausch, E. Duwenig, and R. Reski. 2002b. Promoter subfragments of the sugar beet V-type H^+-ATPase subunit c isoform drive the expression of transgenes in the moss *Physcomitrella patens*. Submitted.

Kiessling J., S. Kruse, S.A. Rensing, K. Harter, E.L. Decker, and R. Reski. 2000. Visualization of a cytoskeleton-like FtsZ network in chloroplasts. J. Cell Biol. 151:945-950.

Koprivova A., A.J. Meyer, G. Schween, C. Herschbach, R. Reski, and S. Kopriva. 2002. Functional knockout of the adenosine 5′phosphosulfate reductase gene in *Physcomitrella patens* revives an old route of sulfate assimilation. Submitted.

Machuka J., S. Bashiardes, E. Ruben, K. Spooner, A. Cuming, C. Knight, and D. Cove. 1999. Sequence analysis of expressed sequence tags from an ABA-treated cDNA library identifies stress response genes in the moss *Physcomitrella patens*. Plant Cell Physiol 40:378-387.

Rensing S.A., S. Rombauts, A. Hohe, D. Lang, E. Duwenig, P. Rouze, Y. Van de Peer, and R. Reski. 2002. The transcriptome of the moss *Physcomitrella patens*: comparative analysis reveals a rich source of new genes. Submitted.

Reski R. 1998. Physcomitrella and Arabidopsis: the David and Goliath of reverse genetics. Trends Plant Sci 3:209-210.

Reski R. 1999. Molecular genetics of Physcomitrella. Planta 208:301-309.

Reski R. 2002. Rings and networks: the amazing complexity of FtsZ in chloroplasts. Trends Plant Sci. 7:103-105.

Reski R., M. Faust, X.-H. Wang, M. Wehe, and W.O. Abel. 1994. Genome analysis of the moss *Physcomitrella patens* (Hedw.) B.S.G. Mol. Gen. Genet. 244:352-359.

Reski R., S. Reynolds, M. Wehe, T. Kleber-Janke, and S. Kruse. 1998. Moss (*Physcomitrella patens*) expressed sequence tags include several sequences which are novel for plants. Bot. Acta 111:143-149.

Richter U., J. Kiessling, B. Hedtke, E. Decker, R. Reski, T. Börner, and A. Weihe. 2002. Phage-type RNA polymerases of *Physcomitrella patens*: two enzymes with dual targeting to mitochondria and plastids. Gene, in press.

Schipper O., D. Schaefer, R. Reski, and A. Fleming. 2002. Expansins in the bryophyte *Physcomitrella patens*. Plant Mol. Biol., in press.

Schlink K., and R. Reski. 2002. Reliable method for preparation of high-quality DNA from the moss *Physcomitrella patens*. Submitted.

Schween G., A. Hohe, A. Koprivova, and R. Reski. 2002a. Effects of nutrients, cell density and culture techniques on protoplast regeneration and early protonema development in *Physcomitrella patens*. Submitted.

Schween G., S. Fleig, and R. Reski. 2002b. High-throughput-PCR screen of 15,000 transgenic *Physcomitrella* plants. Plant Mol. Biol. Rep. 20:43-47.

Sperling P., T. Egener, J. Lucht, R. Reski, P. Cirpus, and E. Heinz. 2002. Identification of a Δ5-fatty acid desaturase from *Physcomitrella patens*. Adv. Res. Plant Lipids, in press.

Strepp R., S. Scholz, S. Kruse, V. Speth, and R. Reski. 1998. Plant nuclear gene knockout reveals a role in plastid division for the homolog of the bacterial cell division protein FtsZ, an ancestral tubulin. Proc. Natl. Acad. Sci. USA 95:4368-4373.

Zank T.K., U. Zähringer, C. Beckmann, G. Pohnert, W. Boland, H. Holtorf, R. Reski, J. Lerchl, and E. Heinz. 2002. Cloning and functional characterisation of an enzyme involved in the elongation of Δ6-polyunsaturated fatty acids from the moss *Physcomitrella patens*. Plant J., in press.

CONTROL OF GENE EXPRESSION BY HISTONE DEACETYLASES

Brian Miki[1] and Keqiang Wu[2]
[1]Eastern Cereal and Oilseed Research Centre, Agriculture and Agri-Food Canada, Ottawa, Ontario K1A 0C6 Canada (e-mail: mikib@em.agr.ca)
[2]Department of Biology, West Virginia University, PO Box 6057, Morgantown WV 26506-6057 USA (email: kewu@mail.wvu.edu)

Keywords Arabidopsis, gene repression, histone deacetylases, promoters

1. INTRODUCTION

In eukaryotes, the strength of the interactions between the basic tails of the core histones and DNA can be modulated through post-translational modifications, such as acetylation (Strahl and Allis, 2000). Acetylation is catalyzed by histone acetyltransferases (HATs) and deacetylation is catalyzed by histone deacetylases (HDACs). Both enzyme classes exist in different multiprotein complexes and establish the state of histone acetylation that is correlated with the state of transcriptional activity and heterochromatinization (Ahringer, 2000; Lusser et al., 2001). In pea, histones were found to be hyperacetylated at the *Pet E* gene promoter, contributing to an open conformation but only when the gene is transcriptionally active (Chua et al., 2001). The HAT, Gcn5, was found to be targeted to the promoters of cold-responsive genes by the transcription factor, CBF-1, which modulates induction (Stockinger et al., 2001). Histone deacetylation was correlated with silencing of one parental set of rRNA genes in interspecific hybrids, a phenomenon known as nucleolar dominance (Chen and Pikaard, 1997). Corepressor complexes containing Polycomb-group proteins have been shown to play a role in regulating the spatial patterns of expression of floral and embryonic homeotic genes and it is believed that histone deacetylases play a key role in these complexes (Eshed et al., 1999; Ogas et al., 1999).

2. PLANT HISTONE DEACETYLASES

Plants generally possess 4 classes of HDAC. Of these, three are commonly found among diverse organisms (RPD3, HDA1, SIR2) and the other, HD2, is specific to plants (reviewed by Lusser et al., 2001). They are coded by

I. K. Vasil (ed.), Plant Biotechnology 2002 and Beyond, 211-214.

Table 1. HDAC genes identified from Arabidopsis and Maize

Genes and origins	Accession #	Putative function	Publications
Arabidopsis			
RPD3-type			
HDA1(AtRPD3A)	AL035538	Pleiotropic	Wu et al. 2000a Tian & Chen 2001
HDA2	AF149413	-	
HDA6(AtRPD3B)	AB008265	Transgene silencing	Murfett et al. 2001
HDA7	AB023031	-	
HDA9	AL138652	-	
HDA1-type			
HDA5	AB006696	-	
HDA8	AC006932	-	
HDA14	AL035678	-	
HDA15	AB026658	-	
HDA18	AT5G61070	-	
SIR2-type			
HDA12	AB009050	-	
HDA16	AL391712	-	
HD2-type			
AtHD2A(HDA3)	AC002534	Seed development	Wu et al. 2000b
AtHD2B(HDA4)	AB006699	-	Wu et al. 2000b
AtHD2C(HDA11)	AL162506	-	Dangl et al. 2001
AtHD2D(HDA13)	AC005824	-	Dangl et al. 2001
Maize			
RPD3-type			
HDA101	AF035815	-	
HDA102	AI438666	-	
HDA108(HD1b)	AF045473	-	
HDA1-type			
HDA109	AW216192	-	
HDA110	AW231694	-	
SIR2-type			
HDA113	AI734474	-	
HD2-type			
HD2a(HDA105)	U82815	-	Lusser et al. 1997
HD2b(HDA103)	AF254072	-	Dangl et al. 2001
HD2c(HDA104)	AF254073	-	Dangl et al. 2001
HDA106(HDA115)	AW126465	-	
HDA114	AI622604	-	

small gene families in plants such as *Arabidopsis* and maize (Table 1). It is likely that they are involved in different biological functions. Studies in which the *Arabidopsis* RPD3-type, HDA1, was repressed using antisense constructs indicated pleitropic effects on the transition to floral development, homeotic effects on floral organ differentiation and the shoot apical meristem (Tian and Chen, 2001; Wu et al., 2000a). Another homologue, HDA6, is involved in the silencing of auxin-responsive promoters (Murfett et al.,

2001). RPD3 family members are expressed throughout the plant where they may perform a range of functions.

In contrast, the plant-specific HDAC, HD2 is believed to be involved in the silencing of rRNA genes in maize embryos because it was localized specifically to the nucleolus (Lusser et al., 1997). However, in *Arabidopsis,* the homologue AtHD2A appears to be involved in seed development as repression using antisense technology generated seed abortion without any other visible pleitropic effects (Wu et al., 2000b). Furthermore, expression of HD2 appears to be selective in reproductive organs. *In situ* hybridization studies indicate that AtHD2A is expressed in all of the major seed tissues (unpublished data). Patterns such as these could be important for repressing patterns of gene expression associated with other developmental programs. Microarray analysis of transcript profiles also revealed that AtHD2B mRNA was elevated in ectopic embryos developing on the cotyledons of transgenic *Arabidopsis* seedlings transformed with the transcription factor BABYBOOM (BBM) (unpublished data). The data suggests that the HD2 family plays a major role in seed and embryo development in *Arabidopsis*.

3. USE OF HDAC FOR REPRESSING TARGET GENES

HDAC plays a fundamental role in gene repression in plants and has been linked to gene silencing involving methylation (reviewed in Finnegan, 2001). Studies from our lab showed that members of both the AtRPD3 and AtHD2 families can interact directly with promoters to silence genes in transgenic plants (Wu et al., 2000a, b). This was demonstrated by fusing the plant HDAC genes to the DNA binding domain of the yeast GAL4 gene to create a fusion protein that could be targeted to promoters of transgenes that contained the GAL4 enhancer sequences. Significant repression of the transgene was demonstrated. Furthermore, deletion analysis of the HD2 gene revealed the domains essential for activity (Wu et al., 2000b). These findings raised the possibility of producing a technology that could be used for silencing transgenes and resident genes within plants by using transcription factors to target HDAC activity to the promoters of the genes.

It was possible to demonstrate both constitutive and organ-specific repression of a target reporter gene. Transgenic plants were constructed with the GUS reporter gene driven by a constitutive promoter with GAL4 enhancers. GAL4/HD2 fusions were expressed separately using constitutive and seed-specific promoters. When the genes were crossed into the same lines repression of the reporter gene paralleled the pattern of expression of the fusion protein. This demonstrated our experimental ability to control repression and the feasibility for new strategies for silencing specific plant genes and transgenes. Targeting systems can be developed using plant transcription factor specificities and the range of HDAC activities can be expanded as described in Table 1. The full extent of the possibilities will be revealed once the plant HDACs have been more fully characterized.

4. REFERENCES

Ahringer, J. 2000. NuRD and Sin3 histone deacetylase complexes in development. Trends Genet. 16: 351-356.

Chen, Z.L., and C.S. Pikkard. 1997. Epigenetic silencing of RNA polymerase I transcription: a role for DNA methylation and histone modification in nucleolar dominance. Genes and Develop. 11: 2124-2136.

Chua Y.L., Brown, A.P.C., and J.C. Gray. 2001. Targeted histone acetylation and altered nuclease accessibility over short regions of the pea plastocyanin gene. Plant Cell 13: 599-612.

Dangl, M., Brosch, G., Haas, H., Loidl, P. and A. Lusser. 2001. Comparative analysis of HD2 type histone deacetylases in higher plants. Planta 213: 280-285.

Eshed, Y., Baum, S.F., and J.L. Bowman. 1999. Distinct mechanisms promote polarity establishment in carpels of *Arabidopsis*. Cell 99: 199-209.

Finnegan, E.J. 2001. Is plant gene expression regulated globally? Trends Genet. 17: 361-365.

Lusser, A., Brosch, G., Loidl, A., Haas, H., and P. Loidl. 1997. Identification of maize histone deacetylase HD2 as an acidic nucleolar phosphoprotein. Science 277:88-91.

Lusser, A., Kölle, D., and P. Loidl. 2001. Histone acetylation: lessons from the plant kingdom. Trends Plant Sci. 6: 59-65.

Murfett, J., Wang, X.-J., Hagen, G., and T.J. Guilfoyle. 2001. Identification of Arabidopsis histone deacetylase HDA6 mutants that affect transgene expression. Plant Cell 13: 1047-1061.

Ogas, J., Kaufmann, S., Henderson, J. and C. Somerville. 1999. PICKLE is a CHD3 chromatin-remodelling factor that regulates the transition from embryonic to vegetative development in *Arabidopsis*. Proc. Natl. Acad. Sci. USA96:13839-13844.

Strahl, B.D., and C.D. Allis. 2000. The language of covalent histone modifications. Nature 403:41-45.

Stockinger, E.J., Mao, Y., Regier, M.K., Triezenberg, S.J., and M.F. Thomashow. 2001. Transcriptional adaptor and histone acetyltransferase proteins in *Arabidopsis* and their interactions with CBF1, a transcriptional activator involved in cold-regulated gene expression. Nucleic Acids Res. 29: 1524-1533.

Tian, L., and Z.J. Chen. 2001. Blocking histone acetylation in *Arabidopsis* induces pleiotropic effects on plant gene regulation and development. Proc. Natl. Acad. Sci. USA 98: 200-205.

Wu, K., Malik, K., Tian, L., Brown, D., and B. Miki. 2000a. Functional analysis of a RPD3 histone deacetylase homologue in *Arabidopsis thaliana*. Plant Mol. Biol. 44: 167-176.

Wu, K., Tian, L., Malik, K., Brown, D., and B. Miki. 2000b. Functional analysis of HD2 histone deacetylase homologues in *Arabidopsis thaliana*. Plant J. 22: 19-27.

SITE-SPECIFIC GENE STACKING METHOD

David W. Ow
Plant Gene Expression Center, USDA/UC Berkeley, Albany, CA 94710; USA (email: ow@pgec.ars.usda.gov)

1. INTRODUCTION

Recombinase-mediated gene targeting has been achieved in a number of plant species (for review, see Ow, 2002). The general scheme requires a first recombination site to be introduced into the genome to serve as the target site for the subsequent insertion of a second DNA molecule. With plants, a current limitation is that the target site can only be generated at random chromosome locations. Reports to date show that recombinase-directed site-specific integration can place a single-copy non-rearranged DNA fragment into the target site at a practical frequency. Moreover, a high percentage of the insertions express the transgene at a predictable and reproducible level (Day et al., 2000). This means that once a suitable target line is found, it can be used for the subsequent delivery of trait genes. However, the current methods do not provide a convenient way to append additional trangenes to the target locus once the first insertion event is obtained. In this paper, a strategy is described that permits the sequential and repeated delivery of new DNA to the genomic target, as might be expected if a transgenic plant line were to be improved over time through the sequential addition of new transgenic traits. Appending DNA onto existing target sites justifies the initial investment in screening for suitable chromosome locations. The clustering of desirable traits also facilitates the introgression of large gene sets to field cultivars.

2. GENE STACKING STRATEGY

The idea of gene stacking rests on a concept that the integrating DNA brings along a different recombination site, such that after insertion of the new recombination site into the genome, the new recombination site then becomes the new target for the next round of integration. While some recombination systems catalyze freely reversible reactions, others do not. Instead, the substrate sites, typically known as *attB* and *attP*, are not identical. This necessitates that the product sites generated from an *attB* x *attP* reaction, *attL* and *attR*, are dissimilar in sequence to *attB* and *attP*. The recombination enzyme that promotes the *attB* x *attP* reaction, often referred to as the

I. K. Vasil (ed.), Plant Biotechnology 2002 and Beyond, 215-218.

integrase, by itself does not recombine *attL* x *attR*. The lack of a readily reversible reaction gives a distinct advantage for employing such a system in DNA integration since integrated molecules are stable. Most importantly, an irreversible system permits a novel gene stacking strategy that is not achievable using only freely reversible systems. In fact, this is the underlying reason for this laboratory's interest in the φC31 recombination system (Thomason et al., 2001).

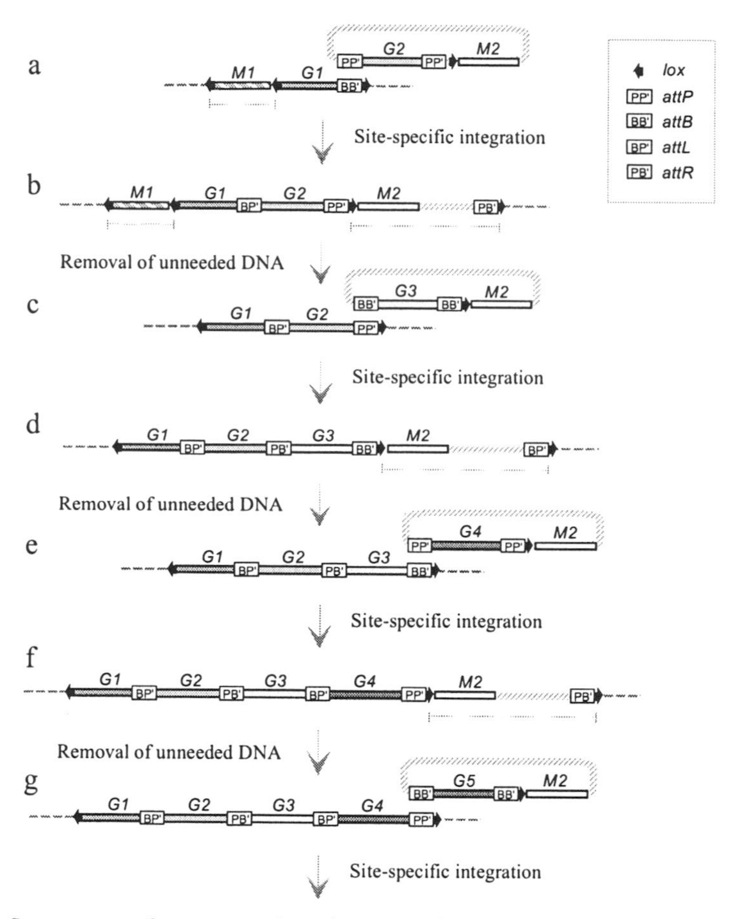

Figure 1. Strategy to stack transgenes through site-specific integration.

Figure 1 shows a strategy to stack genes sequentially using a non-reversible system (such as the φC31 system) along with a reversible system (such as the Cre-*lox* system). Shown are BB', PP', BP' and PB' as *attB*, *attP*, *attL* and *attR*, respectively, filled arrowhead as *lox* site, *G1*, *G2*, *G3*, *G4*, *G5*, as trait genes, and *M1*, *M2* as marker genes (gene promoters and terminators not shown). The process begins with a single copy trait gene linked to a marker: *lox-M1-lox-G1*-BB'-(inverted *lox*). The single copy locus may be obtained by molecular screening. Alternatively, a complex multicopy integration pattern may be resolved by Cre-*lox* site-specific recombination into a single copy state (Srivastava et al., 1999). If a resolution-based strategy were used, the marker *M1* would have been deleted, leaving a configuration consisting of *lox-G1*-BB' (inverted *lox*). To append *G2* to the *G1* locus, the integrating plasmid with the structure PP'-*G2*-PP'-*lox-M2* recombines with the genomic BB' target (Fig. 1a). The integrase can be provided, for example, by transient expression from a cotransformed plasmid. Since either PP' can recombine with the single BB', two different integration structures would arise that are distinguishable by molecular analysis. Figure 1b shows only the structure useful for further stacking, consisting of *lox-M1-lox-G1*-BP'-*G2*-PP'-*lox-M2*-plasmid backbone-PB'-(inverted *lox*). The Cre recombinase is introduced into the system to remove the unneeded DNA (indicated by dotted lines). The resulting structure becomes *lox-G1*-BP'-*G2*-PP'-(inverted *lox*). To stack *G3*, the construct BB'-*G3*-BB'-*lox-M2* is introduced (Fig. 1c). Analogous to the previous steps, the genome has only a single PP' site to recombine with either of the BB' sites on the plasmid. Recombination with the *G3* upstream site produces the structure shown in Figure 1d. After removing the unneeded DNA, the locus containing *G1*, *G2*, and *G3* is ready for the stacking of *G4* (Fig. 1e). In another variation, sets of inverted *attB* and *attP* sites, rather than sets of directly oriented sites, can also be used. The sequence of events is analogous to those described for Fig. 1.

There are several features worth noting. First, the vector for delivery of *G4* is the same as the vector for delivery of *G2*. Likewise, the vector for delivery of *G5* (Fig. 1g) is the same as the vector for delivery of *G3*. In principle, the stacking process can be repeated indefinitely, alternating between the uses of two simple vectors. Second, the stacking of *G2* onward requires only a single marker gene, and if *M1* is first removed, a single marker can be used throughout. This bypasses the need to continually develop new selectable markers. Third, the trait genes, such as *G1*, *G2* and so on, should not be narrowly interpreted as a single promoter-coding region-terminator fragment. Not only could each DNA fragment be composed of multiple transgenes, but could also include border DNA that insulate its (their) expression from surrounding regulatory elements. This may be useful when clustering transgenes that bring with them dominant *cis*-regulatory elements.

3. FUTURE PROSPECTS

The stacking strategy described above requires the use of one non-reversible site-specific recombination system, such as the ϕC31 system, and one freely reversible system, such as the Cre-*lox*, the FLP-*FRT* or the R-*RS* system. Other systems with similar properties may also be developed for this use. The gene stacking protocol takes into consideration the issue of selectable markers in commercial products (Ow, 2000). Avoiding the use of antibiotics resistance genes is possible, but alternative markers may not necessarily be free of public scrutiny either. For genes that are not relevant to the intended traits to be introduced, a prudent approach in dealing with the controversy is to just get rid of them. Hence, DNA deletion steps are used to eliminate as much as possible the DNA not needed for an engineered trait. Only short recombination sequences are necessarily co-introduced along with the trait genes, but most become non-recombinogenic BP' or PB' sites.

The immediate task ahead is to test the efficacy of the stacking strategy. Providing that it be successful, suitable target lines in crop plants would need to be generated. This could be a major undertaking given the large number of different crop plants where this technology may be applicable. A concerted effort by interested parties would be much more preferable to independent efforts. How a target site is constructed dictates future stacking options. If engineered with common elements, they can be shared among research and commercial communities.

4. REFERENCES

Day, C.D., Lee, E., Kobayashi, J., Holappa, L.D., Albert, H. and Ow, D.W. 2000. Transgene integration into the same chromosomal location can produce alleles that express at a predictable level, or alleles that are differentially silenced. Genes & Dev. 14: 2869-2880.

Ow, D. 2000. Marker Genes. Joint FAO/WHO Expert Consultation on Foods Derived from Biotechnology. http://www.who.int/fsf/GMfood/Consultation_May2000/Biotech_00_14.pdf.

Ow, D.W. 2002. Recombinase-directed plant transformation for the post genomic era. Plant Mol. Biol. 48: 183-200.

Thomason, L.C., Calendar, R. and Ow, D.W. 2001. Gene insertion and replacement in *Schizosacchromyces pombe* mediated by the *Streptomyces* bacteriophage ϕC31 site-specific recombination system. Mol. Genet. Genomics 265: 1031-1038.

Srivastava, V., Anderson, O.A. and Ow, D.W. 1999. Single-copy transgenic wheat generated through the resolution of complex integration patterns. Proc. Natl. Acad. Sci. (USA) 96: 11117-11121.

CHEMICAL CONTROL OF TRANSGENE EXPRESSION IN TRANSGENIC PLANTS: BASIC RESEARCH AND BIOTECHNOLOGICAL APPLICATIONS

Nam-Hai Chua
Laboratory of Plant Molecular Biology, Rockefeller University, 1230 York Avenue, New York, NY 10021, USA (email: chua@rockefeller.edu)

1. INTRODUCTION

Investigations of plant gene functions through reverse genetics often entail the use of a constitutive promoter to manipulate the expression level of a target gene. Any phenotypic changes in transgenic plants can then be correlated with alterations in target gene expression levels. Whereas this approach may work for most genes, for a number of genes that play an essential role in plant growth and development, constitutive overexpression of their cDNAs or constitutive silencing of their expression often lead to lethality thus providing little information on the gene function. To circumvent this lethality problem, inducible systems for chemical control of gene expression were developed so that transgene expression can be turned on or off at will by applying the appropriate chemical inducer at the desired time.

1.1. Two Chemical Inducible Gene Control Systems

Several chemical inducible systems for control of transgene expression with different mechanisms of action have been developed (cf. Zuo and Chua, 2000). In our lab, we have developed two such systems based on the well-established concept that eukaryotic transcription activators are modular in nature and can be reconstituted using a DNA-binding domain and an activation domain almost from any sources. In addition, this type of chimeric transcription activator can be placed under chemical control by appending to it a regulatory domain from a steroid hormone receptor. Based on this principle, we have developed two systems designated as GVG (Aoyama and Chia, 1997) and XVE (Zuo et al., 2000), responsive to glucocorticoid and beta-estradiol, respectively (Fig. 1).

I. K. Vasil (ed.), Plant Biotechnology 2002 and Beyond, 219-223.

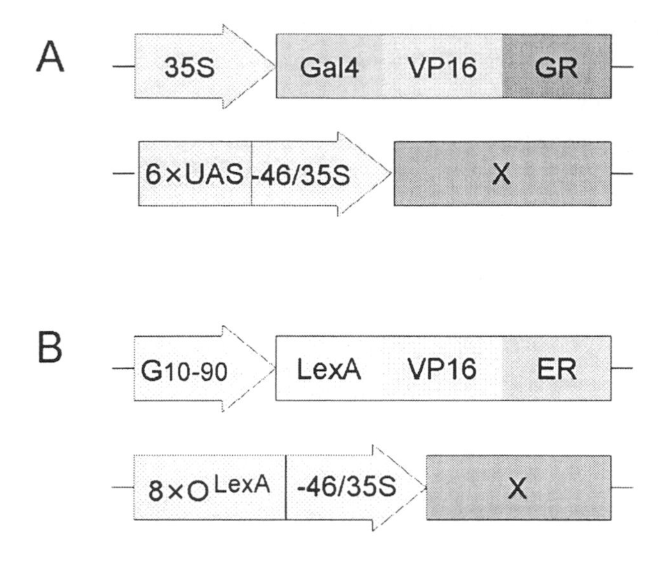

Fig 1. Schematic diagrams of the GVG and the XVE vector. A. In the first transcription unit, a 35S promoter is used to express the chimeric transcription factor, GVG, which is composed of the DNA binding domain of Gal4, the VP16 activation domain, and the regulatory domain of the glucocorticoid receptor. The second transcription unit consists of 6 copies of the upstream activating sequence (UAS) fused to the minimal 35S promoter (-46 to + 9) which is used to express target gene X. B. In the first transcription unit, a synthetic G1090 promoter is used to express the chimeric transcription factor, XVE, which is composed of the DNA binding domain of LexA, the VP16 activation domain, and the regulatory domain of the estrogen receptor. The second transcription unit consists of 8 copies of the lexA operator sequence (used as a cis-element) fused to the minimal 35S promoter (-46 to +9) which is used to express the target gene X.

Both the GVG and the XVE system provide regulation at the transcriptional level. Two transcriptional units are used. In the first transcription unit, a constitutive or tissue-specific promoter is used to express the chimeric transcription factor, GVG or XVE.

The second transcription unit consists of multiple copies of the cognate cis-acting element (UAS for GVG and LEX A binding sites for XVE) placed upstream of a minimal 35S promoter that is used to express any downstream target gene. After synthesis, the GVG or the XVE transcription activator is anchored in the cytoplasm by binding to HSP 90. Upon addition of the appropriate inducer, this interaction is disrupted and the released GVG or XVE factor can then translocate into the nucleus to activate genes with the cognate cis elements.

We have successfully used the GVG and the XVE system for controlled gene expression and gene silencing. In addition, we have used these systems for the following applications.

1.2. Marker-Free Transformation

Current methods of plant transformation require the use of an antibiotic-resistance marker to select for transgenic regenerants. We have attempted to develop alternate strategies to screen for transgenic regenerants without the use of such marker genes. One approach is to employ genes whose overexpression can promote shoot regeneration from explants in the absence of any plant hormones (cf. Zuo et al., 2002b). In this case, transformation events are screened as shoot regeneration events as the regenerated shoots can only be derived from transformed cells that have incorporated the shoot-promoting gene. Non-transformed cells in the explants are unable to regenerate because of the absence of the appropriate plant hormones. However, unregulated overexpression of such genes often causes developmental abnormality. For example, although overexpression of the Agrobacterium *ipt* gene can induce shoot regeneration the regenerated shoots are abnormal, root growth is severely inhibited, and the plants are infertile. To overcome these problems, we placed the *ipt* gene under the control of the GVG system so that once the transformants are identified, expression of the *ipt* gene can be turned off by inducer withdrawal (Kunkel et al., 1999). Tobacco leaf discs were inoculated with Agrobacteria carrying a GVG-*ipt* construct and incubated on MS medium without any auxin or cytokinin. In the presence but not the absence of the dexamethasone inducer many abnormal shoots emerged from the inoculated leaf discs. The shoots, which were presumably transformed, were excised and placed in MS medium without the chemical inducer but with auxin to stimulate root growth. After several weeks, normal shoots and roots emerged from these explants and more than 90% of them proved to be transformed as shown by molecular analysis. The transgenes were shown to transmit to subsequent generation as a dominant genetic trait (Kunkel et al., 1999).

In addition to the Agrobacterium *ipt* gene, we have used the Arabidopsis *CKI1* and *STM* genes and obtained essentially the same results (Niu, Zuo, and Chua, unpublished results)

1.3. Isolation Of Plant Genes That Promote Shoot Regeneration Or Somatic Embryogenesis

Our success in using shoot-promoting genes for marker-free transformation prompted us to search for Arabidopsis genes that have such a function. We have devised two different strategies to isolate such genes. In the first, we constructed a normalized Arabidopsis cDNA library under the control of a 35S promoter. Agrobacteria carrying this library were used to inoculate Arabidopsis root explants treated with auxin but without cytokinin. Under normal circumstances, no shoot should emerge from such root cultures. After inoculation with the Agrobacteria library, however, green calli or abnormal shoots appeared after 3-4 weeks. The cDNA sequences were retrieved from such explants and re-tested for their alleged function but using the inducible XVE system. Using this method, we have isolated the *ESR 1* gene that can enhance shoot regeneration in the absence as well as the presence of cytokinin (Banno et al., 2001).

In the second strategy, we inoculated root explants with a chemical inducible activation tagging XVE vector and incubated the explants in the presence of the beta-estradiol inducer (Zuo et al., 2002a) Regenerated shoots were excised and cultured on MS medium without the inducer to turn off expression of the tagged endogenous gene. The shoots were allowed to root and grown into mature plants. Progeny seeds were then re-tested for the inducible phenotype. Once the phenotype was confirmed, the tagged sequences were retrieved by PCR. Using this method, we have isolated several genes that can promote shoot regeneration and at least one gene that can promote somatic embryogenesis (Zuo et al., 2002a).

1.4. Inducible System For DNA Recombination

The XVE system can be combined with cre/lox to obtain a DNA recombination system that is regulated by beta-estradiol. This system, designated as the CLX system, has been successfully used to remove a kanamycin- resistance marker gene from transgenic Arabidopsis plants (Zuo et al., 2001). The CLX vector can also be used to make genetic chimera by activation, in specific tissues, of a WT gene that has been rendered inactive by a stuffer DNA sequence placed between the gene promoter and the coding sequence. When transformed into a mutant plant, this method can be used to generate one or more leaves carrying a WT gene in the context of a mutant plant.

The CLX vector can also be used for conditional genetic complementation in order to assess the role of a gene at different developmental stages. For example, a gene *(XYZ)* may play a role during embryogenesis as well as in adult stages of the plant. Because a *xyz* mutant would be blocked in embryogenesis, the role of the *XYZ* gene in other stages of plant development cannot be investigated. One can complement the *xyz* mutant with the *XYZ* gene inserted within the loxP sites of the CLX vector. This construct should complement the *xyz* mutation and rescue the embryo lethal phenotype. After seedlings of the complemented lines have germinated, they can be treated with the inducer to remove the *XYZ* gene through DNA recombination, and the phenotype of the *xyz* mutant at later developmental stages can then be investigated.

2. REFERENCES

Banno, H., Y. Ikeda, Q.-W. Niu, and N.-H. Chua 2001 Overexpression of Arabidopsis ERS1 Induces Initiation of Shoot Regeneration. Plant Cell 13: 2609-2618.

Kunkel, T., Q.-W. Niu, Y.-S. Chan, and N.-H. Chua 1999 Inducible isopentenyl transferase as a high-efficiency marker for plant transformation. Nature Biotech. 17: 916-919.

Zuo, J. and N.-H. Chua 2000 Chemical-inducible systems for regulated expression of plant genes. Curr. Opin. Biotech. 11: 146-151.

Zuo, J., Q.-W. Niu, and N.-H. Chua 2000 An estrogen receptor-based transactivator XVE mediates highly inducible gene expression in transgenic plants. Plant J. 24: 265-273.

Zuo, J., Q.-W. Niu,. G. Frugis, and N.-H. Chua 2002a The WUSCHEL gene promotes vegetative-to-embryonic transition in Arabidopsis. Plant J. 30: 1-12.

Zuo, J., Q.-W. Niu, Y. Ikeda, and N.-H. Chua 2002b Marker-free transformation: increasing transformation frequency by use of regeneration-promoting genes. Curr. Opin. Biotech. 13: 173-180.

Zuo, J. Q.-W. Niu,. S. G. Møller, and N.-H Chua 2001 Chemical-regulated, site-specific DNA excision in transgenic plants. Nature Biotech. 19:157-161.

CRE/*LOX* MEDIATED MARKER GENE EXCISION IN TRANSGENIC CROP PLANTS

Larry Gilbertson[1], Waly Dioh[1], Prince Addae[1], Joanne Ekena[2], Greg Keithly[1], Mark Neuman[1], Virginia Peschke[1], Mike Petersen[2], Chris Samuelson[1], Shubha Subbarao[1], Liping Wei[1], Wanggen Zhang[1], and Ken Barton[1]
[1]Monsanto Company, 700 Chesterfield Pkwy, Chesterfield MO 63198, [2]Monsanto Company, Agracetus Campus, 8520 University Ave. Middleton WI, 53562. (email: larry.a.gilbertson@monsanto.com)

Keywords Cre recombinase, Cre/lox, marker removal, antibiotic resistance marker, ARM

1. INTRODUCTION

Selectable marker genes are used in most plant transformation processes. Frequently, however, the selectable marker gene does not confer useful traits on the transgenic plant, and there are a number of advantages to removing the selectable marker gene after regeneration of the plant, including reducing the overall complexity of the transgene array, removal of redundant genetic elements, and improved regulatory and public acceptance, especially when antibiotic resistance marker genes are used as selectable markers. A number of site-specific recombinases of prokaryotic or yeast origin have been shown to function in transgenic plants for marker removal, including Cre/*lox* from bacteriophage P1 (Hoess and Abremski, 1990). We chose the Cre/*lox* system to develop technologies for marker removal in commercially important crop species.

There are a several strategies by which Cre/*lox* can be used to remove marker genes from transgenic plants. In all strategies the marker gene is flanked by directly repeated *lox* sites, and excision occurs when Cre activity is present. The strategies differ in how Cre function is delivered. In one strategy, the plants in which the marker gene is flanked by *lox* sites can be crossed with plants that express Cre activity (herein referred to as the "crossing strategy"). In this case, marker excision occurs in the F1 progeny, followed by loss of the *cre* gene by genetic segregation in the F2 generation (Russell et al., 1992; Dale and Ow, 1991). In another strategy, the *cre* gene can be included on the DNA segment that is flanked by *lox* sites. In this strategy, herein referred to as the "autoexcision strategy", it is critical that the *cre* gene be regulated such

I. K. Vasil (ed.), Plant Biotechnology 2002 and Beyond, 225-228.

that expression is activated after the marker gene is no longer needed. Autoexcision has recently been demonstrated in *Arabidopsis thaliana* using a chemically inducible promoter (Zuo et al., 2001).

We have tested the crossing and autoexcision strategies to remove an *npt*II gene from transgenic maize, wheat, soybean, and cotton, and found that both can efficiently and completely remove a marker gene.

2. RESULTS AND DISCUSSION

Transgenic maize, wheat, soybean and cotton plants were generated containing a *lox* "reporter" construct that allows visual assays for the excision of the marker gene. Cre-mediated excision of the *npt*II gene results in expression of the *gfp* gene. These "lox plants" were crossed with plants that express the *cre* transgene. In corn and wheat, Cre was expressed by the promoter from the rice *Actin1* gene (McElroy et al., 1990). In soybean and cotton, Cre was expressed by either the CaMV enhanced 35S promoter (Kay et al., 1987) or the FMV 34S promoter (Sanger et al., 1990).

Excision was found to be quite efficient in maize and wheat. In maize, a variety of crosses were made between multiple independent R0 Cre lines and independent R0 or R1 lox lines. In nearly all cases the progeny exhibited gfp expression in the ratio expected assuming Mendelian segregation of the Cre and lox transgenes and assuming complete excision at the populational level. Southern blot analysis was carried out to confirm that excision was complete and precise. The Southern results correlated well with GFP segregation data. Ten independent excision events representing 10 independent lox lines were cloned by PCR and sequenced to confirm that the excision was precise at the DNA level.

In order to test if marker excision is heritable, F1 plants that were selfed from a number of the crosses were selected and examined for GFP activity in F2 progeny. GFP expression segregated 3:1, as expected if excision is heritable, in nearly all F2 populations, indicating that marker excision was generally complete in the F1 plant, including in the male and female germ lines, and that marker removal in transgenic plants is stable.

Results from the crosses between Cre and lox lines in soybean and cotton indicated that, while Cre-mediated marker excision occurs frequently in F1 progeny, it is less complete at the plant level, evidenced by incomplete GFP fluorescence in somatic tissues, and corroborated by Southern blot analysis. F2 analysis showed that heritable excision did occur, albeit at lower frequencies than in maize.

To test the autoexcision strategy in maize, we inserted a *cre* cassette between the *lox* sites. The *cre* gene was expressed by an inducible heat shock protein promoter (HSP17.5E) from soybean (Ainley and Key, 1990). Transgenic callus tissue was divided into two groups, one of which was treated with a heat shock, and the other left untreated. Marker excision, indicated by GFP fluorescence and confirmed by Southern blot, occurred only in the calli that were treated with the heat shock and the plants derived from those calli. When these plants were carried to the next generation, marker free plants were obtained at the frequency expected if marker excision was complete in the R0 calli/plants.

After Cre-mediated marker removal occurs, one *lox* site, typically a wild type *loxP* site, is left in the chromosome. Modified *lox* sites with base pair substitutions in the Cre binding site can be used to generate a less active *lox* site as a product of excision (Albert et al., 1995). We designed and tested a series of modified *lox* sites in maize using the crossing strategy. In general, *lox* sites with 3-4 base pair substitutions reduce the efficiency of excision, but generate a relatively inactive *lox* site as a product of excision.

The availability of an efficient Cre/*lox* mediated marker excision tool in maize provided an opportunity to assess the effect of the presence of the selectable marker gene cassette on the expression of an adjacent gene, and the role that gene orientation plays in this effect. Three *lox* constructs were generated and used to transform maize. The three constructs have the same gene cassettes [*uid* (GUS) and *npt*II flanked by *lox* sites] but differ in the relative orientations of the gene cassettes. Transgenic maize plants were obtained with these three constructs and crossed with hemizygous Cre plants to excise the marker gene. F1 progeny contained both excised and unexcised plants. Both types of plants were crossed by wild type maize, and *uid* expression levels were measured and compared between excised and unexcised populations. For two of the constructs we observed little to no difference in *uid* expression between excised and unexcised plants. For one construct, however, excision invariably resulted in a change in *uid* gene expression. In this construct the marker gene (*npt*II) was upstream of the *uid* gene cassette. These results demonstrate the utility of the Cre/*lox* system for addressing basic questions in gene expression.

3. REFERENCES

Ainley, W. M. and Key, J. L. 1990. Development of a heat shock inducible expression cassette for plants: characterization of parameters for its use in transient expression assays. Plant Mol. Biol. 14, 949-967.

Albert, H., Dale, E. C., Lee, E., and Ow, D. W. 1995. Site-specific integration of DNA into wild-type and mutant lox sites placed in the plant genome. Plant J. 7, 649-659.

Dale, E. C. and Ow, D. W. 1991. Gene transfer with subsequent removal of the selection gene from the host genome. Proc. Natl. Acad. Sci. U. S. A 88, 10558-10562.

Hoess,R. and Abremski,K. 1990. The Cre-lox recombination system. In Nucleic Acids and Molecular Biology, (F.Eckstein and D.M.J.Lilley, eds.) Berlin: Springer-Verlag, pp. 99-109.

Kay, R., Chan, A., Daly, M., and McPherson, J. 1987. Duplication of CaMV 35S Promoter Sequences Creates a Strong Enhancer for Plant Genes. Science 236, 1299-1302.

McElroy, D., Zhang, W., Cao, J., and Wu, R. 1990. Isolation of an efficient actin promoter for use in rice transformation. Plant Cell 2, 163-171.

Russell, S. H., Hoopes, J. L., and Odell, J. T. 1992. Directed excision of a transgene from the plant genome. Mol. Gen. Genet. 234, 49-59.

Sanger, M., Daubert, S., and Goodman, R. M. 1990. Characteristics of a strong promoter from figwort mosaic virus: comparison with the analogous 35S promoter from cauliflower mosaic virus and the regulated mannopine synthase promoter. Plant Mol. Biol. 14, 433-443.

Zuo, J., Niu, Q. W., Moller, S. G., and Chua, N. H. 2001. Chemical-regulated, site-specific DNA excision in transgenic plants. Nat. Biotechnol. 19, 157-161.

MARKER GENE ELIMINATION FROM TRANSGENIC SUGARBEET BY A CHEMICALLY REGULATED *CRE-LOX* SYSTEM

H. Wang[1*], J. Kraus[1], J. Dettendorfer[1], N.-H. Chua[2], and R. Nehls[1]
[1]PLANTA GmbH, Grimsehlstr.31, 37574 Einbeck, Germany
[2]Laboratory of Plant Molecular Biology, The Rockefeller University, New York, NY10021
*Corresponding author (h.wang@kws.de)

Keywords marker gene elimination, XVE, estradiol, *cre-lox*, sugarbeet

1. INTRODUCTION

The absence of selection markers is a highly desired feature of transgenic crops considering market acceptance but also to enable multiple transformation. However, the current strategies for obtaining marker-free plants either require time consuming sexual crosses or depend on highly efficient transformation protocols (Hohn et al., 2001) which are either not preferred or not available for sugarbeet. Recently, Zuo et al. (2001) developed a chemical-regulated marker elimination system in *Arabidopsis* with which marker-free transgenic plants may be obtained directly from *in vitro* culture. This system is based on the introduction of an XVE inducible system for controlling *cre-lox*-mediated recombination. The marker elimination process is initiated by the addition of an external inducer (estradiol) which activates the estrogen receptor-based fusion transactivator XVE. This consequently enables the expression of *cre,* resulting in the excision of the marker gene flanked by two *lox* sites through recombination. We report here of the successful application of this chemical-regulated *cre-lox* marker elimination system in sugarbeet.

2. RESULTS AND DISCUSSION

2.1. XVE Inducible System Functions Efficiently In Sugarbeet

One of the prerequisites for the utility of the XVE-*cre-lox* mediated marker elimination in sugarbeet is the inductive capability of the XVE inducible system. Therefore, plasmid DNA of pER8-*gfp* in which *gfp* is under the control of XVE has been transferred to sugarbeet leaves via particle bombardment. Upon treatment with 20μM estradiol such leaf explants

I. K. Vasil (ed.), Plant Biotechnology 2002 and Beyond, 229-231.

showed clearly visible green fluorescent spots, whereas leaves bombarded with blank particles or with construct DNA but not treated with estradiol showed no sign of green fluorescence.

The inductive capability was further confirmed in pER8-*gfp* stable transgenic plants obtained via *Agrobacterium*-mediated transformation. Again, only upon estradiol treatment of transgenic tissues, bright green fluorescence has been observed in all tissues of transgenic shoots including adventitious shoot meristems. The induction of XVE has also been detected via RT-PCR analysis of *gfp* gene expression. RT-PCR analysis of non-treated plants, however, also showed *gfp* expression in some of the plants (7 out of 15). This gives evidence of some leakiness in the control of the system in sugarbeet, albeit below the detection level of *gfp* fluorescence.

As a conclusion, the XVE system is capable to function in sugarbeet with acceptable deficiencies in induction control.

2.2. Marker Gene Elimination From Transgenic Sugarbeet Can Be Achieved By Using A Combined XVE Cre-Lox System

The excision of marker genes has been evaluated utilizing the gene construct, pCLX-*gfp*, in which the *gfp* coding sequence is separated from its promoter by a *lox*-flanked sequence comprising of a marker gene (*nptII*) and a XVE-controlled *cre* gene. The excision of these *lox*-flanked sequences would bring *gfp* close to its promoter and eventually result in *gfp* expression. This event can be confirmed by conducting PCR with a pair of specially designed primers, P1-P4 (Zuo et al., 2001).

Transgenic shoots bearing pCLX-*gfp* were produced by *Agrobacterium*-mediated transformation. While growing on regeneration medium containing 20μM estradiol, apical meristems with strong green fluorescence have been observed (Fig. 1), whereas no distinct fluorescence could be detected in tissues grown on medium without estradiol. The precise excision of *lox*-flanked sequences has been further confirmed by PCR analysis with P1- P4 (Fig. 2). These results clearly indicate the occurrence of marker gene elimination induced by estradiol. More importantly, it has been demonstrated that XVE-controlled *cre* activation can be directed to meristems and thus to the germline which greatly facilitates the creation of homogeneously marker-free progeny. Not surprisingly, it was found through PCR analysis that some transgenic plants (3/16) also showed marker elimination without prior estradiol treatment. The analysis of sexual progeny is under way. No aberrant phenotypes are detected in plants transplanted to the greenhouse.

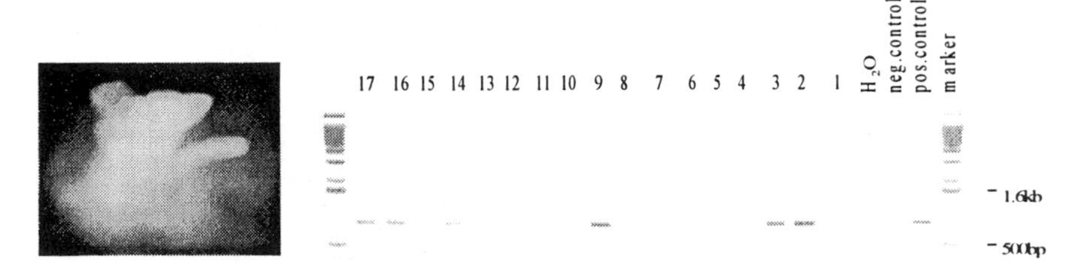

Figure 1. gpf expression indicates marker gene Elimination in shoot Meristem.

Figure 2. PCR analysis: presence of the 0.7kb PCR fragment confirms marker gene excision.

Based on the results shown above, we conclude that the XVE-controlled *cre-lox* recombination system is an efficient tool for the removal of undesired marker genes in sugarbeet.

3. REFERENCES

Hohn B., Levy A.A. and H.Puchta. 2001. Elimination of selection markers from transgenic plants. Curr. Opin. in Biotech. 12: 139-143

Zuo J., Niu Q-W., Moller SG. and N-H. Chua. 2001. Chemical-regulated, site-specific DNA excision in transgenic plant. Nature Biotechnology 19: 157-161

HOMOLOGOUS RECOMBINATION ALLOWS EFFICIENT ISOLATION OF MARKER-FREE TRANSPLASTOMIC PLANTS

Anil Day[1], Mikhajlo K. Zubko[1], Vasumathi Kode[1], Elisabeth A. Mudd[1] and Siriluck Iamtham[1]

[1]School of Biological Sciences, 3.614 Stopford Building, Oxford Road, University of Manchester, Manchester M13 9PT, UK. (e-mail: anil.day@man.ac.uk)

Keywords Clean gene, biosafety, GM crops, antibiotic- resistance, plastid transformation,

1. INTRODUCTION

Gene transfer technologies allow the rapid introduction of new traits into plants. The technology is reliant on bacterial marker genes for cloning trait genes in *Escherichia coli* plasmids, and plant marker genes for identifying transformed plant cells. Marker genes have served their purpose once transgenic plants are obtained but they are often retained (Fig 1A). Antibiotic resistance genes are commonly used as selectable markers and their use in transgenic research has been criticised due to the theoretical risk of their acquisition by pathogenic bacteria. Within the European Economic Community, directive 2001/18/EC requires the gradual elimination of antibiotic resistance markers, which might have adverse effects on human health and the environment, from genetically manipulated organisms by the end of 2004 for commercial releases and the end of 2008 for research purposes. Compliance with the directive requires the development of efficient procedures for removing antibiotic resistance genes from transgenic crops or the use of alternative marker genes. A gene excision strategy has the advantages of eliminating the need for risk assessments on marker genes in transgenic crops and allowing multiple trait genes to be combined in the same plant by repeat transformations with the most efficient marker gene. Removing excess foreign DNA from transgenic crops has the additional benefit of focusing attention on the important trait genes.

2. HOMOLOGOUS RECOMBINATION IN PLASTIDS

Exploitation of native plant enzymes that mediate homologous recombination provides an ideal opportunity for developing clean gene transformation

I. K. Vasil (ed.), Plant Biotechnology 2002 and Beyond, 233-235.

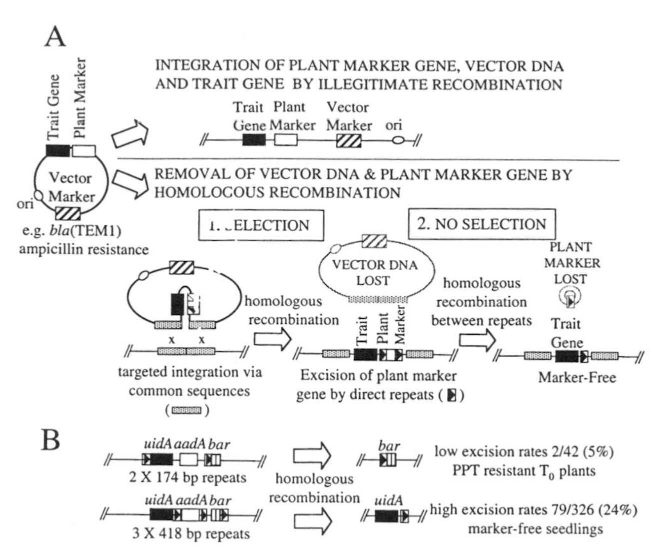

Figure 1. A) Homologous recombination removes the excess DNA integrated by illegitimate recombination B) Multiple direct repeats promote excision of marker genes in plastids

methods. Homology based integration allows precise targeting of foreign genes and elimination of vector sequences (Fig. 1A). Once transgenic plants have been isolated, marker genes can be excised by recombination acting on direct repeats. Use of native plant enzymes eliminates the need for foreign site-specific DNA recombinases and their target sites. The scheme requires relatively high rates of homologous recombination to succeed. This situation is found in plastids, which contain an efficient and relatively unexplored homologous recombination apparatus.

We used engineered direct repeats to excise marker genes from plastids (Iamtham and Day, 2000). The procedure is simple and is based on first selecting an excisable marker gene in transgenic plastids. In the second step, selection is relaxed to allow accumulation of marker-free plastids following homology-based marker gene excision. The frequency of marker gene excision is influenced by the number of direct repeats in a construct (Fig. 1B) and must be sufficiently high for effective generation of marker-free plastids but not too high to allow marker retention when selecting for plastid transformants. The time point at which the selective agent is removed allows the appearance of large numbers of marker-free plastids, cells and organs. The procedure is efficient (24% of seedlings were marker-free) and allows several excision pathways from a single integration event. We isolated tobacco plastid transformants containing the *bar* or *uidA* genes, free of the *aadA* marker (Fig 1B). Excision leaves behind a solo engineered repeat. The use of new direct repeats at each transformation/excision cycle will prevent undesirable recombination events with old solo repeats. Exposure to gamma radiation, which promotes recombination, allowed the isolation of a marker-free plant from a relatively stable two repeat construct. Efficient removal of marker genes, combined with high levels of gene expression, make plastids suitable for biopharming novel products in plants.

3. REFERENCE

Iamtham S., and A. Day (2000) Removal of antibiotic resistance genes from transgenic tobacco plastids. Nature Biotechnol. 18: 1172-1176.

EVALUATION OF CONSTITUTIVE CESTRUM YELLOW LEAF CURLING VIRUS PROMOTER IN MAIZE AND TOMATO

Masha Kononova[1], Livia Stavolone[2] and Thomas Hohn[2]
[1]Syngenta Global Gene Expression Optimization Team, RTP, USA; [2]Friedrich Miescher-Institute, Basel, Switzerland (email: maria.kononova@syngenta.com)

Keywords Cestrum yellow leaf curling virus, constitutive promoter, spatial expression, transgenic maize, transgenic tomato

1. SUMMARY OF PROMOTER EXPRESSION ANALYSIS

We have cloned and evaluated two versions of a novel, strong and constitutive promoter from Cestrum yellow leaf-curling virus (CmYLCV) called CmpC (short- 346bp) and CmpS (longer- 400bp), which can be used for regulating transgene expression in a wide variety of plant species. CmYLCV belongs to the Caulimoviridae family and was first reported in *Cestrum parqui* from the Solanaceae by Ragozzino (1974). Recently, CmYLCV was cloned and seven open reading frames were identified in the genomic sequence (Hohn et al., 2001). To evaluate the utility of the CmYLCV promoter to drive expression of heterologous genes in plants, two versions of the full-length transcript promoter were cloned in front of the GUS, CAT and FP reporter genes and tested in transient assays in *Nicotiana plumbaginifolia*, *Orichophragmus violaceus* and *Oriza sativa* protoplasts as well as in stably transformed *Zea mays* and *Lycopersicon esculentum*. The transient expression experiments show that, depending on the plant system used, the expression level of CmpC and CmpS promoter fragments are higher than the expression level of the widely used 35S promoter from Cauliflower Mosaic Virus (Hohn et al., 2001) and that the longer promoter fragment is the weaker one. Expression analysis of CmpC and CmpS promoter fragments in stably transformed maize (Figure1) have shown that both fragments are on average ten times greater than the strong constitutive Ubi1 promoter from *Z. mays* (Christensen et al., 1992) and that the expression levels in tomato (Figure 2) are comparable with the strong, constitutive SMAS promoter (Ni et al., 1994). Moreover, the spatial expression analysis has shown that both promoters express in various tissue types, except pollen in both maize and tomato and that both promoter fragments retain their high expression levels through at least two generations. We limited our analysis to the single copy

I. K. Vasil (ed.), Plant Biotechnology 2002 and Beyond, 237-238.

events only (Ingham et al., 2001) since it is a widely accepted idea that the copy number of a transgene affects the expression level in transgenic plants.

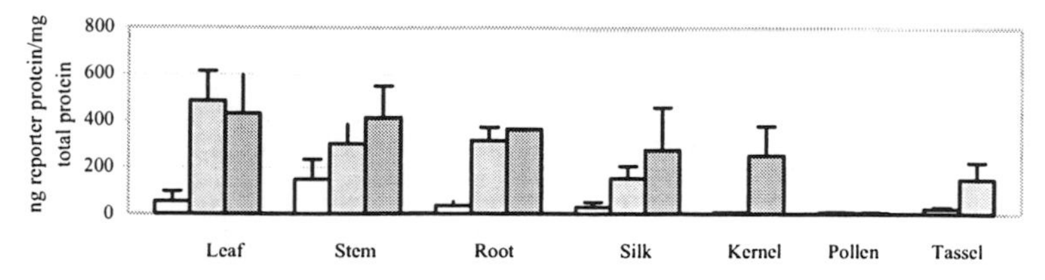

White bars- Maize UBI promoter, light gray bars -Cmp fragment of Cestrum Promoter, dark gray bars-CmpS fragment of Cestrum Promoter

Figure1. Comparative expression levels of Cmp, CmpS and ZmUbi-1 promoters in T0 Z. mays

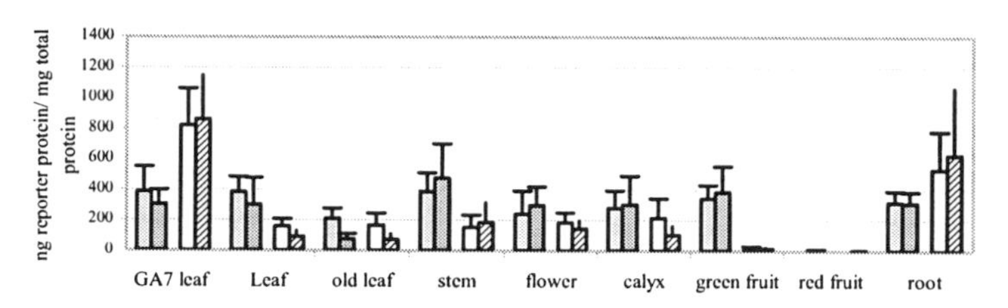

Light gray bars- CMP Promoter linked to FP gene, Dark gray bars –CMPS promoter linked to PMI gene, white bars- SMAS promoter linked to FP gene, striped bars- SMAS promoter linked to PMI

Figure2. Comparative expression levels of Cmp, CmpS and SMAS promoters in T0 L. esculentum

3. REFERENCES

Christensen A., Sharrock R., and Quail P. 1992. Maize polyubiquitin genes: structure, thermal perturbation of expression and transcript splicing, and promoter activity following transfer to protoplasts by electroporation. Plant Mol. Biol. 18: 675-689.

Hohn T., Stavolone L., de Haan P.T., Ligon H.T. and Kononova M. 2001. Cestrum yellow leaf curling virus promoters. Patent: WO 0173087-A.

Ingham DJ, Beer S, Money S, Hansen G. 2001. Quantitative real-time PCR assay for determining transgene copy number in transformed plants. Biotechniques 31: 132.

Ni M., Cui D., Einstein J., Narasimhulu S., Vergara C., and Gelvin S. 1995 Strength and tissue specificity of chimeric promoters derived from the octopine and mannopine synthase genes. Plant J. 7: 651-676.

Ragozzino, A. (1974). Ann. Fac. Sci. Agr. Univ. Napoli IV 8: 249.

HEAT-INDUCED TRANSGENE SILENCING IS CONVEYED BY SIGNAL TRANSFER

Inge Broer[1], Stephanie Walter[1], Sandra Kerbach[2], Stefan Köhne[3] and Katrin Neumann[1]
[1] Universität Rostock; Justus-von-Liebig-Weg 8; D-18051 Rostock; [2] Universität Hamburg; Ohnhorststr. 18; D-22609 Hamburg; [3] WIS-Dezernat; Humboldtstr.; D-29633 Munster (email: inge.broer@biologie.uni-rostock.de)

Keywords Heat stress, transgene inactivation, systemic gene silencing, grafting

1. INTRODUCTION

The stress response of plants is mostly accompanied by the controlled activation and silencing of endogenous plant genes. Transgenes that integrate randomly into the plants genome, are controlled by similar mechanisms like endogenous plant genes. The influence of environmental factors on transgene expression has already been observed in *Petunia hybrida* plants carrying the maize A1 cDNA under control of the *CaMV*35S RNA promoter in 1990. Heat stress and other factors lead to a stable loss of flower pigmentation, accompanied by the methylation of the viral promoter (Meyer et al., 1992). Transgene inactivation directly correlated to high temperatures was described for transgenic *Nia*, *luc*, and *npt*II genes in tobacco (Palauqui and Vaucheret, 1995; Neumann et al., 1997; Conner et al., 1998; Köhne et al., 1998), for the *Ltp*2-*gus* transgene in rice (Morino et al., 1999), and for the *npt*II transgene in *Arabidopsis* (Meza et al., 2001).

The heat-induced inactivation of transgenes will be of importance for the use of transgenic plants in agriculture, hence we analysed the expression of the Phosphinothricin (Pt)-resistance gene *pat* isolated from *Streptomyces viridochromogenes* (Strauch et al., 1988) in *Nicotiana tabacum*. Based on the DNA sequence of the native *pat* gene, three different chimaeric genes were constructed: a) the *pat*40 gene composed of the long (823 bp) 35S promoter, a modified *pat* coding region with an ATG start codon, synthetic 5' and 3' UTR sequences and the *nos* terminator sequence (Broer, I.; 1989), b) the synthetic *pat*S gene composed of a short (534 bp) 35S promoter, a synthetic coding region adapted to the plants codon usage, synthetic 5' and 3'UTR's differing from those present in *pat*40 and the 35S terminator sequence (Eckes et al., 1989), and c) the *pat*43 gene composed of the long 35S promoter, the

I. K. Vasil (ed.), Plant Biotechnology 2002 and Beyond, 239-241.

*pat*S coding region, the *pat*40 5' and 3' UTRs and the *nos* terminator (Köhne et al., 1998). All three genes coding for the same Phosphinothricin-*N*-acetyl-transferase, were transferred to *N. tabacum* SRI and the resulting herbicide resistant plants exposed to elevated temperatures. The loss of herbicide resistance in response to the heat treatment at 37°C was analysed in aseptically grown plants as well as in the green house.

The activity of the GC-rich *pat*40 gene was reduced in 100% of the lines, accompanied by a loss of enzyme activity, the PAT-protein and *pat*40 RNA. After resetting the culture temperature to 25°C, transgene activity was regained. In contrast, the expression of the synthetic AT-rich *pat*S gene was always stable in tobacco plants. The inactivation of the *pat*40 gene was neither correlated with the transgene copy number nor with the methylation of the promoter region, but dependent on the sequence of the transgene itself. Analysing the *pat*43 plants we could demonstrate, that the steady state level of *pat* specific RNA could be stabilized by the presence of the *pat*S coding region, while no PAT protein could be detected in heat treated *pat*43 plants. Hence, the inactivation took place on RNA (*pat*40) as well as on Protein level (*pat*43).

In order to determine whether the heat-induced transgene inactivation is mediated by a mobile signal we performed single-leaf heat-treatments of *N. tabacum* SRI plants carrying the *pat*40 gene, the *pat*43 gene or the *pat*S gene, respectively. In transgenic *pat*40-plants the incubation of one fully expanded leaf at 37°C led to a strong reduction of transgene encoded enzyme activity based on the loss of PAT protein and *pat* –RNA in the heat treated leaf as well as in untreated neighbouring leafs. This reduction of transgene-mediated herbicide resistance decreased with the increasing distance to the heat-treated leaf. *Pat*43-plants displayed the inactivation only on protein level. Hence transgene inactivation is delivered from the heat treated leaf to adjacent untreated ones. As demonstrated for the plants completely incubated at 37°C, no heat-mediated reduction of transgene expression could be observed in single-leaf heat-treated *pat*S-plants on enzyme activity, protein or RNA level.

2. REFERENCES

Broer, I. (1989) Expression des Phosphinothricin-*N*-Acetyltransferase-Gens aus *strepto-myces viridochromogenes* in *Nicotiana tabacum*. Dissertation, University of Bielefeld, Germany.

Conner, A.J., Mlynárová, L., Stiekema, W.J., Nap, J.-P. (1998) Meiotic stability of transgene expression is unaffected by flanking matrix-associated regions. Mol. Breeding 4: 47-58.

Eckes, P., Vijtewaal, B., Donn, G. (1989) Synthetic gene confers resistance to the broad-spectrum herbicide L-Phosphino-thricin in plants. J. Cell Biochem. (Suppl.) 13D: 334 .

Köhne, S., Neumann, K., Pühler, A., Broer, I. (1998) The Heat-treatment Induced Reduction of the *pat* Gene Encoded Herbicide Resistance in *Nicotiana tabacum* is Influenced by the Transgene Sequence. Plant Physiol. 153: 631-642.

Meyer, P., Linn, F., Heidmann, I., Meyer, H., Niedenhof, I., Saedler, H. (1992) Endogenous and environmental factors influence 35S promotor methylation of a maize A1 gene construct in transgenic petunia and its colour phenotype. Mol. Gen. Genet. 231: 345-352.

Meza, T.J., Kamfjord, D., Håkelien, A.-M., Evans, I., Godager, L.H., Mandal, A., Jakob-sen, K.S., Aalen, R.B. (2001) The frequency of silencing in *Arabidopsis thaliana* varies highly between progeny of siblings and can be influenced by environmental factors. Transgenic Res. 10: 53-67.

Morino, K., Olsen, O.-A., Shimamoto, K. (1999) Silencing of an aleurone-specific gene in transgenic rice is caused by rearranged transgene. The Plant Journal 17(3): 275-285.

Neumann, K., Dröge, Laser, W., Köhne, S., Broer, I. (1997) Heat Treatment Results in a Loss of Transgene-Encoded Activities in Several Tobacco Lines. Plant Physiol. 115: 939-947.

Palauqui, J.C., Vaucheret, H. (1995) Field trial analysis of nitrate reductase cosuppression: a comparative study of 38 combinations of transgene loci. Plant Mol. Biol. 29: 149-159.

Strauch, E., Wohlleben, W., Püler, A. (1988) Cloning of a phosphinothricin-*N*-acetyl transferase from *Streptomyces virido-chromogenes* Tü 494 and its expression in *Streptomyces lividans* and *E. coli*. Gene 63: 65-74.

SILENCING OF THE POLLEN GENE *NTP303* CEASES POLLEN TUBE GROWTH *IN VIVO* IN *NICOTIANA TABACUM*

George J. Wullems, Peter F.M. de Groot, Mark de Been and M.M.A. van Herpen
Dept. Molecular Plant Physiology, Nijmegen University, 6525 ED Nijmegen, The Netherlands (e-mail: wullems@sci.kun.nl)

Keywords Pollen gene expression, gene silencing, pollen tube growth

1. INTRODUCTION

The main objective of our study is the elucidation of the molecular and physiological basis for the role of the male reproductive organ, the pollen grain or gametophyte, in the process of fertilization. Although the morphological and chemical aspects of pollen development and pollen tube growth have been studied in great detail, knowledge of the underlying molecular processes is relatively limited. It is known that the high transcriptional activity during pollen development leads to a considerable increase in the total amount of RNA, mRNA and dry weight from the late unicellular microspore stage up to mature pollen. Based on transcription kinetics pollen formation can be divided in two developmental periods, with different sets of genes expressed. Transcripts of the 'early' genes appear soon after meiosis and are reduced or undetectable in mature pollen. Transcripts of 'late' genes are first detected after microspore mitosis and continue to accumulate as pollen matures.

2. RESULTS

In this report, an antisense approach is used to examine the effect of abolishing the production of NTP303 on pollen development and pollen tube growth, in order to gain insight in the function of the protein. We showed that silencing of the *ntp303* gene is accompanied with the failure of transgenic pollen to contribute in a successful fertilisation.

Progeny analysis using *ntp303* transformed plants as male and wild type plants as female revealed that transgene constructs were not transmitted to the progeny, suggesting that NTP303 is involved in effective fertilisation.Within

I. K. Vasil (ed.), Plant Biotechnology 2002 and Beyond, 243-245.

the selected plants the amount of *ntp303* messengers in pollen grains was reduced to 50 % of the wild type level. The fact that the plants could be selected on bases of their non-Mendelian segregation pattern indicates that *in vivo* the function of NTP303 is affected. As transformed pollen is capable of *in vitro* germination, *ntp303* is apparently not involved in hydration or not important for the initial germination. So, NTP303 is involved in a pollen event that occurs during travelling the style or is accomplished in the fertilisation itself.

2.1. Ntp303 RNA

With respect to the ntp303 RNA in doubled haploid transgenic plant lines 417, 449 and 496, the level of ntp303 RNA is strongly reduced as compared to the control line 428. This reduction was shown for all developmental stages analysed. The mRNA kinetics in the silenced lines were similar with the kinetics of the ntp303 RNA during wild type (Weterings et al., 1992, 1995) and control pollen development, although the quantitative level was much lower in the silenced lines and not detectable anymore after 17 hours germination in vitro. In the three silenced lines, the average reduction of the ntp303 RNA in the mature stage was 79 %. After the mature stage, the amount of ntp303 RNA dropped dramatically, absolutely and relatively. In the silenced lines a further reduction of ntp303 RNA from mature to 17 hours after germination of 90 % was observed, whereas in wild type and control pollen after 17 hours germination this RNA was reduced with 42 %.

2.2. NTP303 Protein

*N*tp303 encodes a 69 kDA glycoprotein which is developmentally regulated (Wittink et al. 2000). In order to study the effect of silencing of *ntp303* gene on the NTP303 protein content in the transgenic lines, the levels were determined in the insoluble fractions of extracts from mature transgenic pollen and after 17 hours germination in vitro. It was shown that the NTP303 protein in the transgenic, silenced lines is strongly reduced in the pollen tubes after 17 hours of germination.

2.3. Pollen Tube Growth In *Vitro*

Mature pollen from wild type, control and silenced transgenic lines were cultured in germination medium to follow their germination and pollen tube growth. The germination percentage was similar for either wild type, control and silenced lines. Also the length of the pollen tubes was similar at different times after germination. Also the growth characteristics of the pollen from the silenced lines were similar as those from the wild type and control pollen.

2.4. Pollen Tube Growth I*n Vivo*

To study pollen germination and pollen tube growth *in vivo*, mature styles from wild type plants were pollinated with similar amounts of wild type and silenced pollen. In mature styles, both wild type and control pollen tubes have a growth rate of 1 mm/hour, whereas pollen tubes from silenced pollen have a growth rate of 0.5 mm/hour. In addition, wilt type pollen tubes reach the ovary in 36 hours, whereas pollen tubes from the silenced lines cease their growth after approximately 24 hours after which time they have reached a distance in the style of 10-12 mm.

3. CONCLUSIONS

Our results indicate that transgenic pollen, in which the *ntp303* gene is silenced, do not take part in effective fertilisation since the linked kanamycin-resistance marker is not transmitted to the offspring through the male germ line. The actual disturbance of fertilisation seems to occur during pollen tube growth. The silencing of the NTP303 protein has only effect on pollen tube growth in vivo, not *in vitro*. This suggests that the NTP303 protein is involved in interaction with the presence of factor(s) of the style to give the optimal environment for pollen tube growth to take place. This involvement could be in nourishment, pollen tube guidance or structural support of the pollen tube.

4. REFERENCES

Weterings K. Reijnen W. Aarsen R. van, Kortstee A., Spijkers J., herpen M.van, Schrauwen J. and Wullems G. 1992. Plant Mol. Biol. 18: 1101-1111.

Weterings K., Reijnen W., Wijn G., Heuvel K. van de, Appeldoorn N., Kort G. de, Herpen M. van, Schrauwen J. and Wullems G. 1995. Sex. Plant Reprod. 8: 11-17.

Wittink R., Knuiman B., Derksen J., Capkova V., Twell D., Schrauwen J. and Wullems G. 2000. Sex. Plant Reprod. 12: 276-284.

THE ROLE OF D-TYPE CYCLINS IN PLANT GROWTH AND DEVELOPMENT

Walter Dewitte[1], Yves Deveaux[1], Rachel Huntley[1], Anne Samland[1] and James A.H. Murray[1]
[1]Institute for Biotechnology, Cambridge CB2 1QT, UK (email:j.murray@biotech.cam.ac.uk)

Keywords CyclinD, cell cycle, Rb pathway, division, differentiation

1. INTRODUCTION

Plant morphogenesis is primarily an iterative process. Shoot development, for example, can be conceptualised as the repetitive formation of phytomers, a term used to refer to a unit consisting of a leaf, associated axial meristem and accompanying internode. The patterning and organisation for the formation of a phytomer or floral organs occurs mainly in the shoot apical meristem. Cell proliferation occurs within, but is not restricted to, the apical region, since it is an ongoing process that continues to a certain degree until the formation of the organ is completed. In the formation of organs such as leaves, cell proliferation has therefore to be tightly controlled for proper organ formation (Wyrzykowska et al., 2002).

The molecular mechanisms which integrate cell proliferation with development are currently an exciting topic in plant sciences, since a "black box" remains between our understanding of the action of "patterning" or morphogenetic genes and the cellular controls of the division process. Progression through the cell cycle in plants, as well as in other eukaryotes, is driven by the consecutive action of different CDK complexes. An active complex consists at least two components, a regulatory subunit termed a cyclin, and a catalytic subunit, the cyclin dependent kinase (CDK). Therefore the current challenge lies in unraveling the mechanisms which link developmental signalling to the regulation of CDK activity, particularly those aspects of CDK activity that control decisions to undertake division and the potentially connected aspect of regulating the balance between proliferation and differentiation.

During the last decades a family of D-types cyclins (CycD) has been isolated in plants (Meijer and Murray, 2000). In *Arabidopsis thaliana*, this family consists of 10 members, and can arguably be subdivided into 7 classes

I. K. Vasil (ed.), Plant Biotechnology 2002 and Beyond, 247-253.

(Vandepoele et al., 2002), or three major groups with three orphan outliers (Oakenfull et al., 2002). Substantial evidence now supports the role of CycD/CDK pathway in controlling entry into the cell cycle and indicates that the CycD/CDK pathway is a putative target for molecular mechanisms regulating both cell numbers and differentiation in organ formation.

2. D-TYPE CYCLINS MEDIATE G1 PROGRESSION

After the initial isolation of the first plant D-type cyclins, it became clear that they share the common defining structural features of a cyclin box, a domain involved in CDK binding, and an retinoblastoma protein (Rb) binding motif. Combined with a degree of sequence homology, the presence of an Rb binding motif is normally considered a defining feature of CycDs. By analogy with animal D-type cyclins which interact with and phosphorylated Rb to control G1 progression, it was suggested that plant CycD may act in the G1/S transition according to a generally conserved mechanism in higher eukaryotes. This model would predict that CycD expression responds positively to mitogenic signals promoting cell division, and then targets its cognate kinase activity to the Rb protein. Upon the hyperphosphorylation of Rb, E2F transcription factors are activated which allow or induce expression of genes involved in processes leading to cell cycle progression and S-phase entry.

A considerable body of largely circumstantial evidence supports this model, including the finding that many genes likely to be required for S-phase entry possess E2F-binding sites that at least in a few cases have been shown to be involved in their cell cycle regulation. However, it is not known whether Rb phosphorylation by CycD kinases is either required or sufficient for S-phase entry, although recent evidence suggests that constitutive expression of the *CYCD3;1* gene in Arabidopsis plants (Dewitte et al., submitted) or suspension cultured cells (Samland, A., Menges, M., and Murray, J.A.H., unpublished data) results in a shift in the exit of cells from G1 phase. This would suggest that CycD3;1 kinase activity specifically, or CycD kinase activity generally, is rate limiting for G1 exit. This points to another outstanding question, which is the role or significance of the relatively large number of CycD and E2F-related genes in plants, and whether these play similar roles in different tissues or encode biochemically distinct functions. This point is further discussed in section 4.

Other evidence supporting a role for CycD's in G1 progression and/or the G1/S transition comes largely from work on suspension cultures. In an Arabidopsis cell suspension culture showing partially synchronous re-entry into the cell cycle after sucrose starvation, levels of CycD3- and CycD2-encoding mRNA increased from a low level during early (CycD2) or late G1 phase (CycD3) (Riou-Khamlichi et al., 2000), suggesting that transcriptional control plays a part in the regulation of both gene's activity. Immunoprecipitation experiments with antisera agains CycD2 and CycD3 proteins revealed that these cyclins interacted *in vivo* with the archetypal PSTAIRE-containing CdkA;1 (Healy et al., 2001). Notwithstanding the already mentioned transcriptional regulation of both *CYCD2;1* and *CYCD3;1* genes, striking differences were observed in the regulation of their protein levels and kinase activity. Analysis suggests that CycD3 protein levels and kinase activity largely reflect transcript abundance, although very rapid reduction in CycD3 protein levels is seen on sucrose removal. In contrast, although CycD2 proteins were present in stationary cells, association of CycD2 with CdkA was only detectable in exponential cultures, which is indicative for a supplementary sequestration mechanism for the control of CycD2-associated kinase activity (Riou-Khamlichi et al., 2000; Healy et al., 2001). These results support the idea that for CycD2 and CycD3 there are at least strikingly different modes of regulating their activity, and possibly therefore differences in their biochemical roles.

Furthermore, Arabidopsis D-type cyclins can bind both human and *Zea mays* Rb (Huntley et al., 1998) and CycD2/CdkA complexes phosphorylate tobacco Rb-proteins *in vitro* (Nakagami et al., 1999). In Arabidopsis three E2F genes have been identified (de Jager et al., 2001), and further related genes also exist. E2F proteins form active heterodimeric complexes with DP proteins, and this interaction confers their nuclear translocation and transactivation (De Veylder et al., 2002; Kosugi and Ohashi, 2002).

These findings indicate that all the elements are present for D-type cyclins to mediate G1/S transition in plants via a mechanism conserved at least conceptually with other higher eukaryotes. Our recent development of a synchronous cell suspension for *Arabidopsis* (Menges and Murray, 2002), and the analysis of cell cycle regulated gene expression (Menges et al., in preparation) will provide powerful tools for the further analysis of the control of the cell cycle.

3. D-TYPE CYCLINS INTEGRATE EXTRACELLULAR SIGNALS

The expression of some D-type cyclins responds positively to extracellular signals such as carbohydrates and plant hormones. In sucrose-starved *Arabidopsis* cell suspensions, the expression of *CYCD2;1* increases almost immediately upon sucrose induction whereas *CYCD3;1* expression increases somewhat later in G1 preceding S-phase entry. Interestingly, the induction of *CYCD2;1* and *CYCD3;1* expression by sucrose did not require de novo protein synthesis nor progression through the cell cycle but appeared to involve the action of protein phosphatases (Riou-Khamlichi et al., 2000). Different D-type cyclins appear to respond to different signals. *CYCD2;1* responded to sucrose but not to hormones, whereas *CYCD3;1* responds to cytokinins, brassinosteroids, auxin and giberellic acid with the strongest response towards cytokinins. (reviewed in Oakenfull et al., 2002).

4. D-TYPE CYCLINS ARE PUTATIVE TARGETS FOR DEVELOPMENTAL PATHWAYS

In situ hybridisation analysis of D-type cyclins showed differential expression of the different D-type cyclins. In *Arabidopsis*, *CYCD4;1* transcripts were detected in vascular tissues, developing embryos and were associated with all root pericycle cells involved in the lateral root initiation and their direct progeny (De Veylder et al., 1998). *CYCD3;1* transcripts are detected in all dividing cells such as vegetative and floral meristems, axial buds, vascular tissues and developing leaves but are not present in differentiating tissues such as older leaves and the pith (Dewitte et al., submitted). In roots, *CYCD3;1* expression as detected by a promoter-GUS fusion was associated with the stele cells at the bases of emerging lateral roots, and in more mature roots more generally with the stele tissues (unpublished data). No expression of GUS was seen in the proliferating cells of the root meristem. In shoot apices of *Antirrhinum* clear cut differences in transcript distribibution of different members of the *CYCD3* class were detected. *Antirrhinum CYCD3a* transcripts accumulated in the young developing leaves whereas *CYCD3b* is expressed in all dividing tissues (Gaudin et al., 2000). In summary, these data indicate regulation of D-type cyclin gene transcription conferred by their promoters, suggesting that the expression of at least some *CYCD* genes is not simply proliferation-associated (and hence in all dividing cells), but is subject to tissue-specific or developmental control. Indeed, overexpression of *AINTEGUMENTA* (*ANT*), a transcription factor involved in the control of cell numbers in organs, conferred ectopic expression of *CYCD3;1* (Mizukami and Fisher, 2000), suggesting that CycD3;1 acts downstream of such patterning genes.

Overexpression experiments of D-type cyclins induced significant effects on growth and development. Overexpression of Arabidopsis *CYCD2;1* shortened the G1 phase in tobacco meristems and hence promoted overall growth rate, but did not alter development (Cockcroft et al., 2000). Constitutive overexpression of both *CYCD3;1* and *CYCD1;1* in Arabidopsis increased cell numbers in leaves and perturbed cellular differentiation. Mild effects on leaf cell number and differentiation were caused by overexpression of *CYCD1;1*, and shoots twist along their longitudinal axis.

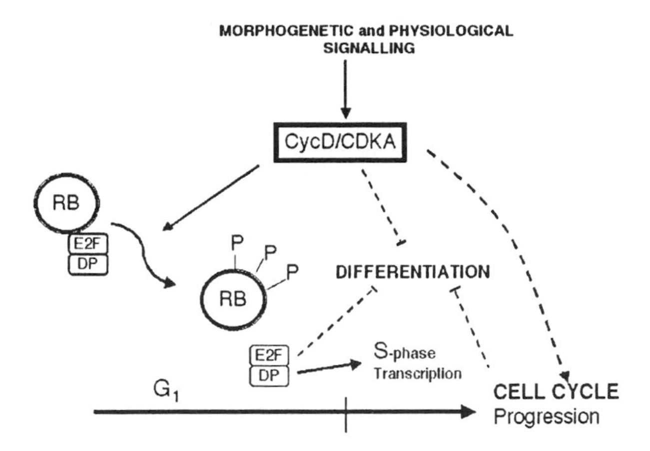

Figure 1. ***A proposed role for D-type cyclins and their cognate kinase complexes in integrating the interlinked processes of cell cycle progression and differentiation in reponse to morphogenetic and physiological signals. D-type cyclins form active complexes with CdkA, and phosphorylate Rb protein. Active E2F/DP heterodimeric factors can then activate cell cycle processes leading to S phase entry. Other interactions with differentiation and cell cycle mechanisms are intriguing aspects to explore.***

Vascular elements were found to differentiate prematurely in pedicels, which could indicate a close relation between cell cycle and differentiation in the formation of vascular elements (Huntley et al., unpublished results). But the effects of increased *CYCD3;1* expression were the most profound. Overexpression of *CYCD3;1* resulted in stunted plants with curled leaves with increased CycD3;1-associated kinase activity. Cell proliferation persisted and differentiation was retarded in leaves and cotelydons of these overexpressors. In shoot apices, differentiation was retarded resulting in a less pronounced border between the shoot apical meristem and surrounding tissues. Furthermore, cells in the shoot apical meristem of *CYCD3;1* overexpressing plants were smaller, indicating a modified relation between cell growth and cell division (Dewitte et al., submitted).

Interestingly, homeostasis of patterning genes was also perturbed in *CYCD3;1* overexpressors, indicating that mechanisms exist which link cell division to the expression of patterning genes, although *ANT* expression is unaltered (see above). This confirms that *CYCD3;1* acts downstream of *ANT* to influence cell number in developing leaves. The effects on differentiation and cell proliferation of *CYCD3* overexpression were also confirmed in tobacco by the use of a chemical inducible expression system (Matole, N.H. and Murray, J.A.H., unpublished results). Application of the inducer dexamethasone resulted in differentiation defects in leaves and flowers in overexpressors of the different members of the *CYCD3* gene family.

5. CONCLUSIONS

The control of cell proliferation in a developmental context is likely to focus primarily on regulation of the G1 phase as the main period of commitment. Results from *CYCD3;1* overexpression also suggests that acceleration through G1 also perturbs normal differentiation, indicating the further importance of this phase in differentitation decisions for cells. During this period D-type cyclins interact with Rb proteins, E2F (and their partner proteins DP) to control the speed and behaviour of cells as they transit this critical phase (Figure 1). D-type cyclins are rate limiting components of this process and are thus good candidates not only to mediate physiological signals but also to interact with developmental controls.

6. REFERENCES

Cockroft, C.E, den Boer, B.G.W., Healy, S.J.M .and Murray, J.A.H. (2000) Cyclin D control of growth rate in plants. Nature 405:575-579.

De Jager S.M., Menges M., Bauer U.-M. and Murray J.A.H. 2001. *Arabidopsis* E2F1 binds a sequence present in the promotor of S-Phase-regulated gene *ATCDC6* and is member of a multigene family with differential activities. Plant Mol. Biol. 47:555-568.

De Veylder L., Beeckman T., Beemster G.T.S., De Almeida Engler J., Ormenese S., Maes S., Naudts M., Van der Scheuren E., Jacqmard A., Engler G. and Inzé D. 2002. Control of proliferation, endoreduplication and differentiation by the Arabidopsis E2Fa-Dpa transcription factor. EMBO J. 21:1-9.

De Veylder, L., de Almeida, E. J., Burssens, S., Manevski, A., Lescure, B., Van Montagu, M., Engler, G. and Inzé, D. 1999. A new D-type cylin of *Arabidopsis thaliana* expressed during lateral root primordia formation. Planta 208:453-462.

Gaudin V., Luness P, Fobert P.R., Towers M., Riou-Khamlichi C., Murray J.A.H., Coen E. and Doonan J. 2000. The expression of D-type cyclin genes defines distinct developmental zones in snapdragon apical meristems and is locally regulated by the cycloidea gene. Plant Physiol. 122 :1137-1148.

Healy J.M.S., Menges M., Doonan J. and Murray J.A.H. 2001. The Arabidopsis D-Type cylins CycD2 and CycD3 both interact *in vivo* with the PSTAIRE Cyclin –dependent Kinase Cdc2a but are differentially controlled. J. Biol. Chem. 276 :7041-7047.

Huntley, R., Healy, S., Freeman, D., Lavender, P., de Jager S., Greenwood, J., Makker, J., Walker, E., Jackman,M., Xie, Q., Bannister, A.J., Kouzarides, T., Gutiérrez C., Doonan, J.H. and Murray, J.A.H. 1998. The maize retinoblastoma protein homoloque ZmRB-1 is regulated during leaf development and displays conserved interactions with G1/S regulators and plant cyclin D (CycD) proteins. Plant Mol. Biol. 37 :155-169.

Kosugi S. and Ohanhi Y. 2002. Interaction of the Arabidopsis E2F and DP proteins confers their concomitant nuclear translocation and transactivation. Plant Physiol. 128:833-843.

Meijer M. and Murray J.A.H. 2000 The role and regulation of D-type cyclins in the plant cell cycle. Plant Mol. Biol. 43:621-633.

Menges M and Murray J.A.H. 2002. Synchronous *Arabidopsis* suspension cultures for analysis of cell-cycle gene activity. Plant J. 26:203-212.

Mizukami Y. and Fisher R.L. 2000. Plant organ Size control: AINTEGUMENTA regulates growth and cell numbers during organogenesis. PNAS 97: 942-947.

Nakagami, H., Sekine, M,. Murakaml, H. and Shinmyo, A. 1999. Tobacco retinoblastoma-related protein phosphorylated by a cyclin dependent kinase complex with cdc2/cyclin D *in vitro*. Plant J. 18: 243-252.

Oakenfull E.A., Riou-Khamlichi C. and Murray J.A.H. 2002. Plant D-type cyclins (CycDs) and the control of G1 progression. Phil. Trans. R. Soc. Lond. In Press

Riou-Khamlichi C., Menges M., Healy J.M.S and Murray J.A.H. 2000. Sugar Control of the plant cell cycle: Differential Regulation of *Arabidopsis* D-type cyclin Gene expression. Mol. Cell. Biol. 20:4513-4521.

Vandepoele K., Raes J., De Veylder L., Rouzé P., Rombauts S. and Inzé D. 2002. Genome-Wide analysis of core cell cycle genes in Arabidopsis. Plant Cell 14:903-916.

Wyrzykowska J., Pien S. , Shen W.H. and Fleming A.J. 2002. Manipulation of leaf shape by modulation of cell division . Development 129:957-964.

USING GENES THAT STIMULATE THE CELL CYCLE TO IMPROVE MAIZE TRANSFORMATION

Bill Gordon-Kamm[1], Yumin Tao[1], Brian Dilkes[2], Keith Lowe[1], George Hoerster[1], Xifan Sun[1], Margit Ross[1], Laura Church[1], Chris Bunde[1], Jeff Farrell[1], Patrea Hill[1], Sheila Maddock[1], Jane Snyder[1], Ricardo Dante[2], Dennis Bidney[1], Ben Bowen[3], Pete John[4], and Brian Larkins[2]
[1]Pioneer Hi-Bred, International, Inc., Johnston, IA 50131; [2]Dept. of Plant Science, Univ. of Arizona, Tuscon, AZ; [3]Lynx Therapeutics, Inc, Hayward, CA; [4]Dept. of Plant Science, ANU, Canberra (William.Gordon-Kamm@Pioneer.com)

Keywords cell cycle, transformation, maize, cyclin-D, RepA

1. INTRODUCTION

The cell cycle's impact on plant transformation has been investigated by various groups, showing that S-phase (Villemont et al., 1997), M-phase (Okada et al., 1986) or both (Meyer et al., 1985) appears to be correlated with increased transformation frequencies. While such studies show that cell cycle progression influences transformation, no methods have been reported that stimulate transformation by expressing cell cycle genes.

2. IMPROVING MAIZE TRANSFORMATION

We tested various cell cycle genes for their impact on maize transformation. In all experiments, DNA was delivered *via* particle bombardment into maize Hi-II immature embryos. All treatments included *UBI::moPAT~GFP::pinII* (a fusion between maize-optimized bialaphos resistance gene and maize-optimized GFP) plus-or-minus the cell cycle genes. Genes tested included maize cell cycle genes (*Cyclin-A*, *cyclin-B*, *Cyclin-D*) a *Schizosaccharomyces pombe cdc25* and another gene known to bind retinoblastoma, the wheat dwarf virus RepA (Figure 1). Of the maize genes, only the *CycD*'s contain a sequence encoding a "LXCXE" Rb-binding domain, and members of this family produced the highest transformation frequencies (three maize *CycD* family members showed similar transformation frequencies - results for two *CycD* genes shown in Figure 2). The cdc25 gene from budding yeast (*Schizosaccharomyces pombe*) also increased transformation, but less than *CycD*. *ZmCycD+* calli retained their morphogenetic competence, were readily regenerable, and produced healthy, fertile plants with Mendelian inheritance.

I. K. Vasil (ed.), Plant Biotechnology 2002 and Beyond, 255-258.

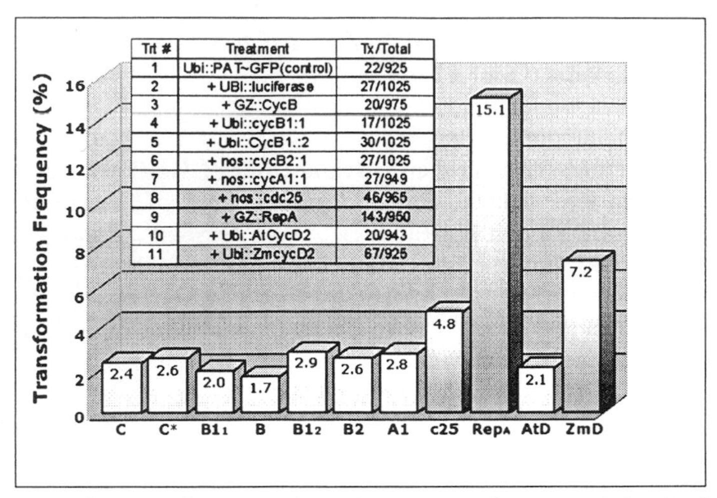

Figure 1. Transformation frequencies for Hi-II immature embryos particle bombarded with the Ubi:PAT~GFP:pinII construct alone (C), with a UBI::luciferase::pinII construct (C) or with each of the various cell cycle genes. For each treatment, approximately 40 replicates with 25 embryos/replicate were done. Transformation frequencies (bars) and the total number of transformants per total number of embryos bombarded (Tx/Total) are shown. Of the cell cycle stimulatory genes tested, three appeared to increase transformation frequency, SpCdc25, ZmCycD and WDV RepA. CycD2 from Arabidopsis did not improve maize transformation relative to control values.*

As with the CycD genes, plants over-expressing the various maize cell cycle genes appeared normal, and transmitted the respective transgene cassettes to progeny in Mendelian fashion. Although fertile with normal inheritance, the *Spcdc25* plants exhibited a dwarf phenotype. Of the genes tested to date, RepA showed the greatest stimulation of transformation frequency. Similar to CycD genes, WDV RepA contains an LxCxE Rb-binding motif (Xie et al., 1995).

Figure 3 shows two trends observed when RepA was included in experiments. Using either a moderate (nos) or strong (Ubiquitin) promoter to drive RepA expression, it was observed that i) transformation frequency increased with promoter strength (1.6% in control; 22.5% for nos::RepA, and 27.5% for Ubi::RepA), and ii) callus growth over the first 14 days appeared to increase with promoter strength. Readily detectable RepA expression was observed in Westerns (in calli and leaf samples) and transgene copy number (based on Southerns) showed a typical range of integrated copies relative to published values in maize (i.e. Register et al., 1994). At the highest expression levels (some Ubi::RepA events) it was difficult to regenerate plants, but plants were readily regenerated at moderate and low expression levels (some Ubi::RepA events and all of the nos::RepA events). Plants were easily regenerated from nos::RepA events, were phenotypically normal, were fertile and transmitted the Ubi:moPAT~GFPm / nos::RepA transgenic locus (and expression) in Mendelian fashion. Surprisingly, for segregating progeny

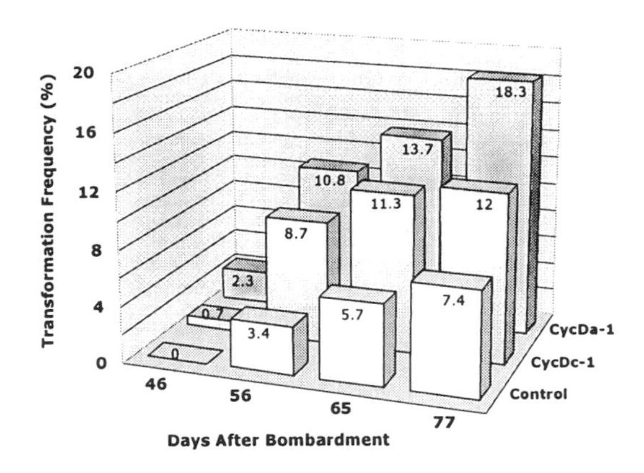

Figure 2. Number of macroscopic GFP-expressing, bialaphos resistant calli recovered at increasing intervals after particle bombardment. An equal number of immature embryos were bombarded in each treatment – with Ubi::PAT~GFP::pinII alone (control), with the control plasmid + Ubi::ZmCycDc-1::pinII, or with the control plasmid + Ubi::ZmCycDa-1::pinII. Both CycD genes increased the final transformation frequency relative to the control, and visible calli were recovered earlier in the selection process (indicative of stimulated growth).

(either RepA+ or wild-type) we found that RepA+ embryos retained this trait of "elevated transformation". Thus, these embryos could be "re-transformed" at high efficiency with a new set of transgenes (for example agronomic genes) and later segregated away from the agronomic locus. These results demonstrate that single cell cycle genes can be used to substantially improve maize transformation.

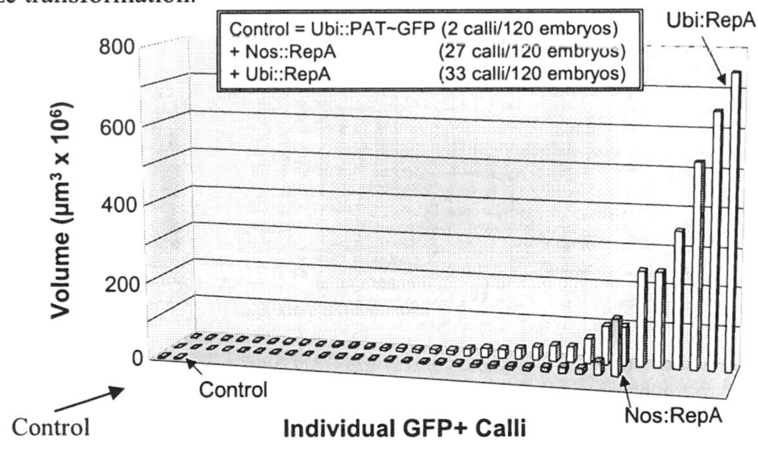

Figure 3. Volume of GFP+ multicellular clusters 14 days after particle bombardment with Ubi::PAT~GFP::pinII alone (control), or with the addition of nos::RepA::pinII or Ubi::RepA::pinII. The calli in the control were growing at a typical rate for early transgenic Hi-II calli, while growth of transformed calli in the nos::RepA and Ubi::RepA treatments was greatly stimulated. Mean volumes for these three treatments were 2.3, 11.0 and 100.4(μm^3 x 10^6), respectively. For each GFP-expressing callus, two perpendicular measurements were taken, averaged to obtain a diameter, and used to calculate volume ($V = 4/3\pi r^3$).

3. REFERENCES

Meyer P., Wlagenbach E., Bussmann K., Hombrecher G., and H. Staedler. 1985. Synchronized tobacco protoplasts are efficiently transformed by DNA. MGG 210:513-518.

Okada K, Takebe I, and T. Nagata. 1986. Expression and integration of genes introduced into highly synchronized plant protoplasts. MGG 205:398-403.

Register JC 3rd, Peterson DJ, Bell PJ, Bullock WP, Evans IJ, Frame B, Greenland AJ, Higgs NS, Jepson I, Jiao S, et al. 1994. Structure and function of selectable and non-selectable transgenes in maize after introduction by particle bombardment. Plant Mol Biol 25:951-61.

Villemont E., Dubois F., Sanwan R.S., Vasseur G., Bourgeois Y, and B.S. Sangwan-Norreel. 1997. Role of the host cell cycle in the *Agrobacterium*-mediated genetic transformation of *Petunia*: evidence of an S-phase control mechanism for T-DNA transfer. Planta 201:160-172.

Xie Q, Suarez-Lopez P, Gutierrez C. 1995. Identification and analysis of a retinoblastoma binding motif in the replication protein of a plant DNA virus: requirement for efficient viral DNA replication. EMBO J 14:4073-82.

THE CDK INHIBITOR ICK1 AFFECTS CELL DIVISION, PLANT GROWTH AND MORPHOGENESIS

Hong Wang[1], Yongming Zhou[1,2], Susan Gilmer[2], Ann Cleary[3], Pete John[3], Steve Whitwell[1], and Larry Fowke[2]
[1]Saskatoon Research Center, AAFC, Saskatoon SK, S7N 0X2, Canada; [2]Department of Biology, University of Saskatchewan, Saskatoon SK, S7N 5E2, Canada; [3]Plant Cell Biology Group, RSBS, Australian National University, Canberra, Australia (email: larry.fowke@usask.ca)

Keywords CDK inhibitor, ICK1, cell division, altered plant growth and development

1. INTRODUCTION

The cell cycle in plants and animals is regulated by cyclin-dependent protein kinases (CDKs), which are activated by association with cyclins and can be inhibited by the action of small protein inhibitors (Mironov et al., 1999). ICK1, the first plant CDK inhibitor, was isolated from *Arabidopsis thaliana* (Wang et al., 1997,1998). Although sharing similarity in a small region with known animal inhibitors, the protein encoded by *ICK1* exhibits a unique structure and recombinant ICK specifically inhibits activity of p13^{suc1}-bound plant CDK *in vitro*. Different approaches were used to establish the role of ICK1 in controlling cell division, plant growth and morphogenesis.

2. MICROINJECTION

Microinjection of ICK1 into individual dividing *Tradescantia virginiana* stamen hair cells during late prophase and prometaphase resulted in an increase in the time from nuclear envelope breakdown to anaphase in a manner dependent on time of injection and load (Cleary et al., 2002). The results suggest that CDKs are important for mitotic progression and that ICK1 is capable of inhibiting cell division.

I. K. Vasil (ed.), Plant Biotechnology 2002 and Beyond, 259-260.

3. OVER-EXPRESSION

Transgenic *Arabidopsis thaliana* plants were generated expressing *ICK1* driven by the cauliflower mosaic 35S promoter (Wang et al., 2000). Growth of transgenic plants was inhibited with some being less than 10% the size of wild-type plants. There were also modifications in plant morphology such as shape and serration of leaves and petals. Increased *ICK1* expression resulted in reduced CDK activity and a reduced number of cells in these plants. The inhibition of cell division by ICK1 thus had profound effects on both plant growth and development.

4. TARGETED EXPRESSION

To determine the effects of ICK1 expression on particular organs and cells, *ICK1* was expressed in *Brassica* plants by using the two tissue-specific *Arabidopsis AP3* and *Brassica Bgp1* promoters (Zhou et al., 2002). Transgenic *AP3-ICK1* plants had modified flowers with petals of reduced size, novel shapes (e.g. tubules, filaments) and often no petals. In some *Bgp1-ICK1* plants, pollen development was affected.

The CDK inhibitor ICK1 slows cell division when injected into dividing plant cells. Transgenic plants over-expressing *ICK1* exhibit reduced CDK activity, reduced growth as a result of fewer cells and altered morphology.

6. REFERENCES

Cleary A.L., Fowke L.C., Wang H. and John P.C.L. 2002. The effect of ICK1, a plant cyclin-dependent kinase inhibitor, on mitosis in living plant cells. Plant Cell Rep. 20: 814-820.

Mironov V., De Veylder L., Van Montagu M. and Inze D. 1999. Cyclin-dependent kinases and cell division in plants – the nexus. Plant Cell 11: 509-521.

Wang H., Fowke L.C. and Crosby W.L. 1997. A plant cyclin-dependent kinase inhibitor gene. Nature 386: 451-452.

Wang H., Qi Q., Schorr P., Cutler A.J., Crosby W.L. and Fowke L.C. 1998. ICK1, a cyclin-dependent protein kinase inhibitor from *Arabidopsis thaliana* interacts with both Cdc2a and CycD3, and its expression is induced by abscisic acid. Plant J. 15: 501-510.

Wang H., Zhou Y., Gilmer S., Whitwell S. and Fowke L.C. 2000. Expression of the cyclin-dependent kinase inhibitor ICK1 affects cell division, plant growth and morphology. Plant J. 24: 613-623.

Zhou Y., Wang H., Gilmer S., Whitwell S., Keller W. and Fowke L.C. 2002. Control of petal and pollen development by the plant cyclin-dependent kinase inhibitor ICK1 in transgenic *Brassica* plants. Planta (in press).

ARABIDOPSIS CDC2A AND CYCLIN GENE PROMOTER::*GUSA* CONSTRUCTS AS MARKERS OF CELL GROWTH AND DIVISION IN HETEROLOGOUS PLANTS

Mark R. Fowler[1], Nigel W. Scott[1], Abdul R. Milan[1,2], Alex C. McCormac[1,3], Manoj K. Mishra[1,4], Cui-Ying Shao[1,5], Jiang Xiaocheng[1], Chun-Lai Zhang[1], Yan Zhou[1], Malcolm C. Elliott[1] and Adrian Slater[1]
[1]The Norman Borlaug Institute for Plant Science Research, De Montfort University, Scraptoft, Leicester, LE7 9SU, UK (e-mail: melliott@dmu.ac.uk)
[2]Horticulture Research Centre, MARDI, P.O. Box 12301, 50774 Kuala Lumpur, Malaysia
[3]School of Biological Sciences-Division of Cell Science, Southampton University, Bassett Crescent East, Southampton, SO16 7PX, UK.
[4]Indian Coffee Board, Manasagangothri, Mysore 570006, Karnataka, India
[5]Department of Molecular and Cell Biology, Institute of Medical Sciences, University of Aberdeen, Foresterhill, Aberdeen, AB25 2ZD, UK.

Keywords CDK, cell division, *Bvcrk1*, cyclin, plant development

1. INTRODUCTION

Cell growth and division are integrated with a programme of differentiation to determine the form of the plant body. In recent years the molecular controls of cell division and growth have been elucidated. Progression through the cell cycle is controlled by a family of serine/threonine protein kinases termed CDKs (cyclin-dependent kinases) that act in association with regulatory proteins termed cyclins. Specific CDKs associate with specific cyclins in order to bring about both the gene expression changes and the physical modifications that are necessary to complete the cell cycle. The nature and interactions of CDKs and cyclins have been studied in detail in mammalian and yeast cells but there is still relatively little information about such systems in plants. Much of the information is drawn by inference. Little is known about the coupling of cell division with cell growth and the integration of these processes into development.

Several classes of CDK (termed CDKA to CDKF) have been identified in plants. Members of the CDKA class have a conserved PSTAIRE epitope and the ability to complement yeast mutants. They are presumed to be *bona fide* CDKs. Members of the CDKB group resemble *bona fide* CDKs but they do not have a perfectly conserved PSTAIRE epitope nor do they complement

I. K. Vasil (ed.), Plant Biotechnology 2002 and Beyond, 261-262.

yeast mutants. The other classes contain both plant specific sequences and ones that resemble mammalian CDK-like sequences. The role played by these other classes of CDK in cell cycle control is not well defined. Plants contain relatively few fully characterised, genuine, CDKs. However, they do contain multiple cyclins, categorised as A-, B- or D-type. Plant A- and B-type cyclins appear to be "mitotic" cyclins, whereas D-type cyclins appear to regulate progression through or into the G_1-phase and entry into the S-phase. The expression of *CDC2a* is proposed to correlate with "competence for division" as opposed to mitotic activity itself, whereas the expression of B-type cyclins is correlated with active cell division.

Arabidopsis cyclin and cyclin-dependent kinase promoter::*gusA* fusions have been used to investigate cell division during development of heterologous plants (tobacco, rice, alfalfa and carrot). The expression patterns of these constructs have been compared with those of a novel CDK-like gene (*Bvcrkl*) promoter::*gusA* fusion. *Bvcrkl* encodes a C-class CDK characterised by a KFMARE epitope and long N- and C-terminal extensions.

2. RESULTS and DISCUSSION

During the development of transgenic tobacco, alfalfa and carrot plants, all three *Arabidopsis* promoters were active in both primary and secondary meristems. *CycB1*;1 activity was restricted to tissues with active cell division. However, *CDC2a* and *CycA2*;1 activity was also noted in tissues that did not contain mitotically active cells. This difference in expression pattern was obvious when promoter activity was analysed in carrot storage roots. All three promoters were active in the vascular cambium, but *CycB1*;1 was only active in individual cells of the meristem. *CycA2*;1 was expressed throughout the inner xylem and outer phloem tissues. The *CDC2a* promoter was active in phloem and periderm. Analyses of promoter activity in diverse plant tissues, both *in vivo* and *in vitro,* indicated that *CDC2a* and *CycA2*;1 expression was more closely correlated with cell growth than with competence for division. Signals regulating the activity of the three promoters were investigated during somatic embryogenesis in alfalfa. The auxin 2,4-D, which is required for cell proliferation in this system, was found to induce only the *CycB1*;1 promoter. The activity of this promoter was inhibited by treatment of the tissue with hydroxyurea but it was not inhibited by treatment with oryzalin, which arrests cells in the M phase. The promoter activity of *Bvcrkl* matched the expression pattern of the *CycB1*;1 promoter more closely than that of *CDC2a*; expression was generally associated with tissues in which active cell division was taking place.

The utility of cell division cycle gene promoter::*gusA* constructs for analyses of growth and development has been confirmed.

LEAFY COTYLEDON GENES AND THE CONTROL OF EMBRYO DEVELOPMENT

John J. Harada[1,2], Sandra L. Stone[1], Raymond W. Kwong[1,2], Hye-seung Lee[1,2], Linda W. Kwong[1], and Julie Pelletier[1]
[1]Section of Plant Biology, Division of Biological Sciences, University of California, One Shields Avenue, Davis, CA 95616, USA (e-mail: jjharada@ucdavis.edu); [2]Graduate Group in Plant Biology, University of California, One Shields Avenue, Davis, CA 95616

Keywords Arabidopsis, B3 domain, CCAAT binding factor, somatic embryogenesis

1. INTRODUCTION

Zygotic embryogenesis begins with the double fertilization event in which the egg cell of the female gametophyte fuses with one sperm nucleus to form the zygote, and the central cell fuses with another sperm nucleus to form the endosperm mother cell (Russell, 1993). The single-celled zygote then undergoes a series of differentiation events, resulting in the formation of a mature embryo. The basic body plan of the plant is established during the early morphogenesis phase of embryogenesis. During this period, regional specification events establish morphological domains within the developing embryo, the polarity of the embryo is expressed as a shoot-root axis, the embryonic tissue and organ systems are formed, and the rudimentary shoot and root apices develop (Goldberg et al., 1994; Jurgens, 2001; West and Harada, 1993). In seed plants, this early embryonic period is followed by the maturation phase in which the embryo acquires the ability to withstand desiccation, storage reserves in the form of proteins, lipids, and starch accumulate in the embryo and/or endosperm, and the embryo becomes metabolically quiescent as a result of desiccation (Bewley, 1997; Harada, 1997; Koornneef and Karssen, 1994). The seed generally remains in a quiescent state until environmental conditions signal the embryo to germinate.

A striking characteristic of plants is that several cell types other than a zygote have the capacity to form embryos. Somatic cells treated with hormones, usually auxin, often can be induced to undergo embryogenesis (Dodeman et al., 1997; Zimmerman, 1993). Microspores can be induced to switch from pollen development to embryogenesis (Reynolds, 1997). A variety of cells in the ovule can undergo asexual developmental pathways to yield embryos

I. K. Vasil (ed.), Plant Biotechnology 2002 and Beyond, 263-268.

through a suite of processes known as apomixis (Koltunow, 1993; Koltunow et al., 1995). Although morphological development appears to follow similar pathways in these different forms of embryogenesis, little is known at a mechanistic level about the processes that initiate embryo development. Nor is it known if common mechanisms are employed in these different embryonic pathways to shift cells to an embryonic fate.

2. ARABIDOPSIS *LEAFY COTYLEDON* GENES ARE ESSENTIAL REGULATORS OF EMBRYO DEVELOPMENT

To address questions about the cellular processes that underlie the initiation of embryo development, we have focused on the *LEAFY COTYLEDON* (*LEC*) genes of Arabidopsis, *LEC1, LEC2*, and *FUSCA3* (*FUS3*). The phenotypes resulting from mutations that initially identified these genes provided strong evidence that they play critical roles in embryogenesis (reviewed by Harada, 2001, and Baumlein et al., 1994; Keith et al., 1994; Meinke, 1992; Meinke et al., 1994; West et al., 1994). First, *LEC* genes are required early in embryogenesis to maintain cell fate, specifically of suspensor cells. Second, cotyledons of *lec* mutants have trichomes on their surfaces and, thus, exhibit leaf-like characteristics. This finding suggests that the *LEC* genes normally function in the specification of cotyledon identity. Third, the *LEC* genes function in the initiation and/or maintenance of the maturation phase. *lec* mutant embryos are intolerant of desiccation, to different degrees, and they are defective in the accumulation of storage proteins and lipids. Fourth, most likely because of their role in maintaining the maturation phase, *LEC* genes are also required to inhibit germination during embryogenesis. *lec* mutants exhibit morphological and molecular characteristics of germinating seedlings, indicating the heterochronic nature of the mutations.

The patterns of *LEC* gene expression suggest that the genes function predominately during embryogenesis. *LEC1* mRNA accumulates seed-specifically in developing embryos and endosperm and is not detected in vegetative organs (Lotan et al., 1998). *LEC2* and *FUS3* mRNAs accumulate primarily in developing embryos and is detected at very low levels in vegetative organs (Luerssen et al., 1998; Stone et al., 2001). Restriction of *LEC* gene expression primarily to the period of seed development explains the observation that defects induced by *lec* mutations are limited to embryo development.

Consequences of expressing the *LEC* genes ectopically revealed new insights about their role in embryogenesis. *LEC1* expressed under control of the *35S* promoter of cauliflower mosaic virus had dramatic effects on embryo development (Lotan et al., 1998). Seedlings expressing the *LEC1* gene retain embryonic characteristics in that cotyledons remain fleshy and do not expand, hypocotyls and roots do not extend, and organs that emerge at the position of leaves have characteristics of cotyledons. Although most *35S::LEC1* plants do not develop beyond the seedling stage, some vegetatively growing plants can be obtained. In rare cases, somatic embryos develop on the surface of leaves. Thus, *LEC1* establishes an embryonic environment that is sufficient to induce somatic embryo formation. Ectopic *LEC2* expression is also able to induce somatic embryo formation (Stone et al., 2001). However, *35S::LEC2* seedlings did not arrest in their development but, rather, are prolific in producing somatic embryos, cotyledon-like organs, and vegetative organs. Because both of these genes are sufficient to confer embryogenic competence to vegetative cells, we hypothesize that *LEC1* and *LEC2* operate early in embryogenesis to induce embryo formation. We also suggest that *LEC1* and *LEC2* normally serve complementary roles during embryogenesis (Stone et al., 2001), with *LEC1* functioning to establish an embryonic environment in seeds and *LEC2* promoting cell proliferation within the embryo. The exact role of these genes in embryogenesis remains to be determined.

3. *LEAFY COTYLEDON* GENES ENCODE TRANSCRIPTION FACTORS

Consistent with the pleiotropic effects of the *lec* mutations, cloning of the *LEC* genes revealed that they encode regulatory proteins. LEC1 shares substantial sequence identity with HAP3 subunits of CCAAT binding transcription factor (CBF, Lotan et al., 1998). CBFs are heterooligomeric transcription factors that consist of either three or four subunits, depending on the organism (reviewed by Maity and de Crombrugghe, 1998; Mantovani, 1999). In yeast, CBF activates a specific set of genes required for mitochondrial respiration. In mammals, CBFs serve a general role by enhancing the transcriptional efficiency of large numbers of genes. Genes encoding other CBF subunits, HAP2 and HAP5, are also found in Arabidopsis, suggesting that the transcription factor operates in plants (Edwards et al., 1998; Gusmaroli et al., 2001; H. Lee, M. Kim, and J.J. Harada, unpublished results).

LEC2 and FUS3 both belong to a class of proteins that possess B3 domains (Luerssen et al., 1998; Stone et al., 2001). Proteins with B3 domains constitute a large family of transcription factors that in Arabidopsis include AUXIN RESPONSE FACTOR (Ulmasov et al., 1997), MONOPTEROS (Hardtke and Berleth, 1998), and RELATED TO ABI3/VP1 (Kagaya et al., 1999). The B3 domain appears to be a plant-specific motif that is responsible for the binding of DNA by these proteins (Ezcurra et al., 2000; Kagaya et al., 1999; Suzuki et al., 1997). Several B3 domain proteins have been shown in functional assays to be transcription factors, including FUS3 (Kagaya et al., 1999; McCarty et al., 1989; Reidt et al., 2000; Ulmasov et al., 1997). Among the large family of Arabidopsis B3 domain transcription factors, LEC2, FUS3, and ABA INSENSITIVE3 (ABI3) are the most closely related, suggesting that LEC2 is also a transcription factor.

Together, the findings summarized above suggest that the *LEC* genes serve as central regulators of embryogenesis. Moreover, LEC1 and LEC2 are sufficient to induce embryo formation, implicating a role for these proteins in the earliest stages of embryo development. Given that the LEC genes all encode transcription factors, we conclude that these regulatory proteins control the transcription of genes required for key processes in embryogenesis. The target genes regulated by LEC1, LEC2, and FUS3 and their precise roles in embryogenesis remain to be identified.

4. REFERENCES

Baumlein, H., Misera, S., Luerben, H., Kolle, K., Horstmann, C., Wobus, U., and Muller, A. J. 1994. The *FUS3* gene of *Arabidopsis thaliana* is a regulator of gene expression during late embryogenesis. Plant J. 6:379-387.

Bewley, J. D. 1997. Seed germination and dormancy. Plant Cell 9:1055-1066.

Dodeman, V. L., Ducreux, G., and Kreis, M. 1997. Zygotic embryogenesis versus somatic embryogenesis. J. Exp. Bot. 48:1493-1509.

Edwards, D., Murray, J. A. H., and Smith, A. G. 1998. Multiple genes encoding the conserved CCAAT-box transcription factor complex are expressed in *Arabidopsis*. Plant Physiol. 117:1015-1022.

Ezcurra, I., Wycliffe, P., Nehlin, L., Ellestrom, M., and Rask, L. 2000. Transactivation of the *Brassica napus* napin promoter by ABI3 requires interaction of the conserved B2 and B3 domains of ABI3 with different cis-elements: B2 mediates activation through an ABRE, whereas B3 interacts with an RY/G-box. Plant J. 24:57-66.

Goldberg, R. B., de Paiva, G., and Yadegari, R. 1994. Plant embryogenesis: zygote to seed. Science 266:605-614.

Gusmaroli, G., Tonelli, C., and Mantovani, R. 2001. Regulation of the CCAAT-binding NF-Y subunits in *Arabidopsis thaliana*. Gene 264:173-185.

Harada, J. J. 1997. Seed maturation and control of germination. In: *Advances in Cellular and Molecular Biology of Plants, Volume 4, Cellular and Molecular Biology of Seed Development.* (B. A. Larkins, and I. K. Vasil, eds.), Kluwer Academic Publishers, Dordrecht, pp. 545-592.

Harada, J. J. 2001. Role of *Arabidopsis* LEAFY COTYLEDON genes in seed development. J. Plant Physiol. 158:405-409.

Hardtke, C. S., and Berleth, T. 1998. The *Arabidopsis* gene *MONOPTEROS* encodes a transcription factor mediating embryo axis formation and vascular development. EMBO J.

17:1405-1411.

Jurgens, G. 2001. Apical-basal pattern formation in *Arabidopsis* embryogenesis. EMBO J. 20:3609-3616.

Kagaya, Y., Ohmiya, K., and Hattori, T. 1999. RAV1, a novel DNA-binding protein, binds to bipartite recognition sequence through two distinct DNA-binding domains uniquely found in higher plants. Nuc. Acids Res. 27:470-478.

Keith, K., Kraml, M., Dengler, N. G., and McCourt, P. 1994. *fusca3*: a heterochronic mutation affecting late embryo development in *Arabidopsis*. Plant Cell 6:589-600.

Koltunow, A. M. 1993. Apomixis: Embryo sacs and embryos formed without meiosis or fertilization in ovules. Plant Cell 5:1425-1437.

Koltunow, A. M., Bicknell, R. A., and Chaudhury, A. M. 1995. Apomixis: Molecular strategies for the generation of genetically identical seeds without fertilization. Plant Physiol. 108:1345-1352.

Koornneef, M., and Karssen, C. M. 1994. Seed Dormancy and Germination, in: *Arabidopsis* (E. M. Meyerowitz, and C. R. Somerville, eds.), Cold Spring Harbor Laboratory Press, Cold Spring Harbor, pp. 313-334.

Lotan, T., Ohto, M., Yee, K. M., West, M. A. L., Lo, R., Kwong, R. W., Yamagishi, K., Fischer, R. L., Goldberg, R. B., and Harada, J. J. 1998. *Arabidopsis* LEAFY COTYLEDON1 is sufficient to induce embryo development in vegetative cells. Cell 93:1195-1205.

Luerssen, H., Kirik, V., Herrmann, P., and Misera, S. 1998. *FUSCA3* encodes a protein with a conserved VP1/ABI3-like B3 domain which is of functional importance for the regulation of seed maturation in *Arabidopsis thaliana.*. Plant J. 15:755-764.

Maity, S. N., and de Crombrugghe, B. 1998. Role of the CCAAT-binding protein CBF/NF-Y in transcription. Trends Biochem. Sci. 23:174-178.

Mantovani, R.. 1999. The molecular biology of the CCAAT-binding factor NF-Y. Gene 239:15-27.

McCarty, D. R., Carson, C. B., Stinard, P. S., and Robertson, D. S. 1989. Molecular analysis of *viviparous-1*: an abscisic acid-insensitive mutant maize. Plant Cell 1:523-532.

Meinke, D. W. 1992. A homoeotic mutant of *Arabidopsis thaliana* with leafy cotyledons. Science 258:1647-1650.

Meinke, D. W., Franzmann, L. H., Nickle, T. C., and Yeung, E. C. 1994. *leafy cotyledon* mutants *of Arabidopsis*. Plant Cell 6:1049-1064.

Reidt, W., Wohlfarth, T., Ellerstroem, M., Czihal, A., Tewes, A., Ezcurra, I., Rask, L., and Baumlein, H. 2000. Gene regulation during late embryogenesis: The RY motif of maturation-specific gene promoters is a direct target of the FUS3 gene product. *Plant J.* 21:401-408.

Reynolds, T. L. 1997. Pollen embryogenesis. Plant Mol. Biol. 33:1-10.

Russell, S. D. 1993. The egg cell: Development and role in fertilization and early embryogenesis. Plant Cell 5:1349-1359.

Stone, S. L., Kwong, L. W., Yee, K. M., Pelletier, J., Lepiniec, L., Fischer, R. L., Goldberg, R.

B., and Harada, J. J. 2001. LEAFY COTYLEDON2 encodes a B3 domain transcription factor that induces embryo development. Proc. Natl. Acad. Sci. USA 98:11806-11811.

Suzuki, M., Kao, C. Y., and McCarty, D. R. 1997. The conserved B3 domain of VIVIPAROUS1 has a cooperative DNA binding activity. Plant Cell 9:799-807.

Ulmasov, T., Hagen, G., and Guilfoyle, T. J. 1997. ARF1, a transcription factor that binds to auxin response elements. Science 276:1865-1868.

West, M. A., and Harada, J. J. 1993. Embryogenesis in higher plants: An overview. Plant Cell 5:1361-1369.

West, M. A. L., Matsudaira Yee, K. L., Danao, J., Zimmerman, J. L., Fischer, R. L., Goldberg, R. B., and Harada, J. J. 1994. *LEAFY COTYLEDON1* is an essential regulator of late embryogenesis and cotyledon identity in *Arabidopsis*. Plant Cell 6:1731-1745.

Zimmerman, J. L. 1993. Somatic embryogenesis: A model for early development in higher plants. Plant Cell 5:1411-1423.

THE ROLE OF THE *ARABIDOPSIS* SOMATIC EMBRYOGENESIS RECEPTOR-LIKE KINASE 1 (AtSERK1) GENE IN EMBRYOGENIC COMPETENCE

Sacco de Vries, Khalid Shah, Ingrid Rienties, Vered Raz, Valerie Hecht and Jenny Russinova
Laboratory of Molecular Biology, Wageningen University, Dreijenlaan 3, 6703 HA Wageningen, The Netherlands (email: sacco.devries@mac.mb.wau.nl)

Keywords AtSERK1, KAPP, embryogenesis, FRET

1. INTRODUCTION

We have reported the isolation of the *Arabidopsis* SOMATIC EMBRYOGENESIS RECEPTOR-LIKE KINASE 1 (*AtSERK1*) gene and demonstrated its role during establishment of somatic embryogenesis in culture. The *AtSERK1* gene is highly expressed during embryogenic cell formation in culture and during early embryogenesis. The *AtSERK1* gene is first expressed in planta during megasporogenesis in the nucleus of developing ovules, in the functional megaspore, and in all cells of the embryo sac up to fertilization. After fertilization, *AtSERK1* expression is seen in all cells of the developing embryo until the heart stage. After this stage, *AtSERK1* expression is no longer detectable in the embryo or in any part of the developing seed. Low expression is detected in adult vascular tissue. Ectopic expression of the full-length *AtSERK1* cDNA under the control of the cauliflower mosaic virus 35S promoter did not result in any altered plant phenotype. However, seedlings that overexpressed the *AtSERK1* mRNA exhibited a 3- to 4-fold increase in efficiency for initiation of somatic embryogenesis. Thus, an increased AtSERK1 level is sufficient to confer embryogenic competence in culture.

2. FUNCTIONAL ANALYSIS OF THE RECEPTOR

The predicted AtSERK1 protein of 625 amino acids has a calculated molecular mass of 69 kD, and is slightly acidic (predicted pI of 5.25). The amino acid sequence of AtSERK1 shows a high percentage of identity with DcSERK (92%) and shares all the characteristic features of that protein,

I. K. Vasil (ed.), Plant Biotechnology 2002 and Beyond, 269-272.

including the five LRRs, the Pro-rich domain containing the so-called Ser-Pro-Pro (SPP) motif, containing two tandemly repeated SPP sequences, the transmembrane domain, and the kinase domain. Figure 1A shows the hydrophilicity plot for AtSERK1 containing two strongly hydrophobic regions. The first region, spanning residues 1 through 29, meets the conditions defining a signal peptide, with a potential signal peptidase cleavage site between positions 29 and 30. The second hydrophobic region, spanning residues 231 through 276, corresponds to the putative transmembrane domain separating the extracellular part of the protein and the intracellular kinase domain. Directly adjacent to the cleavage site of the putative signal peptide, the AtSERK1 protein contains a Leu-rich domain of 45 amino acids fitting the Leu-zipper (LZ) pattern Lx6Lx6Lx6L. It is surprising that these two domains are not present in DcSERK, which instead contains 28 amino acids that are absent in AtSERK1. The substantial similarity between the two proteins (92%) only begins at position 99 of AtSERK1. The LRR domain of AtSERK1 extends from positions 75 through 194 and is composed of five units. In most LRR receptor kinases, the transmembrane domain immediately follows the LRR domain. However, in AtSERK1, as in DcSERK, a Pro-rich region containing a repeated SPP motif separates these domains. We consider the SPP motif to be one of the hallmarks of the SERK-like RLKs. This motif has been suggested to act as a hinge providing flexibility to the extracellular part of the receptor or as a region for interaction with the cell wall. The intracellular region of AtSERK1 is also similar to DcSERK, containing the 11 subdomains characteristic of the catalytic core of Ser/Thr protein kinases.

To determine the intrinsic biochemical properties of the AtSERK1 protein, we have expressed the intracellular catalytic domain as a glutathione S-transferase (GST) fusion protein in *E. coli*. The AtSERK1-GST fusion protein mainly autophosphorylates on threonine residues (K_m=4x10^{-6}M) and the reaction is Mg^{2++} dependent and inhibited by Mn^{2++}. A lysine to glutamic acid substitution (K330E) in the kinase domain of AtSERK1 abolishes all kinase activity. The active AtSERK1^{kin} can autophosphorylate inactive AtSERK1^{K330E} protein demonstrating an intermolecular mechanism of autophosphorylation. The AtSERK1 kinase protein was modelled using insulin receptor kinase as template. Based on this model, threonine residues in the AtSERK1 activation loop of catalytic sub-domain VIII are predicted to be targets for autophosphorylation. Replacing Thr-468 with either alanine or glutamic acid completely obliterated the AtSERK1 ability to be phosphorylated. Transphosphorylation on MBP and casein showed tyrosine, serine and threonine as targets, demonstrating that AtSERK1 is a dual specificity kinase. The AtSERK1^{T468A} was not able to transphosphorylate, showing that only phosphorylated AtSERK1 kinase can transphosphorylate.

To determine whether the predicted AtSERK1 protein is able to homodimerize, the *AtSERK1* cDNA was fused to two different variants of green fluorescent protein (GFP), a yellow-emitting GFP (YFP) and a cyan-emitting GFP (CFP), and transiently expressed in both plant protoplasts and insect cells. Using confocal laser scanning microscopy (CSLM) it was determined that the AtSERK1-YFP fusion protein is targeted to plasma membranes in both plant and animal cells. The extracellular LRRs, and in particular the N-linked oligosaccharides that are present on them appear to be essential for correct localization of the AtSERK1-YFP protein. The potential for dimerization of the AtSERK1 protein was investigated by measuring the YFP/CFP fluorescence emission ratio using fluorescence spectral imaging microscopy (FSPIM). This ratio will increase due to fluorescence resonance energy transfer (FRET) if the AtSERK1-CFP and AtSERK1-YFP fusion proteins interact. In 15% of the cells the YFP/CFP emission ratio for plasma membrane localized AtSERK1 proteins was enhanced. Yeast protein interaction experiments confirmed the possibility for AtSERK1 homodimerization. Elimination of the extracellular LZ domain reduced the YFP/CFP emission ratio to control levels indicating that without the LZ Domain AtSERK1 is monomeric.

A first step towards elucidation of the AtSERK1-containing signaling complex suggested that the AtSERK1 protein interacts with the kinase associated protein phosphatase (KAPP) in vitro. The kinase interaction (KI) domain of KAPP does not interact with a catalytically inactive kinase mutant. Using mutant AtSERK1 proteins in which Thr-462, Thr-463 and Thr-468 in the A-loop of the AtSERK1 kinase domain were replaced by alanines, we show that phosphorylation status of the receptor is involved in interaction with KAPP. KAPP and AtSERK1 cDNAs were fused to two different variants of GFP, the yellow fluorescent protein (YFP) or the cyan fluorescent protein (CFP). Both KAPP and AtSERK1 proteins are found at the plasma membrane. Our results show that AtSERK1-CFP becomes sequestered into intracellular vesicles when transiently co-expressed with KAPP-YFP proteins. AtSERK1^{T463A}–CFP and AtSERK1$^{3T \rightarrow A}$-CFP proteins were partially sequestered intracellularly in the absence of KAPP-YFP protein, suggesting an active role for KAPP dephosphorylation of threonine residues in the AtSERK1 A-loop in receptor internalization. The interaction between the KAPP-CFP/YFP and AtSERK1-CFP/YFP fusion proteins was investigated with fluorescence spectral imaging microscopy (FSPIM). Our results show that AtSERK1-CFP and KAPP-YFP proteins are co-localized at the plasma membrane but only show fluorescence energy transfer (FRET) indicative of physical interaction in intracellular vesicles. These results suggest that KAPP is an integral part of the AtSERK1 endocytosis mechanism.

Current work employs extensive yeast interaction screens to identify additional components of the AtSERK1 complex, characterization of mutant phenotypes resulting from inactivation of the AtSERK1 gene as well as other members of the family of 5 highly similar receptor kinases present in the

Arabidopsis genome. Finally, we have begun to identify the ligands that activate the AtSERK receptor kinases using a variety of approaches.

3. REFERENCES

Hecht, V. Vielle-Calzada, J.-P., von Recklinghausen, I., Hartog, M., Zwartjes, C., Schmidt, E., Boutilier, K., Grossniklaus, U., de Vries, S. (2001) The *Arabidopsis* somatic embryogenesis receptor kinase gene in ovule and embryo development. Plant Physiol 127: 803–816

Shah, K., Gadella, T.W.J., van Erp, H., de Vries, S.C. (2001) Subcellular localization and oligomerization of the AtSERK1 protein. J. Mol. Biol. 309,:641-655.

Shah, K., Russinova, E., Gadella, T.W.J., Willemse, J., de Vries, S.C. (2002) The *Arabidopsis* kinase associated protein phosphatase controls internalization of the somatic embryogenesis receptor kinase 1. Genes & Development (in press).

Shah, K., Vervoort, J., de Vries, S.C. (2001) Autophosphorylation properties of AtSERK1, a receptor-like kinase from *Arabidopsis thaliana*. J. Biol. Chem. 276: 41263-41269.

ANALYSIS OF PEPTIDE SIGNALLING IN THE EMBRYO SAC OF MAIZE AND WHEAT

Suseno Amien, Stefanie Sprunck, Mihaela Márton, Simone Cordts and Thomas Dresselhaus
Biocenter Klein-Flottbek, AMP II, University of Hamburg, Ohnhorststrasse 18, D-22609 Hamburg, Germany (e-mail: dresselh@botanik.uni-hamburg.de)

Keywords Reproduction, female gametophyte, fertilisation, cell-cell communication, signal peptide, genomics

1. INTRODUCTION

In animal systems, cell-cell communications are mainly mediated by signals such as steroids and peptides. Among these signalling molecules, peptides are the most commonly used, probably because of their diversity. In plants, to date only a few peptides have been identified acting as signalling molecules mediating self incompatibility, cell and growth regulation, defence signalling and root nodulation (reviewed by Matsubayashi et al., 2001; Lindsey et al., 2002). However, the presence of more than 500 genes encoding putative receptor-like kinases (RLKs) in the *Arabidopsis* genome (The *Arabidopsis* Genome Initiative, 2000) supports the idea that peptide activated signalling cascades are also common in plants. For two of the 170 putative members of the LRR-(leucine-rich repeat)-RLK superfamily, the binding ligands have recently been identified as small signalling peptides (Trotochaud et al., 2000; Matsubayashi et al., 2002).

We are interested in cell-cell communication of the cells of the female gametophyte (embryo sac). Maturation of the embryo sac requires a very well balanced cell to cell communication system. From a coenocyte containing eight nuclei, seven cells differentiate and due to their functions and fates are positioned at precise locations within the female gametophyte. Before fertilisation, these cells communicate the position of the female gametes to the male gametes and prevent polyspermy upon fertilisation. The occurrence of fertilisation is communicated to the maternal tissues to provide nutrients for the developing embryo and endosperm. We thus assume that secreted signalling molecules (peptides) produced by the embryo sac play important roles before and after the fertilisation process.

I. K. Vasil (ed.), Plant Biotechnology 2002 and Beyond, 273-277.

2. RESULTS AND DISCUSSION

Isolating peptides from the embryo sac is not possible due to the limited amount of material. Therefore, we are using a genomics based approach to identify peptide genes expressed in the female gametophyte before and after fertilisation. Embryo sac cells of maize and wheat were microdissected before and after fertilisation and first cDNA libraries have been generated. Applying differential screening methods, we searched maize libraries for genes highly upregulated in the embryo sac before or after fertilisation. In addition, thousands of ESTs were generated from wheat libraries. cDNAs and ESTs of both plant species were analysed for the occurrence of genes encoding peptides (27-100 aa). PSORT (http://psort.nibb.ac.jp) and SignalP (http://www.cbs.dtu.dk/-services/SignalP) programs were used to identify signal peptides within protein sequences and WinPep (http://www.ipw.agrl.ethz.ch/~lhennig-/winpep) for transmembrane regions. Among the genes expressed in egg cells, we identified up to now eight classes encoding putative peptides (Table 1).

Table 1. Examples of putative peptides encoded by cDNAs or ESTs isolated from egg cell of maize (Zm) and wheat (Ta).

Clone (gene)	Peptide length[1]	Signal peptide	Homology
ZmES1-4	91-92 aa	+	PCP, SCR, Defensins?
TaEC/D03/12	97 aa	+	OsLTP2 (Lipid transfer protein)
TaEC/F04/11	79 aa	+	OsPSK4 (Phytosulphokine)
ZmTLA1	27 aa	+	-
ZmEC222	78 aa	+	-
ZmEC235	94 aa	-	(Type Ib,membrane?)
TaEC/D10/17	83 aa	-	OsBLE1 (Brassinolide induced)
ZmZ110[2]	42 aa	-	Caudal? (NLS?)

[1]length of putative propeptides
[2]isolated from a maize zygote library

The four genes of the first class (*ZmES1-4; Zea mays* embryo sac) encode cysteine-rich peptides with structural homology to defensins and are specifically expressed only in embryo sac cells (Cordts et al., 2001). This class is similar to PCP and SCR peptides involved in self incompatibility in *Brassica*, which are mediating cell-cell communication between the male gametophyte (pollen) with the female receptive tissue (stigma) (Doughty et al., 1998; Schopfer et al., 1999) and contains a predicted N-terminal signal peptide. ZmES4-GFP fusion protein was detected predominantly in the synergids around the filiform apparatus using conventional fluorescence microscopy (Figure 1). Confocal laser scanning microscopy (CLSM) showed protein localisation also outside the embryo sac at the extracellular space of the micropylar region.

The second class is almost identical to a secreted lipid transfer protein (LTP) from rice. Although the function of plant LTPs is yet unclear, like the PCP/SCR class, these cysteine-rich peptides are characterised by a conserved motif of eight cysteines to form four disulfide bonds. Secreated LTPs are activated after binding hydrophobic molecules and are thought to be involved in plant defence against biotic and abiotic stress as well as developmental regulation including somatic embryogenesis (Blein et al., 2002; Guiderdoni et al., 2002). The third class (*TaEC/F04/11*; *Triticum aestivum* egg cell) encodes a peptide homologous to rice phytosulfokines (PSKs). Mature PSKs were shown to be very small (four to five amino acids, respectively) disulphated peptides involved in cell division (Matsubayashi and Sakagami, 1996).

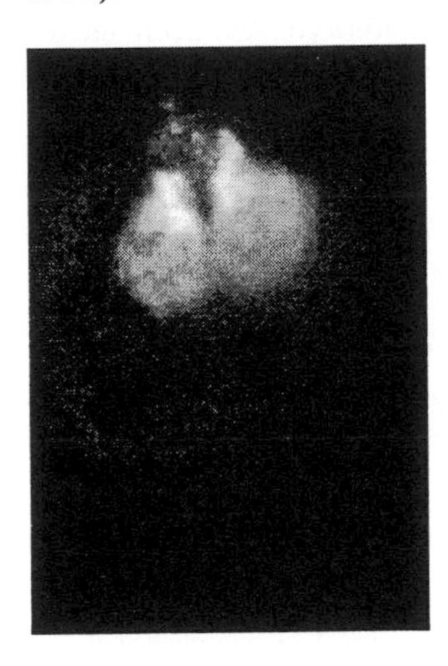

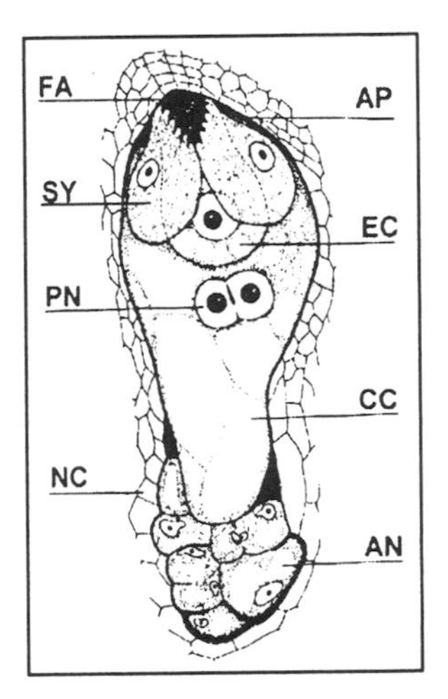

Figure 1. The ZmES4 peptide is localised in the synergids and most prominently around the filiform apparatus. Left: section through the ovule of a transgenic maize line expressing a ZmES4-GFP fusion protein under the control of the ZmES4 promoter. Right: mature embryo sac of maize. Drawing modified after Diboll and Larson (1966); AN: antipodal cells; AP: apical pocket, CC: central cell; EC: egg cell, FA: filiform apparatus, NC: nucellar cells; PN: polar nuclei; SY: synergids.

The next two classes represent up to now undescribed peptides. *ZmTLA1* (*Zea mays* transparent leaf areas) and *ZmEC222* (*Zea mays* egg cell) contain predicted signal peptides at the N-terminus, but homologous sequences to these two classes could not be found in the fully sequenced *Arabidopsis* genome, nor in EST databases. Another class is represented by ZmEC235, which is lacking a signal peptide, but is predicted to be a type 1b membrane protein localised in the plasma membrane. TaEC/D10/17 and ZmZ110 (*Zea mays* zygote) are included in Table 1 as examples for peptides missing signal peptides and transmembrane domains, but displaying interesting homologies to e.g. a Brassinolide induced peptide gene from rice or to the NLS region (nuclear localisation signal) of the homeotic CAUDAL protein from *Drosophila*.

Single cell RT-PCR with dissected cells of the embryo sac, *in situ* hybridisation with ovule sections, but also Northern blots with some 20 tissues were applied to analyse the expression pattern of some of these genes. In order to study their function, we have started to express candidate genes ectopically in *Arabidopsis*, to generate knock-outs in maize using the antisense strategy and to monitor the sub-cellular localisation of some candidate peptides using GFP fusion proteins. E.g. for ZmTLA1 we could show that the fusion protein is secreted and transported in the apoplast.

The functional analysis of the peptide classes described above will provide a much deeper insight into features of signalling peptides in general, but especially into their role during the fertilisation process. We are aiming to identify further classes from other cells of the embryo sac such as the synergids and central cells, but also from zygotes at defined stages after fertilisation. Our focus for future analyses will be on peptide genes expressed in the female gametophyte before and after fertilisation, which are not expressed elsewhere in the plant.

3. REFERENCES

Blein, J.P., P. Coutos-Thévenot, D. Marion, and M. Ponchet. 2002. From elecitins to liprid-transfer ptoteins: a new insight in cell signalling involved in plant defence mechanisms. Trends Plant Sci. 7:293-296.

Cordts S., J. Bantin, P.E. Wittich, R. Kranz, H. Lörz, and T. Dresselhaus. 2001. *ZmES* genes encode peptides with structural homology to defensins and are specifically expressed in the female gametophyte of maize. Plant J. 25:103-114.

Diboll A.G., and D.A. Larson. 1966. An electron microscopic study of the mature megagametophyte in *Zea mays*. Amer. J. Bot. 53:391-402.

Doughty, J., S. Dixon, S.J. Hiscock, A.C. Willis, I.A.P. Parkin, and H.G. Dickinson. 1998. PCP-A1, a defensin-like *Brassica* pollen coat protein that binds the S locus glycoprotein, is the product of gametophytic gene expression. Plant Cell 10:1333-1347.

Guiderdoni, E., M.J. Cordero, F. Vignols, J.M. Garcia-Garrido, M. Lescot, D. Tharreau, D. Meynard, N. Ferriere, J.L. Notteghem, and M. Delseny. 2002. Inducibility by pathogen attack and developmental regulation of the rice *Ltp1* gene. Plant Mol. Biol. 49:683-699.

Lindsey K., S. Casson, and P. Chilley. 2002. Peptides: new signalling molecules in plants. Trends Plant Sci. 7:78-83.

Matsubayashi, Y., M. Ogawa, A. Morita, and Y. Sakagami. 2002. An LRR receptor kinase involved in perception of a peptide plant hormone, phytosulfokine. Science 296:1470-1472.

Matsubaysahi Y., and Y. Sakagami. 1996. Phytosulfokine, sulfated polypeptides that induce the proliferation of single mesophyll cells of *Asparagus officinalis* L. Proc. Natl. Acad. Sci. USA 93:7623-7627.

Matsubaysahi Y., H. Yang, and Y. Sakagami. 2001. Peptide signals and their receptors in higher plants. Trends Plant Sci. 6:573-577.

Schopfer, C.R., M.E. Nasrallah, and J. B. Nasrallah. 1999. The male determinant of self-incompatibility in *Brassica*. Science 286:1697-1700.

The *Arabidopsis* Genome Initiative. 2000. Analysis of the genome sequence of the flowering plant *Arabidopsis thaliana*. Nature 408:796-815.

Trotochaud, A.E., S. Jeong, and S.E. Clark. 2000. CLAVATA3, a multimeric ligand for the CLAVATA1 receptor-kinase. Science 289:613-617.

SOMATIC EMBRYOGENESIS IN *ARABIDOPSIS THALIANA* PROMOTED BY THE WUSCHEL HOMEODOMAIN PROTEIN

Jianru Zuo[1,2], Qi-Wen Niu[1], Giovanna Frugis, and Nam-Hai Chua[3]
Laboratory of Plant Molecular Biology, The Rockefeller University, 1230 York Avenue, New York, NY 10021, USA
[1] These authors contributed equally to this work.
[2] Present address: Institute of Genetics and Developmental Biology, Chinese Academy of Sciences, 917 Datun Road, Beijing 100101, People's Republic of China.
[3] Corresponding author (e-mail: chua@rockvax.rockefeller.edu)

Keywords WUSCHEL/PGA6, somatic embryogenesis, *Arabidopsis*, activation tagging

1. INTRODUCTION

Somatic embryogenesis is a unique pathway for asexual propagation or somatic cloning in plants. The developmental process of somatic embryogenesis shares considerable similarity with that of zygotic embryogenesis, and this is likely due to the conservation in the underpinning cellular and molecular mechanisms between the two processes. Therefore, somatic embryogenesis provides an attractive model system for studying zygotic embryogenesis, particularly because zygotic embryos are encased by maternal tissues and difficult to access by biochemical and molecular tools. Moreover, in biotechnological applications, most economically important crop as well as non-crop plants are regenerated via somatic embryogenesis. In contrast to organogenesis, which requires a high cytokinin to auxin ratio, somatic embryogenesis does not require any external cytokinins, but rather is dependent on high concentrations of 2,4-D, a synthetic chemical that has long been used as a functional analog of auxin. It is generally believed that somatic embryogenesis is mediated by a signaling cascade triggered by external auxin or 2,4-D. However, the signal transduction pathway, particularly the molecular mechanism involved in the transition of a vegetative cell to an embryogenic competent cell, remains largely unexplored.

To dissect the signaling pathway during somatic embryogenesis, we have employed a genetic approach to identify gain-of-function mutations that can promote embryogenic callus formation from *Arabidopsis* root explants.

I. K. Vasil (ed.), *Plant Biotechnology 2002 and Beyond*, 279-281.

Arabidopsis thaliana is known to be a species difficult for somatic embryogenesis. Thus far, embryogenic calli can only be induced from immature embryos of wild-type plants (Wu et al., 1992) or from the *primodia timing* (*pt*) mutant plant (Mordhorst et al., 1998). Therefore, *Arabidopsis* vegetative explants appear to be reliable materials for screening of genetic mutations involved in the vegetative-to-embryogenic transition. In a functional screen using a chemical inducible activation tagging system (the LexA-VP16-Estragen Receptor or XVE system; Zuo et al., 2000), we identified two alleles of an *Arabidopsis* gene *PGA6* (*Plant Growth Activator 6*) whose induced overexpression caused high frequency somatic embryo formation in all tested tissues and organs, without any external plant hormones. Upon inducer withdrawal all these somatic embryos, following a developmental process remarkably similar to that of zygotic embryogenesis, were able to germinate and grow into healthy, fertile plants. These results suggested that *PGA6* is involved in the maintenance of embryonic cell identity, and is able to promote the vegetative-to-embryogenic transition. Molecular and genetic analyses showed that PGA6 was identical to WUSCHEL (WUS), a homeodomain protein previously characterized as a key regulator for specification of meristem cell fate (Mayer et al., 1998). Transgenic plants carrying an estradiol-inducible XVE-*WUS* transgene can phenocopy *pga 6-1* and *pga 6-2*.

At the molecular level, WUS-induced somatic embryo formation was characterized by an ectopic expression of *LEC1*, which is normally expressed only in embryos and seeds (Lotan et al., 1998). Surprisingly, re-induction of WUS expression strongly represses the *LEC1* expression. These observations suggest that the *LEC1* expression in *pga6* explants was not a direct response to *WUS* overexpression but rather a consequence of the *pga6* somatic embryo development. On the other hand, a developmental path redefined by *WUS* overexpression, leads to the repression of *LEC1*, a gene presumably involved in embryo maturation. In summary, our results suggest that WUS/PGA6 plays a critical key role during embryogenesis, presumably by promoting the vegetative-to-embryogenic transition and/or maintaining the identity of the embryonic stem cells. Details of this study and a more comprehensive discussion can be found in Zuo et al. (2002).

2. REFERENCES

Lotan, T., Ohto, M., Yee, K.M., West, M.A., Lo, R., Kwong, R.W., Yamagishi, K., Fischer, R.L., Goldberg, R.B., and Harada, J.J. 1998 *Arabidopsis LEAFY COTYLEDON1* is sufficient to induce embryo development in vegetative cells. Cell 93:1195-1205.

Mayer, K.F., Schoof, H., Haecker, A., Lenhard, M., Jurgens, G., and Laux, T. 1998. Role of *WUSCHEL* in regulating stem cell fate in the *Arabidopsis* shoot meristem. Cell 95:805-815.

Mordhorst, A.P., Voerman, K.J., Hartog, M.V., Meijer, E.A., van Went, J., Koornneef, M., and de Vries, S.C. 1998. Somatic embryogenesis in *Arabidopsis thaliana* is facilitated by mutations in genes repressing meristematic cell divisions. Genetics 149:549-563.

Wu, Y., Haberland, G., Zhou, C., and Koop, H.-U. 1992. Somatic embryogenesis, formation of morphogenetic callus and normal development in zygotic embryos of *Arabidopsis thaliana in vitro.* Protoplasma 169:89-96.

Zuo, J., Niu, Q., Frugis, G., and Chua, N.-H. 2002. The WUSCHEL gene promotes vegetative-to-embryonic transition in *Arabidopsis*. Plant J. in press.

Zuo, J., Niu, Q., and Chua, N.-H. 2000. An estrogen receptor-based transactivator XVE mediates highly inducible gene expression in plants. Plant J. 24:265-273.

MAIZE *LEC1* IMPROVES TRANSFORMATION IN BOTH MAIZE AND WHEAT

Keith Lowe, George Hoerster, Xifan Sun, Sonriza Rasco-Gaunt, Paul Lazerri, Sam Ellis, Shane Abbitt, Kimberly Glassman and Bill Gordon-Kamm
Pioneer Hi-Bred Int'l Inc. 7300 NW 62nd Ave. PO Box 1004 Johnston, IA 50131 (e-mail: keith.lowe@pioneer.com)

Keywords *LEC1*, transformation, re-transformation

1. INTRODUCTION

Since ectopic expression of *LEC1* in *Arabidopsis* can lead to adventive formation of embryo-like structures (Lotan et al., 1998), it was hypothesized that ectopic expression of the maize *LEC1* (Lowe et al., 2001) gene could be used to induce somatic embryogenesis in maize and wheat leading to an improved culture response and higher transformation frequencies.

When *LEC1* is used with a screenable/visible marker (UBI::moPAT~moGFP::pinII), we've shown that transformation frequencies in maize Hi-II germplasm and maize elite lines are increased. Measuring callus volume three weeks post-bombardment also showed that growth rates increased as promoter strength driving *LEC1* increased (see table.1). Since ectopic *LEC1* expression greatly improves primary transformation, experiments were undertaken to determine if *LEC1* could be used to improve re-transformation. In these experiments embryos were taken from T0 plants transformed with *LEC1* linked to GFP. These segregating embryos were then transformed via particle gun or *Agrobacterium*, separated based on GFP expression (linked to *LEC1*) and selected for herbicide resistance. The transformed *LEC1*/GFP+ segregates transformed at a much higher frequencies than the wild type segregates suggesting that this strategy can be used to make recalcitrant maize inbreds more amenable to transformation.

Typically transformed cells are initially compromised and require antibiotic or herbicide selection to enhance their recovery by killing the rapidly growing non-transformed cells. The growth advantage provided by *LEC1* overcomes some of these problems and can be used with a screenable visual marker such as a fluorescent protein to recover transformed cells on non-selective medium. Using the maize *LEC1* gene in recalcitrant wheat genotypes has produced

I. K. Vasil (ed.), Plant Biotechnology 2002 and Beyond, 283-284.

similar results. In wheat *LEC1* increased transformation frequencies and enhanced morphogenetic quality (more embryogenic with easier regeneration of plants) allowing the recovery of transgenic plants without chemical selection. By carefully controlling the expression of the anthocyanin pathway using well characterized maize genes it is possible to provide a easily visualized screenable marker that can be used for identifying transformants. Recently, we have used a combination of the maize *LEC1* gene and CRC, a fusion between the maize R and C1 (Bruce et al., 2000) genes to recover transformants at frequencies comparable to that of bialaphos selection. Controlled expression of *LEC1* and CRC using either tissue-specific or inducible expression provides a non-herbicide method for selecting transformants. This positive growth transformation system also allows one to recover transgenic plants wherein all the transforming DNA components are derived from maize.

Table 1. Total numbers of GFP+ multicellular transformants and size distribution 21 days post-bombardment. From measured diameters for each GFP+ colony, volumes were calculated ($V=3/4\pi r^3$) and tabulated in size classes. Two control treatments include a frame-shifted LEC1 gene (LEC1-FS) driven by a strong promoter(Ubi), and a maize HAP3 homolog behind the maize In2 promoter. As promoter strength driving the maize LEC1 increased (In2<nos<Ubi), the total number of transformants increased and growth rate (based on colony size) also increased.

Size class (mm3)	EXPRESSION CASSETTE				
	In2i: *LEC1*-FS	In2: HAP3	In2: *LEC1*	nos: *LEC1*	Ubi: *LEC1*
1-10	12	13	25	25	25
11-100	1	5	12	22	22
101-1000	2	4	7	21	24
1001-10,000	0	0	3	8	15
Total Txs	**15**	**22**	**47**	**73**	**81**

2. References

Bruce, W., Folkerts, O., Garnaat, C., Crasta, O., Roth, B., and Bowen, B. 2000. Plant Cell. 12: 65-79.

Lotan, T., Ohto,M., Yee,K.M., West,M.A., Lo, R., Kwong,R.W., Yamagishi,K., Fischer, R.L., Goldberg,J.B. and Harada, J.J. 1998. Cell 93: 1195-1205.

Lowe,K.S., and Gordon-Kamm,W.J. 2001. Genbank accession number AF410176.

ANTISENSE SUPPRESSION OF A CYTOKININ-BINDING PROTEIN FROM *PETUNIA* CAUSES EXCESSIVE BRANCHING AND REDUCES ADVENTITIOUS SHOOT BUD INDUCTION IN VITRO

Prakash P. Kumar[1], B. Dhinoth Kumar and Shoba Ranganathan
Department of Biological Sciences, National University of Singapore, Singapore 117543.
([1]corresponding author e-mail: dbskumar@nus.edu.sg)

1. INTRODUCTION

We have characterized a cDNA coding for a putative cytokinin binding protein (*PETCBP*) from *Petunia hybrida*. The derived amino acid sequence of PETCBP shows high homology to S-adenosyl L-homocysteine hydrolase (SAHH). SAHH is involved in regulating methylation and several homologs are known from plants and animals. We used the crystal structure of human SAHH as the template for PETCBP model building. Antisense suppression of *PETCBP* indicated a significant delay and decline in the number of adventitious shoot buds initiated from leaf explants compared to wild type explants. The antisense suppression lines exhibited excessive branching phenotype and a delayed flowering response compared to the wild type. Our results indicate that *PETCBP* plays an important role in plant development and manipulation of homologs of this gene may help in improving leafy vegetables.

2. RESULTS AND DISCUSSION

Cytokinins regulate many plant physiological events such as nutrient metabolism, expansion and senescence of leaves and lateral branching. Also, the central dogma of plant tissue culture accepts that cytokinins act with auxins to promote cell division and shoot initiation in vitro. Despite the wealth of knowledge concerning the physiological effects of cytokinins, the molecular mechanisms underlying cytokinin signal perception and mode of action remain largely unknown. Many cytokinin binding proteins (CBP) have been identified in plants.

Petunia hybrida cv Dazzler shoots maintained by nodal cuttings on phytohormone-free Murashige and Skoog medium served as the explant source. Leaf disks (10 mm x 5 mm) from fully expanded leaves were

I. K. Vasil (ed.), Plant Biotechnology 2002 and Beyond, 285-287.

cultured on shoot induction medium (SI) [1mg/l benzyl adenine, 100 μg/l α-naphthaleneacetic acid and 30 g/l sucrose] (Horsch et al. 1985). The same procedure was used for recovering transgenic shoots.

We prepared a cDNA library with RNA from 7-day-old explants of *Petunia* (Ausubel et al. 1993). A literature search indicated that *CKI1* of *Arabidopsis* (Kakimoto 1996) might be a prime candidate for a gene regulating shoot regeneration. Therefore, we designed a pair of degenerate primers corresponding to the conserved region of *CKI1*. PCR amplification of the cDNA library under low stringency using the primers yielded a 640 bp fragment, which was used as a probe to screen the original cDNA library. Out of 18 positive cDNA clones isolated, 6 showed high similarity to CBP from tobacco and several other species. The longest clone (1.24 kb) was used for further analysis. This was only a partial length cDNA as revealed by the transcript size on an RNA gel blot, and the missing ~500 bp at the 5' end was cloned by 5'-RACE. Genomic Southern blot analysis showed that it exists as a single copy gene in *Petunia*. RNA gel blot analysis revealed a basal level of expression on day 0 (original leaf explants) and maximum expression was in the leaf explants cultured on SI medium for 7 days. The maximal expression precedes the initiation of shoot primordia, indicating a correlation with shoot bud induction. The stem, flower buds and flowers also expressed *PETCBP*.

We observed a significant delay in shoot induction and a decrease in the number of shoot buds produced per leaf explant derived from the antisense petunia transgenic lines harboring a 700 bp gene-specific 3' fragment of *PETCBP* cDNA. The wild-type explants produced 35±2 shoot buds per explant, while the numbers ranged from 3 to 21 shoot buds per explant among several antisense lines. RNA gel blot analysis showed a strong negative correlation between shoot regeneration and the level of *PETCBP* expression. Further, the antisense petunia lines flowered in the pots only after 6 months, whereas the control plants produced flowers within 2 months after transfer. The control plants showed sparse branching during the early stages, but the antisense lines exhibited excessive branching giving them a bushy appearance right from the early stages of growth.

Temporal regulation of expression of *SAHH* in *Medicago* stem was observed, suggesting a role for the gene in shoot development (Abrahams et al., 1995). Antisense suppression of *SAHH* in tobacco resulted in dwarfing, and some of the lines had double flowers (with petaloid staminodes) along with a reduction in genomic DNA methylation (Tanaka et al., 1997).

Sequence analysis of PETCBP revealed a strong homology to SAHH, which catalyzes the reversible hydrolysis of S-adenosyl-L-homocysteine (SAH), a methyl transferase inhibitor. SAHH controls biological methylation reactions by the regulation of intracellular S-adenosyl-methionine/SAH ratio. The role of DNA methylation in plant development and differentiation has been shown in several species including *Arabidopsis* and *Petunia*. Thus, shoot regeneration was inhibited when *Petunia* explants were cultured on SI medium containing either 5-azacytidine or 5-aza-2'-deoxycytidine (Prakash and Kumar, 1997). The present observations suggest that such inhibition in regeneration might involve *PETCBP*.

Several homologs of SAHH are known from plants and animals. The crystal structures of SAHH from human and rat (1A7A & 1B3R, respectively) have been solved. They form homotetramers. We used the human SAHH (chain A of the ternary complex 1A7A) as the template for PETCBP model building. The model was built iteratively using the program MODELLER. Most of the plant CBP lack the N-terminal loop and the first helix (αA2) compared to mammalian SAHH. There is a 28-residue deletion at the N-terminus of PETCBP compared with the template structure and its first residue (Met1) aligns with Met29 of 1A7A, which is at the start of the first helix of the catalytic domain. Additionally, there are 2 insertions (8 and 17 amino acids) in the solvent exposed loops.

Our study shows that *PETCBP* is associated with shoot bud induction in petunia. Furthermore, use of this gene or its orthologs in other species may benefit developmental enhancement efforts. For example, the antisense suppression lines that exhibited excessive branching along with a delay in flowering indicate that this gene is a potential candidate for improving leafy vegetable species where such phenotype is desirable.

3. REFERENCES

Abrahams S., Hayes C.M., and Watson J.M. 1995. Expression patterns of 3 genes in the stem of lucerne (*Medicago sativa*). Plant Mol. Biol. 27:513-528.

Ausubel F.M., Brent R., Kingston R.E., Moore D.D., Seidman J.G., Smith J.A., Struhl K., Albright L.M., Coen D.M., Varki A., and Janssen K. 1993. Current Protocols in Molecular Biology. Current Protocols, USA.

Horsch R.B., Fry J.E., Hoffmann N., Eicholtz D., Rogers S.G., and Fraley R.T. 1985. A simple and general method for transferring genes into plants. Science 227: 1229-1231.

Kakimoto T. 1996. CKI1, a histidine kinase homolog implicated in cytokinin signal transduction. Science 274: 982-985.

Prakash A.P., and Kumar P.P. 1997. Inhibition of shoot induction by 5-azacytidine and 5-aza-2'-deoxycytidine in *Petunia* involves DNA hypomethylation. Plant Cell Rep. 16:719-724.

Tanaka H., Masuta C., Uehara K., Kataoka J., Koiwai A., and Noma M. 1997. Morphological changes and hypomethylation of DNA in transgenic tobacco expressing antisense RNA of the S-adenosyl-L-homocysteine hydrolase gene. Plant Mol. Biol. 35:981-986.

MOLECULAR MARKERS ASSOCIATED WITH PLANT REGENERATION FROM SHOOT MERISTEM CULTURES DERIVED FROM GERMINATED CORN (*ZEA MAYS* L.) SEEDS

Wenbin Li, Jie Liu, Pat Masilamany, Jeff H. Taylor, Genlou Sun, Manilal William, and K. Peter Pauls
Biotechnology Division, Department of Plant Agriculture, University of Guelph, Guelph, ON, Canada N1G 2W1 (email: ppauls@uoguelph.ca)

A number of studies have shown that regeneration from immature embryo and anther cultures is genetically determined in corn and is controlled by a few genes (Willman et al., 1989). In addition, several RFLPs markers associated with regeneration from these cultures were identified (Dufour et al., 2001; Armstrong et al., 1992; Wan et al., 1992). Somatic embryo formation from immature embryos is restricted to a few genotypes but the regeneration from corn shoot tip cultures initiated from seedlings occurs in a broad range of genotypes (Li et al., 2002). No information exists regarding genetic determination of regeneration from this tissue culture system. The current study was initiated to investigate the type and number of genes controlling regeneration in shoot tip cultures and to identify a molecular marker linked to the trait.

Forty-five North America corn genotypes were screened for regeneration in cultures initiated from the shoot tip meristems of 4- day-old seedlings. The explants were cultured on modified shoot-tip meristem culture medium (MS basal medium plus 4 g/l L-proline, 3% sucrose, 1 mg/l 2,4-D and 2mg/l BAP; Li et al., 2002) in continuous light (30-40 $\mu Em^{-2} s^{-1}$) at 25 C. Of the 45 genotypes that were tested, 98% produced multi-shoots at some frequency. According to the frequency distribution of explants producing multi-shoots the genotypes could be divided into four response groups, namely: 0-20%, 30-50%, 60-90% and 100% explants with shoots. The finding that there were discrete categories of responsiveness suggested that the ability to regenerate multi-shoots from seedling shoot meristem explants may be controlled by a few genes rather than many genes. In the screen CG-44 was identified as a nonresponder (0% explants produced shoots), whereas CG-37 was a high responder (100% explants produced shoots) (Fig. 1).

I. K. Vasil (ed.), Plant Biotechnology 2002 and Beyond, 289-292.

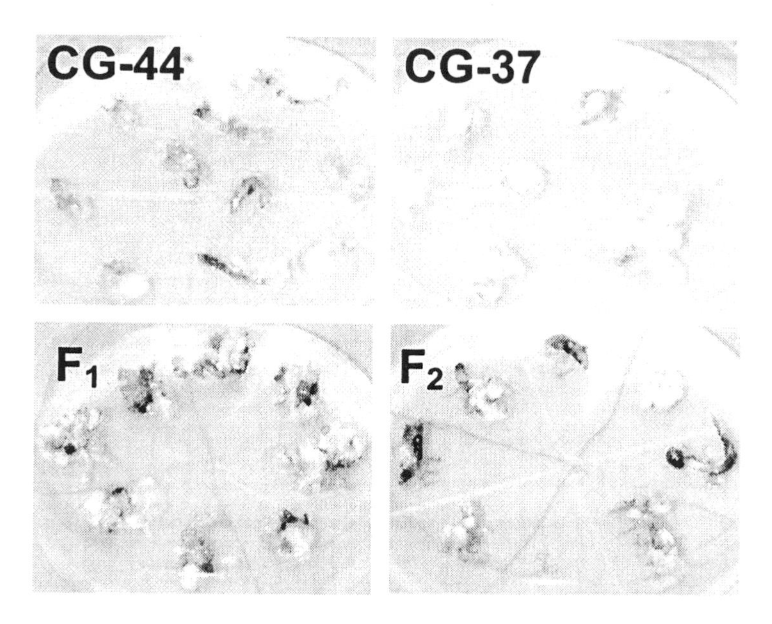

Figure 1. Shoot regeneration from shoot tip explants from CG-37, CG-44 and F2 lines obtained from a cross between CG-37 and CG-44.

From a screen of 101 RAPD markers, produced by PCR amplification of DNA isolated from the 45 genotypes (with QIAGEN Dneasy 96 Plant Test Kit) and 10-mer random primers (obtained from UBC), 5 markers were found to be significantly associated (P = 0.05 or 0.01) with regeneration based on the percentage explants that produced multi-shoots (Table 1).

To further test these associations a segregating population was created by crossing CG-37 with CG-44 and selfing the F_1. Regeneration in cultures produced from the 150 F_2s was evaluated on the basis of the number of shoots produced per explant. DNA was isolated from culture samples and scored for RAPD markers. The distribution of shoot regeneration efficiency appeared to fall into discrete categories including non-responders, individuals that responded like the highly regenerable parent (CG-37) and individuals that produced twice as many shoots per explant as CG-37 (Fig. 2). The heritability for this trait was estimated to be 0.7. These results suggest that the regeneration in this system is controlled by a few genes. In addition, a significant correlation between the number of multi-shoots and callus size was observed.

A RAPD marker (BC603-1,700) associated with multishoot production was identified. This marker accounted for 18% of the variability in multi-shoot production frequency from corn shoot tip explants in this population. The 1,700 bp marker was cloned and sequenced and is ready to be converted into a SCAR marker.

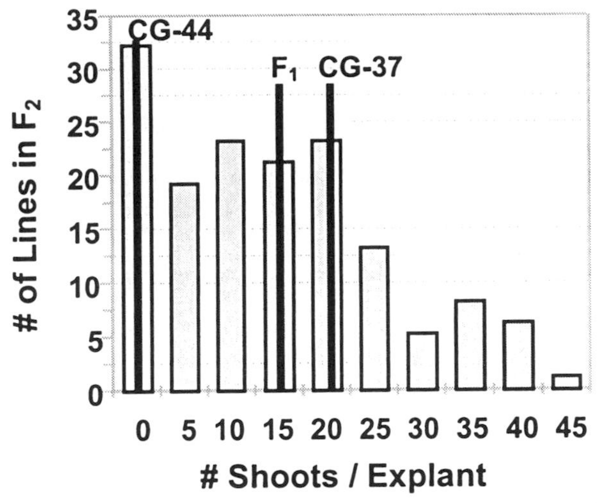

Figure 2. Regeneration responses of explants from F_2 of CG-37 x CG-44. Lines indicate responses of parents and F_1.

Table 1. Presence or absence of RAPD markers associated with regeneration from corn shoot tip explants in a collection of North American inbred corn lines.

		High Regeneration (>25%) Inbreds																Low Regen-eration (<25%) Inbreds				Marker Frequency in high regeneration	Marker Frequency in low regeneration	Difference	P
	Inbreds	CG39	CG33	CG103	CG59	CG62	CG102	CG69	CG99	CG74	CG88	CG89	CG90	CG91	CG94	CG95	CG97	CG96	CG72	CG64	CG36				
Markers	Regen-eration %	44	49	82	85	83	76	100	86	100	100	48	63	48	70	60	74	25	2	20	12				
625.>2000		1	1	1	1	1	1	1	1	1	1	1	1	1	1	1	0	0	0	1	0	0.93	0.25	0.68	3.07**
302.500		0	0	0	0	1	1	1	0	1	0	0	0	0	0	0	0	1	1	1	1	0.25	1.00	-0.75	2.78**
631.800		0	1	0	0	0	0	1	0	0	0	1	0	0	0	1	0	1	1	1	0	0.25	0.75	-0.50	1.92*
646.1800		1	1	1	0	1	0	0	0	0	1	0	1	1	0	1	1	0	0	0	0	0.56	0	0.56	2.08*
603.250		0	0	0	0	0	0	1	0	0	1	1	0	1	0	0	1	1	1	1	1	0.31	1.00	-0.69	2.54**

**P< 0.01
*P< 0.05

REFERENCES

Armstrong CL, J Romero-Severson and TK Hodges (1992) Theor Appl Genet. 84:755-762.

Dufour P, C Johnsson, S Antoine-Michard (2001) Theor Appl Genet 102: 993-1001.

Li W, P Masilamany, KJ Kasha and KP Pauls (2002) In vitro Cell Devel Biol. 37: 1-9.

Wan Y, TR Rocheford and JM Wildholm (1992) Theor Appl Genet 85: 360-365.

Willman MR, SM Schroll and TK Hodges (1989) In vitro Cell Devel Biol. 25: 95-100.

DEVELOPMENT OF AN AUTOMATED IMAGE COLLECTION SYSTEM FOR GENERATING TIME-LAPSE ANIMATIONS OF PLANT TISSUE GROWTH AND GREEN FLUORESCENT PROTEIN GENE EXPRESSION

Marco T. Buenrostro-Nava, Peter P. Ling and John J. Finer
OARDC/The Ohio State University, Wooster, OH 44691, USA (e-mail: finer.1@osu.edu)

Keywords Embryogenesis, automation, gfp, gene expression

1. INTRODUCTION

Robotics systems have been widely used by the industry to perform tasks that may be hazardous, time consuming, or impossible to perform by humans. In the area of plant developmental biology, these systems have been used to gather information on how plants grow and develop under different environmental conditions. Digital imaging permits the non-destructive evaluation of both plant growth and their response to different environments. Images can either be analyzed as they are acquired (real time) or stored for subsequent analysis. Digital imaging has recently been used for the analysis of gene expression using the jellyfish Green Fluorescent Protein (GFP). GFP detection requires no exogenous cofactors or destructive assays (Haseloff and Siemering, 1998), which greatly simplifies reporter gene detection in living organisms.

Collection and assembly of digital images over time permit the generation of digital time-lapse animations of plant growth. Although the general approach of time-lapse animation has occasionally been utilized to monitor plant "movement" over the past few decades, digital imaging and computer animation is a more recent development; with applications in both education and basic research.

2. MATERIALS AND METHODS

The automated system consisted of a camera mounted on a microscope, with an XY belt-driven positioning table (Arrick Robotics Inc., Hurst, Texas)

I. K. Vasil (ed.), Plant Biotechnology 2002 and Beyond, 293-295.

containing a 40 x 40 cm Plexiglas platform. The platform consisted of 16 perforations of 5.4 cm diameter, which were evenly distributed in four rows and four columns. For image collection, a SPOT-RT camera (Diagnostic Instruments Inc., Sterling Heights, Michigan) was mounted on a Leica MZFLIII stereomicroscope (Leica, Heerbrugg, Switzerland). To maintain the subject materials under aseptic conditions, the automated system was placed in a laminar air-flow hood.

In order to create the automated plant growth and GFP gene expression analysis system described above, several modifications were made to the software provided with each system to create custom applications to precisely control both the XY table and the SPOT-RT camera. First, a C base code for the MD2 software (Arrick Robotics Inc.), provided with the XY table, was modified using the Visual C++® v6.0 program (Microsoft Co.) to run under a Windows® environment. Secondly, to be able to control the camera to obtain sequential images at different time intervals between two images, new applications were written using the Visual C++® program. All of these algorithms were integrated into a single program.

3. RESULTS AND DISCUSSION

Growth of a GFP-expressing *Agrobacterium* (Finer and Finer, 2000) on inoculated soybean tissue was tracked using the automated system. *Agrobacterium* showed an initial fluorescence 12 h after inoculation (Figure 1). The total area covered by *Agrobacterium* was much larger in the treatments where *Agrobacterium* was inoculated onto embryogenic tissue (compared to inoculation on medium without plant tissue), suggesting that the plant tissue stimulated bacterial growth. Animation of bacterial growth showed that the bacterial "colony" grew mostly by spreading from the outside but internal colonial rings could also be seen expanding, suggesting that growth of the colony occurred throughout.

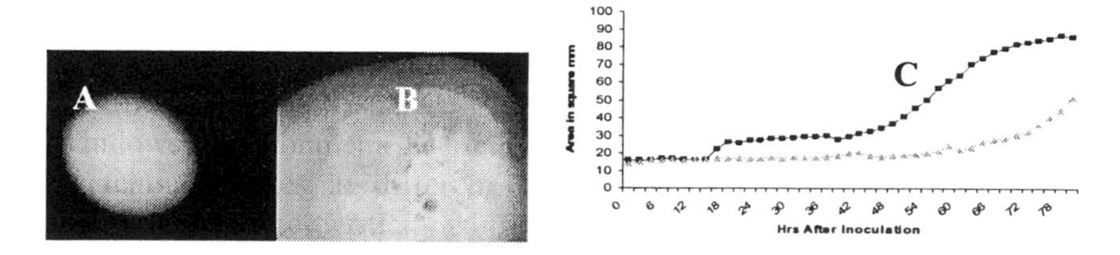

Figure 1. Growth of GFP-expressing Agrobacterium in the absence (A) or presence (B) of embryogenic soybean tissue. C - Image analysis of GFP-expressing Agrobacterium in the absence (▲) or presence (■) of embryogenic soybean tissue.

For evaluation of transient expression, two different *gfp* genes were introduced into lima bean cotyledons via particle bombardment. Four hrs after bombardment, GFP was first detected with the *sgfp-TYG* gene, reaching a maximum expression at 24 hrs after bombardment (Figure 2). For the *mgfp5-ER* gene, low intensity GFP foci appeared between 8 and 10 hrs after bombardment, becoming more apparent at 11 h after bombardment, and finally reaching a maximum expression at 25 hrs after bombardment. The delay in expression of the *mgfp5-ER* gene suggests a time requirement for protein processing due to ER targeting. This difference in GFP fluorescence has been previously reported in soybean embryogenic tissue (Ponappa et al., 1999) and sugar cane calli (Elliott et al., 1999). Animation of transient expression showed in great detail the kinetics of transient expression of both GFP constructions. For both the *mgfp5-ER* and *sgfp-TYG* genes, some of the GFP foci that were first observed gave rise to the highest eventual GFP expression of individual GFP foci. Some of the GFP-expressing cells abruptly stopped fluorescing, suggesting cell death or a very rapid decline in gene expression.

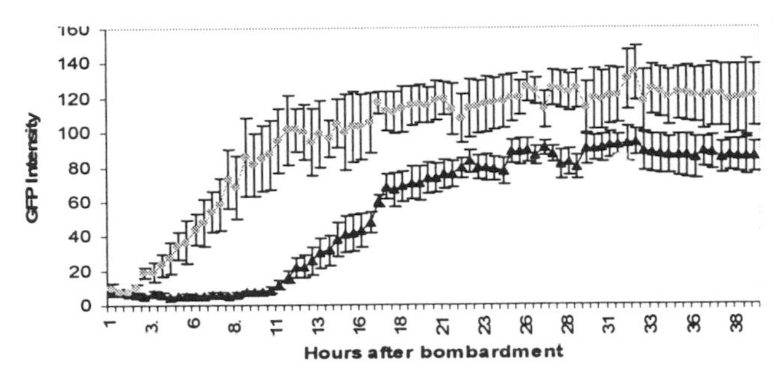

Figure 2. Transient GFP expression of the mgfp5-ER (▲) and sgfp-TYG (◈) genes in lima beans.

4. REFERENCES

Elliott AR, Campbell JA, Dugdale B, Brettell RIS, Grof CPL (1999) Green-fluorescent protein facilitates rapid *in vivo* detection of genetically transformed plant cells. Plant Cell Reports 18:707-714

Finer KR, Finer JJ (2000) Use of *Agrobacterium* expressing green fluorescent protein to evaluate colonization of sonication-assisted *Agrobacterium*-mediated transformation-treated soybean cotyledons. Lett Appl Microbiol 30:406-410

Haseloff J, Siemering KR (1998) The uses of green fluorescent protein in plants. In Green Fluorescent Protein: Properties, Application, and Protocols., a. S. K. Martin Chalfie, ed. (Wiley-Liss, Inc.), pp. 191 - 219.

Ponappa T, Brzozowski AE, Finer JJ (1999) Transient expression and stable transformation of soybean using the jellyfish green fluorescent protein. Plant Cell Reports 19:6-12

APPLICATIONS OF ETHYLENE TECHNOLOGY TO CLIMACTERIC FRUITS: A PROGRESS REPORT

Harry J. Klee
Horticultural Sciences and Plant Molecular and Cellular Biology Program, University of Florida, Gainesville, FL 32611-0690, USA (e-mail: hjklee@mail.ifas.ufl.edu)

Keywords transgenic plants, ETR1, ACC synthase, ACC deaminase

1. INTRODUCTION

The large losses that occur to stored perishable fruits and vegetables justify the significant effort that researchers have placed into the problem. Historically, efforts have focused on controlling postharvest handling and storage conditions of existing materials. In parallel, breeders have attempted to develop varieties with longer keeping properties. These latter efforts have focused principally out of necessity on firmness, with the trend being toward developing fruits that are harder and resist handling better. In recent years, biotechnology has been utilized to complement the existing tools available to breeders and seed companies. Most of the early efforts of biotechnology were focused on tomato. This is because of the wealth of molecular and physiological knowledge that has been amassed as well as the relative ease of genetic manipulation and the simple fact that tomato is a major horticultural crop around the world.

Tomato is a climacteric fruit. That is, at the onset of ripening, fruits enter a phase where they produce an autocatalytic burst of ethylene synthesis that is accompanied by a corresponding increase in the rate of respiration. The ethylene that is produced is essential for ripening to occur and blocking either ethylene synthesis or perception prevents ripening. This mode of ripening is conserved in many important fruits including banana, apple, peach, mango, papaya and melon. Thus, technology developed for control of ethylene in tomato should be directly applicable to many commercially produced species. This paper summarizes the status of our knowledge in controlling ethylene synthesis and perception. While most of the effort has been focused on tomato, the lessons learned will greatly facilitate manipulation of these other crops.

I. K. Vasil (ed.), Plant Biotechnology 2002 and Beyond, 297-303.

2. THE AVAILABLE TOOLS

Of all the plant hormones, our knowledge of ethylene synthesis and perception is arguably the greatest. The regulation of its synthesis is well established. The genes encoding the two committed enzymes, ACC synthase (ACS) and ACC oxidase (ACO) have been cloned and much is known about their regulation during development and fruit ripening, in particular (Oeller et al., 1991; Theologis, 1993; Barry et al., 1996; Wang et al., 2002). These enzymes sequentially convert the ubiquitous precursor S-adenosylmethionine to 1-aminocyclopropane and on to ethylene. Although expression of the genes for both enzymes is highly regulated, it is generally established that the first enzyme, ACS, is the rate-limiting step in synthesis of ethylene.

The mechanisms of ethylene perception are not well understood but many of the elements of the signal transduction pathway have been identified by mutation analysis (Wang et al., 2002). It is clear that a family of genes encode a set of at least partially redundant receptor molecules. In Arabidopsis, where the receptor was first identified, there are five structurally diverged proteins in the receptor family (Chang et al., 1993; Hua et al., 1995, 1998). In tomato, the family consists of six proteins (Wilkinson et al., 1995; Zhou et al., 1996a,b; Lashbrook et al., 1998; Tieman and Klee, 1999). Although one Arabidopsis receptor, ETR1 is clearly a histidine protein kinase (Schaller and Bleecker, 1995), it is not clear how the receptors function. What is clear for purposes of this discussion is that dominant mutations in the receptor are capable of conferring ethylene insensitivity in a plant expressing the mutant form of the protein. The first characterized receptor mutant, *etr1-1*, confers complete insensitivity to ethylene in Arabidopsis (Chang et al., 1993). Likewise, a dominant mutation in one of the tomato receptors, NR, has been shown to confer ethylene insensitivity in that plant. The *Nr* mutant plants have fruits that do not ripen, even upon exposure to ethylene (Wilkinson et al., 1995). A number of genes lying downstream of the receptor are also capable of conferring ethylene insensitivity as well. However, these genes are recessive and it is the loss of function that confers insensitivity. These genes include *EIN2* and *EIN3*. The former appears to be a single gene that is essential for ethylene signaling while the latter is encoded by a family. There are several genes related to EIN3 in both Arabidopsis (Solano et al., 1998) and tomato. In tomato, we have identified three genes in the EIN3 family and have shown that all three are partially redundant (Tieman et al., 2001).

3. MANIPULATION OF ETHYLENE RESPONSES

As a first step to manipulation of ethylene responses, one must consider whether the appropriate target is synthesis or perception of the hormone.

Each approach has advantages and disadvantages. The answer to which approach is best lies in the nature of the final product. The first question is whether the product needs to respond to ethylene. In the case of climacteric fruits the answer is a very clear yes. If the fruits do not respond to ethylene, they will not ripen and will not be suitable for sale. Thus, on the surface, the answer for climacteric fruits should be that it is most desirable to regulate ethylene synthesis. The ideal product, then, is a fruit that makes less ethylene than wild type and is significantly slower to ripen. But that fruit, when the time is appropriate, can be ripened on demand by exogenous application of ethylene gas. The next question that must be asked is how much reduction in ethylene production do we want? This answer is not always obvious. It will ultimately depend on the commercial systems in place. Our work with tomato indicated that reduction sin ethylene synthesis up to 90% had very little effect on the rate of ripening (H. Klee, unpublished). But from 90% up to 100%, there is a linear relationship with ripening rate. Thus, a fruit that is 100% inhibited will never ripen while a fruit that is 95% inhibited will take substantially longer than a control to ripen but will ultimately achieve a state of ripeness. This is important because the fruit that is 100% inhibited might seem like an ideal product, being capable of storage for months. But such a fruit also requires continuous expose to ethylene for upwards of ten days to achieve a fully ripe stage. The production systems are not set up to handle this large scale, continuous ethylene application and wholesaler are not likely to adopt such a product. But a hypothetical product that will initiate a degree of ethylene synthesis upon exposure to a large concentration of exogenous ethylene (growers typically employ at least 100 ppm), will continue to ripen after it is removed from the gas room. This is probably closer to the ideal product even though it won't last as long as the 100% inhibited product. A further consideration is that stored fruits, even if they do not ripen, will continue to respire. We and others have produced fruits whose appearance is perfect for months (Hamilton et al., 1990; Klee et al., 1991; Oeller et al., 1991; Good et al., 1994). But flavor is not maintained over that period. The sugars and acids in stored fruits, so important for flavor, are consumed in a very few weeks. What is left is a perfect-appearing, rather bland tomato that would not encourage repeat purchases (H. Klee, unpublished). In the end, it will be necessary to achieve a range of inhibition and go through a process of trials to determine the best level of inhibition for a given fruit crop.

How is inhibition of ethylene synthesis best achieved? There are genes that will act in a dominant manner to degrade precursors of ethylene. Success has been reported with tissue specific expression of a S-adenosylmethionine (SAM) hydrolase gene (Good et al., 1994). The product of this gene degrades the precursor of ethylene, .SAM. Since SAM is a general cofactor of many enzymes in a cell, expression must be high and limited spatially and temporally to ripening fruits. A different gene, ACC deaminase targets degradation of the first committed ethylene precursor, ACC. This is converted to the non-toxic □-ketobutyric acid. Expression of ACC deaminase in tomato has also resulted in fruits with greatly extended shelf life (Klee et al., 1991). In theory, since ACC is a precursor only to ethylene, tissue specific expression is less critical. The major advantage of this approach to ethylene control is tat one gene serves all purposes. The genes that have been shown to work in tomato, for example, should also work well in other, unrelated plant species. One does not need extensive knowledge of the endogenous ethylene synthesis system in order to achieve the desired end product. One issue that arises with both of these enzymes is that they compete with the endogenous ethylene biosynthetic enzymes for substrate. Since ACS and ACO have relatively high affinities for their substrates, very high levels of the degrading enzymes must be achieved for ethylene control. In practical terms, this means that a large number of transgenic plants may need to be produced and screened to identify those lines that achieve 90%+ reduction in ethylene synthesis.

For reduction in expression of endogenous ACS and ACO genes, several techniques are available including antisense RNA expression, cosuppression and RNAi. All of these techniques have been demonstrated to work effectively in controlling expression of a range of genes. We and others have successfully achieved almost complete shutoff of ethylene synthesis using antisense ACS and ACO genes. The technique works effectively and reliably. However, in order to use this approach, one must have all of the necessary tools and a good understanding of how the target organism regulates ethylene synthesis during ripening. In order to shut down ethylene synthesis, the right genes need to be targeted. Since the target genes are members of families, and ethylene synthesis may be controlled by a subset of these genes, knowledge of expression patterns is essential to designing an effective strategy. As an example, ACS in tomato is encoded by a large family of ten or more genes whose sequences, at the extreme, are less than 50% identical (Oetiker et al., 1997). Only two of these genes are responsible for the climacteric ethylene synthesis associated with ripening. Thus, targeting the right genes for shutoff may not be straightforward without some knowledge of gene expression. If that knowledge is available, however, the strategy is highly effective and straightforward.

4. POTENTIAL ISSUES

What has become apparent over the last several years is that shutting down ethylene synthesis can have some negative consequences. Ethylene is an important regulator of many developmental and environmental responses. When ethylene is removed, unpleasant things can happen to a plant. For example, we have shown that ethylene is essential for formation of adventitious roots (Clark et al., 1999). Ethylene insensitive tomato and petunia plants cannot be propagated by stem cuttings. This can be a major shortcoming in vegetatively propagated species. Probably a more important problem, however, relates to pathogen responses. Plants that have been engineered for constitutive reduction in either ethylene sensitivity of synthesis exhibit altered pathogen responses. In some instances, disease symptoms are reduced in these plants (Bent et al., 1992; Lund et al., 1998), while in others, susceptibility is substantially increased (Knoester et al., 1998; Thomma et al., 1999). In extreme cases, plants become susceptible to pathogens that are normally non-hosts (D. Clark and H. Klee, unpublished observations). Clearly, increased susceptibility to pathogens is a serious issue that will preclude commercialization of some engineered plants.

Is there a way around these issues? Most likely, yes. The answer would appear to lie in targeted expression of trangenes. If the desired trait is delayed fruit ripening, one need only reduce ethylene in ripening fruits. Such a targeted approach should permit normal ethylene responses throughout the rest of the plant. Targeting should be possible via tissue specific transcriptional promoters. It may also be possible via targeting specific members of the biosynthetic gene family. While both of these approaches require a level of sophistication in strategies, they are possible. If the value of the product is high enough, the investment will be warranted. Although the widespread introduction of extended shelf life *rin* hybrids has made transgene-based approaches in tomato redundant, there are plenty of opportunities for enhanced shelf life fruits remaining. We believe that the value is there and are optimistic about the future.

5. ACKNOWLEDGEMENTS

The author wishes to thank the many collaborators who have been involved in engineering delayed ripening tomatoes both at Monsanto and the University of Florida. In particular, the efforts of Keith Kretzmer, Tasneem Rangwala and Glenn Austin at Monsanto and David Clark at UF have been invaluable.

6. REFERENCES

Barry, C., Blume, B., Bouzayen, M., Cooper, W., Hamilton, A., and Grierson, D. 1996. Differential expression of the 1-aminocyclopropane-1-carboxylate oxidase gene family of tomato. Plant J. 9: 525-535.

Bent, A., Innes, R., Ecker, J., and Staskawitz, B. 1992. Disease development in ethylene-insensitive *Arabidopsis thaliana* infected with virulent and avirulent *Pseudomonas* and *Xanthomonas* pathogens. Mol.Plant-Microbe Interact. 5: 372-378.

Chang, C., Kwok, S. F., Bleecker, A. B., and Meyerowitz, E. M. 1993. *Arabidopsis* ethylene-response gene *ETR1*: similarity of products to two-component regulators. Science 262: 539-

544.

Clark, D. G., Gubrium, E. K., Barrett, J. E., Nell, T. A., and Klee, H. J. 1999.Root formation in ethylene-insensitive plants. Plant Physiol. 121: 53-59 .

Good, X., Kellogg, J. A., Wagoner, W., Langhoff, D., Matsumura, W., and Bestwick, R. K. 1994.Reduced ethylene synthesis by transgenic tomatoes expressing S- adenosylmethionine hydrolase. Plant Mol. Biol. 26: 781-790 .

Hamilton, A. J., Lycett, G. W., and Grierson, D. 1990. Antisense gene that inhibits synthesis of the hormone ethylene in transgenic plants. Nature 346: 284-287.

Hua, J., Chang, C., Sun, Q., and Meyerowitz, E. M. 1995. Ethylene insensitivity conferred by *Arabidopsis ERS* gene. Science 269: 1712-1714.

Hua, J., Sakai, H., Nourizadeh, S., Chen, Q., Bleecker, A., Ecker, J., and Meyerowitz, E. 1998. *EIN4* and *ERS2* are members of the putative ethylene receptor gene family in Arabidopsis. Plant Cell 10: 1321-1332.

Klee, H. J., Hayford, M. B., Kretzmer, K. A., Barry, G. F., and Kishore, G. M. 1991. Control of ethylene synthesis by expression of a bacterial enzyme in transgenic tomato plants. Plant Cell 3: 1187-1193.

Knoester, M., Van Loon, L. C., Van Den Heuvel, J., Hennig, J., Bol, J. F., and Linthorst, H. J. M. 1998. Ethylene-insensitive tobacco lacks nonhost resistance against soil-borne fungi. Proc. Nat. Acad. Sci. USA 95: 1933-1937 .

Lashbrook, C., Tieman, D., and Klee, H. 1998. Differential regulation of the tomato *ETR* gene family throughout plant development. Plant J. 15: 243-252.

Lund, S. T., Stall, R. E., and Klee, H. J. 1998.Ethylene regulates the susceptible response to pathogen infection in tomato. Plant Cell 10: 371-382 .

Oeller, P. W., Min-Wong, L., Taylor, L. P., Pike, D. A., and Theologis, A. 1991. Reversible inhibition of tomato fruit senescence by antisense RNA. Science 254: 437-439.

Oetiker, J., Olson, D., Shiu, O., and Yang, S. F. 1997. Differential induction of seven 1-aminocyclopropane-1-carboxylate synthase genes by elicitor in suspension cultures of tomato (*Lycopersicon esculentum*). Plant Molec. Biol. 34: 275-286.

Schaller, G. E. and Bleecker, A. 1995. Ethylene binding sites generated in yeast expressing the *Arabidopsis* ETR1 gene. Science 270: 1809-1811.

Solano, R., Stepanova, A., Chao, Q. M., and Ecker, J. R. 1998.Nuclear events in ethylene signaling: a transcriptional cascade mediated by ETHYLENE-INSENSITIVE3 and ETHYLENE-RESPONSE-FACTOR1. Genes & Development 12: 3703-3714 .

Theologis, A. 1993. One rotten apple spoils the whole bushel: the role of ethylene in fruit ripening. Cell 70: 181-184.

Thomma, B, Eggermont, K., Tierens, K, and Broekaert, W. F. 1999. Requirement of Functional Ethylene-Insensitive 2 Gene for Efficient Resistance of Arabidopsis to Infection by Botrytis Cinerea. Plant Physiol 121: 1093-1101 .

Tieman, D. and Klee, H. 1999. Differential expression of two novel members of the tomato ethylene-receptor family . Plant Physiol. 120: 165-172.

Tieman, D. M., Ciardi, J. A., Taylor, M. G., and Klee, H. J. 2001. Members of the Tomato *Leeil* (*Ein3*-Like) Gene Family Are Functionally Redundant and Regulate Ethylene Responses Throughout Plant Development. Plant J. 26: 47-58 .

Wang, K. L. C., Li, H., and Ecker, J. R. 2002. Ethylene biosynthesis and signaling networks. Plant Cell 14: S131-S151.

Wilkinson, J. Q., Lanahan, M. B., Yen, H.-C., Giovannoni, J. J., and Klee, H. J. 1995. An ethylene-inducible component of signal transduction encoded by *Never-ripe*. Science 270: 1807-1809.

Zhou, D., Kalaitzis, P., Mattoo, A., and Tucker, M. 1996. The mRNA for an ETR1 homologue in tomato is constitutively expressed in vegetative and reproductive tissues. Plant Molec. Biol. 30: 1331-1338.

Zhou, D., Mattoo, A., and Tucker, M. 1996. Molecular cloning of a tomato cDNA encoding an ethylene receptor. Plant Physiol. 110: 1435-1436.

MODIFYING THE AMINO ACID COMPOSITION OF GRAINS USING GENE TECHNOLOGY

Nicholas D. Hagan, Linda M. Tabe, Lisa Molvig and T.J.V. Higgins
CSIRO Plant Industry, GPO Box 1600, Canberra, ACT 2601, Australia (e-mail: TJ.Higgins@csiro.au)

Keywords Essential amino acids, methionine, lysine, rumen-stable, storage proteins, rice, lupin

Animals cannot synthesise 10 of the 20 amino acids needed for protein production and must obtain these "essential" amino acids from their diet. Although cereals and legumes are major sources of protein for humans and livestock, individually these crops do not supply the full complement of essential amino acids. Cereal grains are deficient in lysine whilst legume grains are deficient in the sulfur containing amino acids cysteine and methionine. Traditional plant breeding has attempted to improve the balance of essential amino acids in seed proteins. Good progress has been made in breeding high lysine corn, but breeding of high methionine legumes has met with only limited success. More recently, gene technology (GT) has been used to introduce new metabolic enzymes or storage proteins into cereals, legumes and other plants and has shown potential in addressing the nutritional deficiencies in these crops (Falco et al., 1995; Altenbach et al., 1992; Pickardt et al., 1995; Molvig et al., 1997).

Accumulation of foreign lysine- or methionine-rich proteins can be limited by the rate of biosynthesis of these amino acids in the plant. In order to increase the availability of limiting amino acids, one approach has been to deregulate the particular biosynthetic pathways. Recently, GM technology has been used to increase flux through amino acid biosynthetic pathways by introducing feedback insensitive versions of the key enzymes involved. Transgenic canola and soybean seeds were produced with increased lysine by circumventing the normal feedback regulation of two enzymes in the lysine biosynthetic pathway, aspartokinase (AK) and dihydrodipicolinic acid synthase (DHPS) (Falco et al., 1995). Plant DHPS is feedback inhibited by lysine whilst AK is feedback inhibited by either lysine or threonine. Transgenic canola plants, containing a feedback-insensitive bacterial homologue of DHPS, showed a doubling of total seed lysine (Falco et al., 1995). Transgenic soybean plants with feedback-insensitive bacterial homologues of both DHPS and AK showed as much as a five-fold increase in

I. K. Vasil (ed.), Plant Biotechnology 2002 and Beyond, 305-308.

total seed lysine (Falco et al., 1995). In the sulfur accumulation pathway, key control points are catalysed by serine acetyl transferase (SAT) and O-acetylserine (thiol) lyase (OASTL). SAT converts serine to O-acetylserine and is an important enzyme in regulating the input of nitrogen into the sulfur assimilation pathway. OASTL is a multi-subunit enzyme that catalyzes the formation of cysteine from O-acetylserine and sulfide. Feedback insensitive versions of these enzymes have now been isolated and research is now focused on the manipulation of plant cysteine and methionine biosynthesis using these enzymes.

Another GT approach to improve nutritional quality has been to express genes for lysine or methionine-rich proteins from other plants. Few lysine-rich storage proteins have been identified but number of seed storage proteins have been characterised that are rich in methionine. The most extensively studied of these have been the Brazil nut (*Bertholletia excelsa*) 2S albumin containing 18% methionine and 8% cysteine residues and sunflower (*Helianthus annuus*) seed albumin (SSA) containing 16% methionine and 8% cysteine residues. The genes for both these proteins have been cloned and introduced into a variety of plants (reviewed in Tabe and Higgins, 1998). In our laboratory, the gene for SSA has been introduced into legume and cereal species including narrow leaf lupin (Molvig et al., 1997) and rice (Hagan et al., unpublished) for expression in the seed.

The SSA gene was placed under the control of the pea vicilin promoter and introduced into narrow leaf lupin (*Lupinus angustifolius*), a grain legume that is low in seed methionine but is commonly fed to stock because of its high level of protein and fibre. SSA accounted for 5% of total seed protein in the transgenic lupin and resulted in a doubling of seed methionine and an overall increase in sulfur-containing amino acids of 20% (Molvig et al., 1997). In chicken feeding trials it was shown that when the high methionine lupins were used at 25% inclusion in the diet it was possible to reduce the level of added methionine by 22% without reducing animal growth (Ravindran et al., 2002). Grain containing the SSA protein also has the potential to be very useful as feed for ruminant animals as SSA has been demonstrated to be resistant to degradation in the rumen. This allows more sulfur-containing amino acids to be delivered to the small intestine of the animal, bypassing incorporation into microbial protein. In feeding trials with sheep, the high methionine lupins gave an 8% increase in live weight gain and 8% increase in wool growth, compared to non-transgenic seeds (White et al., 2001). At present, the high methionine lupin is at an advanced stage of evaluation for commercial release.

Although cereal grains are not deficient in the sulfur containing amino acids, the SSA gene was introduced into rice to gauge its effects on the levels of methionine, cysteine and the endogenous sulfur containing storage proteins. Transgenic rice expressing SSA at a level of approximately 7% of total protein did not show any increase in sulfur amino acids or total protein. However, significant changes in the protein profile of the seeds were observed. Endogenous sulfur-rich storage proteins were down regulated in the SSA transgenic seed, presumably because SSA had sequestered most of the available sulfur. In addition, sulfur-poor storage proteins were up regulated, allowing the seed to maintain a constant level of total protein (Hagan et al., unpublished).

In recent years, efficient promoters have been identified that have enabled sulfur-rich proteins to be expressed at high levels in transgenic plants. However, there are a variety of responses between species, which means that even a high level of expression doesn't necessarily result in a large increase in the desired amino acid. Good examples of this can be seen in transgenic lupins and rice expressing the gene for the SSA protein. Lupins expressing SSA at 5% of total protein show a doubling of seed methionine whilst rice expressing SSA at 7% of total protein showed no increase in seed methionine. Evidence from both legume and cereals demonstrate that high levels of transgene expression may lead to altered levels of endogenous proteins which may or may not affect the sulfur amino acid content of the seed protein (for review see Tabe et al., 2002).

Despite current limitations, the introduction of high-lysine or high-methionine proteins combined with the approach of using biosynthetic pathway manipulation shows great promise in creating food that is able to supply a balanced level of all the essential amino acids.

REFERENCES

Altenbach S.B., Chiung-Chi K., Staraci L.C., Pearson K.W., Wainwright C., Georgescu and Townsend J. (1992). Accumulation of a Brazil nut albumin in seeds of transgenic canola results in enhanced levels of seed protein methionine. Plant Mol. Biol. 18: 235-245.

Falco S.C., Guida T., Mauvais J., Sanders C., Ward R.T. and Webber P. (1995) Transgenic canola and soybean seeds with increased lysine. Biotech. 13:577-582.

Molvig L., Tabe L., Eggum B.O., Moore A., Craig S., Spencer D. and Higgins T.J.V. (1997). Enhanced methionine levels and increased nutritive value of seeds of transgenic lupins (*Lupinus angustifolius* L.) expressing a sunflower seed albumin gene. Proc. Natl. Acad. Sci. USA 94: 8393-8398

Pickardt T., Saalbach I., Waddell D., Meixner M., Müntz K. and Schieder O. (1995). Seed specific expression of the 2S albumin gene from Brazil nut (*Bertholletia excelsa*) in transgenic *Vicia narbonensis*. Mol. Breed. 1: 295-301.

Ravindran V., Tabe L.M., Molvig L., Higgins T.J.V. and Bryden W.L. (2002). Nutritional evaluation of transgenic high-methionine lupins (*Lupinus angustifolius* L) with broiler chickens. J. Sci. Food. Agric. 82: 280-285

Roesler K.R. and Rao A.G. (1999). Conformation and stability of barley chymotrypsin inhibitor-2 (CI-2) mutants containing multiple lysine substitutions. Protein Eng. 12: 967-973.

Tabe L. and Higgins T.J.V. (1998). Engineering plant protein composition for improved nutrition. Trends in Plant Science 3: 282-286

Tabe L., Hagan N. and Higgins T.J.V. (2002). Plasticity of seed protein composition in response to nitrogen and sulfur availability. Current Opinion in Plant Biology 5: 212-217

White C.L., Tabe L.M., Dove H., Hamblin J., Young P., Phillips N., Taylor R., Gulati S., Ashes J. and Higgins T.J.V. (2001). Transgenic lupin seed containing sunflower albumin has a higher nutritional value for sheep than seed from the non-transgenic parent. J. Sci. Food Agric. 81: 147-154

ENGINEERING OF APOMIXIS IN CROP PLANTS: WHAT CAN WE LEARN FROM SEXUAL MODEL SYSTEMS?

Ueli Grossniklaus[1,2], James M. Moore[1,2,3], Vladimir Brukhin[1], Jacqueline Gheyselinck[1], Ramamurthy Baskar[1,2], Jean-Philippe Vielle-Calzada[2,4], Célia Baroux[1], Damian R. Page[1], and Charles Spillane[1,2]
[1]Institute of Plant Biology, University of Zürich, Zollikerstrasse 107, CH-8008 Zürich, Switzerland (e-mail: grossnik@botinst.unizh.ch);[2]Cold Spring Harbor Laboratory, 1 Bungtown Road, Cold Spring Harbor, NY 11724, USA;[3]Graduate Program in Genetics, State University of New York, Stony Brook, NY 11794, USA;[4]Present Address: CINVESTAV-Irapuato, Plant Biotechnology Unit, CP 36500, Irapuato, GTO, México

Keywords Apomixis, biotechnology, enhancer detection, genomic imprinting, insertional mutagenesis, sexual reproduction

1. INTRODUCTION

The development of apomixis technology in crop plants is a desirable goal. Apomixis is the asexual reproduction through seeds, which occurs in over 400 flowering plants (Nogler, 1984). The introduction of clonal reproduction to crop plants will allow the indefinite propagation of any desirable genotype (including that of heterozygous F1 hybrids) and will completely transform current breeding and seed production strategies. Developmental aspects of apomixis (Koltunow, 1993; Grossniklaus, 2001; Spillane et al., 2001), its genetic control (Savidan, 2000; Grossniklaus et al., 2001a; Grimanelli et al., 2001), and its potential use in agriculture (Koltunow et al., 1995; Hanna et al., 1998; Jefferson and Bicknell, 1996; Thoenissen, 2001) have been extensively reviewed. Here, we provide a short summary of developmental and genetic aspects and report on our program using sexual model systems to identify genes and promoters relevant to the engineering of apomixis.

2. DEVELOPMENTAL GENETICS OF APOMIXIS

Apomixis occurs in more than 40 angiosperm families suggesting a polyphyletic evolutionary origin. The modes of apomixis are diverse (Crane, 2001) but from a developmental perspective, they can all be viewed as a deregulation of key steps of the sexual developmental program (Koltunow, 1993; Vielle-Calzada et al., 1996; Grossniklaus et al., 1998; Grossniklaus,

I. K. Vasil (ed.), Plant Biotechnology 2002 and Beyond, 309-314.

2001). The different types of apomixis vary in the time point or the cell type where this deregulation occurs. Naturally occurring apomixis differs from sexual reproduction in three major elements (Grossniklaus, 2001; Spillane et al., 2001). These are (i) the absence or alteration of meiosis preventing reduction (apomeiosis), (ii) the activation of the egg cell to form an embryo in the absence of fertilization (parthenogenesis), and (iii) the formation of endosperm with (pseudogamous) or without (autonomous) fertilization. The development of functional endosperm in apomicts relies on alterations of embryo sac development and/or fertilization or the circumvention of seed abortion due to parental effects (imprinting barriers) (Grossniklaus et al., 1998, 2001b; Savidan, 2000; Grossniklaus, 2001; Grimanelli et al., .2001).

While apomixis may be controlled by a single master-regulatory locus in some species (Savidan, 2001), it is controlled by several loci in others (Grossniklaus et al., 2001a). Therefore, the genetic basis for apomixis may be quite diverse involving a variety of genes or even the combination of entire divergent genomes in hybrids (Carman, 2001; Grimanelli et al., 2001). Epigenetic changes in gene expression may have also contributed to the evolution of apomixis (Spillane et al., 2001). Therefore, it seems promising to engineer apomixis by combining its elements, the genetic control of each of which can be addressed separately.

Some of the developmental processes that are deregulated during apomixis are functions of the sporophyte, e.g. the determination of a nucellar cell that initiates embryo sac development (functional megaspore in a sexual and aposporic initial cell in an apomictic plant). Most of them, however, are functions of the female gametophyte: embryo sac initiation, the control of nuclear division cycles, the process of double fertilization, endosperm and embryo initiation, and the epigenetic makeup that differentiates between maternal and paternal genomes. These processes are poorly understood at the genetic and molecular level. A better understanding of the molecular basis underlying these aspects of sexual reproduction will provide genetic and molecular tools for the development of apomixis technology in crop plants.

3. SEXUAL MODEL SYSTEMS AS A SOURCE OF GENES FOR THE ENGINEERING OF APOMIXIS

The exploitation of sexual model systems to elucidate the genetic control of developmental steps that are deregulated during apomictic reproduction is at the center of our research program. We use both *Arabidopsis thaliana* and *Zea mays* (maize) as genetic model systems in a multifacetted effort to identify genes and regulatory elements that can be used as molecular tools for engineering elements of apomixis. In maize, our genetic approach relies on Robertson's *Mutator*, a powerful insertional mutagen (Benetzen, 1996), to identify mutants relevant to apomixis (Grossniklaus, 2001). In *Arabidopsis*, we use the *Ds* transponson-based enhancer detection system developed by Sundaresan et al. (1997). In addition to serving as an insertional mutagen, enhancer detection allows the identification of genes specifically expressed in particular cells and tissues or at developmental stages when deregulation of the sexual program is likely to occur in apomicts. Enhancer detection relies on a mobile genetic element (here a *Ds* transposon) carrying a reporter gene (here the *uidA* gene encoding β-glucuronidase or GUS) under the control of a minimal promoter. If the transposon inserts in the neighborhood of tissue- and/or stage-specific cis-regulatory elements (e.g. enhancers) GUS will be expressed in a pattern controlled by these elements. Usually, the detected elements control a nearby gene expressed in the same way as GUS. Thus, genes can be identified based on their pattern of expression. Enhancer detection is currently the most powerful tool to isolate genes expressed in specific tissues including single cells, for instance the functional megaspore or the mature egg cell; cell types playing crucial roles in plant reproduction. Gene traps are a variation of this method (Springer, 2000).

We use enhancer detector and gene trap elements to identify genes that are relevant to sexual reproduction and may provide molecular tools for the engineering of apomixis. We have generated a library of close to 7000 lines (transposants), each of which carries such an element at a random position in the genome. The transposants were subjected to four different screens (Figure 1) aimed at identifying genes affecting megasporogenesis, which could be relevant to the engineering of apomeiosis, as well as megagametogenesis and fertilization, which may provide new insights to modify various functions of the female gametophyte that are altered in apomictic species (parthenogenesis and endosperm formation). Given the lack of knowledge concerning the genetic control of these developmental processes, the isolation of gametophytic mutants is of crucial importance for the engineering of apomixis. The two phenotypic screens identify female sterile mutants, some of which affect megasporogenesis, and gametophytic mutants disrupting embryo sac development (e.g. Moore et al., 1997) or double fertilization. The expression screens aim at the identification of genes that are either expressed in young ovules during megasporogenesis (e.g. Vielle-Calzada et al., 1998) or in mature ovules when cell specification and double fertilization occur (e.g. Vielle-Calzada et al., 2000). These screens also provide regulatory elements that will be used to misexpress candidate genes at a specific time or in specific tissues in order to deregulate the sexual process towards apomixis (Grossniklaus et al., 1998; Grossniklaus, 2001).

A second major goal is the elucidation of the mechanisms that control genomic imprinting. Mutants affecting this process will allow imprinting barriers to the introduction of apomixis to be overcome. To identify mutants that affect the imprinting process, we make use of the gametophytic maternal

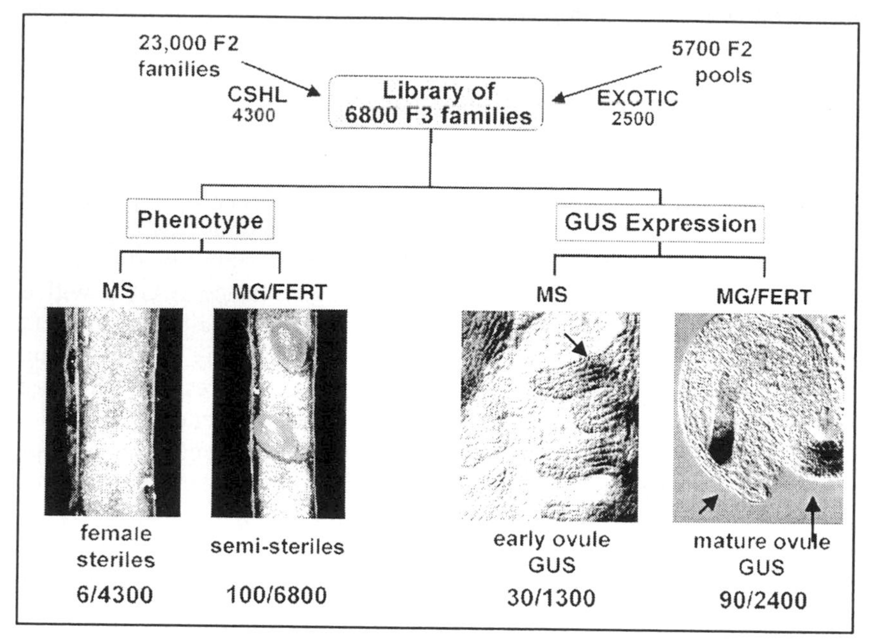

Figure 1. Genetic screens for mutants affecting steps in sexual reproduction that are relevant to the engineering of apomixis. The four screens are based on a library of 6800 transposants that were either screened by phenotype for insertional mutants affecting plant reproduction or by staining for GUS expression patterns in relevant tissues (see text for details). Phenotypic and expression screens targeted megasporogenesis (MS) as well as megagametogenesis and fertilization (MG/FERT). The pictures on the left show a sterile and a semi-sterile (megagametyophyte lethal) mutant, in which all or half of the ovules (white structures) do not initiate seed development, respectively. The pictures on the right show GUS expression at the basis of early ovule primordia or in the embryo sac (strong in egg apparatus) and funiculus of a mature ovule, respectively (arrowheads). The numbers below each panel indicate approximately how many positive transposants were identified in a given screen.

effect mutation *medea* (*mea*) (Grossniklaus et al., 1998). *MEA* was shown to be regulated by genomic imprinting (Vielle-Calzada et al, 1999). In a heterozygous plant, all the seeds derived from a mutant embryo sac abort irrespective of the paternal contribution, because the paternally inherited *MEA* allele is not expressed during seed development. The resulting 50% seed abortion serves as the basis for a second-site modifier screen for mutants with less (suppressors) or more (enhancers) seed abortion. Modifier loci can represent upstream regulators, MEA-interacting proteins, or downstream target genes. But a subset of them will affect the imprinting process itself and could be involved in the establishment, maintenance or resetting of the epigenetic mark providing the molecular basis for genomic imprinting. We have used a variety of approaches to identify modifier mutants, including chemical mutagenesis, screening for natural variants in the *Arabidopsis* gene pool, and testing candidate loci for genetic interactions with *mea* (e.g. Vielle-Calzada et al., 1999). In the long term, a better understanding of the mechanisms controlling megasporogenesis, embryo sac development, double fertilization and genomic imprinting will provide the molecular tools for the engineering of apomixis in sexual crops.

4. ACKNOWLEDGEMENTS

This contribution is a summary of our activities in apomixis research and, as such, largely cites our own work. We apologize to our colleagues who have not been cited due to limited space. We thank Wendy Gagliano, Marilu Hoeppner, Alison Coluccio and Julie Thomas for technical assistance. We are indebted to our former colleagues Sundaresan Venkatesan, Robert Martienssen and Hong Ma for developing the enhancer detection system and providing starter lines. The presented projects were supported by the Cold Spring Harbor Laboratory President's Council, grant 98-35304-6412 of the NRI Competitive Grants Program of the US Department of Agriculture, grant MCB-9723948 of the National Science Foundation (USA), grant 99.0306 of the Bundesamt für Forschung und Wissenschaft as part of the EXOTIC Project QLG2-CT-1999-00351 of the European Union, Pioneer Hi-Bred International, the Novartis Forschungsstiftung and the Kanton of Zürich. D.P. and C.B. are recipients of a Roche Research Foundation fellowship, J-P. V-C. was supported by a postdoctoral fellowship of the Fonds National Suisse de la Recherche Scientifique, and U.G. was supported by EMBO, HFSP, the Janggen-Poehn Stiftung and a Searle Scholarship.

5. REFERENCES

Bennetzen J.L. 1996. The *Mutator* transposable element system of maize. Curr. Top. Microbiol. Immunol. 204:195-229.

Carman J.G. 2001. The gene effect: genome collisions and apomixis. The Flowering of Apomixis: From Mechanisms to Genetic Engineering. Y. Savidan, J.G. Carman, and T. Dresselhaus (eds.). CIMMYT, IRD, European Commission DG VI, México. Pp 95-110.

Crane C. 2001. Classification of apomictic mechanisms. The Flowering of Apomixis: From Mechanisms to Genetic Engineering. Y. Savidan, J.G. Carman, and T. Dresselhaus (eds.). CIMMYT, IRD, European Commission DG VI, México. Pp 24-43.

Grimanelli D., O. Leblanc, E. Perotti, and U. Grossniklaus. 2001. Developmental genetics of gametophytic apomixis. Trends Gen. 17:597-604.

Grossniklaus U. 2001 From sexuality to apomixis: molecular and genetic approaches. The Flowering of Apomixis: From Mechanisms to Genetic Engineering. Y. Savidan, J.G. Carman, and T. Dresselhaus (eds.) CIMMYT, IRD, European Commission DG VI. México. Pp. 168-211.

Grossniklaus U., J.M. Moore, and W.B. Gagliano. 1998a. Molecular and genetic approaches to understanding and engineering apomixis: Arabidopsis as a powerful tool. Advances in Hybrid Rice Technology. Proceedings of the 3rd International Symposium on Hybrid Rice, 1996. S.S. Virmani, E.D. Siddiq, and K. Muralidharan (eds.) International Rice Research Institute, Manila. Pp. 187-212.

Grossniklaus U., J-P. Vielle-Calzada, M.A. Hoeppner, and W.B. Gagliano. 1998b. Maternal control of embryogenesis by *MEDEA*, a *Polycomb* group gene in *Arabidopsis*. Science 280:446-450.

Grossniklaus U., G.A. Nogler, and P. van Dijk. 2001a. How to avoid sex: The genetic control of gametophytic apomixis. Plant Cell 13:1491-1498.

Grossniklaus U., C. Spillane, D.R. Page, and C. Koehler. 2001b. Genomic imprinting and seed development: endosperm formation with and without sex. Curr. Opin. Plant. Biol. 4:21-27.

Hanna W.W., D. Roche, and P. Ozias-Akins. 1998. Use of apomixis in crop improvement. Advances in Hybrid Rice Technology. Proceedings of the 3rd International Symposium on Hybrid Rice, 1996. S.S. Virmani, E.D. Siddiq, and K. Muralidharan (eds.) International Rice Research Institute, Manila. Pp. 283-296.

Jefferson R.A., and R. Bicknell. 1996. The potential impacts of apomixis: a molecular genetics approach. The Impact of Plant Molecular Genetics. B.W.S. Sobral (Ed.). Birkhäuser, Boston. Pp. 87-101.

Koltunow A.M. 1993. Apomixis: embryo sacs and embryos formed without meiosis or fertilization in ovules. Plant Cell 5:1425-1437.

Moore J.M., J-P. Vielle-Calzada, W.B. Gagliano, and U.Grossniklaus. 1997 Genetic characterization of *hadad*, a mutant disrupting female gametogenesis in *Arabidopsis thaliana*. Cold Spring Harb. Symp. Quant. Biol. 62:35-47.

Nogler G.A. 1984. Gametophytic apomixis. Embryology of angiosperms. B.M. Johri (Ed.) Springer, Berlin. Pp. 475-518.

Savidan Y.H. 2000. Apomixis: Genetics and Breeding. Plant Breed. Rev. 18:13-86.

Savidan Y.H. 2001. Gametophytic apomixis: a successful mutation of the female gametogenesis. Current Trends in the Embryology of Angiosperms S.S. Bhojwani and W.Y. Soh (eds.). Kluwer Academic Publishers, The Neterlands. Pp. 419-433.

Spillane C., A. Steimer, and U. Grossniklaus. 2001. Apomixis in agriculture: the quest for clonal seeds. Sex. Plant Reprod. 14:179-187.

Springer P.S. 2000. Gene traps: tools for plant development and genomics. Plant Cell 12:1007-1020.

Sundaresan V., P. Springer, T. Volpe, S. Haward, J.D. Jones, C. Dean, H. Ma, and R. Martienssen. 1995. Patterns of gene action in plant development revealed by enhancer trap and gene trap transposable elements. Genes Dev. 9:1797-1810.

Thoenissen G.H. 2001. Feeding the world in the 21st century: plant breeding, biotechnology and the potential role of apomixis. The Flowering of Apomixis: From Mechanisms to Genetic Engineering. Y. Savidan, J.G. Carman, and T. Dresselhaus (eds.) CIMMYT, IRD, European Commission DG VI. México. Pp. 1-7.

Vielle-Calzada J.P., C.F. Crane, and D.M. Stelly. 1996. Apomixis: The asexual revolution. Science 274: 1322-1323.

Vielle-Calzada J.P.,J.M. Moore, W.B. Gagliano, and U. Grossniklaus. 1998. Altering sexual development in *Arabidopsis*. J. Plant Biol. 41:73-81.

Vielle-Calzada J-P., J. Thomas, C. Spillane, A. Coluccio, M.A. Hoeppner, and U. Grossniklaus. 1999. Maintenance of genomic imprinting at the Arabidopsis medea locus requires zygotic DDM1 activity. Genes Dev. 13:2971-2982.

Vielle-Calzada J.P., R. Baskar, and U. Grossniklaus. 2000. Delayed activation of the paternal genome during seed development. Nature 404:91-94.

OVEREXPRESSION OF *ARABIDOPSIS DWARF4* IN TOMATO INCREASES BRANCHING AND FRUIT NUMBER

Zhihong Chen Cook[a], Julissa Sosa[a], Shozo Fujioka[b] and Kenneth A. Feldmann[a]
[a]Ceres, Inc. 3007 Malibu Canyon Road, Malibu, California 90265
[b]RIKEN (The Institute of Physical and Chemical Research), Wako-shi, Saitama 351-0198, Japan (e-mail: zchen@ceres-inc.com)

Keywords DWF4, brassinosteroids, overexpression, tomato transformation

1. INTRODUCTION

Brasssinosteroids (BRs) are a group of plant steroid hormones which mediate many important processes in plant growth and development including stimulation of cell division, cell elongation, vascular differentiation, and stress responses. BR-deficient and BR-insensitive mutants have been identified in Arabidopsis, pea, rice and tomato. BR mutants have a very characteristic dwarf phenotype in all of these species. Genes for many of these mutants have been cloned and characterized. The Arabidopsis *DWF4* gene, for example, which catalyzes the C22 oxidation steps in the BR biosynthetic pathway is the rate limiting step (Choe et al. 1998). Overexpression of *DWF4* in Arabidopsis resulted in a 35% increase in inflorescence height, and a 59% increase in seed yield (Choe et al. 2001). Tomato (*Lycopersicon esculentum*) is an important horticultural crop and an excellent model system for biochemical and genetic analysis of plant growth and development. Although the counterpart of the *DWF4* gene in tomato has not been cloned, we expected that introducing the Arabidopsis *DWF4* gene into tomato would lead to increased yield, similar to Arabidopsis.

2. RESULTS AND DISCUSSION

2.1. Transformation of Tomato

Tomato, cv micro-tom, was transformed with a binary vector containing the *Arabidopsis thaliana DWF4* gene under the control of a constitutive promoter by means of Agrobacterium-mediated transformation. In the control vector *DWF4* was replaced by a histone-targeted yellow fluorescence protein (YFP)

I. K. Vasil (ed.), Plant Biotechnology 2002 and Beyond, 315-318.

gene construct. The bar gene, driven by a constitutive promoter was used for selection. In the control transgenic lines, expression of YFP was observed in the nuclei of all plant tissues, except pollen. Tomato overexpression of *DWF4* (TOD) lines were confirmed by PCR.

2.2. Analysis of TOD T1 Transgenic Plants

TOD primary transformants could be divided into two categories depending on their height: TOD-tall and TOD-short. TOD-tall lines had much elongated internodes resulting in taller plants compared to the controls, while TOD-short had heights similar to the control (data not shown). However, mean branch number in both types of plants was about 150% greater than controls after 3 months in the greenhouse. Most strikingly, fruit number in both types of plants, at 4 months of age, was more than doubled. The increase in branch number suggests that an increase in BR levels may cause the release of suppression of the axillary buds. A similar effect occurs after external application of cytokinins. Transcripts of *DWF4* in TOD lines were detected by RT-PCR in leaves, roots, stems, flower buds, fruits and fruit exocarp. Analysis of the young leaves of various TOD lines revealed that the intensities of the RT-PCR bands were correlated with strong and weak phenotypes. With a moderately expressed constitutive promoter, fruit size and seed number per fruit in the TOD lines was similar to control lines. Most TOD lines had an extended growth period of flowering and gave rise to fruits for 7 months, while control plants lived only 4 months.

2.3. Comparison of BRs Profiles Between TOD and Control Plants

Leaf samples were analyzed for endogenous BRs by GC-SIM. This analysis showed that the levels of compounds downstream of DWF4 were increased about 2 fold in the TOD leaf samples, compared to the control samples. This was very similar to the profile observed in Arabidopsis lines over-expressing *DWF4*. Nomura et al. (2001) showed that the profiles of BRs in the late C-6 oxidation pathway in tomato are very similar to Arabidopsis, suggesting common biosynthetic control mechanisms in these two species.

2.4. Stability of TOD Progeny

Four lines, with strong phenotypes for each type of TOD (tall and short), segregating for single gene insertions were selected for further evaluation in subsequent generations. The TOD-tall and TOD-short lines both continued to show phenotypes similar to the primary transformant including height differences and increased branch and fruit number (Figs. 1, 2). Not surprisingly, there were no visible differences in phenotypes of the heterozygous and homozygous plants.

Figure 1. Phenotypes of T2 TOD plants in later developmental stages.

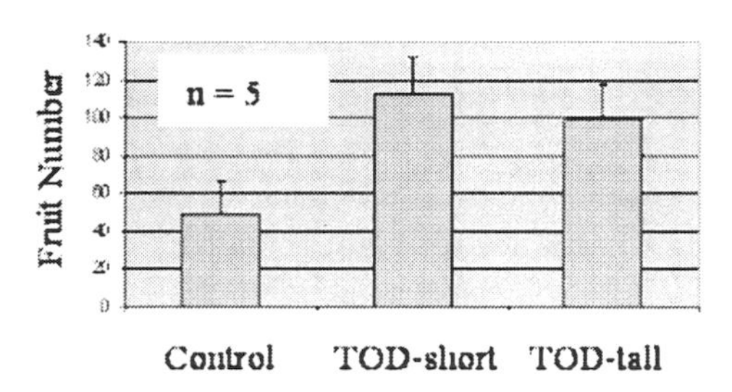

Figure 2. Increased fruit number in T2 TOD plants

2.5. Observation on Elongated Tissue

TOD-tall plants have longer hypocotyls, leaf petioles, internodes and inflorescence petioles than TOD-short plants, while root length was not significantly different. TOD-short plants were very similar in size to control plants but much more highly branched. Hand-dissected stems showed that cells in TOD-tall internodes were more elongated, approximately 2 fold, than those of TOD-short plants. These differences may be the result of differences in insertion sites between TOD-short and TOD-tall plants. Conversely, BRs may regulate cell elongation via unrelated, or synergistic, biochemical pathways.

3. CONCLUSIONS

Constitutive overexpression of Arabidopsis *DWF4*, encoding an enzyme in BR biosynthesis, in tomato resulted in the doubling of branch and fruit numbers. This result suggests that the *DWF4* gene may be a useful tool for manipulating yield in tomato.

4. REFERENCES

Choe S., Dilkes B. P., Fujioka S., Takatsuto S., Sakurai A. and Feldmann K.A. 1998. the DWF4 gene of Arabidopsis encodes a cytochrome p450 that mediates multiple 22a-hydroxylation steps in brassinosteroid biosynthesis. Plant Cell 10: 231-243.

Choe S., Jujioka S., Noguchi T.l., Takatsuto S., Yoshida S. and Feldmann K.A. 2001. Overexpression of DWARF4 in the brassinosteroid biosynthetic pathway results in increased vegetative growth and seed yield in Arabidopsis. Plant J. 26(6): 573-582.

Nomura T., Sato T., Bishop G.J., Kamiya Y., Takatsuto S. and Yokota T. 2001. Accumulation of 6-deoxocastasterone in Arabidopsis, pea and tomato is suggestive of common rate-limiting steps in brassinosteroid biosynthesis. Phytochemistry 57: 171-178.

QUALITY OF FRUIT OF LYTIC PROTEIN TRANSGENIC APPLE LINES WITH ENHANCED RESISTANCE TO FIRE BLIGHT

H.S. Aldwinckle, E.E. Borejsza-Wysocka, and J.L. Norelli
Dept. of Plant Pathology, Cornell University, Geneva, NY 14456 and USDA, ARS, Appalachian Fruit Research Station, Kearneysville, WV 25430 USA (email: HSA1@cornell.edu)

Good crops of fruit were obtained on lytic protein-transgenic Royal Gala and Galaxy apple trees in 2000 and 2001. The transgenic trees included Royal Gala lines transgenic for attacin, avian lysozyme, and SB37 cecropin, and Galaxy fruit transgenic for attacin. In order to prevent transgenic pollen pollinating bearing trees near the field trial, large netting structures, each covering 2 rows of flowering transgenic trees, were erected. Flowers were manually pollinated with Idared pollen under the netting, and excellent fruit set was obtained. Two weeks after all flowers had undergone petal fall, the netting was removed and fruit developed normally. Fruit were manually thinned, where necessary, to a maximum crop load based on trunk cross sectional area. However, because of light flowering on some trees, crop loads were not completely uniform. Fruits were harvested from transgenic and non-transgenic Royal Gala and Galaxy trees in the second week of September each year, and placed in cold storage. All transgenic and non-transgenic fruit were graded for weight and color on an automatic grading line, and later pressure tested for firmness without skin, and assayed for soluble solids and titratable acidity. Lines with a degree of increased resistance to fire blight had color and size data that fell within the range of the non-transgenic Royal Gala and Galaxy controls. The data indicate the need for a more intensive trial of the most resistant lines.

Abstract only – no manuscript received.

I. K. Vasil (ed.), Plant Biotechnology 2002 and Beyond, 319.

TRANSFORMATION OF *BRASSICA NAPUS* WITH cDNAS ENCODING PROTEINS THAT STIMULATE *IN VITRO* TRIACYLGLYCEROL BIOSYNTHESIS

Randall J. Weselake[1], William B. Wiehler[1], Nii A. Patterson[2], Cory L. Nykiforuk[1], Katherine Cianflone[3], Maurice M. Moloney[2] and André Laroche[4]

[1]Department of Chemistry & Biochemistry, University of Lethbridge, Alberta, Canada T1K 3M4 (e-mail: weselake@uleth.ca); [2]Department of Biological Sciences, University of Calgary, Alberta, Canada T2N 1N4 ; [3]Mike Rosenbloom Labortatory for Cardiovascular Research, McGill University Health Centre, Montréal, Quebec, Canada H3A 1A1; [4]Agriculture & Agri-Food Canada Lethbridge Research Centre, Alberta T1J 4B1

Keywords Seed oil biosynthesis; DGAT; ASP; BSA

Human acylation stimulating protein (ASP) and bovine serum albumin (BSA) have been shown to stimulate diacylglycerol acyltransferase activity in microsomes from cultures of *Brassica napus* (Weselake et al., 2000; Little et al., 1994). In the current study, *B. napus* was transformed with cDNA encoding ASP or BSA to evaluate possible effects on seed oil biosynthesis.

Agrobacterium tumefaciens with a construct containing cDNA encoding ASP or BSA, under the control of an oleosin promoter, was used to transform *B. napus* L. cv Westar. Plants were grown in a greenhouse. cDNA encoding BSA was a gift from Dr. E.W. Holowachuk of the Mary Imogene Bassett Hospital Research Institute, Cooperstown, NY. Lipid extraction and analysis on T_1 seed was according to Weselake et al. (1993).

cDNAs encoding ASP or BSA were detected in leaf DNA and transcripts were detected in developing seeds (data not shown). ASP 8 and ASP 10 contained significantly elevated lipid content on a per mature seed basis (P = 0.03 & 0.04, respectively), which appeared to be related to an increase in seed size (Figure 1). ASP 3 and ASP 14 exhibited decreases in 18:2 (P = 0.05) and α-18:3 content (P = 0.04), respectively (Table 1). BSA 11 exhibited a significant increase in 18:2 and α-18:3 content (P = 0.04 & 0.02, respectively). These observations suggest that expression of ASP or BSA cDNA in *B. napus* could be used to alter lipid metabolism, and possibly seed development in ASP plants.

I. K. Vasil (ed.), Plant Biotechnology 2002 and Beyond, 321-322.

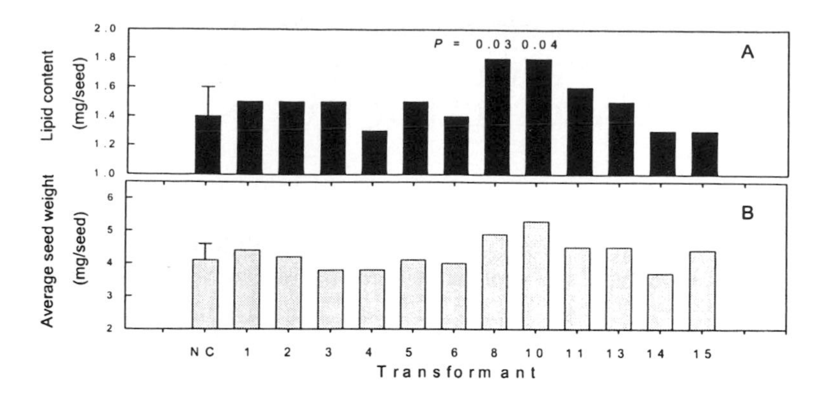

Figure 1. Lipid content (A) and seed weight (B) of mature seed from ASP plants. NC = negative control (averages ± SD for 6 plants transformed with the null vector).

*Table 1. Fatty acid composition (% wt) of total acyl lipids in mature seeds of control (NC) and transgenic plants. * P < 0.05*

Transformant	16:0	16:1	18:0	18:1	18:2	18:3	20:0	20:1
NC	4.4	0.3	3.4	65.7	15.8	5.3	1.0	1.2
ASP 3	3.8	0.2	3.7	70.5	13.3*	4.1	1.0	1.2
ASP 14	4.2	0.3	4.0	68.6	14.3	3.6*	1.1	1.2
BSA 11	4.7	0.3	4.0	59.4	18.6*	7.6*	1.1	1.1

ACKNOWLEDGEMENT

This work was supported by the Alberta Agricultural Research Institute.

REFERENCES

Little D., Weselake R., Pomeroy K., Furukawa-Stoffer T., and Bagu J. 1994. Solubilization and characterization of diacylglycerol acyltransferase from microspore-derived cultures of oilseed rape. Biochem. J. 304:951-958.

Weselake R., Kazala C., Cianflone K., Boehr D., Middleton C., Rennie C., Laroche A., and Recnik I. 2000. Human acylation stimulating protein enhances triacylglycerol biosynthesis in plant microsomes. FEBS Lett. 481:189-192.

Weselake R., Pomeroy K., Furukawa T., Golden J., Little D., and Laroche A. 1993. Developmental profile of diacylglycerol acyltransferase in maturing seeds of oilseed rape and safflower and microspore-derived cultures of oilseed rape. Plant Physiol. 102:565-571.

TOWARDS TRANSFORMATION, REGENERATION AND SCREENING OF PAPAYA CONTAINING ANTISENSE ACC SYNTHASE GENE

Pablito M. Magdalita[1], Antonio C. Laurena[1], Bessie M. Yabut-Perez[1], Maribel M. Zaporteza[1], Evelyn Mae Tecson-Mendoza[1], Violeta N. Villegas[1] and Jimmy R. Botella[2]

[1]Institute of Plant Breeding, College of Agriculture, University of the Philippines at Los Baños, College, Laguna, 4031, The Philippines (e-mail: pab@lgn.csi.com.ph); [2]The Plant Genetic Engineering Laboratory, Department of Botany, The University of Queensland, Brisbane, 4072, Australia

Keywords transformation, microprojectile bombardment, somatic embryos, antisense ACC synthase

1. INTRODUCTION

The papaya (*Carica papaya* L.) is usually being harvested when they are about 25% ripe or when a tinge of yellow color appears on the skin. It takes only one to two weeks before the fruit ripens completely from the time of harvest. Losses due to postharvest diseases in some species reach up to about 60% of annual production. The two most important qualities required for efficient marketing of fruits are the taste and the overall appearance. The flexibility in marketing is determined mainly by the rate of fruit ripening. An extended ripening phase would prolong its shelf-life, thus allowing the fruit to be shipped to distant markets without spoilage, thereby increasing the potential target markets, and enabling the fruit to reach the market in better condition. The papaya is a climacteric fruit and ripening is in part, being controlled by the simple hydrocarbon ethylene. The rate-limiting enzyme of the ethylene biosynthetic pathway is ACC synthase. This enzyme has been the target in engineering the ethylene biosynthesis pathway using the antisense technology. This technology for instance has been used to reduce translation of either ACC synthase or ACC oxidase by antisense RNA that blocks ethylene production in tomato, thereby delaying fruit ripening (Oeller et al., 1991; Hamilton et al., 1991). Two ACC synthase genes expressed during fruit ripening have been cloned from papaya (Mason and Botella, 1997). These genes could be used for transforming papaya plants that could constitutively express an antisense copy of the ACC synthase gene. In this study, we aimed to apply antisense technology and the existing transformation protocol to produce transgenic Solo papayas.

I. K. Vasil (ed.), Plant Biotechnology 2002 and Beyond, 323-327.

2. MATERIALS AND METHODS

2.1. Somatic Embryogenesis

Immature green Davao Solo papaya fruits (90-120 day old) were surface-sterilized with 70% ethanol. The embryos were isolated and cultured on half-strength MS salts and vitamins (Murashige and Skoog, 1962) supplemented with 10 mg L^{-1} 2,4-D, 400 mg L^{-1} glutamine, 50 mg L^{-1} myo-inositol, 10 mg L^{-1} thiamine-HCl, 30 g L^{-1} sucrose and 0.3% phytagel, pH 5.7. Highly embryogenic cultures were selected and about 7-12 somatic embryo clumps were spread using a sterile spatula onto a filter paper overlayed on somatic embryo induction medium following the procedures of Gonsalves et al. (1997).

2.2. Transient Expression

Two parameters (pressure of the helium blast and distance of the target tissue) for microprojectile bombardment were studied. In experiment 1, the effect of four different pressure levels (1000, 1500, 1800 and 2000 kPa) on transient gene expression was tested. Prior to bombardment, the somatic embryos were placed for 3-4 hours on high osmotic medium and then bombarded with the plasmid DNA of pBI 121 using a particle inflow gun. The bombarded embryos were transferred onto a recovery medium and incubated for 48 hours. After this period, GUS activity was assayed histochemically by incubating the embryos in X-gluc solution overnight at 37°C. Transient expression was assayed 12 hours after incubation and measured as total blue foci count per shot area. In a second experiment, three distances (17.5 as control, 15.0 and 12.5 cm.) of the target somatic embryos from the filter containing the microprojectiles were tested.

2.3. Stable Transformation

The primary somatic embryos were spread, placed on an osmotic medium and then bombarded with the plasmid DNA of the antisense ACC synthase gene. All bombarded embryos were placed for 7 days on a recovery medium and after this period, the embryos were transferred onto a pre-selection medium containing half strength MS plus 150 mg L^{-1} kanamycin sulphate. All cultures were incubated for 2-3 months onto this medium. After this period, the embryos were transferred onto full selection medium described above, except that the kanamycin level was increased to 300 mg L^{-1}. The putatively transformed embryos were germinated on a regeneration medium containing half strength De Fossard (DF; De Fossard et al., 1974) plus 0.25 μM each of BAP and NAA and 10 μM GA_3. The plantlets were transferred onto a growing medium containing full strength DF salts and vitamins plus 30 g L^{-1} sucrose and grown for several months.

2.4. Acclimatization and Potting-out

The regenerated plantlets were taken out of the culture vessels and then they were planted in plastic cups containing a mixture of sterile garden soil and coir dust. They were placed inside a humidity cabinet in a Physical Containment Glasshouse Level 2 (BL 2) a the Institute of Plant Breeding. The plantlets were kept inside the cabinet for two weeks and later transferred into pots.

2.5 Screening for Sex Identification

The genomic DNA of 14 independent putative transformed lines was extracted. Primers specific for sex identification consisting of T1F, T1R, W11F and W11R were used. A 25 μL reaction (1x PCR buffer, 1.5 mM Mg Cl_2, 0.152 mM dNTP, 0.176 μM primers, 1 unit Taq DNA polymerase, 3.0 μL genomic DNA) was subjected to an initial denaturation of 95°C for 3 minutes followed by 25 cycles of denaturation at 95°C for 1 minute, annealing at 58°C for 1 minute, and extension at 72°C for 2 minutes and then 7 minutes with a hold at 4°C. The PCR product was electrophoresed in 1.2% agarose gel, stained with ethidium bromide and photographed under UV using a Photodocumentation System. In addition, the plants are presently being screened for the presence of the introduced gene by Southern blotting and for expression by Nothern analysis.

3. RESULTS AND DISCUSSION

Solo papaya zygotic embryos successfully produced somatic embryos after culture onto the somatic embryo induction medium. The somatic embryos usually originated from the apical dome but in some cases, they originated from the cotyledonary leaves or the suspensor. In experiment 1, it was found that the highest number of blue spots averaging 833, an indication of GUS expression, occurred when the somatic embryos were bombarded using pressure levels of 1,500-1,800 kPa while the target somatic embryos were ca. 17.5 cm away from the target. A lower pressure level ie. 1,000 kPa gave low level of transient expression. Experiment 2 aimed to get high levels of GUS expression of about 1,000 blue spots on the bombarded somatic embryos using a lower pressure. It was found that blue spots on somatic embryos were highest using a pressure level of 1,000 kPa when the target was 12.5 cm away from the filter unit. Both parameters are dependent on each other for effecting transient expression (Able et al., 2001) In general, a lower pressure level is preferred over a higher pressure level. This is because a higher pressure could cause mechanical damage to the embryos when heavy metal particles (tungsten) penetrate either the cell wall, plasmalemma and other internal structures (Hunold et al., 1994). For stable transformation, a total of 160 bombardments using the antisense ACC synthase construct and controls were made. After one-month incubation of the bombarded somatic embryos on the pre-selection medium, their growth started to decrease. A further 2-3 months incubation on the full selection medium caused death of many untransformed somatic embryos. Initially, the embryos and calli stopped growing, gradually they became light yellow in color, and later they became bleached, and then they eventually died. In contrast, the putatively transformed somatic embryos and embryogenic calli were yellowish or golden yellow in color. These putatively transformed tissues grew despite the presence of kanamycin in the selection medium. These tissues started to form small germinating embryo clumps and plantlets about 8-18 months after microprojectile bombardment. The germinating embryos were initially greenish in color and later formed shoots with or without roots. The regenerated plants surviving kanamycin selection grew slowly during the first three months of growth, but later grew vigorously in culture. Weak-looking plantlets did not grow properly and were discarded. The strong putative transgenic plants upon potting-out in the Physical Containment Glasshouse survived, but some died due to transplanting shock. Twenty one putative transgenic plants survived out of the 32 plants planted in soil. These plants continued to grow vigorously and are presently at the vegetative stage of growth. Out of 14 independent putative transgenic lines tested, 9 lines showed 2 bands while 5 lines showed 1 band, indicating that they are hermaphrodite and female lines, respectively. None of the lines tested showed no band indicating that there are no male lines.

4. ACKNOWLEDGEMENTS

The authors thank the Australian Centre for International Agricultural Research (ACIAR), the Philippine Council for Agriculture, Forestry and Natural Resources and Development (PCARRD) and the Department of Science and Technology (DOST) of the Philippine government for funding this research.

5. REFERENCES

Able, J.A., Rathus, C. and Godwin, I.D. (2001). The investigation of optimal bombardment parameters for transient and stable transgene expression in sorghum. In Vitro Cell. Dev. Biol. - Plant 37:314-348.

De Fossard, R.A., Myint, A. and Lee, E.C.M. (1974). A broad spectrum tissue culture experiment with tobacco (*Nicotiana tabacum* L.) pith tissue callus. Physiol. Plant. 31:125-130.

Gonsalves, C., Cai, W., Tennant, P. and Gonsalves, D. (1997). Effective development of papaya ringspot virus resistant papaya with unstranslatable coat protein gene using a modified microprojectile transformation method. Acta Hortic. 461:311-314.

Hamilton, A.J., Bouzayen, M. and Grierson, D. (1991). Identification of a tomato gene for the ethylene forming enzyme by expression in yeast. Proc. Nat. Acad. Sci. USA 88:7434-7437.

Hunold, R., Bronner, R. and Hahne, G. (1994). Early events in microprojectile bombardment: cell viability and particle location. Plant J. 5:593-604.

Mason, M.G. and Botella, J.R. (1997). Identification and characterisation of two 1-aminocyclopropane-1-carboxylic (ACC) synthase cDNAs expressed during papaya (*Carica papaya*) fruit ripening. Aust. J. Plant Physiol. 24:239-244.

Murashige, T. and Skoog, F. (1962). A revised medium for rapid growth and bioassays with tobacco tissue cultures. Physiol. Plant. 15:473-497.

Oeller, P.W., Min-Wong, L., Taylor, L.P., Pike, D.A. and Theologis, A. (1991). Reversible inhibition of tomato fruit senescence by antisense RNA. Science 254:437-439.

FUNCTIONAL GENOMICS TO ISOLATE GENES INVOLVED IN FRAGRANCE PRODUCTION FOR GENETIC ENGINEERING OF SCENT IN FLOWERS

Efraim Lewinsohn[1], Moshe Shalit[1,2], David Gang[3], Noa Lavid[1], Einat Bar[1], David Weiss[2], Alexander Vainstein[2], Zach Adam[2], Dani Zamir[2], Natalia Dudareva[4], Michele Zaccai[5], James E. Simon[6], and Eran Pichersky[3]

[1]Newe Ya'ar Research Center, ARO, Ramat Yishay, Israel; (email: twefraim@volcani.agri.gov.il); [2]The Faculty of Agriculture, The Hebrew University of Jerusalem, Israel; [3]Dept. of Molecular, Cellular and Developmental Biology, University of Michigan, Ann Arbor, MI, USA; [4]Dept. of Horticulture, Purdue University, West Lafayette, IN, USA; [5]The Institutes for Applied Research, Ben Gurion University of the Negev, Beersheva, Israel; [6] New Use Agriculture and Natural Plant Products, Rutgers University, New Brunswick, NJ, USA.

Keywords Flower scent, rose, genomics, functional expression, *O*-methyltransferases

1. INTRODUCTION

Scents of flowers are usually made of mixtures of hundreds (or even thousands) of volatile compounds, normally emitted from flowers to attract pollinators. Different varieties of the same plant may emit a completely different array of compounds (Vainstein et al., 2001). Not only the presence or absence of an individual component might affect a particular flower scent, at times the same compound might have an agreeable or disagreeable scent depending on its concentration. Most of the research in flower scent has been aimed at elucidating the chemical structures of key scent components and in attempting their chemical synthesis for use in the perfumery and cosmetics industries. Despite the vast number of chemical structures involved, the large majority of scent compounds are biosynthesized by a surprisingly small number of metabolic pathways. These metabolic pathways are often ubiquitous, and specialization has developed through small but important modifications of ancestral genes and pathways (Pichersky and Gang, 2000).

I. K. Vasil (ed.), Plant Biotechnology 2002 and Beyond, 329-332.

2. THE GENOMIC APPROACH TO IDENTIFY GENES INVOLVED IN SCENT FORMATION IN ROSES

We have initiated a project aimed at utilizing genomic methodologies to discover many of the genes involved in the formation of the unique aroma of roses. Two varieties were compared: "Golden Gate", a yellow-petal variety with almost no scent, but with a relatively long shelf-life and "Fragrant Cloud" a red-petal variety with a profound and intense smell, but very short shelf-life. cDNA libraries derived from mRNA obtained from the petals of each of the two varieties were generated. Sequence information for more than 3,500 EST (expressed sequence tags) was obtained. (Guterman et al., 2001). The sequence information obtained was cross examined with the pattern of microarray expression, in a search for genes that were expressed in coordination to the appearance of particular scent compounds. Also, sequence information was utilized to identify genes that are similar to known genes from other plants involved in scent production, or genes similar to those with a putative biochemical function that might be related to the formation of scent volatiles. Such predictions afforded the identification of novel genes involved in scent production in roses. Full-length clones were expressed in *E. coli* utilizing expression vectors. The bacterial cells were induced with IPTG to generate sufficient amounts of the recombinant native proteins and *in vitro* enzymatic assays were adapted to detect possible enzymatic activities of the gene products generated. Putative substrates in a variety of assay conditions were offered to the recombinant proteins, and the products formed *in vitro* have been identified utilizing GC-MS methodologies. Two novel *O*-methyltransferases, that accept phenolic volatiles as substrates, have been thus identified (Fig. 1). The genes are related to other plant methyltransferases involved in methylation of other phenolic substrates. Their levels of expression correlate well with the availability of enzymatic activity and the emission of the methylated products during flower development. The two genes are very similar to each other, differing only in 13 amino acids, yet, they display distinct substrate specificities towards the phenolic substrate acceptor. One of the genes (OOMT1) primarily accepts orcinol to generate its monomethyl ether (Fig. 1), while OOMT2 primarily accepts orcinol monomethyl ether to generate orcinol dimethyl ether, a compound emitted by the flowers. Other genes identified using this approach include a sesquiterpene synthase, and an acetyltransferase, whose gene products are able to enzymatically produce the sesquiterpene Germacrene D and geranyl acetate respectively (Table 1). Other genes of yet unknown biochemical function, but possibly involved in the formation of the unique scent of roses have also been noted and are currently been examined.

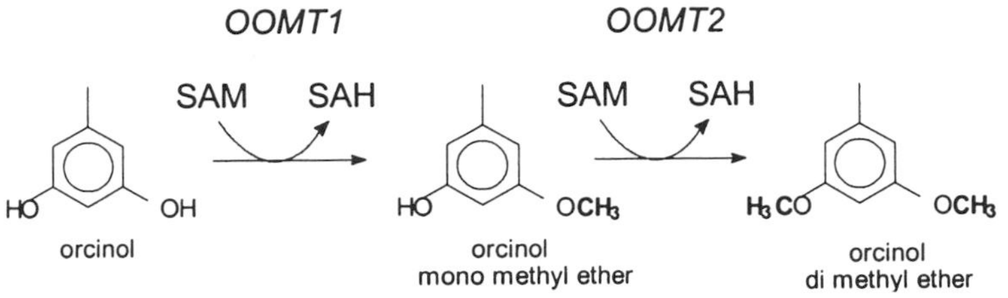

Fig. 1. Enzymatic activities found in the gene products of clones OOMT1 and OOMT2, isolated from rose petals (Lavid et al., 2002). The genes are involved in the formation of orcinol dimethyl ether, a volatile emitted from roses.

3. GENETIC ENGINEERING TO MODIFY AROMA AND SCENT

S-linalool is a non-cyclic monoterpene alcohol found in the scents of many flowers. A gene that codes for *S*-linalool synthase was isolated from the flower tissues of *Clarkia breweri* (Onagraceae), a wild Californian plant. It has been ectopically expressed in tomato fruits, as well as in carnation and petunia flowers in attempts to modify their aroma. *S*-linalool was accumulated in transgenic tomato fruit, but not in controls (Lewinsohn et al., 2001). Thus the diversion of the endogenous terpenoid pathway (committed mainly to lycopene biosynthesis) to the production of volatiles was demonstrated. The transgenic fruits also accumulated 8-hydroxylinalool, a derivative formed by endogenous, yet unidentified enzymes. In transgenic carnation flowers, linalool and linalool oxide were emitted by transgenic plants (Lavy et al., 2002). In petunia, the linalool produced was converted to the scent-less glucoside derivative (Lücker et al., 2001). Other genes derived from *C. breweri* and other flowers are being ectopically expressed in lisianthus flowers in attempts to modify their scent.

Table 1. Putative biochemical roles of EST's isolated from rose petals

Clone	Computer-prediction	Enzymatic activity[a] Substrate	Product
RSS1	sesquiterpene synthase	FPP	Germacrene D
AAT1	alcohol acetyltransferase	Geraniol	Geranyl acetate
OOMT1	O-methyltransferase	Orcinol	Orcinol mono-methyl ether
OOMT2	O-methyltransferase	Orcinol mono-methyl ether	Orcinol dimethyl ether

[a]in cell-free extracts derived from *E.coli* overexpressing the putative clone.

4. REFERENCES

Guterman, I., Dafny-Yelin, M., Shalit, M., Emanuel, M., Shaham, M., Piestun, D. Zuker, A., Ovadis, M., Lavy, M, Lavid, N., Lewinsohn, E., Pichersky, E., Vainstein, A., Zamir, D., Adam, Z., and Weiss, D. 2001. An integrated genomic approach to discovering fragrance-related genes in rose petals. Flowering Newsletter 32: 31-37.

Lavid, N., Wang, J., Shalit, M., Guterman, I., Bar, E., Beuerle, T., Menda, N., Shafir, S., Zamir, D., Adam, Z., Vainstein, A., Weiss, D., Pichersky, E., and Lewinsohn, E. 2002. *O*-Methyltransferases involved in the biosynthesis of volatile phenolic derivatives in rose petals. (*Submitted*).

Lavy, M., Zuker, A., Lewinsohn, E., Larkov, O., Ravid, U., Vainstein, A., and Weiss, D. 2002. Linalool and linalool oxide production in transgenic carnation flowers expressing the *Clarkia breweri* linalool synthase gene. Mol. Breed. (In Press).

Lewinsohn, E., Schalechet, F., Wilkinson, J., Matsui, K., Tadmor, K., Nam, K.H., Amar, O., Lastochkin, E., Larkov, O., Ravid, U., Hiatt, W., Gepstein, S., and Pichersky, E. 2001. Enhanced levels of the aroma and flavor compound *S*-linalool by metabolic engineering of the terpenoid pathway in tomato fruits. Plant Physiol. 127: 1256-1265.

Lücker, J., Bowmeester, J., Schwab, W., Blaas, .J, van der Plas, L.H.W., and Verhoeven, H.A. 2001. Expression of *Clarkia S*-linalool synthase in transgenic petunia plants results in the accumulation of S-linalyl-β-D-glucopyranoside. Plant J. 27: 315-324.

Pichersky, E., and Gang, D.R.. 2000. Genetics and biochemistry of secondary metabolites in plants: an evolutionary perspective. Trends Plant Sci. 10:439-445.

Vainstein, A., Lewinsohn, E., Pichersky, E., and Weiss, D. 2001. Floral fragrance – new inroads into an old commodity. Plant Physiol. 127: 1383-1389.

FLORIGENE FLOWERS: FROM LABORATORY TO MARKET

Chin-yi Lu, Stephen F. Chandler, John G. Mason and Filippa Brugliera
16 Gipps Street, Collingwood, Melbourne VIC 3066, Australia (clu@florigene.com.au)

Keywords Flowers, color modification, vase life, transformation

1. INTRODUCTION

Plant biotechnology has opened up new ways for the production of crops with improved traits. It is also a useful tool for the breeding of flowers. The floriculture industry is driven by novelty. New varieties are most easily distinguished by new color but plant and flower form, variegation, fragrance, longevity, hardiness and resistance to insects and pests are also important.

New flower varieties are traditionally created by breeders using hybridization. However, conventional breeding programs have limitations and gene technology offers avenues for the development of new varieties:

- Introduction of genes from widely diverse germplasm between genera or even families. This could never be achieved by conventional means.
- Directed improvement of specific varieties for specific traits (such as color or disease resistance).
- A much-reduced time frame for variety development.

2. HISTORY OF FLORIGENE

Florigene was established in 1986, as Calgene Pacific. The company's primary objectives have been the application of gene technology to the floriculture industry. In 1990 Calgene Pacific entered into a joint venture with Suntory of Japan for development of novel color flowers. Suntory and Florigene have since maintained a research and commercial relationship (International Flower Developments:IFD).

In 1993 Calgene Pacific bought Florigene BV of the Netherlands. As a result Calgene Pacific was named Florigene Ltd and the Dutch operation became Florigene Europe BV. Early in 2000 NuFarm, an Australian based,

I. K. Vasil (ed.), Plant Biotechnology 2002 and Beyond, 333-336.

multinational chemical company purchased Florigene Ltd. After this all research and product development activities were consolidated in Australia.

3. FLORIGENE'S RESEARCH PROGRAMS

3.1. Color Modification

Flower colors are mainly due to flavonoids, carotenoids and betalains. Flavonoids are the most common flower pigments and provide flower colors from yellow to red, purple and blue (Tanaka et al., 1998). The most important class of flavonoids are the anthocyanins, which contribute to red and blue flowers. The flavonoid biosynthetic pathway has been elucidated (Holton and Cornish, 1995) and factors involved in the determination of flower color have been studied extensively (Mol et al., 1999). Florigene has isolated and patented key genes of the pathway (Holton et al., 1993; Brugliera et al., 1994). Selected pathway genes have been introduced into carnation, rose and gerbera to create novel colored flowers.

3.2. Enhanced Vase Life

The senescence of some cut flowers is due to the synthesis of ethylene.The final stages of ethylene biosynthesis are:

S-adenosyl methionine —ACC synthase→ 1-aminocyclopropane-1-carboxylic acid —ACC oxidase→ Ethylene

The carnation industry uses preservative solutions containing silver ions, which are toxic to humans and a skin irritant, to extend flower vase life. Using co-suppression technology, we have developed varieties of carnation without ACC synthase activity and so do not produce ethylene. Thus, flowers have an enhanced vase life without the use of chemical preservatives.

3.3. Transformation

Cultivar selection for transformation is based on the suitability for transformation (reasonable regeneration and gene transfer frequency, acceptable genetic background) and the commercial value of the cultivar. It is critical to select a cultivar that is well established in the global marketplace and has excellent grower characteristics.

Florigene/IFD have developed somatic embryogenesis and/or organogenesis regeneration systems for many ornamental plants including petunia, chrysanthemum, gerbera, African violet, lily, rose and carnation (Lu and Chandler, 1995). Transformation systems have been established using *Agrobacterium* and in some cases the particle gun. Various selective agents (kanamycin, G-418, hygromycin, chlorsulfuron and phosphinothricin) were used to select for transformed cells and recover transgenic plants. Over 20,000 separate transformation events have been generated in the course of the last 15 years.

4. FLORIGENE'S PRODUCT DEVELOPMENT PROGRAM

Critical to Florigene's efforts to develop new cultivars is an understanding of the demand for quality in the flower industry. Relationships with conventional breeders provide germplasm and assistance in trial evaluations.

Transgenic events are rejected if they have poor flower form, whatever the color of the flowers. Other factors taken into account are the stability of color, flower size and general health of the transgenic plant. A small group of promising lines, with the desirable phenotypes, are also analyzed for gene integration pattern. Any lines that have integration of DNA outside of the T-DNA borders are eliminated. Trials are required not only to assess the suitability of the transgenic for further roll out but also to determine the best growing climate for new varieties. Trials provide an opportunity to assess productivity, disease resistance, color stability, vase life and flower form.

Regulatory approvals are pivotal to product roll out. Regulatory approvals are required in both the production countries and the consuming countries. Much of the data collected from trials can be used for all countries. In parallel with trials and securing regulatory approvals, we prepare to roll out product. To speed up the time to market, propagation (bulking up) of planting material is carried out, within the constraints of regulatory approvals.

5. FLORIGENE'S PRODUCT IN THE MARKET PLACE

Florigene has sold transgenic carnation flowers for six years. Flowers were first released in Australia, followed by Japan and USA. Our products are from two carnation types – the spray, which has a branching stem with flowers on each branch, and the standard - which is a single stem with a single large flower. The first Florigene products were mauve-violet sprays –

Florigene Moondust™ and Florigene Moonshadow™ . These are still sold on the Australian, US and Japanese markets.

In 2000 Florigene launched four standard carnations – all modified for novel color, and representing a new range of mauve-violet flowers. These carnations (Florigene Moonaqua™, Florigene Moonshade™, Florigene Moonlite™ and Florigene Moonvista™) are largely grown in South America for export to USA and Japan. Plants are also grown in Australia for the Australian market. In 2001 Florigene became the first company in the world, in collaboration with our Australian partner Tesselaar, to sell genetically modified plants by mail order to the general public for home garden use.

6. ACKNOWLEDGEMENTS

We would like to thank all employees from Florigene and Florigene Europe for their contributions to our research and product development.

7. REFERENCES

Holton, T.A., F. Brugliera, D.R. Lester, Y. Tanaka, C.D. Hyland, J.G.T. Menting, C. Lu, E. Farcy, T.W. Stevenson and E.C. Cornish. 1993. Cloning and expression of cytochrome P450 genes controlling flower color. Nature 366:276-279.

Brugliera F., T.A. Holton, T.W. Stevenson, E. Farcy, C.Lu and E.C. Cornish. 1994. Isolation and characterization of a cDNA clone corresponding to the *Rt* locus of *Petunia hybrida*. Plant J. 5:81-92.

Holton T.A., and E.C. Cornish. 1995. Genetics and biochemistry of anthocyanin biosynthesis. Plant Cell 7:1071-1083.

Lu, C., and S.F. Chandler. 1995. Genetic transformation of *Dianthus caryophyllus* (Carnation). Biotechnology in Agriculture and Forestry 34. Y.P.S. Bajaj (Ed.). Springer-Verlag Germany. Pp. 156-170.

Mol, J., E. Grotewold and R. Koes. 1999. How genes paint flowers. Plant Biotechnology and In Vitro Biology in the 21st Century. A. Altman, M. Ziv and S. Izhar (eds.). Kluwer Academic Publishers. Holland. Pp. 597-600.

Tanaka, Y., S. Tsuda and T. Kusumi. 1998. Metabolic engineering to modify flower color. Plant Cell Physiol. 39(11):1119-1136.

BIOTECHNOLOGY OF FLORICULTURE CROPS – SCIENTIFIC QUESTIONS AND REAL WORLD ANSWERS

David G. Clark[1], **Holly Loucas, Kenichi Shibuya, Beverly Underwood, Kristin Barry and Jason Jandrew**
[1]University of Florida, Environmental Horticulture Department, Gainesville, FL 32611-0670, USA (e-mail: geranium@ufl.edu)

Keywords Dwarfism, flower senescence, leaf senescence, phytohormones

1. INTRODUCTION

The floriculture crop industry is technically diverse and is characterized by the use of hundreds of different plant species. Due to the large amount of genetic diversity used by this industry, there are often very complex issues that arise during crop production and during postharvest handling throughout the wholesale and retail markets. Crop production practices are often very complex, and may be complicated by the demands for precise crop timing, and the different cultural requirements of different cultivars. After the crops is produced, optimal conditions for postharvest shipping and handling are difficult to maintain, and this can lead to subsequent poor quality once the crop has arrived at the retail market. In addition to the demands of complex production and marketing systems, the demand from consumers for a constant supply of new and interesting flowering plants with unique characteristics continues to increase. Needs for product quality and availability and subsequent garden performance have driven the availability of new flowering crops to unprecedented levels over the last 5-10 years. With such a vast array of issues in floriculture crops, there is a great deal of potential for using some of these crops as model systems to study the potential for use of transgenic plants with improved horticultural characteristics. In many cases, biotechnology applications are proving to be very difficult, but there have been several advances made with engineering a wide variety of genetic traits in floriculture crops. There have also been significant gains made in cloning important genes that are proving to be involved with biological processes that scientists hope to manipulate in ornamental crops in the future.

I. K. Vasil (ed.), Plant Biotechnology 2002 and Beyond, 337-342.

2. MANIPULATION OF PHYTOHORMONE SYNTHESIS AND PERCEPTION

There has been much research attention given to the manipulation of phytohormone synthesis and perception in floriculture crops due to the wide range of physiological processes they participate in. Vast amounts of money are spent on growth regulating chemicals that are applied to plants to control the synthesis of gibberellic acid (GA) and subsequent plant height during crop production. In many potted flowering crops and bedding plant crops, three to five applications of growth regulator per crop adds a significant amount of production cost in both labor and chemicals. Thus, the manipulation of endogenous GA concentration in plants through genetic engineering has the potential to produce ornamental and flowering plants with a diverse array of reduced-height phenotypes. The initial cloning and characterization of the GA5 gene from spinach, which encodes gibberellin 20-oxidase (Wu et al., 1996) provided the first practical molecular tool needed to develop plants with altered endogenous GA levels. Over-expression of GA 20-oxidase in *Arabidopsis* leads to increased endogenous GA production, and subsequent stem elongation, while expression of antisense GA20-oxidase leads to reduced GA levels and reduced rates of stem elongation (Coles et al., 1999). Similar experiments focused on the expression of antisense GA2-oxidase in transgenic petunias have produced similar phenotypes, suggesting that the control of endogenous GA levels may be applicable to important floriculture crops as well (Clark, unpublished data). It is very likely that traditional breeder selection for the appropriate plant height among transgenic lines will be critical for applying these technologies to any given crop, and breeders should be able to select any particular plant height they desire. One extremely important application for dwarfing technology will likely be observed in plant species used in the turfgrass industry. Lawn grasses (*Agrostis sp*) requiring less labor for maintenance due to slower growth have been developed. In low management areas, these "low mow" grasses have a great deal of potential for saving money on labor costs and equipment maintenance, and could help cut down on lawn mower pollution and use of fossil fuels. It is likely that these dwarf turfgrasses will be important for homeowners and municipal or roadside situations, but it is not likely that these grasses will be used to any great extent on golf courses or sports turf facilities because they may not grow fast enough to cover the damage inflicted by constant use.

Hormonal control of plant morphology is also being pursued in floriculture crops with the goal of engineering plants with delayed leaf senescence. Natural leaf senescence in many plants is characterized by lower leaf yellowing or chlorosis as nutrients and other components of the cells are degraded (esp. chlorophyll). In floriculture crops, leaf chlorosis can cause a decreased aesthetic appearance of ornamental plants and thus a decrease in the salability of those plants or poor garden performance.

One way to prevent leaf senescence is through the manipulation of cytokinin synthesis. Cytokinins are an important class of phytohormones that influence numerous aspects of plant growth and development, and have been shown to delay and, in some cases, reverse the leaf senescence process (Gan and Amasino, 1996). After many years of physiological research on cytokinins, the molecular mechanisms of plant cytokinin biosynthesis and perception are just now being elucidated. Although experiments focused on the manipulation of endogenous cytokinin synthesis in transgenic plants with endogenous plant genes have been lacking, a gene from *Agrobacterium tumefaciens* that encodes the isopentenyl transferase *(IPT)* enzyme has been available for use in transgenic plant research for a number of years. *IPT* is encoded on the Ti (tumor inducing) plasmid of *Agrobacterium tumefaciens* (Akiyoshi et al., 1984; Barry et al., 1984) and catalyses the condensation of dimethylallylpyrophosphate (DMAPP) and 5'AMP to form isopentenyladenosine 5'-phostphate ([9R-5'P]iP), which is then quickly converted to different cytokinins. Attempts have been made to use this gene under the control of various promoters but the results of such experiments were often complicated by abnormal growth patterns of the plants due to the lack of control of gene expression in a temporal or spatial specific manner. In an attempt to overcome the problems associated with past work on transgenic plants overproducing cytokinins, Gan and Amasino (1995) developed a genetic construct that utilized the highly senescence specific promoter from SAG 12 (P_{SAG-12}) to drive IPT expression. This construct had three important features due to P_{SAG-12} specificity: temporal regulation, spatial regulation, and quantitative regulation (Gan and Amasino, 1995; Gan and Amasino, 1996). When leaf senescence is triggered, P_{SAG-12} activates transcription of IPT, leading to the subsequent production of functional *IPT* enzyme. The enzyme then catalyzes cytokinin production, which in turn delays senescence. Without senescence signals, the P_{SAG-12} promoter attenuates IPT transcription and subsequent enzyme production, thus providing autoregulatory control of cytokinin synthesis. Transgenic P_{SAG-12}–*IPT* tobacco plants displayed a "normal" growth habit except that leaf senescence was inhibited and there was a significant increase in flower number, increased biomass due to the presence of lower leaves, and increased seed-yield (Gan and Amasino, 1995). Since this time, transgenic P_{SAG-12} lettuce plants (McCabe et al., 2001) *Nicotiana alata* (Schroeder et al., 2001) and petunias (Clark et al., in press) with similar delayed leaf senescence phenotypes have been produced. The P_{SAG-12} promoter has also been used to drive expression of the *KNOTTED-1* mutant gene from *Maize* in transgenic tobacco plants to confer a delayed leaf senescence phenotype (Ori et al., 1999), thus extending the utility of delayed leaf senescence technologies. Preliminary experiments on transgenic P_{SAG-12} :*KNOTTED-1* petunias with delayed leaf senescence has shown that this approach may actually produce plants with more desirable phenotypes and fewer undesirable side-effects than with $P_{SAG\text{-}12}$–*IPT* (Clark et al., submitted). It is likely that horticultural performance studies will show that breeder selection of transgenic plants under a wide range of selection criteria will be essential to providing the growers with the data required to produce these plants effectively on a commercial scale.

One hormone of particular interest in the ornamental plant industry is ethylene. Ethylene is involved in many physiological processes in plants including fruit ripening, petal senescence, abscission, and seed germination. Floral senescence is of particular interest because spoilage of many important floriculture crops is known to occur due to ethylene gas in the postharvest environment. Since ethylene is such a significant problem in both potted flowering crops and cut flowers there have been several attempts to produce chemical control methods for both ethylene synthesis and sensitivity. Ethylene sensitivity has long been managed in floriculture crops through use of silver thiosulfate (STS), which makes plant tissue insensitive to ethylene, but also has environmental downsides that appear to be restricting its commercial use. Another chemical approach gaining popularity amongst the industry is with the use of 1-MCP (1-methylcyclopropene), which is a compound that blocks the ethylene receptor protein and makes plant tissue insensitive to ethylene (Serek et al., 1995). Typically applied as a gas, MCP provides a means by which large amounts of tissue can be treated for a short amount of time. Although this compound has proven to be effective in some crops, there is some difficulty with use of 1-MCP in crops that continue to produce new ethylene receptor proteins through development during postharvest transit, thus limiting the residual effect of the compound (Cameron and Reid, 2001).

The method of ethylene control receiving the most research attention in terms of genetic engineering has been at the level of ethylene perception. Wilkinson et al. (1997) transformed petunia with a dominant mutant *Arabidopsis* ethylene receptor, etr1-1, under the control of a constitutive Cauliflower Mosaic Virus 35S (CaMV35S) promoter to produce ethylene-insensitivity throughout the whole plant. These petunias had significantly delayed floral senescence compared to wild-type plants, but had physiological side effects such as decreased adventitious root formation, increased disease susceptibility, and delayed fruit ripening (Wilkinson et al., 1997; Clark et al., 1999; Gubrium et al., 2000). Currently, experiments focused to reduce expression of other genes in the ethylene signal transduction pathway (namely ein2 and ein3) are being conducted to determine if there are other potential candidates that may be more feasible for use in altering ethylene sensitivity for commercial purposes. Regardless of the method used to achieve ethylene insensitivity, poor horticultural performance characteristics severely limit the commercial utility of ethylene-insensitive plants. Thus it is clear that the key to use of ethylene sensitivity manipulation will lie with the selection of promoters used to drive transcription of any transgene. Research utilizing the tools of functional genomics will soon lead to the isolation of temporally and spatially regulated genes in floral crops. This approach will prove invaluable in isolating the upstream promoter elements needed to drive gene expression in a very specific manner. Once these elements are located, they can be used to drive the expression of proven transgenes and efficiently tested in transgenic plants. It will be imperative for breeders to conduct a significant amount of field and greenhouse trailing of these plants in order to completely eliminate the possibility of negative side effects.

3. REFERENCES

Akiyoshi, D.E., Klee, H., Amasino, R.M., Nester, E.W., and M.P., Gordon. 1984. T-DNA of *Agrobacterium tumefaciens* encodes an enzyme of cytokinin biosynthesis. Proc. Nat. Acad. Sci. USA 81:5994-5998.

Barry, G.F., Rodgers, S.G., Fraley, R.T. and Brand, L. 1984. Identification of a cloned cytokinin biosynthetic gene. Proc. Nat. Acad. Sci. USA 81:4776-4780.

Cameron, A.C., and Reid, M.S. 2001. 1-MCP blocks ethylene-induced petal abscission of *Pelargonium peltatum* but the effect is transient. Postharv. Biol. Technol. 22:169-177.

Clark, D.G., Gubrium, E.K., Klee, H.J., Barrett, J.E., and Nell, T.A. 1999. Root formation in ethylene insensitive plants. Plant Physiol. 121:53-59.

Clark, D.G., Dervinis, C., Barrett, J.E., and Klee, H.J. Horticultural performance of transgenic P*sag12-IPT* petunias. J. Amer. Soc. Hort. Sci. (In Press).

Coles, J.P., Phillips, A.L., Croker, S.J., Garcia-Lepe, R., Lewis, M.J. and Hedden, P. 1999. Modification of gibberellin production and plant development in *Arabidopsis* by sense and antisense expression of gibberellin 20-oxidase genes. Plant J. 17:547-556.

Gan, S., and Amasino, R.M. 1995. Inhibition of leaf senescence by autoregulated production of cytokinin. Science 270:1986-1988.

Gan, S., and R.M. Amasino. 1996. Cytokinins in plant senescence: From spray and pray to clone and play. BioEssays 18:557-565.

Gubrium, E.K., Clark, D.G., Barrett, J.E. and Nell, T.A. 2000. Horticultural performance of transgenic ethylene insensitive petunias. J. Amer. Soc. Hort. Sci. 125:277-281.

McCabe, M.S., Garratt, L.C., Schepers, F., Jordi, W.J.R.M., Stoopen, G.M., Davelaar, E., van Rhijn, J.H.A., Power, J.B. and Davey, M.R. 2001. Effects of P_{SAG12}-*IPT* gene expression on development and senescence in transgenic lettuce. Plant Physiol. 127:505-516.

Ori, N., Juarez, M.T., Jackson, D., Yamaguchi, J., Banowetz, G.M. and Hake, S. 1999. Leaf senescence is delayed in tobacco plants expressing the maize homeobox gene knotted1 under the control of a senescence-activated promoter. Plant Cell 11:1073-1080.

Schroeder, K.R., Stimart, D.P. and Nordheim, E.V. 2001. Response of *Nicotiana alata* to insertion of an autoregulated senescence-inhibtion gene. J. Amer. Soc. Hort. Sci. 125:523-530.

Serek, M., Tamari, G., Sisler, E.C. and Borochov, A. 1995. Inhibition of ethylene-induced cellular senescence symptoms by 1-methylcyclopropene, a new inhibitor of ethylene action. Physiol. Plant. 94:229-232.

Whitehead, C.S., Halevy, A.H. and Reid, M.S. 1984. Roles of ethylene and 1-aminocyclopropane-1-carboxylic acid in pollination and wound-induced senescence of *Petunia hybrida* flowers. Physiol. Plant. 61:643-648.

Wilkinson, J.Q., Lanahan, M.B., Clark, D.G., Bleecker, A.B., Chang, C., Meyerowitz, E.M., and Klee, H.J. 1997. A dominant mutant receptor from *Arabidopsis* confers ethylene insensitivity in heterologous plants. Nature Biotechnol. 15:444-447.

Wu, K., Li, L., Gage, D.A. and Zeevaart, J.A.D. 1996. Molecular cloning and photoperiod-regulated expression of gibberellin 20-oxidase from the long-day plant spinach. Plant Physiol. 110:547-554.

GENE EXPRESSION IN SPACE BIOLOGY EXPERIMENTS

Anna-Lisa Paul[1,2] and Robert J. Ferl[1]
[1]Department of Horticultural Sciences and the Biotechnology Program, University of Florida, Gainesville, FL 32611-0690, USA (robferl@ufl.edu); [2] (alp@ufl.edu)

Keywords Spaceflight, arabidopsis, transgenic, GUS, GFP, microgravity, signal transduction, hypoxia, Adh

1. INTRODUCTION

As humans continue to explore beyond the confines of our own planet we are faced with a variety of challenges to facilitate these explorations. There are, of course, the mechanical engineering challenges that encompass the problems of getting us off the planet and out to orbital habitats such as the shuttle and the space station or other planetary surfaces such as the moon or Mars. In addition there biological and environmental engineering challenges that encompass supporting humans within these habitats. The first stage of environmental engineering embraced by the space program was to simply make a closed system of refreshed air and waste disposal supported by food and other rations that astronauts brought with them. This is largely still the way of it; however, there is increasing interest in augmenting this basic approach with biological systems that can help in the environmental requirements of refreshing the air, metabolizing wastes and providing a portion of nourishment as we endeavor to mount longer missions in space. Plants are central to any life support system that contains biological components and thus understanding the metabolic challenges that may face plant growth and development in space is central to this effort. Early studies concerning the physiological responses of plants in a spaceflight environment clearly illustrate that plants are metabolically stressed during spaceflight when compared to their ground control counterparts (Ferl et al., 2002 and references therein). Thus, the successful incorporation of plants into orbiting habitats will require a keen understanding of the basis for the spaceflight stress-response in plants. At least one apparent link in this stress response is known. Plants exhibiting spaceflight-associated stress appear to exhibit some of the hallmarks of hypoxia. To explore this phenomenon, we and others have turned initial focus to a key enzyme in the hypoxic stress pathway, alcohol dehydrogenase (Adh) (Porterfield et al., 1997; Paul et al., 2001).

I. K. Vasil (ed.), Plant Biotechnology 2002 and Beyond, 343-346.

Mounting a spaceflight experiment requires complex hardware and systems, and experiments need to be tailored to the constraints imposed by these systems. Considerations are primarily made for weight, size and energy consumption. The model plant *Arabidopsis thaliana* (arabidopsis) lends itself well to spaceflight experiments as it is small, grows fast and can be easily genetically engineered with transgene reporters.

2. THE TAGES SYSTEM

The question of how plants respond to a spaceflight environment can be addressed at many levels, but we have chosen to ask the question at the level of gene expression. The first generation of spaceflight experiments employing genetically engineered arabidopsis plants was the TAGES PGIM-01 experiment that flew on STS-93 in the summer of 1999. TAGES (Transgenic Arabidopsis Gene Expression System) experiments employed arabidopsis plants that had been engineered with a reporter gene composed of the Adh gene promoter coupled to the coding region of the gene encoding β-glucuronidase (GUS). The plants were grown in specialized hardware known as the Plant Growth Facility (PGF) which is composed of six small Plant Growth Chambers (PGC's) (Figure 1).

Figure 1. An example of a PGC unit containing six transgenic arabidopsis plants growing in nutrient agar tubes. A square Petri plate of nutrient agar supports a set of vertically grown seedings.

The PGF is a middeck locker facility that provides a controlled environment designed minimize environmental stress outside of the features of spaceflight under investigation. As mentioned, expression of the Adh gene seems to be associated with the adaptations plants undergo to cope with the perceived "stress" of spaceflight. The Adh/GUS transgene responds just as the native Adh gene, the only difference being that the Adh/GUS reporter gene provides an indication of the level and distribution of gene product in the tissues in which it is expressed. The Adh/GUS gene product creates an intense blue histochemical stain when incubated with 5-bromo-4-chloro-3-indolyl-beta-D-glucuronic acid (X-Gluc) substrate (Paul and Ferl, 2002) (Figure 2).

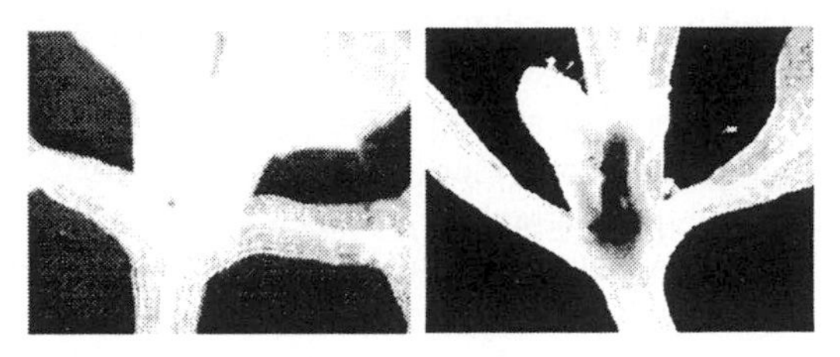

Figure 2. The control plant on the left does not express Adh/GUS. The plant on the right has been hypoxically stressed, and expresses the transgene in the shoot apex (darkly stained area).

The patterns of Adh/GUS gene expression seen in the TAGES PGIM-01 flight experiment confirmed that Adh was indeed induced by some aspect of spaceflight at the transcriptional level, yet the patterns of gene expression were not consistent with what one would expect of hypoxically stressed plants (Paul et al. 2001). Although the roots of the flight plants expressed Adh/GUS in a pattern that was very similar to the pattern seen in plants exposed to root zone hypoxia, the shoots of the flight plants were completely clear of Adh/GUS expression. Normally, plants whose roots are deliberately flooded to mimic root-zone hypoxia show Adh/GUS expression in the shoots as well as the roots. The example shown in Figure 2 is, in fact, the aerial portion of a deliberately root-flooded ground control. These results indicate that while some aspect of spaceflight induced Adh expression the response does not follow the patterns defined by terrestrial models. It is possible that the response is in the plant, it is not clear that the plant is responding to true hypoxia; rather the response may be an inappropriate adaptation of the hypoxic stress metabolism to respond to an unfamiliar stress, and that is why the patterns of expression differ.

Another possibility is that the signal from the root to the shoot has been disrupted by a feature of spaceflight. The aerial portion of the plant shown expressing Adh/GUS in Figure 2 is fully aerobic, yet the hypoxic signal from the flooded root is sufficient to induce Adh/GUS expression in the shoot as well. If calcium ion mediated signal transduction is disrupted in a plant experiencing root-zone hypoxia, the signal is not transduced to the shoot, and the shoot does not express Adh.

3. GFP - THE NEXT GENERATION OF TRANSGENE

Reporter gene technology is a powerful tool of molecular investigation, yet as effective as a reporter as GUS is, it has one major drawback; the plants must be fixed and stained to see tissue specific localization of gene expression. The desire for a method to observe transgene expression in real-time leads us to the adaptation of Green Fluorescent Protein (GFP) for a new generation of reporter genes. GFP fluoresces when excited by blue light and is therefore nondestructive. Plant tissues are essentially transparent to the excitation wavelength and any tissues that are expressing GFP will emit green light in response to blue light excitation. Fluorescence can be observed both non-destructively in real-time, photographed and then the plants can be allowed to continue to grow for subsequent observations. This feature opens the door to the telemetric collection of gene expression data, which in turn, facilitates taking gene expression data from closed systems in orbit, obviating the need for crew time, and from distant places, such as from a probe on the surface of Mars. The Adh/GFP reporter is induced identically to the Adh/GUS reporter (Manak et al., 2002). Data on the hypoxic induction of Adh/GFP has been collected telemetrically and analyzed from digital photographs in preparation for automated spaceflight analysis of gene expression.

4. REFERENCES

Ferl, R., R. Wheeler, H.G. Levine, and A.L. Paul. 2002. Plants in space. Curr. Opin. Plant Biol. 5: 258-263.

Manak, M.S., A.L. Paul, P.C. Sehnke, and R.J. Ferl. 2002. Remote sensing of gene expression in Planta: transgenic plants as monitors of exogenous stress perception in extraterrestrial environments. Life Support Biosph. Sci. 8: 83-91.

Paul, A.L., C.J. Daugherty, E.A. Bihn, D.K. Chapman, K.L. Norwood, and R.J. Ferl. 2001. Transgene expression patterns indicate that spaceflight affects stress signal perception and transduction in arabidopsis. Plant Physiol. 126: 613-621.

Paul, A.L. and R.J. Ferl. 2002. Molecular Aspects of Stress-Gene Regulation During Spaceflight. J. Plant Growth Regul. [epub ahead of print].

Porterfield, D.M., S.W. Matthews, C.J. Daugherty, and M.E. Musgrave. 1997. Spaceflight exposure effects on transcription, activity, and localization of alcohol dehydrogenase in the roots of *Arabidopsis thaliana*. Plant Physiol. 113: 685-693.

STRESS AND GENOME SHOCK IN DEVELOPING SOMATIC EMBRYOS IN SPACE

A.D. Krikorian
State University of New York at Stony Brook, Stony Brook, New York 11794-5215 USA
(email: Abraham.Krikorian@sunysb.edu)

Keywords Somatic embryogenesis, stress, chromosomes, somaclonal variation

1. INTRODUCTION

The study of somatic embryogenesis of daylily (*Hemerocallis*) during space-flight offers an unexpected framework for understanding the consequences of stressful environments and inadequate optimization of culture conditions. Justification for research on plants in space focuses on the need to understand better how to use them for managing 'Life Support' (Krikorian, 2001). Or, that research using the distinctive features of the space environment such as near-weightlessness can increase understanding of life processes on Earth, or even trigger novel bio-processes (Durzan, 2000). Indeed the presence of the earth's gravitational field is *the* factor that has remained most unchanged throughout evolution (Krikorian and Levine, 1991; Conger et al., 1998). But it is equally arguable that plants have multiple levels of control, including mechanisms to protect them against potentially catastrophic stress, such as the absence of any normally-required gravitational signals, that have evolved through time. There is no evidence suggesting the uncoupling of a single factor such as gravity from the rest of any paradigm addressing plant development. Thus microgravity is but one factor of many that can impinge on plant development (Krikorian, 1996c).

2. SOMATIC EMBRYOS IN SPACE

Embryogenic 'units' of different size and morphological complexity on media that have been "more-" or "less- optimized" using different strategies have comprised the test system (Krikorian, 1996a). Vessels for exposing somatic embryos at different developmental stages range from specially flight-qualified "Plant Cell Culture Chambers" to petri dishes. Some of these allow fixation in space. Introduction of fixative in the course of space flight has allowed definition of events in terms of time and space (Krikorian,

I. K. Vasil (ed.), Plant Biotechnology 2002 and Beyond, 347-350.

1996b). Also, procedures have routinely entailed 'synchronous ground controls' to allow observations as any given experiment has proceeded in space. Analysis after 'recovery' of chromosomes has been carried out by examining materials fixed using colchicine as the cytostatic, or by use of direct fixation. Analyses have relied not only on squashes of randomly selected embryos pulled off their support surfaces but karyotyping, determining of centromeric index, etc. — all of which have been used to quantify results. Assessment of embryogenic response has been by staging and qualitative scoring at the end of an experiment and at weekly intervals up to 6 weeks and also by use of SEM. Germination of somatic embryos and rearing them to rooted plantlets was done by transferring randomly selected embryos of different stages of development to appropriate media.

What we were able to learn from our early earth-based *in vitro* work with embryogenic cells about sensitivity, vulnerability and requirements for responsiveness, especially at specific "phenocritically sensitive" stages of development (Krikorian, 1996c, 2000), eventually provided a rational framework for developing and testing a "space environment-unique stresses to cells" hypothesis. Stated in its barest essentials "Space provides a unique environment that can adversely impact the processes of cell division. The extent to which the adverse events become manifest is dependent on the extent of pre-existing stresses." The more morphologically advanced a system, the less likely it is to suffer catastrophic stress effects in the space environment; the less developed, the greater the vulnerability. That means that intact seedlings and whole plants are predictably less vulnerable whereas systems growing *de novo* like tissue and cell cultured ones are more vulnerable (Krikorian, 1999). In either case, stress effects, like hormonal ones, are additive, or may be synergistic. This means that only when there is a necessary constellation of stresses will damage result. It is interesting to note that my pre-occupation with utilizing the smallest size embryogenic somatic cell units (Krikorian 1999), designated by us as somatic embryo initials or embryocytes, that could advance in their development muddied the track since such units have proven quite sensitive to the space environment. Indeed, later experiments in which both small and larger embryogenic units were utilized, larger ones have shown far less chromosomal damage (Krikorian, 2000). Equally significant has been the evidence that different kinds of hardware can affect the responses. For example, poor ventilation or significant temperature fluctuations within dishes results in moisture condensation, and even liquid accumulation on the surface where somatic embryogenic growth was to have occurred. Water logging, in space as on earth, has an adverse effect on growth of very young somatic embryo initials. Similarly, in cases where the ambient temperature in the space environment was elevated during flight, premature drying out of the systems has induced dormancy and hence less chromosome damage. In short, there is both a biological and physical component to the cytological situations(s) encountered.

3. CONCLUSIONS

1. We have not been able to reproduce closely the damage encountered in space-grown materials; we conclude therefore there are un- or underappreciated subtleties of the space environment which add stress(es) to an otherwise acceptable environment or developmental stage, that is, one that is not as vulnerable. 2.Variations in chromosome number and structure can occur in phenocritically-sensitive embryogenic cells in response to external influences and stresses and signals from impaired osmotic relations and water stress, pH excursion via proton extrusion, and various kinds of mechano-perturbation associated with certain types of substratum and also level of vibration. 3. Marginally-optimized, normally pivotal, environmental parameters imposed on or available to embryogenic cells at different levels of their organization or advancement, will be manifested by chromosomal changes and even the production of developmentally aberrant embryos. 4. The degree to which organisms are able to adapt to or "tolerate" stress "insult" is more or less proportional to the amount of non-living material in their tissue. (Thus, cells with well-developed walls and large vacuoles are to be viewed as more stress-tolerant than those with poorly developed, thin walls and small or few vacuoles.). 5. Meristematic cells are more sensitive than mature, differentiated vegetative cells. Susceptibility is directly proportional to the proliferative capacity and inversely proportional to the degree of differentiation of a tissue in a somatic embryo. 6. Increasing ploidy of plant tissues is one of the modifying factors involved in stress tolerance. Diploids become more sensitive to stress as nuclear volume increases, while polyploids become more resistant as nuclear volume increases. 7. Rate of cell division is a factor in stress-tolerance of developing embryos because stress effects are not compounded in slowly dividing cells because there is time for DNA repair etc., whereas errors do not accumulate in rapidly dividing cells. 8. A threshold appears to exist for each of the above relationships. They are temporally- and stress intensity-dependent. Chromosome damage, loss of viability or failure to progress in somatic embryogenesis are progressive and a critical threshold of stress has to be reached or achieved before a given result is manifested, whether continued development or destruction of the normal pathway or progression. 9. Selective alleviation or unloading of stress can be accomplished by a strategy of optimization of the test system and growing conditions based on the classical plant physiological principle of “optima and limiting factors”. 10. Cell level stress effects due to failure to reach and maintain critical threshold levels of insult in daughter embryo initial cells from insulted embryos after division are reversible, provided rescue occurs within a critical window of time.

Support of NASA is acknowledged and appreciated.

4. REFERENCES

Conger, B.V., Z. Tomaszewski, J.K. McDaniels, and A. Vasilenko. 1998. Spaceflight reduces somatic embryogenesis in orchardgrass. Plant Cell Environ. 21: 1197-1203.

Durzan, D.J. 2000. Metabolic engineering of plant cells in a space environment. Biotech. Gen. Eng. Rev. 17: 349-383.

Krikorian, A.D. 1996a. Strategies for maintenance of plant cell cultures at "zero" or minimal growth: A perspective on managing cultures in the context of a long-duration experiment in the Space environment. Bot. Rev. 62: 41-108.

Krikorian, A. D. 1996b. Embryogenic somatic cell cultures of daylily (*Hemerocallis*): A system to probe spaceflight-associated mitotic disturbances. H. Suge (Ed.) Plants in space biology. Institute of Genetic Ecology, Tohoku University, Sendai, Japan, Pp 111-126.

Krikorian A.D. 1996bc Space stress and genome shock in developing plant cells. Physiol. Plant. 98: 901-908.

Krikorian, A. D. 1999. Somatic embryos of daylily in space. Advances in Space Research (32nd COSPAR Scientific Assembly, Nagoya, Japan) 23: 1987-1997.

Krikorian, A.D. 2000. Historical insights into some contemporary problems in somatic embryogenesis. Somatic embryogenesis in plants. S. Mohan Jain, Pramod K. Gupta and Ronald J. Newton (Eds.) Kluwer Academic Publishers, Dordrecht, Boston, London. 6: 17-49.

Krikorian, A.D. 2001. Novel applications of plant tissue culture and conventional breeding techniques to space biology research. . From soil to cell- a broad approach to plant life. In: Bender, L. and Kumar, A. (Eds.).Giessen Pp. 97-130. Also, Electronic Library (GEB) http://bibd.uni-giessen.de/ghtm/2001/uni/p010012.htm.

Krikorian, A.D. and H.G. Levine. 1991. Development and growth in space. In: Plant Physiology: A Treatise. F.C. Steward, and R.G.S. Bidwell (Eds.). Academic Press, Orlando.10: Pp. 491-555.

PLANT DEVELOPMENT IN SPACE OR IN SIMULATED MICROGRAVITY

Gérald Perbal
Laboratoire CEMV, Université Pierre et Marie Curie, 4 Place Jussieu 75252 Paris Cedex 05, France (e-mail: Gerald.perbal@snv.jussieu.fr)

Keywords Gravitropism, gravimorphism, microgravity, development

1. INTRODUCTION

In manned spacecraft or satellites, living organisms are subjected to a specific environment whose principal components are microgravity, cosmic rays and cabin atmosphere. As plant morphology is greatly dependent upon gravity on Earth, it was foreseen that the nullification of this factor in space should have led to a great disturbance of plant development. The pioneering experiments (mainly carried out by Russian investigators) have confirmed this point of view (Halstead and Dutcher, 1987) until two of them (cited in Merkys and Laurinavicius, 1990) were able to obtain a complete life cycle (seed-to-seed) in space with *Arabidopsis thaliana*. This experiment has clearly demonstrated that plant growth was possible in near weightlessness, but did not explain the reason why the previous experiments were unsuccessful .

The main causes of the various problems (Kuang et al., 1995; Kuang et al., 1996) encountered in growing plants in space have recently been discovered. They deal mainly with gas exchanges (Musgrave et al., 1997) and a possible hypoxia (Paul et al., 2001) in a microgravity environment. Thus, the direct effects of microgravity on plants remains to be clarified: it is the main goal of this paper.

2. METHODS OF CULTURE OF PLANTS IN SPACE

Basically, it should be possible to use any kind of soil for growing plants in space. However, hydration of a substrate is sometimes not so simple in microgravity conditions. In particular, water distribution in a soil on the ground is not homogenous since percolation occurs, whereas it is homogenous in microgravity. Agarose should be used as a substrate since water distribution should be the same at different levels (Aarrouf et al., 1999b). However it must be noted that when seeds are planted in agarose, they are obviously hydrated and begin to germinate. This implies that at least part of the early development takes place before and during the launch that is in various gravity conditions. Different substrates as foam, sponge, filter

I. K. Vasil (ed.), Plant Biotechnology 2002 and Beyond, 351-357.

paper (porous materials), have been used and water or growth medium were often injected *via* syringes. As an example, we have used small plant growth chambers to grow lentil seedlings. The dry seeds were maintained against a sponge and were hydrated by injecting water with a syringe, *via* an entry tube. The entry tube permitted gas exchanges between the cabin atmosphere and the plant growth chambers (Perbal et al., 1987). Gas composition in the cabin can also be a problem since it can be different from that on Earth. This difficulty can be overcome at least partially by using a 1 *g* control in space ; i.e. to grow plants on a 1 *g* centrifuge in order to compare the microgravity sample to a 1 *g* sample cultivated in the same air conditions. However, in microgravity there is no convection which means that the mixing of gases depends only upon diffusion and for this reason it can be slower than under 1 *g*. In this case, the effect of microgravity can be considered to be indirect since it modifies the physical environment of the plant.

Such problems were not really taken into account in the pioneering experiments and were discovered rather recently (Musgrave et al., 1997) after a critical analysis of the results obtained in space in the frame of different flights and with different species.

In order to determine the role of gravity on plant growth many investigators have also used clinostats, the role of which is to stimulate microgravity by rotating plants about a horizontal axis. The action of gravity becomes omnilateral and the plant grows as if gravity was not present. There are different types of clinostats : one axis clinostat, two axe clinostat and 3 D clinostat. The validity of the clinostat as a simulation of microgravity has been discussed (Salisbury and Wheeler, 1981; Lorenzi et Perbal , 1990; Sievers and Hejnowicz, 1992; Hoson et al., 1997) and it can be said that at least clinorotation increases gas mixing in the plant growth chamber (Albrecht-Buehler, 1991). Moreover, when some plants were cultivated in the same way on earth as in space, their growth was different (Legué et al., 1992).

3. PLANT GROWTH IN SPACE AND IN SIMULATED MICROGRAVITY

Numerous species have been grown in space and it appeared that microgravity had no effect on germination (Halstead and Dutcher, 1987; Kordyium, 1997). However, the direction of growth of the radicle being strongly dependent upon gravity on Earth is in microgravity related to the orientation of the embryo in the seed (Volkmann et al., 1986). The root tip can be subjected to oscillations (Johnsson, 1997) at the very beginning of germination, but its orientation afterward is random to a certain extend (Johnsson et al., 1996). Bending of the root was also observed in simulated microgravity (Hoson, 1994).

The growth of the primary root has been described many times (Kordyum, 1997), but the data seemed to be controversial (Aarrouf et al., 1999b). A careful review of the literature done by Claassen and Spooner (1994) has led to the general conclusion that root growth in microgravity was the same as that on Earth for the first two days, greater for 3-5 days and less for longer periods of culture. This conclusion was reinforced by Hilaire et al. (1996) and by Aarrouf et al. (1999b) with clinostat work. On *Brassica napus* (Aarrouf et al., 1999b), it was shown that the growth of the primary root was greater on the clinostat for 15 days than in the vertical position. However, after 15 days, the primary root stopped growing in simulated microgravity and secondary roots were more numerous and initiated closer to the primary root tip. This indicated that the apical dominance was lost (or strongly lowered) after two weeks on the clinostat.

As gravitropic curvature involved a redistribution of auxin and perhaps abscisic acid, Schulze et al. (1992) examined the IAA and ABA contents in *Zea mays* seedlings grown in space. A significant difference in IAA content was found only in roots. Aarrouf et al. (1999b) have shown that IAA, ABA and zeatin contents were modified in the root system on the clinostat. These authors demonstrated that after 5 and 10 days of growth there were higher contents of IAA and ABA in the primary root of clinorotated *Brassica napus* seedlings. After 25 days the zeatin content became higher in this organ on the clinostat whereas IAA and ABA contents were similar. Aarrouf et al.(1999b) pointed out that the effect of simulated microgravity can be dependent upon the stage of development of the seedlings.

If we can extrapolate the results obtained on clinostat to what occurred in space, it can be concluded that the root system is strongly modified by microgravity surely because of the loss of apical dominance.

In a series of space experiments Krikorian and co-workers (Krikorian and Levine, 1991; Levine and Krikorian, 1992; Krikorian, 1996) have studied chromosome disturbances in roots of several species. The evaluation of the available facts have led them to put forward that indirect effects played a major role in the chromosome aberrations and that plants grown in space were subjected to various stresses.

For the majority of the species investigated, a decrease in the mitotic index was observed. However, mitotic index gives a very poor indication for analysing the cell cycle. Driss-Ecole et al. (1994b) and Yu et al. (1999) have studied the percentage of the various phases of the cell cycle in the lentil root meristem. They have shown that after 28 and 29 h only onc cell cycle was completed and that there were more cells in the G1 and S phases of the second cell cycle on the ground (or in the 1 *g* control in space) than in microgravity. Moreover, the results obtained indicated that there were more cells in the G2 phase of the first cell cycle in microgravity than in 1 *g*.

The slowing down of the mitotic activity in the meristem of the primary root could lead to an early decrease or the removal of apical dominance and to formation of numerous lateral roots described in many investigated species grown in space (Halstead and Dutcher, 1987). Interestingly, Driss-Ecole et al. (1994a) on *Veronica arvensis* and Aarrouf et al. (1999a) on *Brassica napus* have shown a greater biomass of the root system on the clinostat, which was due to greater formation and elongation of secondary roots.According to Aarrouf et al. (1999b), there should be a relationship between the modification of the hormonal balance and the development of the root system in microgravity and in simulated microgravity.

During the early phase of development of shoots, bending was observed in simulated microgravity (Hoson et al., 1995) and in space (Hoson et al., 1999). The data obtained on shoot growth are controversial (Claassen and Spooner, 1994), the most conclusive experiment was conducted in the frame of Salyut 7 (Merkys and Laurinavicius, 1990). Lettuce seedlings were grown in space either in microgravity or on a 1 *g* centrifuge. The growth of the hypocotyls was slightly greater in microgravity than on the 1 g centrifuge, but in both cases less than on the ground. Surely some factors (different from gravity) were responsible for slowing down of growth in space.

Recent work done by Kiss et al. (1998) has demonstrated the role of ethylene in the development of *Arabidopsis thaliana* grown in space. The fact that ethylene was the cause of a slowing down of seedling growth in space could eventually explain the results obtained on lettuce. It must be added that clinostat work also conducted to a greater development of the shoot of *Veronica arvensis* (Driss-Ecole et al., 1994a) and *Brassica napus* (Aarrouf et al., 1999a).

The development of the pegs was studied in Cucumber (Kamada et al., 2000; Fujii et al., 2000) in space and it has been shown that two pegs are formed in microgravity and on the ground when the seedlings are vertically oriented. When the seedlings were horizontally oriented only one peg was observed and this morphogenesis is dependent upon the auxin level in the transition zone between the root and the hypocotyl.

4. CONCLUSIONS

Several attempts to grow plants through a complete life cycle in space were not successful surely because of the lack of gas exchanges in space. Viable seeds were obtained with *Arabidopsis thaliana* (Merkys and Laurinavicius, 1990) and *Brassica napus* (Musgrave et al., 2000) although they were smaller in microgravity and presented some differences in their reserves. The results obtained on the vegetative phase can be summarized as follows. Germination is not perturbed in microgravity although cell cycle in the root meristem can be slowed down even during the first cycle following hydration. In most cases, the root system develops faster in microgravity and in simulated microgravity after the phase of germination. But the primary root seems to stop growing early and a loss of apical dominance occurs leading to greater formation and elongation of the secondary roots. It seems that the shoot is less perturbed in microgravity than the root system (Aarrouf et al., 1999a; Driss-Ecole et al., 1994a). The difference observed in microgravity and in simulated microgravity could be due to the distribution of hormones (IAA and ABA). Recent works using clinostats or in space have also shown that photosynthesis can be perturbed by microgravity or simulated microgravity. Whether these perturbations are direct or indirect effects of microgravity remains to be determined.

5. REFERENCES

Aarrouf J., N. Darbelley, C. Demandre, N. Razafindramboa, and G. Perbal. 1999a. Effect of horizontal clinorotation on the root system development and on lipid breakdown in Rapeseed (*Brassica napus*) seedlings. Plant Cell Physiol. 40(4):396-405.

Aarrouf J., D. Schoëvaërt, R. Maldiney, and G. Perbal. 1999b. Changes in hormonal balance and meristematic activity in primary root tips on the slowly rotating clinostat and their effect on the development of the rapeseed root system. Physiol. Plant. 105:708-718.

Albrecht-Buehler G. 1991. Possible mechanisms of indirect gravity sensing by cells. ASGSB Bull. 4(2):25-34.

Claassen D.E., and S. Spooner. 1994. Impact of altered gravity on aspects of cell biology. Int. Rev. Cytol. 156:301-373.

Driss-Ecole D., A. Cottignies, B. Jeune, F. Corbineau, and G. Perbal. 1994a. Increased mass production of *Veronica arvensis* grown on a slowly rotating clinostat. Environ. Exptl. Bot. 34(3):303-310.

Driss-Ecole D., D. Schoëvaërt, M. Noin, and G. Perbal. 1994b. Densitometric analysis of nuclear DNA content in lentil roots grown in space. Biol. Cell. 81:59-64.

Fujii N, M. Kamada, S. Yamasaki, and H. Takahashi. 2000. Differential accumulation of *Aux/IAA* mRNA during seedling development and gravity response in cucumber (*Cucumis sativus* L.). Plant Mol. Biol. 42:731-740.

Halstead T.W., and F.R. Dutcher. 1987. Plants in space. Annu. Rev. Plant Physiol. 38:317-345.

Hilaire E., V. B. Peterson, J.A. Guikema, and C.S. Brown. 1996. Clinorotation affects morphology and ethylene production in Soybean seedlings. Plant Cell Physiol. 37(7):929-934.

Hoson T. 1994. Automorphogenesis of maize roots under simulated microgravity conditions. Plant Soil 165:309-314.

Hoson T.,K. Soga, R. Mori, M. Saiki, K. Wakabayashi, S. Kamisaka, S. Kamigaichi, S. Aizawa, I. Yoshizaki, C. Mukai, T. Shimazu, K. Fukui, and M. Yamashita. 1999. Morphogenesis of rice and *Arabidopsis* seedlings in space. J. Plant Res. 112:477-486.

Hoson T., S. Kamisaka, Y. Masuda, M. Yamashita, and B. Buchen. 1997. Evaluation of the three-dimensional clinostat as a simulator of weightlessness. Planta 203:S187-S197.

Hoson T., S. Kamisaka, R. Yamamoto, M. Yamashita, and Y. Masuda. 1995. Automorphosis of maize shoots under simulated microgravity on a three-dimensional clinostat. Physiol. Plant. 93:346-351.

Johnsson A. 1997. Circumnutations: results from recent experiments on Earth and in space. Planta 203:S147-S158.

Johnsson A., C. Karlsson, D.K. Chapmann J.D. Braseth, and T.H. Iversen. 1996. Dynamics of root growth in microgravity. J. Biotech. 47:155-165.

Kamada M., N. Fujii, S. Aizawa, S. Kamigaichi, C. Mukai, T. Shimazu, and H. Takahashi. 2000. Control of gravimorphogenesis by auxin: accumulation pattern of *CS-IAA1* mRNA in cucumber seedlings grown in space and on the ground. Planta 211:493-501.

Kiss J.Z., W.J. Katembe, and R.E. Edelmann. 1998. Gravitropism and development of wild-type and starch-deficient mutants of *Arabidopsis* during spaceflight. Physiol. Plant. 102:493-502.

Kordyum E.L. 1997. Biology of plant cells in microgravity and under clinostating. Int. Rev. Cytol. 171:1-79.

Krikorian A.D. 1996. Space stress and genome shock in developing plant cells. Physiol. Plant. 98:901-908.

Krikorian A.D., and H.G. Levine. 1991. Plant Physiology. A treatise. Academic Press Vol.(X) Growth and Development. Pp. 491-555.

Kuang A., M.E. Musgrave, and S.W. Matthews. 1996. Modification of reproductive development in *Arabidopsis thaliana* under spaceflight conditions. Planta 198:588-594.

Kuang A., M.E. Musgrave, S.W. Matthews, D.B. Cummins, and S.C. Tucker. 1995. Pollen and ovule development in *Arabidopsis thaliana* under spaceflight conditions. Amer. J. Bot. 82(5):585-595.

Legué V., F. Yu, D. Driss-Ecole, and G. Perbal. 1992. Cell cycle and differentiation in lentil roots grown on a slowly rotating clinostat. Physiol. Plant. 84:386-392.

Levine H.G., and A.D. Krikorian. 1992. Shoot growth in aseptically cultivated daylily and haplopappus plantlets after a 5-day spaceflight. Physiol. Plant. 86:349-359.

Lorenzi G., and G. Perbal. 1990. Root growth and statocytes polarity in lentil seedling roots grown in microgravity or on a slowly rotating clinostat. Physiol. Plant. 78:532-537.

Merkys A.J., and R.S. Laurinavicius. 1990. Plant growth in space. Fundamentals of Space Biology. M. Asashima and G.M. Malacinski (eds.) Japan Sci. Soc. Press, Tokyo/Springer-Verlag, Berlin. Pp. 69-83.

Musgrave M.E., A. Kuang, and D.M. Porterfield. 1997. Plant reproduction in spaceflight environment. ASGSB Bull., 10:83-90.

Musgrave M.E., A. Kuang, Y. Xiao, S.C. Stout, G.E. Bingham, L.G. Briarty, M.A. Levinskikh, V.N. Sychev, and I.G. Podolski. 2000. Gravity independence of seed-to-seed cycling in ***Brassica rapa***. Planta 210:400-406.

Perbal G., D. Driss-Ecole, J. Rutin, and G. Sallé. 1987. Graviperception of lentil seedling roots grown in space (Spacelab D1 Mission). Physiol. Plant. 70:119-126.

Paul A.-L., C.J. Daugherty, E.A. Bihn, D.K. Chapman, K.L.L. Norwood, and R.J. Ferl. 2001. Transgene expression patterns indicate that spaceflight affects stress signal perception and transduction in *Arabidopsis*. Plant Physiol. 126:613-621.

Salisbury F.B., and R.M. Wheeler. 1981. Interpreting plant responses to clinostating. I Mechanical stresses and ethylene. Plant Physiol. 67:677-685.

Schulze A., P.J. Jensen, M. Desrosiers, J.G. Buta, and R.S. Bandurski. 1992. Studies on the growth and indole-3-acetic acid and abscisic acid content of *Zea mays* seedlings grown in microgravity. 100:692-698.

Sievers A., and Z. Hejnowicz. 1992. How well does the clinostat mimic the effect of microgravity on plant cells and organs? ASGSB Bull. 5(2):69-75.

Volkmann D., H.M. Behrens, and A. Sievers. 1986. Development and gravity sensing of Cress roots under microgravity. Naturwis. 73:438-441.

Yu F., D. Driss-Ecole, J. Rembur, V. Legué, and G. Perbal. 1999. Effect of microgravity on the cell cycle in the lentil root. Physiol. Plant. 105:171-178.

MORPHOGENESIS, HYDROTROPISM, AND DISTRIBUTION OF AUXIN SIGNALS IN CUCUMBER SEEDLINGS GROWN IN MICROGRAVITY

Hideyuki Takahashi, Motoshi Kamada, Yuko Saito, Aakie Kobayashi, Atsushi Higashitani and Nobuharu Fujii
Graduate School of Life Sciences, Tohoku University, Katahira, Aoba-ku, Sendai 980-8577, Japan (e-mail: hideyuki@ige.tohoku.ac.jp)

Keywords Auxin, cucumber, gravity, hydrotropism, microgravity, peg

1. INTRODUCTION

For pulling the seed coat out, seedlings of most cucurbits develop a protuberance, peg, on the concave side of the bending transition zone between hypocotyl and root when seeds were placed in a horizontal position for germination (Darwin and Darwin, 1880; Takahashi, 1997). It has been argued whether gravity is a cause for the lateral placement of a peg in cucurbit seedlings (Wiztzum and Gersani, 1975; Takahashi, 1997). Auxin appears to play an important role in developing the peg (Wiztzum and Gersani, 1975; Takahashi, 1997). To verify this hypothesis, we conducted a spaceflight experiment and examined the roles of gravity and auxin in peg formation of cucumber seedlings. Furthermore, we studied hydrotropism and its regulation by auxin in cucumber roots because the seedlings grown in microgravity appeared to become responsive to moisture gradient exhibiting hydrotropism without gravitropic interference.

2. MODIFICATION OF PEG FORMATION IN MICROGRAVITY

On the ground, seedlings developed a peg on the lower side of the transition zone when seeds were germinated in a horizontal position (Fig. 1A). However, space-grown seedlings developed a peg on each side of the transition zone as the ground control seedlings in a vertical position did (Fig. 1B,C). Thus, cucumber seedlings potentially develop a peg on each side of the cotyledonary plane of the transition zone. The result implies that on the ground peg formation on the upper side of the transition zone in a horizontal

I. K. Vasil (ed.), Plant Biotechnology 2002 and Beyond, 359-362.

position is suppressed in response to gravity while its formation is not suppressed on the lower side (Takahashi et al., 2000).

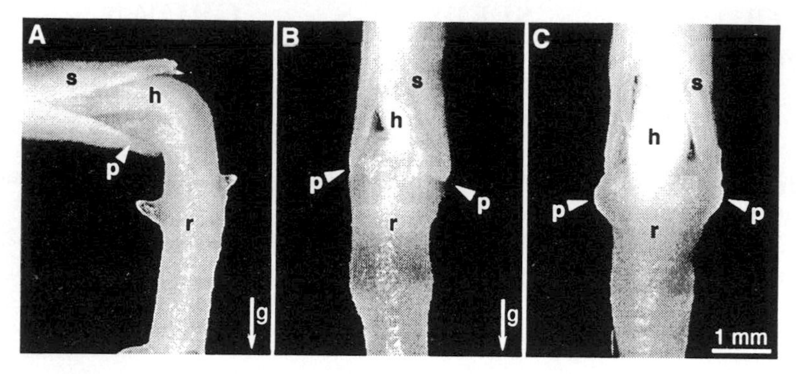

Figure 1. Cucumber seedlings were grown in space and on the ground, aeroponically for 70 h. A and B, seeds of cucumber (cv. Shinfushinari-jibai) were germinated in a horizontal and vertical positions, respectively. C, Seeds were germinated in micrograavity. p (arrow head), peg; h, hypocotyl; r, root; s, seed coat.

3. MODIFICATION OF AUXIN TRANSPORT AND DISTRIBUTION BY GRAVITY FOR PEG FORMATION

3.1. Expression Pattern of An Auxin-inducible Gene, *CS-IAA1*

When seeds were germinated in microgravity or in a vertical position, mRNA of an auxin-inducible gene, *CS-IAA1*, substantially accumulated in the transition zone and root (Fig. 2). There was no detectable difference in the accumulation of *CS-IAA1* mRNA between both sides of the cotyledonary planes. On the other hand, there was a differential accumulation of *CS-IAA1* mRNA in the transition zone in a horizontal position. We observed much accumulation of *CS-IAA1* mRNA in the lower side. This differential accumulation of *CS-IAA1* occurred mainly due to the decrease in the mRNA on the upper side of the transition zone (Kamada et al., 2000).

3.2. Distribution of Endogenous Auxin

The content of free IAA was lower and conjugated IAA was more abundant on the upper side of the transition zone of the gravistimulated seedlings compared with the lower side. These results support the idea that a decrease in auxin level due to a modification of auxin transport or metabolism causes the suppression of peg formation on the upper side of the transition zone in a horizontal position. Cucumber seedlings treated with auxin transport inhibitors exhibited agravitropic growth and developed a peg on each side of

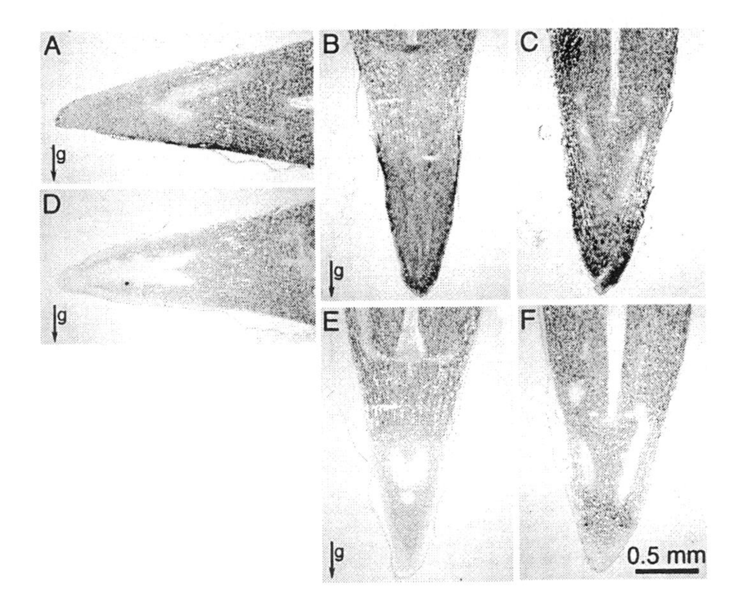

Figure 2. Accumulation pattern of CS-IAA1 mRNA was analyzed by in situ hybridization in cucumber seedlings grown for 29 h in space and on the ground. A-C, CS-IAA1 anti-sense probes; D-F, CS-IAA1 sense probes. Seeds were germinated in a horizontal position (A, D), in a vertical position (B, E) or in microgravity (C, F).

the transition zone, accompanying a transient increase in *CS-IAA1* mRNA. By analyzing the expression of *CS-PIN1* and *CS-AUX1*, we found that auxin carriers play an important role in controlling cytosolic auxin concentration for *CS-IAA* mRNA accumulation and peg formation (unpublished data).

4. HYDROTROPISM AND AUXIN IN MICROGRAVITY

Our spaceflight experiment suggested that in microgravity cucumber roots exhibited positive hydrotropism in the absence of gravitropic response. Cucumber roots rotated on a clinostat did display hydrotropism and accumulated *CS-IAA1* mRNA differentially: it was abundant on the concave side of the bending roots compared with that of the convex side (Mizuno et al., 2002). The results suggest that auxin plays a role in hydrotropism and that both hydrostimulation and gravistimulation modify auxin transport for its redistribution.

5. CONCLUSION

Our spaceflight experiment showed that microgravity in space is useful tool for the study of gravimorphogenesis and hydrotropism in which auxin metabolism or transport plays an important role. Although the role of asymmetric distribution of auxin in gravitropism has long been implicated, spaceflight experiments have not paid much attention to auxin dynamics in microgravity. Adopting a suitable experimental system, the gravity-regulated dynamics of auxin and its causal mechanism need to be studied in microgravity.

6. REFERENCES

Darwin, C., and F. Darwin. 1880. The Power of Movement in Plants. John Murray, London.

Kamada, M., N. Fujii, S. Aizawa, S. Kamigaichi, C. Mukai, T. Shimazu, and H. Takahashi. 2000. Control of gravimorphogenesis by auxin: Accumulation pattern of *CS-IAA1* mRNA in cucumber seedlings grown in space and on the ground. Planta 211: 493-501.

Mizuno, H., A. Kobayashi, N. Fujii, M. Yamashita, and H. Takahashi. 2002. Hydrotropic response and expression pattern of auxin-inducible gene, CS-IAA1, in the primary roots of clinorotated cucumber seedlings. Plant Cell Physiol. 43: 793-801.

Takahashi, H. 1997. Gravimorphogenesis: gravity-regulated formation of the peg in cucumber seedlings. Planta 203: S164-169.

Takahashi, H., M. Kamada, Y. Yamazaki, N. Fujii, A. Higashitani, S. Aizawa, I. Yoshizaki, S. Kamigaichi, C. Mukai, T. Shimazu, and K. Fukui. 2000. Morphogenesis in cucumber seedlings is negatively controlled by gravity. Planta 210: 515-518.

Wiztzum, A., and M. Gersani. 1975. The role of polar movement of IAA in the development of the peg in *Cucumis sativus* L. Bot. Gaz. 136: 5-16.

THE ASSEMBLY AND POTENTIAL APPLICATIONS OF IMMUNOGLOBULINS EXPRESSED IN TRANSGENIC PLANTS

Pascal M.W. Drake, Daniel M. Chargelegue, Patricia Obregon, Alessandra Prada, Lorenzo Frigerio and Julian Ma

Dept. of Oral Medicine and Pathology, Unit of Immunology, Guy's Hospital, London, UK. (e-mail: julian.ma@kcl.ac.uk)

1. INTRODUCTION

There is increasing interest in transgenic plant technology for the production of a wide range of recombinant proteins for pharmaceutical and industrial uses. There are many potential benefits in using plants as bioreactors for the production of such economically important proteins, such as the similarities between protein synthesis and post-translational modifications in plant and mammalian cells, the possibility of production scale-up to agricultural levels, as well as a number of safety and ethical issues. A consistent problem however, has been the disappointingly low expression levels achieved in many cases. There are several stages at which intervention might help to improve yield. These include the transformation event, transcription efficiency of the foreign gene, mRNA stability, translation of mRNA to recombinant protein, protein folding and assembly, post-translational modifications, intracellular targeting and transport, protein stability and downstream protein purification. But the fact that some recombinant proteins, particularly complex multimeric immunoglobulins, can be routinely expressed and extracted at levels of around 1% of total soluble protein, suggest that the difficulties with other proteins may lie at the post-translational level.

Recombinant proteins have been expressed in a number of plant cell sub-compartments, including the cytosol (Tavladoraki et al., 1993), plastids (De Cosa et al., 2001), the endoplasmic reticulum (Fiedler and Conrad, 1995) and apoplastic space (Hiatt et al., 1989). The cell compartment targeted is clearly important both with respect to protein stability but also for synthesis and

I. K. Vasil (ed.), Plant Biotechnology 2002 and Beyond, 363-370.

folding. Generally, cytosolic expression has resulted in the lowest yield of functional protein, whilst targeting proteins for secretion has resulted in the highest. More recently, it was demonstrated that proteins produced in the endomembrane system can often be stabilised by retention in the ER using a C-terminal peptide retention sequence (Fiedler and Conrad, 1995). The endoplasmic reticulum (ER) is the gateway of the secretory pathway (Muntz, 1998) . The folding and assembly of newly synthesised proteins targeted through the ER is a complex process with stringent quality control mechanisms and involves interactions with a battery of chaperones and enzymes (reviewed in Ellgaard et al., 1999). In mammalian and plant cells, the best characterised chaperone is BiP (binding protein), a lumenal endoplasmic reticulum (ER) resident member of the heat shock protein 70 family of stress proteins (Haas and Wabl, 1983). BiP has been identified in various mammals (Haas and Meo, 1988; Munro and Pelham, 1986; Ting et al., 1987; Wooden et al.,1988), yeast (Normington et al., 1989; Rose et al.,. 1989; Normington et al., 1989) and plants (Denecke et al., 1991; Fontes et al., 1991). By binding to newly synthesised polypeptides, BiP is thought to stabilise partially folded intermediates during folding and assist in the assembly of protein oligomers (Gething, 1999). BiP also has other functions in protein translocation into the ER, prevention and dissolution of protein aggregates and retention of misfolded or unassembled subunit proteins (Gething, 1999; Nishikawa et al., 2001).

In mammalian cells, the interactions between immunoglobulin chains and chaperones have been partially characterised. Newly synthesised heavy and light chains associate with BiP immediately after synthesis (Knittler and Haas, 1992; Vanhove et al., 2001). The interaction is brief but BiP displays a strong preference for early folding intermediates over mature molecules. Thereafter, a relay of chaperones is likely to be involved (Melnick and Argon, 1995), including GRP94 and possibly GRP170 (Melnick et al., 1994), as well as protein disulphide isomerase (PDI) (Roth and Pierce, 1987).

Relatively little is known about ER chaperones in plants. Limited evidence is available for the interaction of BiP with newly-synthesised endogenous and defective polypeptides (Pedrazzini et al., 1994; Vitale and Denecke, 1999), but until recently, the role of BiP in expression of recombinant mammalian proteins was not understood. Plant BiP shares approximately 69% homology with mammalian BiP at the amino acid level (Denecke et al., 1991), this compares favourably with yeast BiP which has 64% overall homology with murine BiP. In tobacco, BiP mRNA expression is normally highest in tissues containing rapidly dividing cells or those that are involved in secretion (Denecke et al., 1991), whereas in maize, BiP is expressed most abundantly in endosperm development (Fontes et al., 1991). It has been demonstrated that ER chaperones such as BiP are required in high quantities to promote efficient expression of recombinant proteins in plants, and under certain circumstances, they may become a limiting factor in protein expression (Leborgne-Castel et al., 1999). However, vegetative parts of plants do not express chaperones at high levels (Denecke et al., 1995). Nevertheless, the synthesis, folding and assembly of full length immunoglobulins (Ig) in plants can be extremely efficient, resulting in expression levels of between 1-5% of total plant protein in leaf tissue (Frigerio et al., 2000; Hiatt et al., 1989; Ma et al., 1995), that compare favourably with mammalian hybridoma cell culture.

It was previously demonstrated that efficient assembly and expression of immunoglobulins in plants could only be achieved by using a leader sequence to target the recombinant immunoglobulin proteins to the ER and the secretory pathway (Hiatt et al., 1989) and we have recently demonstrated the role of the plant BiP homologue with folding and assembly of Ig light and heavy chains (Nuttall et al., submitted). We believe now that the involvement of ER-resident chaperones promotes processing and expression of immunoglobulin molecules in plants.

The demonstration that the BiP pathway in plants is functional for mammalian proteins, provides a rationale for the use of plants as an expression system and is significant in demonstrating the suitability and potential versatility of transgenic plants for producing a variety of mammalian proteins. Protein translocation and folding in the ER can be one of the rate limiting steps in protein secretion, and the presence of protein chaperones is important for high efficiency turnover, leading to high levels of production. It has been reported that the overexpression of BiP (and PDI) in yeast cells greatly improves the efficiency of folding and secretion of single chain antibody fragments (Shusta et al., 1998). Likewise, when BiP is overexpressed in transgenic plants, it is able to alleviate ER stress induced by tunicamycin (Leborgne-Castel et al., 1999). It will be very interesting to test the effects of BiP overexpression in plants expressing model immunoglobulins. The passage of proteins through the secretory pathway is a complicated process (Vitale and Galili, 2001) and it is clear that BiP will not be the only chaperone involved. Plant homologues of GRP 94, and protein disulphide isomerase (Shorrosh and Dixon, 1991) have been identified, and it will also be important to establish their role in Ig assembly.

Another aspect of recombinant protein stability in transgenic plant expression systems, is the ultimate fate of the expressed protein and site of accumulation. Early evidence suggested that immunoglobulins were secreted into the apoplastic space, a relatively stable environment for protein accumulation (Hiatt et al., 1989). IgG antibodies are certainly secreted extracellularly (Frigerio et al., 2000), but somewhat surprisingly, they also appear to be excreted out of the plant. This mirrors recent findings with the recombinant green fluorescent protein (GFP), bacterial xylanase and human placental alkaline phosphatase (SEAP) expressed in plants, which were rhizosecreted in transgenic tobacco (Borisjuk et al., 1999). All three secreted proteins retained their biological activity and accumulated in higher amounts in the medium than in root tissue. In these and other cases reported so far, the expressed molecule has been a single protein with no assembly requirements but to our knowledge, rhizosecretion has not been reported for large multimeric complexes such as an immunoglobulin. We have however shown that a functional full length immunoglobulin complex was secreted from *N. tabacum* roots. We were able to use direct nitrocellulose root blots to demonstrate antibody rhizosecretion. ELISA of plant medium around the roots of IgG plants also indicated that the antibody was functional, as immune complexes were formed between rhizosecreted antibody and exogenous antigen that was added to the medium. Thus, in contrast to previously described limitations in cell wall porosity (Carpita et al., 1979), full-size IgG antibodies are able to cross the cell wall, despite having a molecular mass of approx. 150 Kd. This apparent anomaly may be due to immunoglobulins being highly flexible molecules, not only through the central hinge region, but also as a result of the domain structure which allows movement in most directions throughout the molecule (Burton, 1990; Lesk and Chothia, 1988).

Secretion is not the only possible fate for antibodies in plants. Secretory IgA, a much larger multimeric complex is correctly and efficiently assembled, yet is largely retained in the ER (Frigerio et al., 2000). However, this does not appear to represent an unfolded protein response, as the antibody is fully functional and assembled, and accumulates to even higher levels that IgG, reaching 5-8% of total soluble protein (Ma et al., 1998). The reason for SIgA retention in the ER is still unclear, however, this may represent a favourable strategy for immunoglobulin expression, in view of the high levels of antibody accumulation.

We have also studied the targeting of IgG antibodies to the plant cell membrane (Vine et al., 2001), using a highly conserved murine transmembrane sequence derived from mammalian B cells. This comprises an extra 71 amino acid residues at the COOH terminus of the Ig heavy chain, arranged in 3 domains, a 17 residue acidic extracellular portion, a 26 residue hydrophobic intra-membrane portion and a 28 residue hydrophilic intracellular portion (Yamawaki-Kataoka et al., 1982). Here we demonstrated assembly of functional antibody by co-expressing the antibody light chain with the membrane anchored heavy chain, and that the antibody is expressed at the plant cell membrane. Rather surprisingly, when exogenous antigen was added to the root medium of transgenic plants grown in hydroponic culture, these plants have shown the ability to sequester antigen in the leaves by formation of an immune complex with cell membrane-retained antibody (Drake et al., FASEB in press).

Overall, we have shown two possible strategies for IgG expression in plants, and both give rise to possible applications. IgG secretion from plants could be used to simplify the purification of recombinant antibodies. The extraction of recombinant proteins from plant tissues can be expensive and time consuming involving plant harvesting, tissue maceration and subsequent protein purification. Protease release during sample homogenization and protein extraction may also result in antibody fragmentation (Ma et al., 1994). In addition, the process of extraction and purification does not allow the continuous production of antibody over the lifetime of the plant. Plants naturally rhizosecrete a variety of compounds which may have a role in nodulation, mycorrhizal colonization, growth inhibition of neighbouring plants, acquisition of nutrients from soil and in defence against toxic metals (Gleba et al., 1999). Single chain Fv and monoclonal antibody γ heavy chain were recovered from the surrounding growth medium of genetically modified tobacco cell suspensions (Firek et al., 1993; Magnuson et al., 1996) and *Agrobacterium rhizogenes* derived hairy roots of tobacco were used to secrete assembled full-length murine IgG1 into surrounding medium (Wongsamuth and Doran, 1997). However, cell suspension cultures can frequently be low yielding and unstable (Borisjuk et al., 1999) whilst hairy root culture requires expensive bioreactors and cannot exploit the autotrophic capacities of the whole plant. Secretion-based systems involving whole plants in hydroponic culture may offer an opportunity to by-pass these steps.

A further application could be the rhizosecretion of appropriate antibodies in the open aquatic environment to reduce the bioavailability of environmental pollutants by neutralising biologically active epitopes. Alternatively, the establishment of rhizosecreting hydroponic cultures within a waste-water treatment plant could provide a decontamination strategy, as immune complexes could also be removed from the aqueous environment by simple precipitation strategies.

In planta antigen sequestration demonstrated by membrane retained IgG could also be used against virtually any pollutant to which monoclonal antibodies can be generated, although further studies will be necessary to extend these observations to a relevant non-protein pollutant. However, the choice of two phytoremediation strategies would increase the number of possible applications of this technology extending its use to both soil and surface water environments. In addition, it is notable that a protein antigen was stabilised by complexing with membrane antibody in plant leaves for up to 72 hours. Thus this technology may have a further application in phytomining, the extraction and concentration of valuable soil components.

2. REFERENCES

Borisjuk, N.V., Borisjuk, L.G., Logendra, S., Petersen, F., Gleba, Y., and Raskin, I. (1999). Production of recombinant proteins in plant root exudates. Nature Biotech. 17: 466-469.

Burton, D.R. (1990). Antibody: the flexible adaptor molecule. Trends Biochem. Sci. 15: 64-69.

Carpita, N., Sabularse, D., Montezinos, D., and Delmer, D. P. (1979). Determination of the pore size of cell walls of living plant cells. Science 205: 1144-1147.

De Cosa, B., Moar, W., Lee, S.B., Miller, M., and Daniell, H. (2001). Overexpression of the Bt cry2Aa2 operon in chloroplasts leads to formation of insecticidal crystals. Nature Biotechnol. 19: 71-74.

Denecke, J., Carlsson, L.E., Vidal, S., Hoglund, A.S., Ek, B., van Zeijl, M.J., Sinjorgo, K.M., and Palva, E.T. (1995). The tobacco homolog of mammalian calreticulin is present in protein complexes in vivo. Plant Cell 7: 91-406.

Denecke, J., Goldman, M.H., Demolder, J., Seurinck, J., and Botterman, J. (1991). The tobacco luminal binding protein is encoded by a multigene family Plant Cell 3: 1025-1035.

Ellgaard, L., Molinari, M., and Helenius, A. (1999). Setting the standards: quality control in the secretory pathway. Science 286: 1882-1888.

Fiedler, U., and Conrad, U. (1995). High level production and long term storage of engineered antibodies in transgenic tobacco seeds. Bio/Technology 10:1090-1094.

Firek, S., Draper, J., Owen, M.R., Gandecha, A., Cockburn, B., and Whitelam, G.C. (1993). Secretion of a functional single-chain Fv protein in transgenic tobacco plants and cell suspension cultures. Plant Mol. Biol. 23: 861-870.

Fontes, E.B.P., Shank, B.B., Wrobel, R.L., Moose, S.P., O'Brian, G.R., Wurtzel, E.T., and Boston, R.S. (1991). Characterization of an immunoglobulin binding-protein homolog in the maize floury-2 endosperm mutant. Plant Cell 3: 483-496.

Frigerio, L., Vine, N.D., Pedrazzini, E., Hein, M.B., Wang, F., Ma, J.K., and Vitale, A. (2000). Assembly, secretion, and vacuolar delivery of a hybrid immunoglobulin in plants. Plant Physiol. 123: 1483-1494.

Gething, M.J. (1999). Role and regulation of the ER chaperone BiP. Semin. Cell Dev. Biol. 10: 465-472.

Gleba, D., Borisjuk, N.V., Borisjuk, L.G., Kneer, R., Poulev, A., Skarzhinskaya, M., Dushenkov, S., Logendra, S., Gleba, Y.Y., and Raskin, I. (1999). Use of plant roots for phytoremediation and molecular farming. Proc. Nat. Acad. Sci. USA 96: 5973-5977.

Haas, I.G., and Meo, T. (1988). cDNA cloning of the immunoglobulin heavy chain binding protein. Proc. Nat. Acad. Sci. USA 85: 2250-2254.

Haas, I.G., and Wabl, M. (1983). Immunoglobulin heavy chain binding protein. Nature 306: 387-389.

Hiatt, A.C., Cafferkey, R., and Bowdish, K. (1989). Production of antibodies in transgenic plants. Nature 342: 76-78.

Knittler, M.R., and Haas, I.G. (1992). Interaction of BiP with newly synthesized immunoglobulin light chain molecules: cycles of sequential binding and release. EMBO J. 11: 1573-1581.

Leborgne-Castel, N., Jelitto-Van Dooren, E.P., Crofts, A.J., and Denecke, J. (1999). Overexpression of BiP in tobacco alleviates endoplasmic reticulum stress. Plant Cell 11: 459-470.

Lesk, A.M., and Chothia, C. (1988). Elbow motion in the immunoglobulins involves a molecular ball-and-socket joint. Nature 335: 188-190.

Ma, J.K.-C., Hiatt, A., Hein, M.B., Vine, N., Wang, F., Stabila, P., van Dolleweerd, C., Mostov, K., and Lehner, T. (1995). Generation and assembly of secretory antibodies in plants. Science 268: 716-719.

Ma, J.K.-C., Lehner, T., Stabila, P., Fux, C.I., and Hiatt, A. (1994). Assembly of monoclonal antibodies with IgG1 and IgA heavy chain domains in transgenic tobacco plants. Eur. J. Immunol. 24: 131-138.

Ma, J.K., Hikmat, B.Y., Wycoff, K., Vine, N.D., Chargelegue, D., Yu, L., Hein, M.B., and Lehner, T. (1998). Characterization of a recombinant plant monoclonal secretory antibody and preventive immunotherapy in humans [see comments]. Nature Medicine 4: 601-606.

Magnuson, N.S., Linzmaier, P.M., Gao, J.W., Reeves, R., An, G., and Lee, J.M. (1996). Enhanced recovery of a secreted mammalian protein from suspension culture of genetically modified tobacco cells. Protein Exp. Purif. 7: 220-228.

Melnick, J., and Argon, Y. (1995). Molecular chaperones and the biosynthesis of antigen receptors. Immunology Today 16: 243-250.

Melnick, J., Dul, J.L., and Argon, Y. (1994). Sequential interaction of the chaperones BiP and GRP94 with immunoglobulin chains in the endoplasmic reticulum. Nature 370: 373-375.

Munro, S., and Pelham, H.R. (1986). An Hsp70-like protein in the ER: identity with the 78 kd glucose-regulated protein and immunoglobulin heavy chain binding protein. Cell 46: 291-300.

Muntz, K. (1998). Deposition of storage proteins. Plant Mol. Biol. 38: 77-99.

Nishikawa, S.I., Fewell, S.W., Kato, Y., Brodsky, J.L., and Endo, T. (2001). Molecular chaperones in the yeast endoplasmic reticulum maintain the solubility of proteins for retrotranslocation and degradation. J Cell Biol. 153: 1061-1070.

Normington, K., Kohno, K., Kozutsumi, Y., Gething, M.J., and Sambrook, J. (1989). S. cerevisiae encodes an essential protein homologous in sequence and function to mammalian BiP. Cell 57: 1223-1236.

Pedrazzini, E., Giovinazzo, G., Bollini, R., Ceriotti, A., and Vitale, A. (1994). Binding of BiP to an assembly-defective protein in plant cells. Plant J. 5: 103-110.

Rose, M.D., Misra, L.M., and Vogel, J.P. (1989). KAR2, a karyogamy gene, is the yeast homolog of the mammalian BiP/GRP78 gene. Cell 57: 1211-1221.

Roth, R.A., and Pierce, S.B. (1987). In vivo cross-linking of protein disulfide isomerase to immunoglobulins. Biochemistry 26: 4179-4182.

Shorrosh, B.S., and Dixon, R.A. (1991). Molecular cloning of a putative plant endomembrane protein resembling vertebrate protein disulfide-isomerase and a phosphatidylinositol-specific phospholipase C. *Proc. Nat. Acad. Sci. USA* 88: 10941-10945.

Shusta, E.V., Raines, R.T., Pluckthun, A., and Wittrup, K.D. (1998). Increasing the secretory capacity of *Saccharomyces cerevisiae* for production of single-chain antibody fragments. Nature Biotech. 16: 773-777.

Tavladoraki, P., Benvenuto, E., Trinca, S., De Martinis, D., Cattaneo, A., and Galeffi, P. (1993). Transgenic plants expressing a functional single-chain Fv antibody are specifically protected from virus attack. Nature 366: 469-472.

Ting, J., Wooden, S.K., Kriz, R., Kelleher, K., Kaufman, R.J., and Lee, A.S. (1987). The nucleotide sequence encoding the hamster 78-kDa glucose-regulated protein (GRP78) and its conservation between hamster and rat. Gene 55: 147-152.

Vanhove, M., Usherwood, Y.K., and Hendershot, L.M. (2001). Unassembled Ig heavy chains do not cycle from BiP in vivo but require light chains to trigger their release. Immunity 15: 105-114.

Vine, N.D., Drake, P., Hiatt, A., and Ma, J.K. (2001). Assembly and plasma membrane targeting of recombinant immunoglobulin chains in plants with a murine immunoglobulin transmembrane sequence. Plant Mol. Biol. 45: 159-167.

Vitale, A., and Denecke, J. (1999). The endoplasmic reticulum-gateway of the secretory pathway. Plant Cell 11: 615-628.

Vitale, A., and Galili, G. (2001). The endomembrane system and the problem of protein sorting. Plant Physiol. 125: 115-118.

Wongsamuth, R., and Doran, P.M. (1997). Production of monoclonal antibodies by tobacco hairy roots. Biotechnology and Bioengineering 54: 401-415.

Wooden, S.K., Kapur, R.P., and Lee, A.S. (1988). The organization of the rat GRP78 gene and A23187-induced expression of fusion gene products targeted intracellularly. Exp. Cell Res. 178: 84-92.

Yamawaki-Kataoka, Y., Nakai, S., Miyata, T., and Honjo, T. (1982). Nucleotide sequences of gene segments encoding membrane domains of immunoglobulin gamma chains. Proc. Nat. Acad. Sci. USA 79: 2623-2627.

MEDICAL MOLECULAR PHARMING: EXPRESSION OF ANTIBODIES, BIOPHARMACEUTICALS AND EDIBLE VACCINES VIA THE CHLOROPLAST GENOME

Henry Daniell
Department of Molecular Biology & Microbiology, University of Central Florida, 12722 Research Parkway, Orlando FL 32826-3227, USA (e-mail: daniell@mail.ucf.edu)

Keywords Chloroplast Genetic Engineering, Genetically Modified Crops, Environmental concerns

1. INTRODUCTION

The daily income of nearly one billion people is less than one U.S. dollar. However, the annual cost of interferon therapy for leukemia, metastasizing carcinoma, Karposi sarcoma or viral hepatitis is $26,000. Among 180 million people infected with hepatitis C virus, a large majority of them have severe cirrhosis. Currently there is no vaccine available for hepatitis C. The annual requirement of Insulin like Growth Factor I, per cirrhotic patient is 600 mg and the current cost per mg is $30,000. In contrast, chloroplast-derived biopharmaceuticals should be inexpensive to produce and store, easy to scale up for mass production, and safer than those derived from animals or humans. Chloroplast genetic engineering is an environmentally friendly approach and offers biological containment of transgenes. Expression of thousands of copies of transgenes per cell via chloroplast genomes has yielded the highest level of foreign proteins ever reported in transgenic plants. Chloroplast transgene expression is also free of position effect and gene silencing, frequently encountered in nuclear transgenic plants. Oral delivery of medicine, via food, should eliminate the purification steps that usually account for most of production costs. Edible vaccines may be the only practical solution to combat bioterrorism. The successful engineering of tomato chromoplasts for high level transgene expression in fruits, coupled to hyper-expression of vaccine antigens and the use of plant-derived antibiotic free selectable markers augur well for oral delivery of edible vaccines or biopharmaceuticals that are currently beyond the reach of those who need them most. This article provides recent developments in the expression of antibodies, biopharmaceuticals edible vaccines via the chloroplast genome and also addresses possible environmental advantages.

I. K. Vasil (ed.), Plant Biotechnology 2002 and Beyond, 371-376.

2. PLANT DERIVED BIOPHARMACEUTICALS AND HUMAN PROTEINS

Generally, levels of pharmaceutical proteins produced in transgenic plants have been less than the levels needed for commercial feasibility, if the protein must be purified. Plant-derived recombinant hepatitis B surface antigen (HBsAg) induced only a low level serum antibody response in a small human study, probably reflecting the low level of expression (1-5 ng/g fresh weight) in transgenic lettuce. Recent studies increased the level of expression to 8.35 μg/g fresh tuber (0.000835% fresh weight). Despite such low levels of expression, adequate primary antibody response was observed in animals fed with uncooked potato tubers; but this response was lost when tubers were cooked. Even though Norwalk virus capsid protein expressed in potatoes caused oral immunization when consumed as food, expression levels were too low for large-scale oral administration. Expression of genes encoding several human proteins in transgenic plants has been disappointingly low: e.g. human serum albumin, 0.02%; human protein C, 0.001%; and erythropoietin, 0.0026% of total soluble protein; and human Interferon-β, 0.000017% of fresh weight (Daniell et al., 2001). A synthetic gene coding for the human epidermal growth factor was expressed only up to 0.001% of total soluble protein in transgenic tobacco. Therefore, there is a great need to increase expression levels of human blood proteins in order to enable the commercial production of pharmacologically important proteins in plants.

3. CHLOROPLAST TRANSGENIC SYSTEM

One alternate approach to increase expression levels is to express transgenes via the chloroplast genomes of higher plants. Foreign genes have been integrated into the chloroplast genome of several crop plants, including tobacco, tomato and potato (up to 10,000 copies of transgenes per cell), resulting in accumulation of recombinant proteins several hundred fold higher than nuclear transgenic plants (up to 47% of the total soluble protein, Decosa et al., 2001). Targeted transgene integration into the chloroplast genomes eliminates the "position effect" frequently observed in nuclear transgenic plants resulting from random integration of transgenes (Daniell et al., 2002). In addition, gene silencing has not been observed in transgenic chloroplasts, in spite of extraordinarily high levels of transgene expression, whereas it is a common phenomenon in nuclear transformation (Daniell and Dhingra, 2002). Because of these reasons, expression and accumulation of foreign proteins is uniform in independent chloroplast transgenic lines. It has been shown that multiple genes can be engineered in a single transformation event via the chloroplast genome, regulated by a single promoter; this facilitates coordinated expression of multi-subunit proteins or engineering new pathways. This was demonstrated by successful expression and assembly of monoclonal antibodies (Daniell et al., 2001), or bacterial operons (DeCosa et al., 2001) in transgenic chloroplasts. Yet another advantage is the lack of toxicity of foreign proteins to plant cells when they are compartmentalized within chloroplasts (Lee et al., 2002). Chloroplast genetic engineering is an environmentally friendly approach, minimizing several environmental concerns (Daniell, 2002). Importantly, chloroplasts are able to process eukaryotic proteins, including correct folding and formation of disulfide bridges. Accumulation of large quantities of a fully assembled form of human somatotropin with the correct disulfide bonds (7% total soluble protein) provides strong evidence for hyper-expression and assembly of pharmaceutical proteins using this approach. In addition, binding assays showed that chloroplast-synthesized Cholera toxin β subunit (CTB) binds to the intestinal membrane GM1-ganglioside receptor, confirming correct folding and disulfide bond formation of the plant-derived CTB pentamers (Daniell et al., 2002). Such folding and assembly of foreign proteins should eliminate the need for highly expensive *in vitro* processing of pharmaceutical proteins produced in recombinant organisms. For example, 60% of the total operating cost for the commercial production of human insulin in *E. coli* is associated with *in vitro* processing (formation of disufide bridges and cleavage of methionine).

3.1. Expression of Human Serum Albumin in Transgenic Chloroplasts

Human Serum Albumin (HSA) accounts for 60% of the total protein in blood serum and it is the most widely used intravenous protein. To date, HSA has been produced primarily by the fractionation of blood. In spite of screening of raw material and heat treatment of final product, HSA derived from blood might harbor pathogens. Transgenic chloroplasts can accumulate large amounts of foreign proteins and have been shown to be able to fold properly several recombinant proteins and form disulfide bonds. Therefore, different vectors were constructed and bombarded into tobacco leaves to direct HSA expression in transgenic chloroplasts. Regulation of HSA under the control of a Shine-Dalgarno sequence (SD), 5' *psbA* region or the *cry2Aa2* UTR resulted in different levels of expression in transgenic chloroplasts from seedlings: 0.8, 1.6, 5.9% of HSA in tsp, respectively. On the other hand, a maximum of 0.02, 0.8 and 7.2% of HSA in tsp was observed in transgenic potted plants regulated by SD, *cry2Aa2* UTR or 5' *psbA* region respectively, demonstrating excessive proteolytic degradation, unless compensated by enhanced translation. The psbA-HSA expression was subject to developmental and light regulations, with the lowest expression observed in seedlings and maximal expression (11.1% tsp) under continuous illumination. HSA accumulated into large inclusion bodies, however, this led to gross underestimation of HSA because ELISA could be performed only in partially solubilized plant extracts. In spite of this underestimation, we report here the highest expression of a HSA and 500-fold higher than previous reports of HSA expression in nuclear transgenic plants. Formation of HSA inclusion bodies not only offered protection from proteolytic degradation but also provided a simple method of purification from other cellular proteins by centrifugation. HSA inclusion bodies could be readily solubilized to obtain monomeric form using appropriate reagents. The *cry2Aa2* UTR mediated expression in seedlings and chromoplasts, although as efficient as *psbA* 5' region, is independent of light regulation and should therefore facilitate expression of foreign genes in non-green tissues, thereby enabling oral delivery of pharmaceuticals.

3.2. Optimization of Codon Composition and Regulatory Elements for Expression of the Human Insulin-like Growth Factor-1 in Transgenic Chloroplasts

The human insulin-like growth factor 1 (IGF-1) is a potent multifunctional anabolic hormone produced by the liver. IGF-1 polypeptide is composed of 70 amino acids with a molecular weight of 7.6 kDa and contains three disulfide bonds. IGF-1 is involved in the regulation of cell proliferation and differentiation of a wide variety of cell and tissue types, and plays an important role in tissue renewal and repair. One cirrhotic patient requires 600mg of IGF-1 per year and the cost per mg is $30,000. In the past, IGF-1 has been expressed in E.coli but the protein cannot be produced in the mature form, because E.coli does not form disulfide bonds in the cytoplasm. Transgenic chloroplast technology provides a good solution to recombinant protein production, because of the capability to achieve high expression levels, and the ability to fold and process eukaryotic proteins with disulfide bridges. To increase the expression levels, a synthetic IGF-1 gene with optimized codons for tobacco chloroplast was made. The integration of the IGF-1 gene into the tobacco chloroplast genome was confirmed using PCR and Southern analyses. The IGF-1 protein has been detected in transgenic tobacco chloroplasts by western blot analysis. The ELISA performed on the transgenic lines show expression levels up to 32% IGF-1 of the total soluble protein. This is the highest level of pharmaceutical protein reported in transgenic plants. Further characterization will be presented.

3.3. Expression of Human Interferon in Transgenic Chloroplasts

Currently, recombinant IFNα2b is being produced through an *E. coli* expression system and is being used for the treatment of viral and malignant diseases. However, due to necessary *in vitro* processing and purification, the annual cost of IFNα2b is about $26,000 per year. In addition, up to 20% of patients develop anti-IFNα antibodies, which may be related to the route of administration, dosage parameters or production-based contaminants. Oral delivery of IFNα2b expressed via the chloroplast genome may eliminate some of these side effects. We have generated a recombinant IFNα2b construct containing a polyhistidine purification tag as well as a thrombin cleavage site. This construct has been integrated into the chloroplast genome of a model tobacco as well as a low nicotine variety of tobacco, LAMD-605. Western blots using interferon alpha MAB has shown protein accumulation in both varieties of tobacco. ELISA has been used to quantify the amount of IFNα2b in leaf tissues. Southern blots to verify homoplasmy has been performed. Results of these investigations will be presented.

3.4. Expression of *Bacillus anthrasis* Protective Antigen in Transgenic Chloroplasts for an Improved Vaccine Against Anthrax

Bacillus anthracis is the causative agent for the anthrax disease, which has become a serious threat due to bioterrorism. Large-scale production of immunogenic protective antigen (PA) is required to produce large quantities of vaccines. Expressing *pag* (gene for PA) in plants would eliminate the toxin contaminants associated with the current vaccine. *Pag* was cloned into the universal chloroplast vector along with *psbA* 5'UTR to enhance expression. Earlier in our lab, another *Bacillus* protein utilized ORF1,2 from the *cry2Aa2* operon to fold the insecticidal protein into cuboidal crystals which protected it from proteolytic degradation. Therefore, a second *pag* construct was made with the ORF1,2 and *psbA* 5' UTR to assess this approach for PA. Chloroplast integration of the transgene was confirmed by PCR and protein expression by western blots. Macrophage lysis assays show that chloroplast derived PA is fully functional. Further characterization of the transgenic lines will be done including protein quantification by ELISA.. An alternative approach for an improved anthrax vaccine would be to achieve high level expression in tomato fruit chromoplasts and use of a selectable marker gene from spinach (BADH) instead of antibiotic resistance genes. Results of all investigations will be presented.

3.5 Expression of Monoclonals in transgenic chloroplasts

Owing to their remarkable specificity and therapeutic nature for defined targets, monoclonal antibodies are emerging as therapeutic drugs at a fast rate. The chloroplast genome was chosen for transformation with antibody genes due to tremendously high levels of foreign protein expression; ability to fold, process and assemble foreign proteins with disulfide bridges; simpler purification and transgene containment via maternal inheritance. To enhance translation, a codon optimized gene under the control of specific 5` untranslated regions (UTRs) was used. IgA-G, a humanized, chimeric monoclonal antibody (Guy`s 13) has been successfully synthesized and assembled in transgenic tobacco chloroplasts with disulfide bridges. Guy`s 13 recognizes the surface antigen Streptococcus mutans, the bacteria that causes dental cavities. In this study, integration into the chloroplast genome was confirmed by PCR and southern blot analyses. Western blot analysis revealed the expression of heavy and light chains individually as well as the fully assembled antibody, thereby suggesting the presence of chaperonins for proper protein folding and enzymes for formation of disulfide bonds within transgenic chloroplasts.

4. REFERENCES

Daniell, H. 2002. Gene flow from genetically modified crops: Current and future technologies for transgene containment. Nature Biotechnology 20: 581-586.

Daniell, H., and A. Dhingra. 2002. Multiple gene engineering. Curr. Opin. Biotech. 13: 136-141.

Daniell, H., M.S. Khan and L. Allison. 2002. Milestones in chloroplast genetic engineering: an environmentally friendly era in biotechnology. Trends Plant Sci. 7:84-91.

Daniell, H., S.B. Lee, T. Panchal, and P. O. Wiebe. 2001. Expression and assembly of the native cholera toxin B subunit gene as functional oligomers in transgenic tobacco chloroplasts. J. Mol. Biol. 311: 1001-1009.

Daniell, H., B. Muthukumar and S.B. Lee. 2001. Engineering the chloroplast genome without the use of antibiotic selection. Curr. Gen. 39: 109-116.

Daniell, H., K. Wycoff and S. Streatfield. 2001. Medical Molecular Farming: Production of antibodies, biopharmaceuticals and edible vaccines in plants. Trends Plant Sci. 6: 219-226.

DeCosa, B., W. Moar, S. B. Lee, M. Miller and H. Daniell. 2001. Hyper-expression of the Bt Cry2Aa2 operon in chloroplasts leads to formation of insecticidal crystals. Nature Biotech. 19:71-74.

Lee, S.B., M.O. Byun, and H. Daniell. 2002. Accumulation of trehalose within transgenic chloroplasts confers drought tolerance. Transgen. Res. in press.

PRODUCTION AND APPLICATION OF PROTEINS FROM TRANSGENIC PLANTS

Elizabeth E. Hood, Michael E. Horn, and John A. Howard
ProdiGene® 101 Gateway Blvd. Ste. 100 College Station, TX 77845 USA (email: eehood@prodigene.com)

Keywords Maize, vaccines, industrial enzymes, containment

1. INTRODUCTION

Plant biotechnology, originally developed to improve agricultural products, has also been used successfully to produce heterologous proteins for application in other industries. The number of proteins produced in plants is increasing and is expected to grow rapidly in the future. The types of proteins targeted for this technology include antibodies, food and feed additives, products for animal health and human pharmaceuticals, and industrial enzymes. Some advantages the system offers include: 1. vaccines are edible and can be delivered directly in food or feed, 2. proteins can be produced at a fraction of the cost of competing product delivery systems, 3. products are bio-based, renewable, safe for the environment, easily scaled-up, and 4. products do not harbor human and animal pathogens. The use of transgenic plants for large-scale production of proteins for industrial, pharmaceutical, veterinary and agricultural use has been shown to be commercially feasible and economically advantageous (Hood and Jilka, 1999; Kusnadi et al. 1998a,b; Whitelam et al. 1993; Hood, 2002). Using plants to produce recombinant proteins is an attractive alternative to other expression systems because of the ease and low cost in generating a large biomass and scaling up production, and the ease in storing and purifying proteins from seed.

2. PRODUCTS

2.1. Demonstration Products and Reagents

ProdiGene has commercialized the first protein products from transgenic plants—avidin (Hood et al., 1997) and β-glucuronidase (GUS; Witcher et al., 1998). Avidin has equivalent activity to the native protein and we have achieved expression levels of 0.2% of dry weight of seed (2g/kg). Avidin,

I. K. Vasil (ed.), Plant Biotechnology 2002 and Beyond, 377-382.

first sold in 1997, was generated from transgenic maize plants that express the chicken *avidin* gene. Used primarily as a diagnostic reagent, the glycoprotein has traditionally been obtained for commercial production from chicken egg white. Long-term stability of expression has been a concern for transgenic proteins. However, after 10 generations in the field, expression levels for avidin have not only been maintained but substantially increased through selection and improvements in germplasm (unpublished data). GUS has been produced commercially from transgenic maize plants that express the *E. coli* β-glucuronidase gene. GUS is a large versatile enzyme that is useful in research because of its ability to be detected cytochemically, spectrophotometrically, and fluorometrically.

2.2. Vaccines

The greatest benefit of plants such as maize in vaccine production is that plants can be used as an edible product for the oral delivery of vaccines. Clearly, oral delivery has tremendous advantage in convenience. The elimination of shots and medical assistance, compared to taking a formulated dose of medication in the form of a wafer, can provide a leap forward in vaccine delivery. This also translates into reducing the dependence on the cold chain for transporting vaccines. Since corn is stable at ambient temperatures and keeps recombinant proteins stable in grain for years, there is no need to maintain refrigeration for storage and transport. This will enable vaccines to be sent to continents such as Africa and Asia, which are in need of vaccines but are not adequately equipped with refrigerated storage and transportation infrastructures to ship the products to where they are needed.

ProdiGene's vaccine program has several targets including the *E. coli* heat labile toxin, Lt-B, the hepatitis B virus, the AIDs virus (HIV), and transmissible gastroenteritis virus (TGEV). Clinical trials will be done in 2002 with Lt-B. When orally administered within a transgenic corn matrix to mice, this antigen induced a significant IgA response in the gut (Streatfield et al., 2001, 2002). Additionally, the subunit vaccines being developed for HepB and AIDs will enter into clinical trials within the next two years.

ProdiGene's animal health program has used the TGE virus to demonstrate efficacy of orally administered vaccine antigens. TGEV is a swine virus that can be lethal to newborn piglets. We have expressed high levels of the spike protein from this virus in maize seed. When fed to pigs, this S-antigen-containing grain elicited serum antibody responses, and protected the pigs from infection with the wild type virus (PCT # US 01/01148; Streatfield et al., 2001).

2.3. Industrial Enzymes

We are also developing industrial enzymes in transgenic maize seed. Expressing foreign enzymes at high levels generates unique issues for plant physiology. We have achieved high expression in seed through the use of seed-preferred promoters, targeting to specific sub-cellular locations, and breeding into protective germplasm.

2.3.1. Laccase

Laccase, a blue copper oxidase, generates free radicals that react with other molecules to form or degrade polymers. Care must be taken to express redox enzymes in a protective manner because they can significantly affect metabolism. High expression levels are best achieved by using a seed-preferred promoter and targeting accumulation to the cell wall space. We used constitutive and seed-specific promoters to express the laccase I isozyme from *Trametes versicolor* in transgenic maize (Hood et al., submitted). Although all vectors produced numerous transgenic events, expression levels were substantially higher in the T1 seed from the embryo-preferred promoter and cell wall targeting than with all other vector combinations. These lines have been further developed to yield 0.1% of dry weight as active laccase.

2.3.2. Trypsin

Trypsin, a proteolytic enzyme involved in digestion, is used commercially in cell culture and protein processing, primarily in the pharmaceutical industry. The use of bovine slaughterhouse refuse, the current source for the production of trypsin, is fraught with problems associated with the occurrence of mad cow disease and hoof and mouth disease. Therefore, a non-animal source of trypsin, such as from transgenic maize, is highly desirable.

Producing proteases in heterologous systems is difficult because of the potential for degrading native proteins. Non-toxic proteins can usually be expressed with constitutive promoters, whereas proteins such as proteases that may harm cellular components are usually best expressed in the seed with seed-specific promoters. ProdiGene has recently been granted a broad-based patent (USP #6,087,558) for the production of proteases in transgenic plants that claims expression of any protease in the zymogen form. This is most strategic because achieving high expression levels of a protease in transgenic plants is nearly impossible without expressing zymogens in seed.

We have expressed bovine pancreatic trypsin in transgenic maize plants using constitutive and seed-preferred promoters (Hood and Woodard, in press). Expression levels were much higher for events generated from the trypsinogen gene, the inactive precursor for trypsin, than those from the active trypsin gene. Moreover, enzyme levels were much higher for the seed-targeted trypsinogen than those for trypsinogen expressed with a constitutive promoter. Thus, seed-preferred expression of the zymogen form yields the highest expression levels of a protease in transgenic plants.

3. PRODUCTION AND CONTAINMENT

In addition to achieving high expression levels, another challenge in transgenic plant technology is to ensure containment of transgenic seed. The principal issues driving the establishment of a containment program are food and environmental safety, public perception and regulatory compliance. The most important issue is safety, and because they are interrelated, addressing safety also addresses the other issues. A containment system to ensure safety comprises physical features, biological control, quality assurance and quality control systems that include standard operating procedures, and tracking systems and finally, economic incentives. The properties of production for commodity corn and identity preserved corn are compared in Table 1. Commodity corn is the most common type grown and the production requires no special treatment, whereas specialty corn such as waxy or white corn, is identity preserved and is supported with a grower premium. Identity contained corn, such as that with specialty chemicals like pharmaceuticals, requires the support of contracts, premiums and standard operating procedures for production.

Four areas in production of transgenic seed are targeted for implementation of an identity containment system: grain production; harvesting; transportation and storage; and processing. Although each area has unique steps for containment, cleaning of equipment and inspection and validation of clean out are procedures common to all four areas of production. During grain production, containment requires the support of the agronomist and isolation of the crop. Specific issues that must be addressed to ensure safety include pollen containment procedures, such as physical or genetic male sterility, isolation of the crop from other corn, border rows for trapping pollen, or delays in planting to prevent synchrony in flowering time. Each of these methods prevents cross-pollination to levels greater than 99%. When combined, these methods can result in excellent pollen containment. Storage areas and farm tools must be free of transgenic seed to prevent products approved for industrial or medical use from entering the food supply. In addition, during harvest of the transgenic crop, the fields must be monitored and the grain segregated. Containment programs also specify the use of custom bulk totes for transporting and storing transgenic products, and seed lots should be properly labeled with identity numbers, which are also tracked during processing.

Table 1: Comparison of Different Systems for Growing Maize

	COMMODITY	IDENTITY PRESERVED	PRODIGENE'S ICS_{SM}
Objective	*Maximize production yield*	*Prevent exposure FROM commodity corn*	*Prevent exposure TO or FROM commodity corn*
GENERAL			
SOPs	None	None	Required
Legal contracts	None	Optional	Required
Economic incentives above commodity	N/A	S.15-S.35/bushel	S.50-S1.00/bushel
Typical acreage	75,000,000	1,000,000	1,000
GROWING			
Grower's seed	Purchased	Purchased	Licensed
Location of plots	No restrictions	No restrictions	Pre-approved locations
Containment required	None	None	Regulated
Regulatory approval	None	None	Required
Regular field inspections	None	None	Required
HARVESTING			
Equipment	No special requirements	No special requirements	Dedicated equipment or standardized clean-out required
Transport	No special requirements	Segregated from commodity corn	ProdiGene or its designee must handle
Storage	No special requirements	Segregated from commodity corn	Dedicated storage containers/facilities
PROCESSING			
Requirements beyond minimum specifications	None	None	Regulated by FDA
Waste or by-products	Not regulated	Not regulated	Regulated by USDA

4. SUMMARY

Production of enzymes and pharmaceuticals from transgenic plants requires use of some unique paradigms in genetics, fractionation, extraction and activation. The plant production system is versatile, enabling diverse products. Issues in product development, regulatory, legal and production must be coordinated for success. The industry will require vigilance and self-regulation during development and compliance with government regulations at maturity. The benefits and potential are enormous.

5. LITERATURE CITED

Hood E.E. 2002. From green plants to industrial enzymes. Enzm. Micro. Tech. 30: 279-283.

Hood et al., High level expression of fungal laccase in transgenic maize. (submitted)

Hood E.E., and J. Jilka. 1999. Plant-based production of xenogenic proteins. Curr. Opin. Biotech. 10: 4.

Hood E.E. and S. Woodard 2002. Industrial proteins produced from plants. In Plants as Factories for Protein Production. E. Hood, and J. Howard (eds.). Plenum Publishers. In press.

Hood E.E., D. Witcher, S. Maddock, T. Meyer, C. Baszczynski, M. Bailey, P. Flynn, J. Register, L. Marshall, D. Bond, E. Kulisek, A. Kusnadi, R. Evangelista, Z. Nikolov, C. Wooge, R. Mehigh, R. Hernan, W. Kappel, D. Ritland, L. Chung-Ping, and J. Howard. 1997. Commercial production of avidin frrom transgenic maize: characterization of transformant, production, processing, extraction and purification. Mol. Breed. 3: Pp. 291-306.

Hood E.E., A. Kusnadi, Z. Nikolov, and J. Howard. 1999. Molecular farming of industrial proteins from transgenic maize. Chemicals via Higher Plant Bioengineering. F. Shahidi, P. Kolodziejczyk, J. Whitaker, A. Munguia, G. Fuller (eds.). Plenum Publishers, New York. Pp. 127-147.

Jilka J.M., E. Hood, R. Dose, and J. Howard. 1999. The benefits of proteins produced in transgenic plants. AgBiotechNet 1: 1-4.

Jilka J.M. 2001. Methods and Compositions for Obtaining Disease Protection for Economically Important Animals. PCT # US 01/01148.

Kusnadi A.R., R. Evangelista, E. Hood, J. Howard, and Z. Nikolov. 1998a. Processing of trangenic corn seed and its effect on the recovery of recombinant β-glucuronidase. Biotech. Engin. 60: 44-52.

Kusnadi A.R., E. Hood, D. Witcher, J. Howard, and Z. Nikolov. 1998b. Production and purification of two recombinant proteins from transgenic corn. Biotechnology Progress 14: 147-155.

Streatfield S., H. Daniell, K. Wycoff. 2001. Medical molecular farming: production of antibodies, biopharmaceuticals and edible vaccines in plants. Trends Plant Sci. 6: 219-226.

Streatfield Stephen, J. Mayor, D. Baker, C. Brooks, B. Lamphear, S. Woodard, K. Beifuss, D. Vicuna, L. Massey, M. Horn, D. Delaney, Z. Nikolov, E. Hood, J. Jilka, and J. Howard. 2002. Development of an edible subunit vaccine in corn against enterotoxigenic strains of escherichia coli. In Vitro Cell Dev. Biol.—Plant 38: 11-17.

Whitelam G.C., B. Cockburn, A. Gandecha, and M. Owen. 1993. Heterologous protein production in transgenic plants. Biotechnology 11: 1-29.

Witcher D.R., E. Hood, D. Peterson, M. Bailey, D. Bond, A. Kusnadi, R. Evangelista, Z. Nikolov, C. Wooge, R. Mehigh, W. Kappel, J. Register, and J. Howard. 1998. Commercial production of β-glucuronidase (GUS): A model system for the production of proteins in plants. Mol. Breed. 4: 301-312.

PLANTS AND HUMAN HEALTH: DELIVERY OF VACCINES VIA TRANSGENIC PLANTS

Tsafrir S. Mor and Charles J. Arntzen
Arizona Biomedical Institute and the Plant Biology Department, PO Box 1601, Arizona State University, Tempe, AZ 85287-1601 USA (emails: tsafrir.mor@asu.edu, charles.arntzen@asu.edu)

Keywords mucosal vaccines, infectious diseases

1. THE THEORY BEHIND EDIBLE VACCINES

One of major challenges of biotechnology is to reduce clinical innovations to economically viable practices. Plant-derived edible vaccines were first conceived and are continued to be developed with this prime directive in mind: merging innovations in medical science and plant biology for the creation of efficacious and affordable pharmaceuticals. Since the emergence of the original idea about 10 years ago, it was embraced by a growing number of laboratories in academia and industry and doing justice to the subject matter requires a much more expansive review then can be afforded here (Daniell et al., 2001; Mor et al., 1998; Tacket and Mason, 1999).

Despite notable successes, traditional "Jennerian" vaccine technology has its limitations. These vaccines usually consist of either inactivated or attenuated strains of the pathogen, which are delivered by injection (the oral polio vaccine is an exception). In contrast, many of the currant vaccine development efforts focus on subunit mucosal vaccines.

A "subunit vaccine" refers to a pathogen-derived protein (or even just an immunogenic domain of a protein, "an epitope") that cannot cause disease but can elicit a protective immune response against the pathogen. Very often the subunit vaccine candidate is a recombinant protein made in transgenic production-hosts (such as cultured yeast cells), then purified, and injected into vaccinees to immunize against a specific disease. Subunit vaccines are generally considered safer to produce (eliminating the need to culture pathogenic organisms) and more importantly, to use.

However, immunization by injection (parenteral delivery) rarely results in specific protective immune responses at the mucosal surfaces of the respiratory, gastrointestinal and genito-urinary tracts. Mucosal immune responses represent a first line of defense against most pathogens. In contrast

I. K. Vasil (ed.), Plant Biotechnology 2002 and Beyond, 383-387.

mucosally targeted vaccines achieve stimulation of both the systemic as well as the mucosal immune networks. In addition, mucosal vaccines delivered orally increase safety and compliance by eliminating the need for needles. While subunit vaccines are effective, they depend on expensive infrastructure for their production, distribution and administration, infrastructure, which is either scarce on non-existing in the developing world, where vaccines are needed the most. Combining a cost-effective production system with a safe and efficacious delivery system, plant edible vaccines, provide a compelling solution.

2. THE THEORY IS PUT TO CLINICAL TRIALS

Since 1992, when their seminal paper describing the expression of hepatitis B surface antigen (HBsAg) in tobacco plants was published (Mason et al., 1992), the group headed Arntzen and Mason developed their ideas in a succession of papers characterizing the recombinant product which assembled into virus like particles (VLPs, Mason et al., 1992), and could invoke specific immune responses in mice upon parenteral delivery (Thanavala et al., 1995). To prove that plant-derived HBsAg can stimulate mucosal immune responses via the oral route, the group switched their expression system to potato tubers and optimized it to increase accumulation of the protein in the plant tubers (Richter et al., 2000). Remarkably, these plant materials proved superior to the yeast-derived antigen in both priming and boosting immune responses in mice (Kong et al., 2001; Richter et al., 2000). Along side the efforts on the hepatitis B vaccine front, Mason and Arntzen explored plant expression of other vaccine candidates including the labile toxin B subunit (LT-B) of enterotoxigenic *Escherichia coli* (ETEC) and the capsid protein of Norwalk virus (NVCP). The plant derived proteins correctly assembled into functional oligomers that could elicit the expected immune responses in animals (Haq et al., 1995; Mason et al., 1996, 1998).

Success in mouse experiments provided motivation for conducting Phase I/II clinical trials to test the safety and immunogenicity of plant-produced LT-B, NVCP and HBsAg (Tacket et al., 1998, 2000, and Thanavala, Mason and Arntzen, unpublished). In the three cases tested, humans who consumed raw potato tubers containing tens of microgram amounts of the antigens developed specific serum and more importantly mucosal immune responses. Significantly, the three antigens in these studies come from three very different pathogens including viral (NV and HBV) and bacterial (*E. coli*) enteric (NV and *E.* coli) as well as non-enteric (HBV). The high titers of mucosal and systemic antibodies in most of the volunteers suggest that vaccinated individuals would be protected from infection (Tacket, et al., 1998, 2000). Taken together these results provide the basis for wider-scale clinical trials with these antigens to be conducted with the aid of international agencies in the near future.

Although high titer of mucosal and systemic antibodies following oral vaccination with edible vaccines is an important immune correlate, proof of concept requires the vaccine to provide long lasting protection against a pathogen challenge. Ethical considerations usually preclude clinical trials from directly assaying protection but in a few cases (e.g. Mason et al., 1998), surrogate protection assays were performed. In contrast, working with veterinary vaccines provides researchers an opportunity to assess the degree of immune protection more directly. An excellent example of this approach is represented by a series of papers originating from the group of Borca (Carrillo et al., 2001 and references therein).

3. "SECOND GENERATION" EDIBLE VACCINES

Multicomponent vaccines that provide protection against several pathogens are very desirable. An elegant approach to achieve this goal, based on epitope fusion to both subunits of the cholera toxin (CT), was recently demonstrated by Yu and Langridge (2001). CT provides a scaffold for presentation of protective epitopes of rotavirus and ETEC, acts as a vaccine candidate by its own right and as a mucosal adjuvant devoid of toxicity. The trivalent edible vaccine elicited significant humoral responses, as well as immune memory B cells and T-helper cell responses, important hallmarks of successful immunization (Yu and Langridge, 2001).

Commonly, foreign proteins in plants accumulate to relatively low levels (0.01–2% of total soluble protein). In the clinical trials described above, 100 g of raw potato tubers expressing LT-B of ETEC in three doses had to be consumed in order to overcome digestive losses of the antigen and to elicit a significant immune response (Tacket et al., 1998). Less immunogenic proteins would require even larger doses to be effective. Even with more palatable alternatives to potatoes (e.g. bananas), these accumulation levels may limit the practicality of edible vaccines

Two solutions to overcome this limitation are being explored. First, techniques to enhance antigen accumulation in plant tissues are being explored. These include, optimization of the coding sequence of bacterial or viral genes for expression as plant nuclear genes, and defining the subcellular compartment in which to accumulate the product for optimal quantity and quality. Several laboratories are also developing alternative expression systems to improve accumulation. For example the expression in plastids is advocated by some (Daniell et al., 2001; Ruf et al., 2001). Other systems involve plant viruses for expression of foreign genes (e.g. Nemchinov et al., 2000) or coat-protein fusions (e.g. Modelska et al., 1998) and even viral assisted expression in transgenic plants (Mor et al., 2002).

The second approach is to enhance the immunogenicity of the orally delivered antigens by using mucosal adjuvants. One such approach is making use of bacterial enterotoxins such as CT or LT (e.g. Yu and Langridge, 2001), mammalian and viral immunomodulators (Matoba, Soreq, Arntzen and Mor unpublished) as well as plant-derived secondary metabolites (Joshi and Arntzen, unpublished).

At the doorstep of the 21st century, the fear of a surge in naturally occurring epidemics is heightened by the threat of bio-terrorism. This new reality makes disease prevention through vaccination a necessity in our ever more interconnected world. Any tools we can master and all the tools we can afford will have to be employed. Technical problems and skeptics aside, edible-vaccines have passed the major hurdles of an emerging vaccine technology: We believe production of vaccines in transgenic plants will become an essential component in our disease prevention arsenal.

4. REFERENCES

Carrillo C., A. Wigdorovitz, K. Trono, M.J. Dus Santos, S. Castanon, A.M. Sadir, R. Ordas, J.M. Escribano and M.V. Borca. 2001. Induction of a virus-specific antibody response to foot and mouth disease virus using the structural protein VP1 expressed in transgenic potato plants. Viral Immunol. 14:49-57.

Daniell H., S.J. Streatfield and K. Wycoff. 2001. Medical molecular farming: production of antibodies, biopharmaceuticals and edible vaccines in plants. Trends Plant Sci. 6:219-226.

Haq T.A., H.S. Mason, J.D. Clements and C.J. Arntzen. 1995. Oral Immunization with a recombinant Bacterial antigen produced in transgenic plants. Science 268:714-716.

Kong Q., L. Richter, Y.F. Yang, C.J. Arntzen, H.S. Mason and Y. Thanavala. 2001. Oral immunization with hepatitis B surface antigen expressed in transgenic plants. Proc. Natl. Acad. Sci. U.S.A. 98:11539-11544.

Mason H.S., D.M.K. Lam and C.J. Arntzen. 1992. Expression of hepatitis B surface antigen in transgenic plants. Proc. Natl. Acad. Sci. U.S.A. 89:11745-11749.

Mason H.S., J.M. Ball, J.-J. Shi, X. Jiang, M.K. Estes and C.J. Arntzen. 1996. Expression of Norwalk virus capsid protein in transgenic tobacco and protein and its oral immunogenicity in mice. Proc. Natl. Acad. Sci. U.S.A. 93:5335-5340.

Mason H.S., T.A. Haq, J.D. Clements and C.J. Arntzen. 1998. Edible Vaccine Protects Mice Against *E. coli* Heat-labile Enterotoxin (LT): Potatoes Expressing a Synthetic LT-B Gene. Vaccine 16:1336-1343.

Modelska A., B. Dietzschold, N. Sleysh, F.Z. Fu, K. Steplewski, D.C. Hooper, H. Koprowski and V. Yusibov. 1998. Immunization against rabies with plant-derived antigen. Proc. Natl. Acad. Sci. U.S.A. 95:2481-2485.

Mor T.S., M.A. Gómez-Lim and K.E. Palmer. 1998. Edible vaccines: a concept comes of age. Trends Microbiol. 6:449-453.

Mor T.S., Y.-S. Moon, K.E. Palmer and H.S. Mason. 2002. Geminivirus vectors for high level expression of foreign proteins in plant cells. Biotechnol. Bioeng. in press.

Nemchinov L.G., T.J. Liang, M.M. Rifaat, H.M. Mazyad, A. Hadidi and J.M. Keith. 2000. Development of a plant-derived subunit vaccine candidate against hepatitis C virus. Arch. Virol. 145:2557-2573.

Richter L.J., Y. Thanavala, C.J. Arntzen and H.S. Mason. 2000. Production of hepatitis B surface antigen in transgenic plants for oral immunization. Nature Biotechnol. 18:1167-1171.

Ruf S., M. Hermann, I.J. Berger, H. Carrer and R. Bock. 2001. Stable genetic transformation of tomato plastids and expression of a foreign protein in fruit. Nature Biotechnol. 19:870-875.

Tacket C.O., H.S. Mason, G. Losonsky, J.D. Clements, S.S. Wasserman, M.M. Levine and C.J. Arntzen. 1998. Immunogenicity in humans of a recombinant bacterial-antigen delivered in transgenic potato. Nature Med. 4:607-609.

Tacket C.O. and H.S. Mason. 1999. A review of oral vaccination with transgenic vegetables. Microbes Infect. 1:777-783.

Tacket C.O., H.S. Mason, G. Losonsky, M.K. Estes, M.M. Levine and C.J. Arntzen. 2000. Human immune responses to a novel Norwalk virus vaccine delivered in transgenic potatoes. J. Infect. Dis. 182:302-305.

Thanavala Y., Y.-F. Yang, P. Lyons, H.S. Mason and C.J. Arntzen. 1995. Immunogenicity of transgenic plant-derived hepatitis B surface antigen. Proc. Natl. Acad. Sci. U.S.A. 92:3358-3361.

Yu J. and W.H. Langridge. 2001. A plant-based multicomponent vaccine protects mice from enteric diseases. Nature Biotechnol. 19:548-552.

PROTECTIVE EFFECT OF ORALLY ADMINISTERED HUMAN INTERFERON (HUIFN)-α AGAINST SYSTEMIC *LISTERIA MONOCYTOGENES* INFECTION AND A PRACTICAL ADVANTAGE OF HUIFN-α DERIVED FROM TRANSGENIC POTATO PLANT

Kenji Ohya[1], Takeshi Matsumura[2,3], Noriko Itchoda[3], Kazuhiko Ohashi[1], Misao Onuma[1] and Chihiro Sugimoto[4]

[1]Department of Disease Control, Graduate School of Veterinary Medicine, Hokkaido University, Kita-ku, Sapporo 060-0818, Hokkaido, Japan; [2]Hokkaido Green-Bio Institute, Naganuma, Yubari-gun 069-1311, Hokkaido, Japan (email: sugimoto@abihiro.ac.jp); [3]Research Institute of Biological Resources, National Institute of Advanced Industrial and Science Technology, Toyohira-ku, Sapporo 062-8517, Japan; [4]National Research Center for Protozoan Diseases, Obihiro 080-8555, Hokkaido, Japan

1. ABSTRACT

Type I interferon (IFN- α/β) is the first cytokine used for clinical applications against viral and neoplasmic diseases. Usually it is administrated by subcutaneous and intramuscular injection, but several studies have reported that orally administered Type I IFN is also effective against viral and autoimmune diseases. Therefore, we examined whether orally administered human IFN (HuIFN)-α can augment protection against systemic bacterial infection using *Listeria monocytogenes* infection in mice as an experimental model. Daily oral administrations for 6 days of 1000 international units (IU) of purified natural HuIFN-α reduced bacterial burden in spleen and liver from *L. monocytogenes*-infected mice. This effect was observed in the middle phase of *L. monocytogenes* infection, but not in the early phase of the infection. Effects of oral administration of HuIFN-α expressed in potato plant wcrc also cxamined in this infection model. Daily oral administrations of extracts of the transgenic potato tuber for 6 days decreased bacterial burden in the spleen. Lower doses of HuIFN-α in the extracts (20 IU/mouse/day) exerted a protective effect at almost the same level as the results achieved by the administration of 1000 IU of HuIFN-α. This result may be due to the 'bioencapsulation' effect for HuIFN-α by plant compartmentalization, which is one of the advantages of the plant expression system over other expression systems of recombinant proteins. Our present observation indicates the transgenic plants expressing cytokines can be used as feed/food and their additives in order to enhance natural immune responses in humans and animals.

2. INTRODUCTION

Cytokines play a major role in the homeostatic maintenance of the mucosal immune system where foreign antigens including infectious agents are encountered for initiation of immune response. For example, mammalian milk contains several cytokines and growth factors such as tumor necrosis factor-α and epidermal growth factor (Goldman, 2000). These molecules in milk may support the growth and development of the epithelium and the immune systems of the gastrointestinal tract.

I. K. Vasil (ed.), Plant Biotechnology 2002 and Beyond, 389-391.

Therefore, administration of cytokines to the mucosal surface should be useful as adjuvants or immune stimulatory agents against infectious diseases (Cummins et al., 1999; Rollwagen and Baqar, 1996).

Interferon (IFN)-α, with more than 20 subtypes, is secreted from virus-infected cells and exhibits anti-viral and tumorcidal activity. Type I IFN (IFN-α/β) is the most widely used cytokine for patients with viral and neoplasmic diseases. Recently, it has been reported that type I IFN shows not only anti-viral activity but also plays a key role in the regulation of the systemic immune response such as dendritic cell maturation (Biron, 2001). We have succeeded to express type I IFN in potato plant (Ohya et al., 2001) as a potential source of a plant-derived pharmaceutical protein.

In this study, we examined whether orally administered IFN-α is effective for non-viral pathogens using a bacterial infection model in mice. *Listeria monocytogenes*, a facultative intracellular pathogen, causes human and animal listeriosis such as encephalitis and abortion, and is widely used as a infection model to analyze interactions between host and intracellular pathogens. Potentials of transgenic plants expressing cytokines in enhancing natural immune responses against infectious diseases are discussed .

3. RESULTS AND DISCUSSION

3.1. Orally Administered Huifn-α Reduces L. Monocytogenes Growth

Natural HuIFN-α was diluted in 20 μl of phosphate-buffered saline (PBS) containing 0.03% bovine serum albumin and administered to oral cavity of mice (BALB/c, female, 6 wks) after 6 hrs fast every day till sample collection. After 6 days of the treatment, 5 x 10^5 colony forming unit (CFU) of *L. monocytogenes* EGD strain was inoculated to mice intraperitoneally . Bacterial numbers in spleen and liver were measured after 2 and 4 days post infection (d.p.i.). No difference in the numbers was observed between PBS- and HuIFN-α-treated groups on 2 d.p.i. However, on 4 d.p.i., bacterial burden in spleen of HuIFN-α-treated mice (1000 IU/day) was reduced compared to that in PBS-treated mice ($\log_{10}$ CFU = 3.85 ± 0.25 [IFN] vs. 4.62 ± 0.38 [PBS]; $p < 0.005$) and liver (4.41 ± 0.37 vs. 5.37 ± 0.77; $p < 0.05$), respectively. In the early phase of infection (0 - 2 d.p.i.), *L. monocytogenes* reaches organs s such as spleen and liver, and start to grow in epithelial cells and phagocytes. Furthermore, in the middle phase (2 - 4 d.p.i.) of infection, number of *L. monocytogenes* in organs reaches the maximum level, but phagocytes start to kill intracellular bacteria vigorously during this stage. Hence, orally administered HuIFN-α cannot protect *L. monocytogenes* infection at the early phase, but is effective for rapid elimination of bacteria in organs during the middle phase of infection. Although the mechanism involved in enhancement of host protective systems by oral IFN-α administration is still unknown, this cytokine may be able to potentiate systemic innate immune response at relatively low dose.

3.2. Effects Of Oral Administration Of Huifn-α Expressed In Potato Plant

As we succeeded in expression of bioactive HuIFN-α in transgenic potato plant (Ohya et al., 2001), it was examined whether HuIFN-α derived from the transgenic potato plant shows the same effect as the natural molecule. Forty μl of the plant leaf extract, containing 20 IU HuIFN-α were daily administered to oral cavity of mice. After 6 days of administration of the plant extract, 5 x 10^5 CFU *L. monocytogenes* was inoculated intraperitoneally. Oral administration of the extract of the HuIFN-α expressing potato plant decreased bacterial burden in spleen on 5 d.p.i. compared to mice treated with extract of untransformed potato plant (3.77 ± 0.15 [IFN/potato] vs. 4.24 ± 0.34 [untransformed potato]; $p < 0.025$). Additionally, even low concentration of HuIFN-α (20 IU/mouse/day) exerted the same effect, compared to the protection level that achieved by the treatment with PBS-diluted natural HuIFN-α (1000 IU/mouse/day), which may be due to bioencapsulation with plant cells or tissues.

4. CONCLUSIONS

This study indicated that transgenic plants expressing HuIFN-α or other cytokines can be used as feed/food or their additives in order to potentiate natural immune responses in human and animals.

5. REFERENCES

Biron, C.A. 2001. Interferons alpha and beta as immune regulators--a new look.. Immunity 14: 661-664.

Cummins, J.M., Beilharz, M.W., and Krakowka, S. 1999. Oral use of interferon. J. Interferon Cytokine Res. 19: 853-857.

Goldman, A.S. 2000. Modulation of the gastrointestinal tract of infants by human milk. Interfaces and interactions. An evolutionary perspective. J. Nutr. 130(2S Suppl): p. 426S-431S.

Ohya, K., Matsumura, T., Ohashi, K., Onuma, M., and Sugimoto, C. 2001. Expression of two subtypes of human IFN-alpha in transgenic potato plants. J Interferon Cytokine Res, 21: 595-602.

Rollwagen, F.M. and Baqar, S. 2001. Oral cytokine administration.. Immunol. Today 17: 548-550.

PRODUCTION OF APROTININ IN TRANSGENIC MAIZE SEEDS FOR THE PHARMACEUTICAL AND CELL CULTURE MARKETS

Donna Delaney, Joseph Jilka, Donna Barker, Philip Irwin, Miranda Poage, Susan Woodard, Michael Horn, Amanda Vinas, Kathy Beifuss, Mark Barker, Barry Wiggins, Carol Drees, Robin Harkey, Zivko Nikolov, Elizabeth Hood, and John Howard
ProdiGene, 101 Gateway Blvd., Suite 100, College Station, TX 77845 USA (e-mail: ddelaney@prodigene.com)

Keywords Molecular pharming, aprotinin, maize

1. INTRODUCTION

Aprotinin is a serine protease inhibitor found in several bovine organs that has a number of applications in both the pharmaceutical and cell culture markets. In cell culture it is used as a component of serum-free media to preserve the integrity of recombinant proteins during fermentation and downstream purification. As a pharmaceutical, it is used to reduce blood loss during major surgeries, such as cardiopulmonary bypass surgery, and as a component of fibrin sealant kits for sutureless wound closure. Current commercial production of aprotinin is from bovine lungs with the inherent concern that contamination with pathogens such as BSE may occur. Using transgenic maize as a production vehicle for aprotinin would provide an inexpensive, easily scalable source of the protein that would be free of human pathogens. Transgenic maize plants expressing aprotinin were generated using both particle bombardment and *Agrobacterium*-mediated transformation. In addition, transformations were performed using three different maize genotypes and seven different constructs incorporating three promoters, two targeting sequences and multiple plant transformation units.

2. RESULTS AND DISCUSSION

T1 seeds from each plant were analyzed using an ELISA specific for bovine aprotinin. Expression levels (% total soluble protein, %TSP) for plants produced by *Agrobacterium* (constructs 2-7) were as much as 100-fold higher than for those produced by bombardment (construct 1)

I. K. Vasil (ed.), Plant Biotechnology 2002 and Beyond, 393-394.

(Table 1). Constructs with a promoter that is active mainly in the embryo had higher expression than either constitutive or endosperm promoters. Targeting to the cell wall rather than the cytoplasm resulted in over a 5-fold increase in expression (compare constructs 3 and 4). For construct 4, the average expression of plants in the HiII and HiII X Lan. genotypes was higher than for HiII X SS. However, ears from HiII X SS and HiII X Lan. produced an average of twice as many T1 seeds as HiII alone. The two constructs containing multiple copies of the aprotinin gene (constructs 5 and 6) were significantly different from each other, but not higher than constructs containing only one copy of the gene. Aprotinin purified from transgenic corn seeds is biochemically identical to aprotinin purified from bovine lung tissue. The high expression levels achieved make maize an efficient manufacturing vehicle for commercial production of aprotinin.

Table 1. Expression of aprotinin in transgenic corn seeds from seven different constructs incorporating different promoters, targets, host genotypes and gene copy numbers.

Construct*	Tissue	Target	Host Genotype†	No. of PTU's‡	Highest Exp. Lvl. (%TSP)^x	Average Exp. Lvl. (%TSP)^y
1	constitutive	cell wall	HiII	1	0.1	0.02d
2	constitutive	cytoplasm	HiII	1	1.1	0.27c
3	embryo	cytoplasm	HiII	1	3.1	0.32c
4	embryo	cell wall	HiII	1	8.9	1.75a
4	embryo	cell wall	HiII X SS	1	6.4	1.15b
4	embryo	cell wall	HiII X Lan.	1	2.7	1.98a
5	embryo	cell wall	HiII	2	3.2	2.10a
6	embryo & endosperm	cell wall	HiII	2	3.9	0.56c
7	endosperm	cell wall	HiII	1	0.3	0.12c

* Construct #1 was integrated using particle bombardment, while for construct 2-7 *Agrobacterium* was used.

† Three different host genotypes were used – HiII, HiII crossed with an elite Stiff Stalk inbred (HiII X SS), and HiII crossed with an elite Lancaster inbred (HiII X Lan.).

‡ PTU's = Plant transformation units = No. of copies of the aprotinin gene.

^{x} Expression level (% total soluble protein) of the highest single T1 seed.

^{y} Average expression of T1 seed (%total soluble protein) over all events. Different letters indicate the numbers are significantly different at $\alpha = 0.05$.

PLANT AND BACTERIAL PRODUCTION OF ENGINEERED ANTIBODIES FOR PHARMACOLOGICAL USE IN ONCOLOGY

P. Galeffi, I. Pietraforte, A. Lombardi, F. Novelli, M. Sperandei, P. Giacomini and C. Cantale
ENEA CR Casaccia - UTS BIOTEC, Via Anguillarese 301,00060 Rome, Italy (e-mail: galeffi@casaccia.enea.it)

Keywords Engineered antibodies, Herb2, plant factory, single chain fragment

Genetically engineered plants are investigated to produce therapeutic proteins with many economic and qualitative benefits, including reduced health risks from human/animal pathogen contamination. They give comparatively high yields, production is in seeds or other plant tissues, and are convenient for long periods of storage, and for distribution and management. The use of the existing agricultural infrastructure for the cultivation, harvesting, storage, and processing of transgenic crops would require relatively little capital investment, making the commercial production of biopharmaceuticals an exciting prospect. Several proteins, enzymes and antibodies are produced in plants and used in clinical trials.

Among antibodies, single chain variable fragment (scFv) represent a particularly promising class for biomedical applications. Many scFvs have been expressed in E. coli for several purposes, but only a few scFvs have been tested in the plant system, e.g. a scFv against Hodgkin's lymphoma in tobacco and a scFv-T84.66 against carcinoembryogenic antigens in various plants. The present work was addressed to produce, in *E.coli* and in *Nicotiana tabacum*, a functional scFv, named scFv-800E6, able to recognise the Herb2 antigen related to breast and ovary human cancers, both *in vitro* and *in vivo.*

PCR primers specific for the hybridoma cell line (murine Mab) 800E6 were used to amplify the VH and VL domains that were subsequently mounted into the opportune plasmid cassettes, with a linker consisting in $(GGGGS)_3$, and finally inserted into the two expression vectors.

The usual cloning and transforming protocols were used, with minor changes which took into account the specificity of this system. All the steps were confirmed by nucleotide sequencing and western

I. K. Vasil (ed.), Plant Biotechnology 2002 and Beyond, 395-396.

blotting.The pHEN bacterial expression vector was selected to clone scFv-800E6 in the *E. coli* HB2151 strain. Western blotting and flow cytometry sorting analyses (FACS) were used to confirm the presence and the functionality of the scFv-800E6 in the bacterial protein extracts. The specificity of the bacterial protein extracts was tested using several cell lines including Her-2/neu+ (SK BR3 and MDAmB) and Her-2/neu-(TD 47D, MCF 7, SK, Endo9, Carlini and colo 38). The pBG-cFv-BIN plant expression vector was selected to clone 800E6 scFv in the *Agrobacterium tumefaciens* GV3101 strain. *Agrobacterium* mediated transformation of *Nicotiana tabacum* leaf disks was carried out following standard procedures.

Seventy-four putative transgenic plants were selected for RNA extraction and RT-PCR analysis to assess scFv expression. Forty-two positive plants were used to analyse the protein. Western blotting and FACS analyses were carried out to verify the presence and the functionality of the plant produced protein. Two protein fractions, crude and partially purified, were analysed by FACS demonstrating a certain ability to bind to HER2+ cells (SKBR-3). Frozen sections of mammary tumoural tissues from human surgery biopsies were used to test the ability of the partially purified scFv-800E6 from *E.coli* to specifically bind the antigen, over-expressed on these cell membranes, as a base for possible diagnostic application. The membrane localization of a red fluorescence from a FITC-conjugate anti-mouse IgG assesses the binding capacity of scFv-800E6 to recognize Her-2/*neu* in these clinical section tissues.

In conclusion, this scFv engineered antibody retains its functionality when produced either in *E. coli* or in plants. More work is necessary to optimise the performance, especially when produced in plants. The final yields appear to be low and more attention should be paid to the overall protein recovery in view of medical application.

TOBACCO CHLOROPLASTS AS A PLATFORM FOR VACCINE PRODUCTION

Pal Maliga[1], Hiroshi Kuroda[1], Sylvie Corneille[1], Kerry Lutz[1], Arun K. Azhagiri[1], Zora Svab[1], John Tregoning[1,2], Peter Nixon[2] and Gordon Dougan[2]

[1]Waksman Institute, Rutgers University, Piscataway, NJ, USA (e-mail: maliga@waksman.rutgers.edu) and [2]Department of Biological Sciences and Centre for Molecular Microbiology and Infection, Imperial College of Science and Technology, London SW7 2AY, UK

Keywords Marker elimination system, plastid expression cassettes, plastid transformation vector, tetanus vaccine

1. INTRODUCTION

Initial developments in the field of plant-based vaccines have been limited by low level of immunogen expression from nuclear genes. An alternative method is to express vaccine antigens from the chloroplast genome. Recently, we developed a production system for expression of recombinant proteins in tobacco chloroplasts including vectors, expression cassettes, and a system for marker gene elimination. This development is reviewed here. For general reviews on plastid transformation see (Bock, 2001; Maliga, 2002; Staub, 2002).

2. ENGINEERING FOR HIGH-LEVEL PROTEIN EXPRESSION

Plastid transformation is based on DNA delivery by the biolistic process, homologous recombination between the transformation vector and the plastid genome and subsequent sorting out of transformed genomes. The Plastid Repeat Vector (pPRV) series targets insertions in the *trnV-rps12* intergenic region (Zoubenko et al., 1994). Plastid transgenes are typically expressed in a 5' PL- (for Promoter and Leader) and a 3' T-cassette (for Terminator). The PL cassette includes a promoter and translation control sequences. The T-cassette encodes the mRNA 3'-UTR. Gene construction is facilitated by a gene assembly system that relies on having the same restriction sites at the boundaries of the PL and T cassettes and coding regions (Figure 1). New

I. K. Vasil (ed.), Plant Biotechnology 2002 and Beyond, 397-400.

transgenes can be readily obtained by shuffling regulatory elements and coding regions.

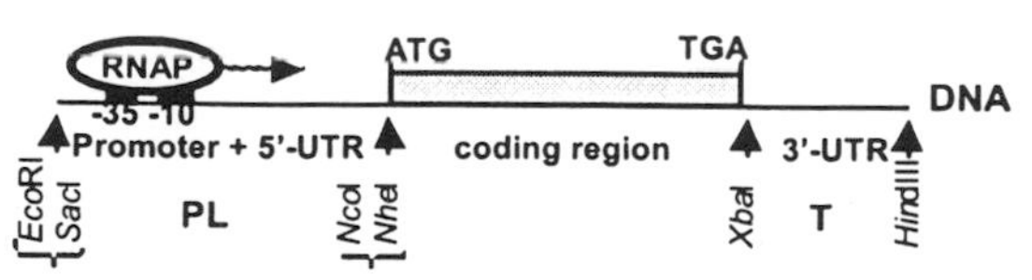

Figure 1. Modular design of plastid PL and T expression cassettes. PL cassette encodes a promoter and 5'-UTR. T cassette encodes the 3'-UTR.

There are a few empiric rules that should be followed to obtain high-level protein accumulation. The first requirement is expression of the transgene from a strong promoter to ensure high levels of mRNA. Our choice is the strong rRNA operon promoter. Targets for engineering are the 5'-UTR and coding region N-terminus. The inclusion of 14 *rbcL* or *atpB* N-terminal amino acids with the cognate 5'-UTR increased NPTII levels from 4.7% to 10.8% (plasmids pHK35 and pHK34) and from 2.5% to 7% (Plasmids pHK31 and pHK30), respectively. Silent mutations downstream of the *rbcL* AUG in the PL cassette of plasmid pHK64 reduced translation efficiency 35-fold in the absence of a change in protein or mRNA stability (Kuroda and Maliga, 2001b). Expression of NPTII from alternative T7 phage translation control signals (Nt-pHK38 and Nt-pHK39) yielded a 7-fold difference in mRNA level and 100-fold difference in protein accumulation. The highest level of NPTII obtained was >23% of TSP (Kuroda and Maliga, 2001a).

To characterize expression of heterologous proteins in plastids, we explored the feasibility of producing a mucosal tetanus vaccine. TetC is a non-toxic 47kDa polypeptide fragment shown to induce a protective immune response. Expression of TetC in *E. coli* was limited by the unfavorable codon bias of the highly AT-rich *Clostridium tetani* coding sequence that could be overcome by expressing a synthetic gene (Makoff et al., 1989). Since proteins expressed in *E. coli* may contain toxic cell wall pyrogens, TetC was expressed in the non-toxic host *S. cerevisiae* (Romanos et al., 1991). The AT-rich *C. tetani* DNA could not be expressed in yeast due to the presence of several fortuitous polyadenylation sites which gave rise to truncated mRNAs. TetC accumulation was obtained from a synthetic (high-GC) gene lacking the polyadenylation sites. However, the yeast-produced TetC was inactive as an immunogen due to glycosylation. We have found that in tobacco chloroplasts both the high-AT bacterial and high-GC synthetic mRNAs are stable. Significant TetC accumulation was obtained from both genes, 25% and 10% of TSP, respectively, proving the versatility of plastids for the expression of both high-AT and high-GC genes. TetC in plastids was not glycosylated, as the enzymatic machinery for glycosylation is localized to the secretory pathway. Immunization of mice with the plastid-produced TetC induced protective levels of TetC antibodies, confirming the potential of chloroplasts for the production of a plant-based mucosal vaccine.

3. BIOSAFETY

Transgenes expressed in plastids are naturally contained, as chloroplasts are not transmitted by pollen (see reviews). In addition, the marker gene can be eliminated by the CRE-*lox* site-specific recombination system (Corneille et al., 2001; Hajdukiewicz et al., 2001)(Figure 2).

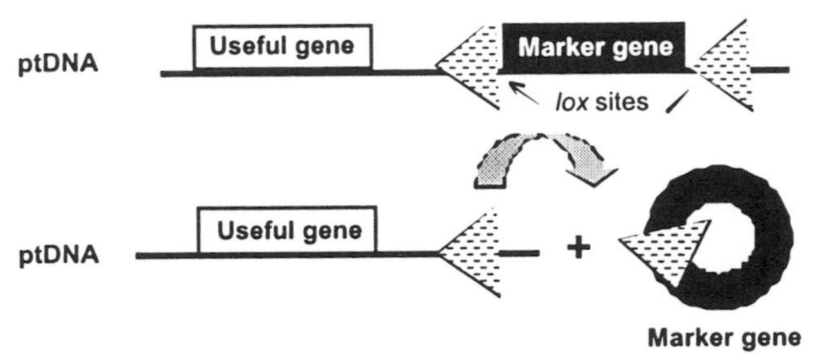

Figure 2. Elimination of marker genes flanked by lox *sites with the CRE site-specific recombinase.*

4. ADVANTAGES OF CHLOROPLASTS FOR VACCINE PRODUCTION

Chloroplasts are a versatile system for vaccine production. It was found that in chloroplasts both the high-AT and high-GC mRNAs are stable. Furthermore, there are no rarely used codons in plastids. Thus, in plastids both bacterial and human genes can be directly expressed without re-synthesis and codon modification. Furthermore, multi-subunit, complex proteins may be expressed from polycistronic mRNAs and proteins with disulfide bridges are folded correctly in chloroplasts. A unique feature of the plastid expression system is the lack of glycosylation (Maliga, 2002; Staub, 2002). Chloroplast-produced vaccines would be inexpensive, easy to produce and more stable to heat, features shared by all plant-based vaccine production systems (Ma, 2000; Walmsley and Arntzen, 2000; Stoger et al., 2002).

5. REFERENCES

Bock R. 2001. Transgenic plastids in basic research and plant biotechnology. J. Mol. Biol. 312:425-438.

Corneille S., K. Lutz, Z. Svab, and P. Maliga. 2001. Efficient elimination of selectable marker genes from the plastid genome by the CRE-*lox* site-specific recombination system. Plant J. 72:171-178.

Hajdukiewicz P.T.J., L. Gilbertson, and J.M. Staub. 2001. Multiple pathways for Cre/*lox*-mediated recombination in plastids. Plant J. 27:161-170.

Kuroda H., and P. Maliga. 2001a. Complementarity of the 16S rRNA penultimate stem with sequences downstream of the AUG destabilizes the plastid mRNAs. Nucleic Acids Res. 29:970-975.

Kuroda H., and P. Maliga. 2001b. Sequences downstream of the translation initiation codon are important determinants of translation efficiency in chloroplasts. Plant Physiol. 125:430-436.

Ma J.K. 2000. Genes, greens, and vaccines. Nat. Biotechnol. 18:1141-1142.

Makoff A.J., M.D. Oxer, M.A. Romanos, N.F. Fairweather, and S. Ballantine. 1989. Expression of tetanus toxin fragment C in *E. coli*: high level expression by removing rare codons. Nucleic Acids Res. 17:10191-10202.

Maliga P. 2002. Engineering the plastid genome of higher plants. Curr. Opin. Plant Biol. 5:164-172.

Romanos M.A., A.J. Makoff, N.F. Fairweather, K.M. Beesley, D.E. Slater, F.B. Rayment, M.M. Payne, and J.J. Clare. 1991. Expression of tetanus toxin fragment C in yeast: gene synthesis is required to eliminate fortuitous polyadenylation sites in AT-rich DNA. Nucleic Acids Res. 19:1461-1467.

Staub J.M. 2002. Expression of recombinant proteins via the plastid genome. In: Parekh SR, Vinci VA (eds) *Handbook of industrial cell culture: mammalian, microbial and plant cells.*, Humana Press Inc., Totowa, NJ. Pp 261-280.

Stoger E., M. Sack, R. Fischer, and P. Christou. 2002. Plantibodies: applications, advantages and bottlenecks. Curr. Opin. Biotechnol. 13:161-166.

Walmsley A.M., and C.J. Arntzen. 2000. Plants for delivery of edible vaccines. Curr. Opin. Biotechnol. 11:126-129.

Zoubenko O.V., L.A. Allison, Z. Svab, and P. Maliga. 1994. Efficient targeting of foreign genes into the tobacco plastid genome. Nucleic Acids Res. 22:3819-3824.

NUTRITIONAL IMPROVEMENT OF RICE TO REDUCE MALNUTRITION IN DEVELOPING COUNTRIES

Ingo Potrykus
Federal Institute of Technology (ETH), Zurich, Switzerland (e-mail: ingo@potrtykus.net)

1. THE SOCIAL CHALLENGE

Malnutrition disorders are the cause for 24,000 deaths per day. Golden Rice represents a genetic engineering concept for development of nutrient-dense staple crops as contribution to reduction of malnutrition in developing countries. Major micronutrient deficiency disorders concern protein/energy, iron/zink, vitamin A, and iodine. These deficiencies are especially severe, where rice is the major staple. Traditional interventions such as distribution, fortification, dietary diversification, and measures against infectious diseases are very helpful in reducing deficiency disorders, but they have not, and they probably can not, solve the problem. Statistics demonstrate that despite enormous efforts in applying these traditional interventions, we are still faced with 2.4 billion iron-deficient women and children, and 400 million vitamin A-deficient children per year. Approaches towards nutrient-dense staple crops offer an alternative opportunity to complement the traditional interventions in a sustainable manner. And genetic engineering has the potential to substantially enhance breeding for nutrient staple crop varieties.

2. THE SCIENTIFIC CHALLENGE

Protein deficiency relates both to the amount and the quality (the content in essential amino acids) of dietary protein. Rice is rich in carbohydrates (energy), but low in protein (and vitamins and minerals) and provides, with a typical daily diet of 300 g, only 10% of the required essential amino-acids. Genetic engineering of an ideal balanced mixture of the missing nine essential amino acids isoleucine, leucine, lysine, methyonine, cysteine, phenylalanine, tyrosine, threonine, tryptophane, and valine, would be beyond our current technical ability. However, thanks to *Asp*-1, a synthetic gene, developed by Jesse Jaynes, and coding for an ideal high-quality storage protein (providing a balanced mixture of all nine essential amino acids), offered the unique opportunity to approach the otherwise unfeasible task. It was, however, completely open, whether rice endosperm would provide for the necessary biochemical background to assemble such a protein if the information would be provided and activated. To test this question, the *Asp-1* gene was placed under endosperm-specific control, linked to an appropriate target sequence for import into the endosperm protein storage vesicles, and

I. K. Vasil (ed.), Plant Biotechnology 2002 and Beyond, 401-406.

transformed into rice (Japonica TP309). Surprisingly and fortunately, the *Asp-1*-transgenic rice plants recovered, according to Western data, accumulate the Asp-1 protein in their endosperm to a range of concentrations, thus providing the mixture of the essential amino acids required. Detailed biochemical analysis has to verify this preliminary assumption, and has to show, whether the concentrations achieved, are nutritionally relevant.

Iron deficiency caused by a rice diet is the consequence of (1) far too low amounts in rice of iron, (2) the presence of an extremely potent inhibitor of iron re-sorption (phytate), and (3) lack of any iron re-sorption-enhancing factors in a vegetative diet. Our genetic engineering task for the endosperm was, therefore, (1) to increase iron content, (2) to reduce the inhibitor, and (3) to add re-sorption-enhancing factors. Transgenic ferritin (from *Phaseolus vulgare*) increased, so far, the iron content by two-fold; a transgenic metallothionin (from *Oryza sativa*) led to a seven-fold increase in an iron re-sorption-enhancing cystein-rich protein, and a transgene coding for a heat-stable phytase (from *Aspergillus fumigatus*) produced high inhibitor-degrading phytase activity, which however, to date, did not maintain its heat-stability character (Lucca et al., 2001). Aiming at phytate degredation after cooking was an essential part of our strategy (to leave the phytate required for germination untouched). We are thus working on alternative strategies and further experiments on the phytate problem are in progress.

Vitamin A-deficiency in rice consuming populations is due to the fact that milled rice is totally devoid of any provitamin A, and that the people concerned are too poor to afford a diversified diet. The situation is especially severe where children are raised on rice gruel. To add provitamin A to rice endosperm required the engineering of a complete biochemical pathway, a task considered unfeasible throughout the course of our experiments. The final success was possible because (a) the complementary expertise of two laboratories (Dr. Peter Beyer, University of Freiburg provided the scientific knowledge and the necessary genes, and my group was specialized in genetic engineering of rice), (b) long-term public funding (from e.g. Swiss agencies and The Rockefeller Foundation), (c) a firm "engineering spirit" to solve the problem, and (d) "good fortune" with the biology of the rice endosperm. The introduction of transgenes for phytoene synthase (*Narcissus*), a phytoene / ξ-carotene double-desaturase (*Erwinia*), and lycopene cyclase (*Narcissus*), to everybody's surprise, completed the biochemical pathway leading from the latest available precursor geranyl-geranly-pyrrhophosphate to β-carotene (pro-vitamin A). Biochemical analysis of the polished rice kernels confirmed that the "golden" endosperm colour was due to varying amounts of provitamin A and further terpenoids of dietary interest (such as lutein and zeaxanthin; Ye et al., 2000). The concentration of 1.6 μg/g may, according to the calculation of an experienced vitamin A nutritionalist (Prof. Robert Russell, Boston), be sufficient to prevent vitamin A-deficiency disorders from a daily diet of 200g of Golden Rice. Nutritional studies with human volunteers, testing this hypothesis are in preparation. Conclusive data will, however, not be available before 2004.

For a "normal" scientist, these scientific successes would have been the completion of their tasks. However, for scientists, motivated to contribute via a humanitarian project to reduction in malnutrition in developing countries, this was only the beginning of a series of further, and rather unusual and exhausting challenges.

3. THE CHALLENGE OF FREE DONATION TO DEVELOPING COUNTRIES

The Golden Rice project had been designed from the beginning as a humanitarian project for poor people in developing countries. To reach this goal and to contribute to relieve from malnutrition in poor populations in developing countries, this scientific success has to be passed on to the subsistence farmers and the urban poor free of charge and limitations. To be in a position to give away the technology, we took care throughout the project to use public funding only. However, independent from our invention (which we could give away) the basic genetic engineering technology used to develop provitamin A-rice had to make use of a great number of patented technologies. "Freedom-to-operate for humanitarian use", the necessary basis for variety development by partner institutions in developing countries, became, therefore, a major undertaking. The inventors solved the problem thanks to an alliance with the ag-biotech industry, which is based on the agreement that transfers the rights for commercial exploitation to ag-biotech industry, which in turn supports the humanitarian project. The difficult problem of definition of "humanitarian project" was solved by defining it "as income from Golden Rice per farmer or trader in developing countries below $ 10,000 p.a." This definition safely includes the target population. Thanks to this agreement the technology is now available via free licences to public research institutions for breeding, variety development and de-novo transformation. Transfer of the technology requires signature on a sub-sub-licence agreement with the inventors (not with ag-biotech industry!). Such agreements have been signed, so far, with IRRI (01) and PhilRice (02) (Philippines), Cuu Long Delta Rice Research Institute (03) (Vietnam), Department of Biotechnology, Delhi (04), DRR, Hyderabad (05), IARI, New Delhi (06), UDSC, New Delhi (07), TNAU, Tamil Nadu (08) (India), Institute of Genetics, Academia Sinica, Beijing (09) and National Key Laboratory of Crop Genetic Improvement, Wuhan (10) (China), Agency for Agricultural Research and Development, Jakarta (11) (Indonesia), and DSIR, Brumeria, Pretoria (12) (South Africa). These institutions constitute the still growing International Humanitarian Golden Rice Network.

4. THE CHALLENGE OF SAFE TECHNOLOGY TRANSFER AND VARIETY DEVELOPMENT

To ensure proper handling of the GMO material, a "Humanitarian Board" has been establsihed, to supervise the choice of partners, to support further improvement, to overlook needs, availability, bio-safety, and socio-economic assessments, to coordinate the activities in the different countries, to support fund raising from public resources, to support deregulation, to facilitate exchange of information, and to mediate information of the public and general support for the humanitarian project. Members of the Board include G.Toenniessen (Rockefeller Foundation), A.Dubock (Syngenta), W.Padolina (IRRI), R.M.Russell (USDA), H.E.Bouis (IFPRI), G.Khush (IRRI), K.Jenny (Indo-Swiss Collaboration in Biotechnology), A.Kratticker (Cornell), and the inventors P.Beyer and I.Potrykus (Chairmen). Variety development in the partner institutions is via backcrossing into, or direct transformation of popular local varieties. Backcrossing from the experimental Japonica rice line into the essential Indica rice lines requires ca. eight generations (or three years). Direct transformation may be faster and has already been achieved into a series of Indica varieties (including the very popular Indica variety IR64) by the Vietnamese and Philippine partner institutions.

5. THE CHALLENGE OF A RADICAL GMO OPPOSITION AND CONSUMER ACCEPTANCE

Golden Rice has, unfortunately, become a key topic in the fight between proponents and opponents of plant biotechnology in food production. A radical GMO opposition is one of the last major stumbling blocks, with the potential to prevent that the poor in developing countries might benefit from the project. Greenpeace and numerous other NGO's are determined to prevent the development and use of Golden Rice. Their major reason for opposition is, obviously, the fact that they see Golden Rice as a "Trojan Horse", opening the road for GMO technology in developing countries. As the opposition has lost, however, with the Golden Rice case, all their standard arguments used so far (Golden Rice e.g. benefits the poor, not industry; it has been developed in the public domain not by industry; it will be available for the poor free of costs and limitations; subsistence farmers can use part of their harvest for the subsequent sowings; Golden Rice cultivation does not require any additional inputs; environmentalists can not conceive of realistic risks to the environment; etc.) and as the public and the media understand the moral dimension of the project, the opposition is in a difficult situation, and is, therefore, trying to bypass this moral dilemma, by claiming that Golden Rice is useless anyhow, because children have to eat 3.75 kg/day. This is definitely wrong, but data to prove the assumption, that, probably, 200g/day will be sufficient, unfortunately, will be available only early in 2004.

6. THE CHALLENGE OF DEREGULATION

It is widely accepted that food derived from transgenic plants must successfully pass all requirements set up by regulatory authorities and we agree in principle. We get, however, more and more the impression that sticking to the existing regulatory framework may be sufficient to severely delay or even prevent the use of Golden Rice for the reduction in malnutrition-caused diseases in poor populations of developing countries. The regulatory framework has become so extensive, and the requirements so sophisticated, that the financial input alone required, becomes rather unrealistic for a "humanitarian" project. A further severe limit is the time factor: Golden Rice is a reality since February 1999. Since Spring 2002 the trait is available in IR64, a popular Indica rice variety grown in many developing countries. New experimental adjustments are, however, required to make Golden Rice more amenable to the regulatory procedure. All this has the consequence that exploitation of the technology is delayed for many more years. In view of the fact that we are not discussing about sensitivities of well-fed European consumers, but about large populations who's life and health depends upon contributions to sustainable reduction in malnutrition (causing 24 000 deaths per day), the question is justified, what level of sophistication in regulation is scientifically and morally justified, and how much of it is just the consequence of weak politicians giving in to the pressure of activists, operating extremely successfully on an emotional level. The Golden Rice Humanitarian Board is determined to perform the humanitarian project at the highest regulatory levels, however the question remains: is it justified to delay use of the technology for many years because of some weak and hypothetical risks, if the consequence is, that many thousands are dying, or have severest health problems (e.g. irreversible blindness), who otherwise could live a healthy and productive life? What is more important to our society - a regulatory framework for minor and mostly hypothetical risks, or life and health of underprivileged human beings? Deregulation should balance risks and benefits! Probably, it is time to re-evaluate the regulatory framework (at least in context with humanitarian projects) to identify those features that are essential and meaningful, and those which are there just in response to political pressure.

We have now, most probably, assembled in one plant nine trans-genes to combine "provitamin A" + "high iron" + "high quality protein" + "insect resistance". Advisors experienced in regulatory affairs inform us, that it will be impossible to receive deregulation for such a transgenic plant - it will be difficult enough to get deregulation for "provitamin A"-rice alone, because the principle of "substantial equivalence" does not apply. Do we really want to ignore the solutions genetic engineering can provide for malnutrition in developing countries because of the sensitivities of a well-fed European minority?

7. REFERENCES

Lucca, P. et al. 2001. Theoret. Appl. Genet. 102:392-396.

Ye, X. et al. 2000. Science 287:303-305.

GENOMIC APPROACH TO ALTERING PHYTOCHEMICALS IN TOMATO FOR HUMAN NUTRITION

Helena Mathews
Exelixis Plant Sciences, 16160 SW Upper Boones Ferry Road, Portland, OR 97224-7744 USA
(email: hmathews@exelixis.com)

Keywords tomato, genomics, activation tagging, mutants, *MYB* transcription factor

1. INTRODUCTION

Phytochemicals are non-nutrient, physiologically active plant components present in relatively small amounts compared to the macronutrients (fats, carbohydrates and proteins). Phytochemicals have recently been the focus of intense research efforts because of their cancer preventive properties. Epidemiological studies have demonstrated that populations consuming phytochemicals through a plant-based diet high in grains, legumes, fruits and vegetables have a markedly reduced incidence of cancer. Only recently have biological scientists begun to identify the mechanism through which phytochemicals reduce cancer risk. Some phytochemicals like the organosulfur compounds in *allium* vegetables such as garlic and onion, detoxify carcinogens and thus help the body to eliminate them. Others such as carotenoids in yellow, red, and green vegetables, function as antioxidants by scavenging free radicals that can attack and damage cellular membranes and DNA. Lycopene in tomatoes is antoher example of a phytochemical that acts as an antioxidant, and has been shown to be especially effective in preventing prostate cancer.

Genomic tools have immense potential to enhance the discovery and manipulation of genes that are involved in the health promoting phytochemicals in plants. Tomato, a phytochemical rich plant has been the subject of our research using the T-DNA mediated activation tagging. Although a large number of plant genes have been sequenced and mapped, functional information is available only for a very small number of them. Activation tagging technology (ACTTAG™) has emerged as a powerful tool in the elucidation of plant gene function (Weigel et al., 2000; Huang et al., 2001; Zubko et al., 2002). It is based on the gene activation by transcriptional enhancers causing ectopic expression of genes in the vicinity of the T-DNA

I. K. Vasil (ed.), Plant Biotechnology 2002 and Beyond, 407-411.

insertion site. ACTTAG™ enables collection of both gain of function and loss of function mutations.

Micro-Tom is a true dwarf tomato (*Lycopersicon esculentum*) plant that was originally developed at the University of Florida (Scott and Harbaugh, 1989). The Micro-Tom genome differs from standard tomato by only two mutations. Originally intended for the decorative horticulture market, its potential for use as a model genetic system was described by Miessner et al. (1997, 1999). Its small size and short generation time along with high transformation rate (Meissner et al., 2000) make it very attractive as a high through put system for the application of activation tagging technology. At Exelixis Plant Sciences we have developed several thousand ACTTAG™ lines and a small fraction of them are analyzed genetically and biochemically. Our preliminary results of these mutants in the context of important phytochemicals are discussed here.

2. MATERIALS AND METHODS

ACTTAG™ mutants were generated in tomato cv. Micro-Tom using *Agrobacterium*-mediated transformation. Seven to ten day old sterile seedlings and one month old plants were used for hypocotyl/shoot tip and stem/leaf explants respectively. A binary vector containing plasmid bluescript, 35S enhancers and antibiotic marker gene *nptII* under the control of a plant promoter (RE4) or Cassava Vein Mottle virus promoter (Verdaguer et al.,1996) was introduced into *Agrobacterium tumefaciens* strains EHA105/101 or GV3101. The transformation procedure followed the established protocol of our lab. The frequency of transformation was calculated as the number of rooted plants in presence of selection (kanamycin) relative to the total number of explants and was expressed as percent points. The rooted plants were out-planted to soil in the Bio-safety greenhouse. The plants were observed twice a week for a period of 3 months for phenotypic variations from wild-type Micro-Tom plants. The tissue samples (leaves and fruits) were frozen for gene isolation and biochemical analysis. Selfed seeds of the T0 generation were raised for the genetic confirmation of the observed variation of the T0 and also the possible knock out traits in second generation. Back crossing to wild plants and T2 generations were carried out for interesting mutant lines.

3. RESULTS AND DISCUSSION

The transformation frequency ranged 30-60% from various explants. Activation tagged transgenics were confirmed at the molecular level by PCR and Southern hybridization for the presence of the plasmid bluescript and/or

35S enhancers. A total of >10,000 activation tagged transgenic lines were grown in the greenhouse. More than 1000 morphological variations (mutants) were observed in T0 and a subset of these mutants were subjected for plasmid rescue and further sequence analysis of the flanking genomic DNA. One of the mutants of relevance to the present discussion is the anthocyanin mutant (ANT1) shown below.

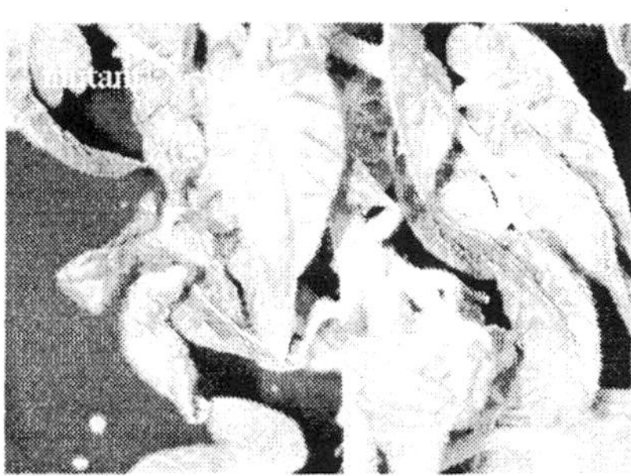

The ANT1 mutant had prominent purple coloration on the leaves, flowers and purple speckles on the fruit. The purple coloration resulted from the over expression of a gene encoding a *MYB* transcription factor. The colorimetric analysis of the ANT1 mutant showed a ten fold increase in the anthocyanin content (0.84 mg in ANT1 versus 0.08 mg per 100g fresh wt of control). The HPLC analysis of the ANT1 mutant showed elevation of several compounds that were undetectable in controls. RTPCR analysis confirmed the over expression of *MYB* transcription factor in the ANT1 mutant plants. Progeny analysis (T1 generation) showed a typical Mendelian inheritance pattern for the purple color, expected of a dominant gene controlled trait. The *ANT1* gene was cloned and introduced into wild tomato and tobacco plants to confirm the gene-to-trait relationship. The *ANT1* transgenic plants of tomato had purple fruits and pigmented trichomes on leaf and stem surface.

A similar pigment mutation by activation tagging has been reported in Arabidopsis – a *MYB* regulator of phenylpropanoid biosynthesis was responsible for upregulated pigment production (Borevitz et al., 2000). In addition to increased anthocyanin content the ripe fruits of ANT1 showed increased levels of carotenoids in comparison to control non-transgenic tomato (Table 1).

Table 1. Carotenoid content of Micro-Tom tomato

Fruit sample	Lutein	Zeaxanthin
Wild type Micro-Tom	3.45	1.48
ANT1 mutant of Micro-Tom	6.34	3.75

The ANT1 mutant is found to have a two fold increase in zeaxanthin, a carotenoid, which together with lutein, is the essential component of the macular pigment in eye (Sandmann 2001).

The genetic mutants have always retained a central role as tools of research on developmental processes in plants. The lack of sufficient mutant populations has been a major obstacle in the advancement of tomato genetics. A number of genes have been isolated in tomato by insertional mutagenesis using targeted tagging of Ac/Ds elements (Bishop et al., 1996; Takken et al., 1998). Meissner et al. (1997, 1999) have described EMS induced mutants with altered pigments, modified leaves and fruits from screening mutagenized population of 9000 M1 and 20,000 M2 plants. *ANT1* perhaps the first tomato gene to be isolated using ACTTAG technology. The overexpression of *ANT1* causes elevated levels of anthocyanins and carotenoids. The identification of the ANT1 and the molecular identification of the responsible gene demonstrates the power of ACTTAG technology in identifying new control points for phytochemical production.

4. REFERENCES

Bishop G.J., K. Harrison and J.D.G. Jones. 1996. The tomato *Dwarf* gene isolated by heterologous transposon taggin encodes the first member of a new cytochrome P_{450} family. Plant Cell 8: 959-969.

Borevitz J.O., Y. Xia, J. Blount, R.A. Dixon and C. Lamb. 2000. Activation tagging identified MYB regulator of phenylpropanoid biosynthesis. Plant Cell 12: 2383-2393.

Huang S., R.E. Cerny, D.S. Bhat and S.M. Brown. 2001. Cloning of an Arabidopsis Patatin-Like gene *STURDY*, by activation T-DNA tagging. Plant Physiol. 125: 573-584.

Meissner R., V. Chague, Q. Zhu, S. Melamed and A.A. Levy. 1999. Mutagenesis and functional genomics in tomato. Meeting on Molecular Biology of tomato. York, UK. July 1999. AB. 2, p. 9.

Meissner R., V. Chague, Q. Zhu, E. Emmanuel, V. Elkind and A.A. Levy. 2000. A high throughput system for transposon tagging and promoter trapping in tomato. Plant J. 22: 265-274.

Meissner R, Y. Jacobson, S. Melamed, S. Levyatuv, S. Gil, A. Ashri, Y. Elkind and A. Levy. 1997. A new model system for tomato genetics. Plant J. 12: 1465-1472.

Muir S.R., G.J. Collins, S. Robinson, S. Hughes, A. Bovy, C.H. Ric De Vos, A.J.V. Tunen and M.E. Verhoeyen. 2001. Overexpression of petunia chalcone synthase isomerase in tomato results in fruit containing increased levels of flavonols. Nature Biotechnology 19: 470-474.

Roemer S., P.D. Fraser, J.W. Kiano, C.A. Shipton, W. Misawa, W. Schuch and P.M. Bramley. 2000. Elevation of the provitamin A content of transgenic tomato plants. Nature Biotechnology 18: 666-669.

Sandmann G. 2001. Genetic manipulation of carotenoid biosynthesis: strategies, problems and achievements. Trends Plant Sci. 6: 14-17.

Scott J.W. and B.K. Harbaugh. 1989. Micro-Tom – a miniature dwarf tomato. Florida Agr. Expt. Sta. Circ. 370:1-6.

Takken F, Schipper D, Nijkamp H and Hille J. 1998. Identification and Ds-tagged isolation of a new gene at the Cf-4 locus of tomato involved in disease resistance to *Cladosporum fulvum* race 5. Plant J. 14: 401-411.

Verdaguer B, de Kochko A, Beachy RN, Fauquet C. 1996. Isolation and expression in transgenic tobacco and rice plants, of the cassava vein mosaic virus (CVMV) promoter. Plant Mol. Biol. 31(6):1129-39.

Weigel D., J.H Ahn, M.A. Blazquez, J.O. Borevitz, S.K. Christensen, C. Fankhauser, C. Ferrandiz, I. Kardailsky, E.J. Malancharuvil, M.M. Neff, J.T. Nguyen, S. Sato, Z.Y. Wang, Y. Xia, , R.A. Dixon, M.J. Harrison, C.J. Lamb, M.F. Yanofsky and J. Chory. 2000. Activation Tagging in Arabidopsis. Plant Physiol. 122: 1003-1013.

Ye X., S. Al-Babili, A. Kloti, J. Zhang, P. Lucca, P. Beyer and I. Potrykus. 2000. Engineering the provitamin A (beta-carotene) biosynthetic pathway into (carotenoid-free) rice endosperm. Science 287: 303-305.

Zubuko E., C.J Adams., I. Machaekova, J. Malbeck., C. Scollan and P. Meyer. 2002. Activation tagging identifies a gene from *Petunia hybrida* responsible for the production of active cytokinins in plants. Plant J. 29: 797-808.

NUTRITIONALLY IMPROVED TRANSGENIC SORGHUM

Zuo-yu Zhao, Kimberly Glassman, Vincent Sewalt, Ning Wang, Mike Miller, Shawn Chang, Teresa Thompson, Sally Catron, Emily Wu, Dennis Bidney, Yilma Kedebe and Rudolf Jung
Trait and Technology Development, Pioneer Hi-Bred International, Inc., A DuPont Company, Johnston, IA50131 USA (e-mail: zuo-yu.zhao@pioneer.com)

Keywords *Agrobacterium*, co-transformation, high lysine, sorghum, transformation

1. INTRODUCTION

Sorghum (*Sorghum bicolor* L.) is the sixth most planted crop in the world, grown on over 100 million acres /year worldwide and currently produces ~60 million metric tons of grain per year. Sorghum is the dietary staple food to over half a billion people in developing countries. However, sorghum grain is low in protein quality due to its low content of essential amino acids, such as lysine. The reliance on sorghum as an important food in regions of Africa and Asia can result in problems associated with malnutrition, especially of children.

Advancements in sorghum tissue culture and transformation research have led to the development of the first efficient technology to genetically transform sorghum by *Agrobacterium* (Zhao et al., 2000). Using this novel technology to improve the nutritional quality of its grain, we introduced into sorghum a high-lysine protein gene that we previously found to be efficacious when transgenically expressed in corn. Sorghum was transformed with a "super-binary" *Agrobacterium* vector (Komari, 1990) containing two unlinked T-DNA cassettes (Komari et al., 1996). One cassette contained the lysine-rich *HT12* gene and another contained a herbicide-resistant *bar* gene as a selectable marker. This co-transformation vector permits the segregation of the marker gene from the trait gene in the progeny of the primary transformants. The elimination of the marker gene is important, because sorghum plants can cross-breed with wild relatives, such as johnsongrass (*Sorghum halepense* L. Pers.) under native conditions (Arriola and Ellstrand, 1996) that could result in herbicide-resistant lines being released into the environment. Of five independent transgenic events that were co-transformed with both genes, three expressed high levels of the HT12 protein in the grain.

I. K. Vasil (ed.), Plant Biotechnology 2002 and Beyond, 413-416.

Several plants from these events were selfed and the progeny was analyzed for segregation of the *HT12* from the *bar*. Preliminary data confirmed the elimination of the *bar* gene in one of the events. This is the first report of successful application of an *Agrobacterium* co-transformation vector in sorghum. Hemizygous seeds in this event showed a 40-60% increase in lysine. A transgenic, lysine-enriched sorghum variety has the potential to directly benefit large populations in developing countries. Assistance in the development and in the release of such a variety would therefore demonstrate to a wide public the possibility of biotechnological solutions to important agricultural and nutritional challenges.

2. SORGHUM TRANSFORMATION

The high lysine analog (HT12 protein) of *Hordeum vulgare* α-hordothionin protein (Rao et al., 1994) contains 44-residue and 12 are lysine residues (27%). The coding sequence for this HT12 protein was under the control of the 27 KD maize gamma zein promoter and terminator (Das et al., 1991). Three copies of this cassette were linked together within one T-DNA borders. A bar coding sequence (Thompson et al., 1987) was controlled by maize ubiquitin promoter (Christensen et al., 1992) and pinII terminator (An et al., 1989) within another T-DNA borders (Figure 1). *Agrobacterium* LBA4404 carrying this vector (PHP16039) was used in this study. Two sorghum lines, P898012 and PHI391 (Pioneer breeding line), were used. Transformation method has been described previously (Zhao et al., 2000).

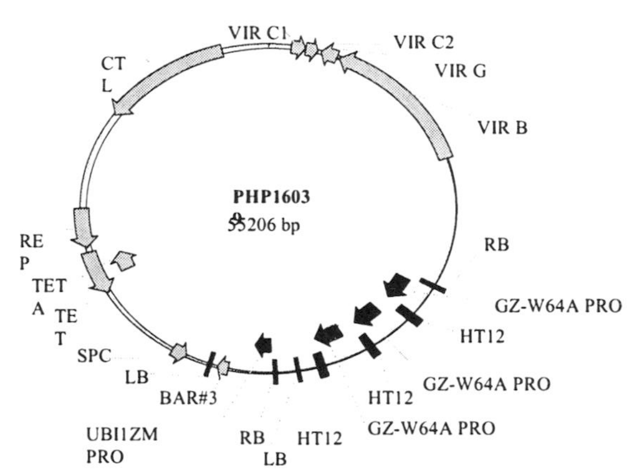

Figure 1. The map of the co-transformation vector.

The results of stable transformation in these two sorghum lines and all analyses performed in these trangenics are listed in Table 1. The analyses include callus PCR of *bar* and *HT12* gene, herbicide (1% Liberty) leaf painting of the T0 plants, DNA Southern blots of the T0 plants to determine the number of inserts of both *bar* and *HT12* gene, as well as *bar* and *HT12* segregation assays in the T1 generation (based on herbicide painting, PCR and DNA Southern blots).

Table 1. The analysis results of sorghum transgenic events.

Event no.	Callus PCR <u>bar</u>	<u>HT-12</u>	Ignite paint T0	No. inserts in T0 <u>*bar*</u>	<u>*HT-12*</u>	Bar/HT-12 segregation in T1 generation
PHI391-1	++	+	2	2		bar segregation
PHI391-2	++	+	2	3		no segregation
PHI391-3	+	-	+	1	0	NA
PHI391-4	+	-	+	1	0	NA
PHI391-5	+	-	NP*	NA	NA	NA
P898012-1	+	+	+	2	1	no segregation
P898012-2	+	+	+	3	3	no segregation
P898012-3	**+**	**+**	**+**	**1**	**1**	**segregation of *bar* from *HT-12***
P898012-4	+	-	+	1	0	NA
P898012-5	+	-	NP*	NA	NA	NA

*NP: we did not send plants to greenhouse. NA: not applied.

HT12 expression in the transgenic sorghum kernels has been evaluated. The analyses were mainly focused on the event, P898012-3, that showed segregation of *bar* from *HT12* in the T1 generation. Protein SDS-PAGE gels showed no visible change in protein profiles between *HT12*(+) and *HT12*(-) kernels. However, Western blot and ELISA assays clearly revealed expression of the HT12 peptide in *HT12* kernels. Lysine content assays further demonstrated an increase of lysine content of about 50% in *HT-12* kernels compared to wild type (Figure 2). Amino acid profiles showed a small reduction of leucine content in *HT-12* kernels with no significant change of other amino acids.

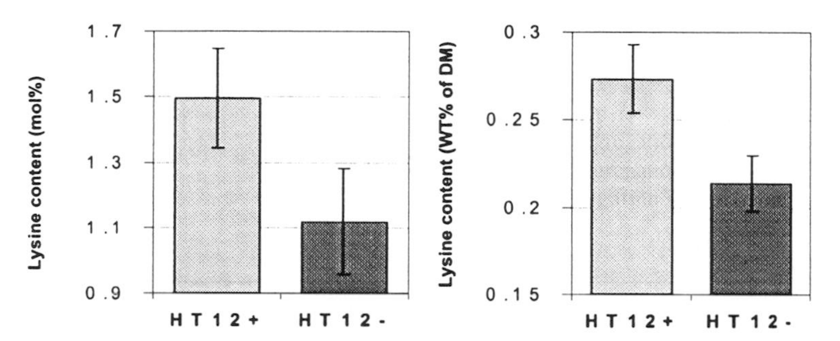

Figure 2. Lysine content assay of sorghum kernels from event P898012-3. Left chart: lysine molar% change and right chart: lysine weight% change in sorghum day meal.

3. REFERENCES

An, G., A. Mitra, H.K. Choi, M.A. Costa, K. An, R.W. Thornburg, and C.A. Ryan. 1989. Functional analysis of the 3' control region of the potato wound-inducible proteinase inhibitor II gene. Plant Cell 1:115-122.

Arriola P.E., and N.C. Ellstrand. 1996. Crop-to-weed gene flow in the genus *Sorghum* (Poaceae): spontaneous interspecific hybridization between johnsongrass, *Sorghum halepense*, and crop sorghum, *S. bicolor*. American J. Botany 83:1153-1160

Christensen, A.H., R.A. Sharrock, and P.H. Quail. 1992. Maize polyubiquitin genes: structure, thermal perturbation of expression and transcript splicing, and promoter activity following transfer to protoplasts by electroporation. Plant Mol. Biol. 18:675-689.

Das, O.P., K. Ward, S. Ray, and J. Messing. 1991. Sequence variation between alleles reveals two types of copy correction at the 27-kDa zein locus of maize. Genomics 11(4):849-856.

Komari T. 1990. Transformation of cultured cells of *Chenopodium quinoa* by binary vectors that carry a fragment of DNA from the virulence region of pTiBo542. Plant Cell Rep. 9:303-306

Komari, T., Y. Hiei, Y. Saito, N. Murai, and T. Kumashiro. 1996. Vectors carrying two separate T-DNAs for co-transformation of higher plants mediated by *Agrobacterium tumefaciens* and segregation of transformants free from selection markers. Plant J. 10:165-174.

Rao, A.G., M. Hassan, and J.C. Hempel. 1994. Structure-function validation of high lysine analogs of α-hordothionin designed by protein modeling. Protein Engineering 7(12):1485-1493.

Thompson, C., N.R. Movva, R. Tizard, R. Crameri, J.E. Davies, M. Lauwereys, and J. Botterman. 1987. Characterization of the herbicide-resistance gene *bar* from *streptomyces hygroscopicus*. EMBO J. 6:2519-2523.

Zhao Z-Y., T. Cai,, L. Tagliani, M. Miller, N. Wang, H. Pang, M. Rudert, S. Schroeder, D. Hondred, J. Seltzel, and D. Pierce. 2000. *Agrobacterium*-mediated sorghum transformation. Plant Mol. Biol. 44:789-798.

IMPROVEMENTS IN THE NUTRITIONAL QUALITY OF THE COTTONSEED

Keerti S. Rathore[1], G. Sunilkumar[1], Lorraine Puckhaber[4], Robert D. Stipanovic[4], Hossen M. Monjur[3], Ernesto Hernandez[3], and C. W. Smith[2]
[1]Institute for Plant Genomics & Biotechnology, Texas A&M University, College Station, TX, 77843 USA (rathore@tamu.edu)
[2]Dept of Soil & Crop Sciences, Texas A&M University, College Station, TX, 77843 USA
[3]Food Protein Research Center, Texas A&M University, College Station, TX, 77843 USA
[4]USDA-ARS, Southern Plains Agricultural Research Center, College Station, TX, 77843 USA

Keywords: Cotton, gossypol, nutritional quality, oleic acid, seed-specific promoter

1. INTRODUCTION

Cottonseed is a by-product of cotton fiber production. With each 100 pounds of fiber, the cotton plant produces approximately 165 pounds of cottonseed. However, it accounts for only 10 to 15% of the total value of a bale of cotton. Nearly 40% of cottonseed is fed directly to ruminant animals and the remainder is used as raw material for the cottonseed processing industry. Following oil extraction, the meal is mainly used as feed for cattle. Cottonseed is composed of ~ 22.5% of high nutritional quality protein. The amount of protein available annually from cottonseed worldwide is sufficient to meet the protein requirements of 350 million people (Lusas and Jividen, 1987). However, it can not be used to feed non-ruminant animals because of the presence of gossypol. Gossypol, a polyphenolic substance, is highly toxic to non-ruminant animals and is detrimental to human health if not removed from the oil. Thus, cottonseed represents a valuable resource that is grossly underutilized, principally due to the presence of gossypol. Elimination of gossypol from the seed will make the meal a valuable source of nutrition for monogastric animals as well as humans. However, this needs to be done in a highly seed-specific manner as gossypol and related terpenoids that are also present in the vegetative parts of the plant are believed to play a role in disease and insect resistance.

Cottonseed oil is another valuable by-product of fiber production. At 1.2 billion pounds per year, cottonseed oil ranks third in total production behind soybean and corn oil in the U.S. The fatty acid composition of this oil is, 24.7% palmitic acid, 17.6% oleic acid and 53.3% linoleic acid (White, 2000).

I. K. Vasil (ed.), Plant Biotechnology 2002 and Beyond, 417-420.

As other major oilseed crops are being modified to improve their fatty acid profile in favor of higher oleic acid levels to enhance nutritional and cooking qualities of their oil (Hitz et al., 1995; Kinney, 1996), it is essential that cottonseed oil composition be improved in order to curb the loss of market share.

We are examining a transgenic approach to improve the nutritional quality of cottonseed by eliminating gossypol from the seed and by altering the fatty acid composition of the oil. Both of these projects involve modification of the biosynthetic pathway by employing the antisense technology to suppress the expression of genes that encode key enzymes. For gossypol reduction the target gene is delta-cadinene synthase. This gene encodes an enzyme that catalyzes cyclization of farnesyldiphosphate to form delta-cadinene, a key intermediate in the biosynthesis of gossypol. Down-regulation of delta-12 desaturase gene is being used to increase oleic acid level and to reduce linoleic acid level in the oil. Delta-12 desaturase catalyzes the reaction that involves addition of a second double bond to oleic acid. The promoter region of cotton alpha-globulin gene B that was functionally characterized and shown to be exclusively expressed in a seed-specific manner in an earlier study (Sunilkumar et al., 2002) was used to drive each of the two antisense transgenes. An efficient transformation method described by Sunilkumar and Rathore (2001) was employed to obtain a large number of independent transgenic lines in each case. T1 and T2 seeds have been analyzed for gossypol levels and thus far fatty acid levels have been examined in T1 seeds.

2. RESULTS

We used a seed-specific promoter (cotton alpha-globulin gene B) to drive the antisense delta-cadinene synthase cDNA clone from *Gossypium hirsutum* to reduce gossypol in cottonseeds. Following transformation of hypocotyl segments of *Gossypium hirsutum*, cv. Coker 312, 60 plants were regenerated from 35 independent transgenic callus lines. Total gossypol was quantified in a pool of 30 T1 seeds from each of the greenhouse-grown primary transformants as described by Stipanovic et al. (1988). Although T1 seeds will be segregating for the transgene, our earlier study (Sunilkumar et al., 2002) had indicated that a pooled sample of 24 T1 seeds can provide an estimation of transgene expression level. Gossypol levels ranged from 0.54% to 1.50% (wt/wt of dry seed kernels) in the T1 seeds from 60 transgenic plants. We selected six lines with the lowest seed-gossypol levels and grew 12 T1 plants from each of these lines to maturity in the greenhouse. In one of the six lines selected, T2 seeds continued to show low gossypol levels. T2 seeds from one of the T1 parent from this line had gossypol levels as low as 0.49%. We used 24 different transgenic lines derived from transformations with various other constructs as control. Mean gossypol value in a sample size of 30 pooled T1 seeds from each of these controls was 1.02%. Thus, we have been able to achieve up to 50% reduction in the seed-gossypol level using antisense-based suppression.

Cotton alpha-globulin gene B promoter was also used to drive the antisense cDNA clone of *G. hirsutum* delta-12 desaturase. Following transformation, 42 plants from 23 lines were regenerated and grown to maturity in the greenhouse. Fatty acid profile in dry seed kernels was analyzed using Gas Chromatography as described by Dahmer et al. (1989). Again a sample size of 30 pooled T1 (segregating) seeds was used for this analysis. The levels of oleic acid (18:1) ranged from 15.5% to 29% in the T1 seeds from different lines. Linoleic acid (18:2) levels ranged from 54.6% to 40.7%, respectively. Lines that exhibited higher levels of oleic acid showed a corresponding reduction in the levels of linoleic acid. The levels of palmitic and stearic acid remained relatively unchanged in the seeds from these lines. The large size of cottonseed allowed us to analyze fatty acid profile at a single seed level. We performed this single seed analysis on seeds from three high-oleic acid lines. Results from this analysis are shown in Table 1. These T1 seeds will be segregating for the transgene. As expected, some seeds had normal complement of these two fatty acids. Some had elevated oleic acid and reduced linoleic acid levels. In some individual seeds, the oleic acid levels were as high as 34% and linoleic acid levels were as low as 32%. Intermediate levels of altered fatty acid profile were found in many seeds.

Table 1. Percent levels of oleic acid (18:1) and linoleic acid (18:2) in the oils from randomly selected individual T1 cottonseeds from three different transgenic lines.

Seed#	Line H50-2		Line H41-1		Line H42-2	
	18:1	**18:2**	**18:1**	**18:2**	**18:1**	**18:2**
1	16.5	48.3	15.7	47.1	15.5	52.7
2	24.4	44.6	18.7	45.9	15.5	50.6
3	25.1	41.6	23.5	42.1	16.3	49.9
4	25.5	41.0	24.7	39.7	17.3	50.8
5	25.8	42.0	25.0	40.9	18.0	47.8
6	26.1	40.5	26.0	38.2	18.8	48.0
7	26.5	40.3	26.0	38.2	18.8	45.6
8	26.9	39.8	26.2	39.0	22.8	42.5
9	27.1	39.4	26.4	39.3	23.8	45.9
10	27.3	39.4	26.6	38.7	24.9	43.0
11	30.0	37.6	27.1	39.6	26.4	41.0
12	30.8	34.3	28.2	35.8	26.5	43.1
13	31.2	34.9	29.2	36.3	30.0	37.5
14	32.4	35.5	30.8	34.3	31.5	33.7
15	34.2	33.2	34.3	32.0	31.5	34.0

In conclusion, results from this study show that antisense-mediated suppression of gene expression can be used to improve the seed quality in cotton.

3. REFERENCES

Dahmer M.L., P.D. Fleming, G.B. Collins, and D.F. Hildebrand. 1989. A rapid screening technique for determining the lipid composition of soybean seeds. J. Amer. Oil Chem. Soc. 66:543-548.

Hitz W.D., N.S. Yadav, R.S. Reiter, C.J. Mauvais, and A.J. Kinney. 1995. Reducing polyunsaturation in oils of transgenic canola and soybean. Plant Lipid Metabolism. J.C. Kader and P. Mazliak (eds.). Kluwer Academic Publishers. Pp. 506-508.

Kinney A.J. 1996. Development of genetically engineered oilseeds. Physiology, Biochemistry and Molecular Biology of Plant Lipids. J.P. Williams, M.U. Khan, and N.W. Lem (eds.). Kluwer Academic Publishers. Pp. 298-300.

Lusas E.W., and G.M. Jividen. 1987. Glandless cottonseed: A review of the first 25 years of processing and utilization research. J. Amer. Oil Chem. Soc. 64:839-854.

Stipanovic R.D., D.W. Altman, D.L. Begin, G.A. Greenblatt, and J.H. Benedict. 1988. Terpenoid aldehydes in upland cottons: analysis by aniline and HPLC methods. J. Agric. Food Chem. 36:509-515.

Sunilkumar G., J.P. Connell, C.W. Smith, A.S. Reddy, and K. S. Rathore. 2002. Cotton alpha-globulin promoter: Isolation and functional characterization in transgenic cotton, Arabidopsis, and tobacco. Transgenic Research - in press.

Sunilkumar G., and K. S. Rathore. 2001. Transgenic cotton: Factors influencing *Agrobacterium*-mediated transformation and regeneration. Molecular Breeding 8:37-52.

White P.G. 2000. Fatty acids in oilseeds (vegetable oils). Fatty acids in foods and their health implications. C.K. Chow (ed.). Marcel Dekker, Inc. Pp. 209-238.

PRODUCTION OF GAMMA LINOLENIC ACID IN SEEDS OF TRANSGENIC SOYBEAN

Tom Clemente[1,2,3], Aiqiu Xing[1], Xingguo Ye[2], Shirley Sato[1], Bruce Schweiger[4] and Anthony Kinney[4]

[1] Center for Biotechnology, University of Nebraska-Lincoln (tclemente1@unl.edu)
[2] Department of Agronomy & Horticulture, University of Nebraska-Lincoln
[3] Plant Science Initiative, University of Nebraska-Lincoln
[4] DuPont Experimental Station, Wilmington, DE

Keywords *Glycine max*, stearidonic acid, nutraceutical, transformation

1. INTRODUCTION

An array of fatty acids has been identified in the plant kingdom. Many of these fatty acids have commercial applications, but cost of goods for some of them are significantly high due to either the production of the fatty acid in the native source may be low and/or intensive cultivation of the plant species is cost-prohibitive. The ability to introduce novel traits into soybean makes it technically feasible to alter oil metabolism for the production of high value/low volume fatty acids cost effectively. One such fatty acid is gamma linolenic acid (GLA). GLA has pharmacological applications in treatment of skin conditions such as eczema and is known to possess some antiviral and anticancer properties (Gill and Valivety, 1997; Horrobin, 1990). Another example of a high value fatty acid that can be cost effectively produced in soybean is stearidonic acid (STA). Like GLA, STA is of interest for the pharmaceutical and nutraceutical industry (Griffiths et al., 1996).

In plants GLA is produced by the conversion of linoleic acid through a single desaturation step via a Δ^6 desaturase (Sayanova et al., 1997). Sayanova et al. (1997) demonstrated that expression of the borage Δ^6 desaturase gene in tobacco resulted in the accumulation of GLA and STA at levels of 13.2% and 9.6%, respectively in the leaves of the transgenic tobacco plants, while in wild-type controls, the presence of these fatty acids was not observed. This provides the rationale that the same strategy could be extended to an oil seed crop such as soybean.

I. K. Vasil (ed.), Plant Biotechnology 2002 and Beyond, 421-424.

2. INTRODUCTION OF THE BORAGE Δ^6 DESATURASE GENE IN SOYBEAN

A cDNA clone of the Δ^6 desatursae gene was generated via RT- PCR from developing floral buds of *B. officinalis* L. The Δ^6 desaturase gene was subsequently subcloned downstream of the seed specific promoter ß-conglycinin. The derived seed-specific cassette was assembled in a two T-DNA binary vector with T-DNA one harboring a *bar* (Thompson et al., 1987) cassette for plant selection and T-DNA two carrying the Δ^6 desaturase cassette. A two T-DNA binary plasmid can be implemented as a strategy to derive marker-free transformants (Komari et al., 1996; Xing et al. 2000). The resultant two T-DNA binary vector is referred to as pPTN331. The binary plasmid pPTN331 was mobilized into *Agrobacterium tumefaciens* strain EHA101. Soybean transformations were conducted with genotypes A3237 (Asgrow Seed Company), Thorne (Ohio State University) and NE3001 (University of Nebraska) using the cotyeldonary-node method following modifications previously described (Zhang et al., 1999; Clemente et al., 2000). Soybean transformants were selected on 5 mg/L glufosinate during the shoot initiation step and 3 mg/L glufosinate during the shoot elongation step. The primary transformants (T_0) were characterized for the presence of both T-DNA elements by Southern blot analysis. The T_0 plants were grown to maturity under greenhouse conditions. Seed chips from up to 16 T_1 seeds from the transformed lines carrying both T-DNA elements were analyzed for fatty acid composition by gas chromatography. The remaining portions of the seed, with the embryonic axis were planted and the developing T_1 individuals subsequently monitored for either a herbicide resistance or susceptible phenotype.

2.1. Preliminary Characterizaion Of Soybean Lines Carrying The Borage Δ^6 Desaturase Gene

A total of 14 lines have been identified in which production of both GLA and STA has been detected in the seed storage lipids (Table 1). Among the soybean lines screened to date three, 414-2, 414-9 and 418-6 had progeny with a GLA/STA/herbicide sensitive phenotype suggesting that these individuals only carry the Δ^6 desaturase T-DNA and thus are marker-free. Molecular analysis is currently being conducted on these putative marker-free individuals to verify the absence of the *bar* T-DNA element in the genome.

A subset of the GLA/STA producing soybean lines and all confirmed marker-free lines will be carried on to homozygousity for evaluation under field conditions in order to monitor the agronomic performance of the GLA/STA soybeans. Moreover, in an attempt to enhance the STA levels in the seed storage lipids we are assembling a two T-DNA binary that will carry in T-

DNA two, the Δ^6 desaturase cassette along with a seed specific cassette harboring the *Arabidopsis* FAD3 gene.

Broadening the germplasm of soybean to produce novel high value fatty acids will potentially expand the market for soybean oil and permit a cost effective exploitation of these and other high value molecules.

Table 1. Seed fatty acid profiles observed in soybean transformants expressing the Δ^6 desaturase gene.

Line	%Plamitic	%Stearic	%Oleic	%Linoleic	%GLA	%Linolenic	%STA
398-4	11.9	4.1	13.4	38.2	**21.2**	7.9	**3.2**
398-6	12.3	4.3	11.9	29.1	**31.7**	6.3	**4.4**
404-1	13.4	3.5	12.0	31.5	**30.4**	5.6	**3.7**
404-7	12.4	3.7	11.7	23.5	**38.4**	5.3	**5.1**
404-11	11.7	4.4	13.5	35.0	**25.4**	6.8	**3.1**
411-5	11.0	4.3	13.5	29.2	**30.7**	6.8	**4.9**
412-2	11.5	4.2	14.6	39.2	**19.5**	8.1	**2.9**
412-4	12.3	3.7	13.5	29.5	**31.0**	5.8	**4.2**
414-2*	11.5	4.4	13.5	43.9	**16.2**	8.1	**2.4**
414-3	11.8	3.4	13.6	29.7	**29.4**	7.5	**4.6**
414-4	12.4	3.5	11.2	28.1	**34.7**	5.6	**4.4**
414-5	11.3	4.7	12.9	48.7	**10.1**	10.6	**1.7**
414-9*	12.0	3.8	11.8	39.8	**21.4**	7.9	**3.3**
418-6*	11.8	3.2	15.3	43.3	**16.7**	7.5	**2.3**
WT	12.0	3.7	13.2	58.2	**ND**	12.9	**ND**

The Line column refers to the soybean transformant designation. Lines followed by * indicate putative marker-free individuals observed among the T plants screened. WT indicates wild type soybean fatty acid profile. Numbers within each of the fatty acid columns indicates the percentage of the respective fatty acid in a representative T sample from the respective soybean line. ND indicates not detected

3. REFERENCES

Clemente, T., B. LaVallee, A. Howe, D. Ward, R. Rozman, P. Hunter, D. Broyles, D. Kasten, and M. Hinchee. 2000. Progeny analysis of glyphosate selected transgenic soybeans derived from *Agrobacterium*-mediated transformation. Crop Sci. 40:797-803.

Gill, I., and R. Valivety. 1997. Polyunsaturated fatty acids: occurrence, biological activities and applications. Trends Biotechn. 15 401-409.

Griffiths, G., E.Y. Brechany, F.M. Jackson, W.W. Christie, S. Styme, and A.K. Stobart. 1996. Distribution and biosynthesis of stearidonic acid in leaves of *Borago officinalis*. Phytochemistry 43:381-386.

Komari, T., Y. Hiei, Y. Saito, N. Murai, and T. Kumashiro. 1996. Vectors carrying two separate T-DNAs for co-transformation of higher plants mediated by *Agrobacterium tumefaciens* and segregation of transformants free from selection markers. Plant J. 10:165-174.

Horrobin, D.F. 1990. Gamma linolenic acid: an intermediate in essential fatty acid metabolism with potential as an ethical pharmaceutical and as a food. Reviews in Contemporary Pharmacotherapy 1:1-45.

Sayanova, O.., M.A. Smith, P. Lapinskas, A.K. Stobart, G. Dobson, W.W. Christie, P.R. Shewry, and J.A. Napier. 1997. Expression of a borage desaturase cDNA containing an N-terminal cytochrome *b5* domain results in the accumulation of high levels of Δ^6-desaturated fatty acids in transgenic tobacco. Proc. Natl. Acad. Sci. USA. 94:4211-4216.

Thompson, C.J., N.R. Movva, R. Tizard, R. Crameri, J.E. Davies, M. Lauwereys, and J. Botterman. 1987. Characterization of the herbicide-resistance gene *bar* from *Streptomyces hygroscopicus*. EMBO J. 6:2519-2523.

Xing, A., Z. Zhang, S. Sato, P. Staswick, and T. Clemente. 2000. The use of the two T-DNA binary system to derive marker-free transgenic soybeans. In Vitro Cell. Dev. Biol.-Plant 36:456-463.

Zhang, Z., A. Xing, P. Staswick, and T. Clemente. 1999. The use of glufosinate as a selective agent in *Agrobacterium*-mediated transformation of soybean. Plant Cell Tiss. Org. Cult. 56:37-46.

TRANSFER AND EXPRESSION OF AN ARTIFICIAL STORAGE PROTEIN (ASP1) GENE IN CASSAVA: TOWARDS IMPROVING NUTRITIVE VALUE OF STORAGE ROOTS

Peng Zhang[1], Jesse M. Jaynes[2], Ingo Potrykus[1], Wilhelm Gruissem[1] and Johanna Puonti-Kaerlas[3]
[1] Institute of Plant Sciences, ETH-Zentrum / LFW E 17, CH-8092 Zürich, Switzerland
[2] Chairman of the NovaTero Foundation, 2417 High Ridge Drive, Raleigh, NC 27606, USA
[3] European Patent Office, D-80298 Munich, Germany (e-mail: zhang.peng@ipw.biol.ethz.ch)

Keywords *Manihot esculenta*, storage root quality, synthetic storage protein, genetic transformation

Cassava (*Manihot esculenta* Crantz) is a staple food of more than 500 million people in the tropics. Its storage roots contain starch up to 85% of their dry weight, but are deficient in protein. People depending heavily on cassava may consequently suffer from qualitative malnutrition, unless they can supplement their diet with protein from other sources. Traditional breeding of cassava is difficult due to irregular flowering and low fertility as well as to low seed set and germination rates of the plants, and attempts to improve the protein content of cassava roots have so far been unsuccessful. Advances in plant genetic engineering now provide an alternative to traditional breeding in improving cassava, such as improved root quality and disease resistance.

In order to increase the nutritional quality of cassava storage roots, a synthetic gene encoding a storage protein (ASP1) rich in essential amino acids (80%) was introduced into embryogenic suspension of cassava via *Agrobacterium*- mediated gene transfer. The ASP1 were designed to have a stable storage protein like structure in plants based on the structurally well studied maize storage zein proteins (Z19 and Z22) (Agros et al., 1982). ASP1 is comprised of four helical repeating monomers, each 20 amino acids long (Fig. 1). The *ASP1* has been transformed into tobacco (Kim et al., 1992) and sweetpotato (Prakash and Egnin, 1997) and showed high protein and essential amino acid levels in the transgenic plants.

I. K. Vasil (ed.), Plant Biotechnology 2002 and Beyond, 425-427.

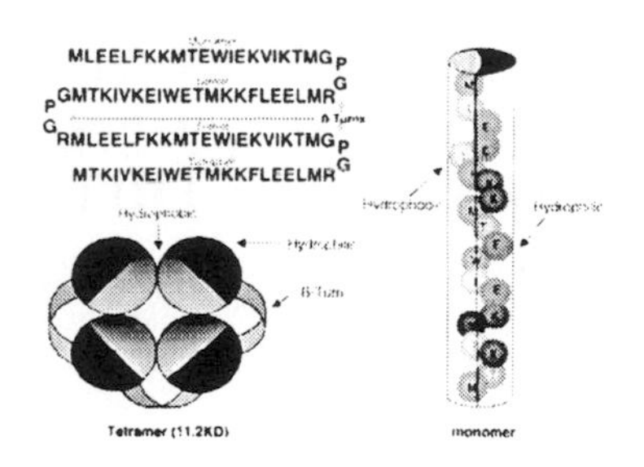

Figure 1. The amino acid sequence and conformation of the ASP1.

To ensure stable expression levels in transgenic plants, the *ASP1* is linked to a KOZAC translational enhancer and driven by the CaMV 35S promoter. Transgenic cassava plants were regenerated from somatic embryo lines derived from hygromycin-resistant friable embryogenic callus lines. Southern and Northern analyses showed the stable integration of *ASP1* in cassava genome (Fig. 2a, b) and its expression at RNA level in transformed plant lines (Fig. 2c). The expression of *ASP1* has been found to correlate with the copy numbers of the transgene. The ASP1 tetramer could be detected in leaves as well as in primary roots of cultured transgenic plants by Western analysis (Fig. 3). These results indicate that the nutritional improvement of cassava storage roots may be achieved by constitutive expression of *ASP1* in transgenic plants. Selected lines are being grown in greenhouse and will be transferred to CIAT or Virgin Island for further analysis.

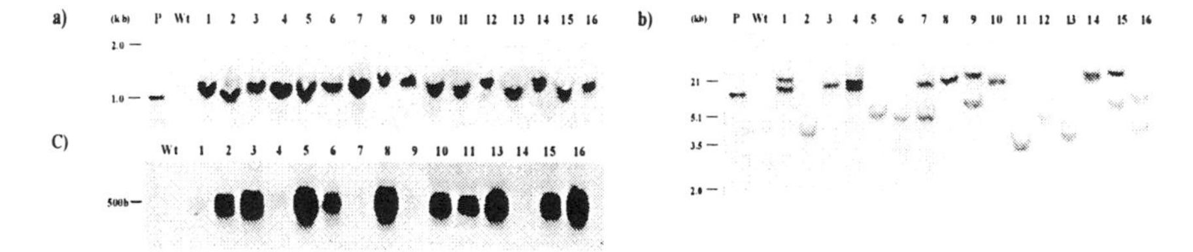

Figure 2. Southern and northern analyses of transgenic cassava plant lines. Genomic DNA was digested with HindIII to release the ASP1 cassette and hybridised with either ASP1 probe (a) or uidA probe (b). Total RNA was hybridised with ASP1 probe (c). P – plasmid control; Wt – wild type control; 1-16 – different transgenic plant lines.

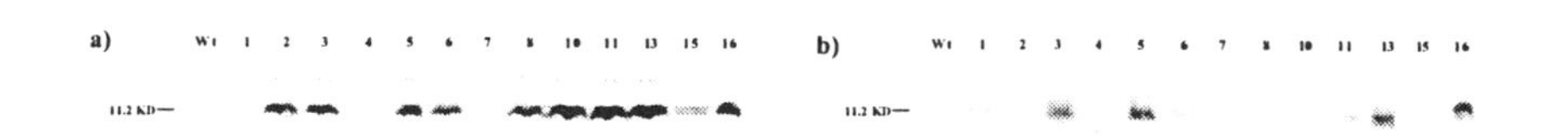

Figure 3. Detection of ASP1 protein in transgenic plant lines by immunoblot analysis in: a) leaves and b) roots. Wt – wild type control; 1-16 – different transgenic plant lines.

REFERENCES

Agros P., Pederson K., Marks D. and Larkins B.A. (1982) A structural model for maize zein proteins. J. Biol. Chem. 257: 9984-9990.

Kim J.H., Cetiner S. and Jaynes J.M. (1992) Enhancing the nutritional quality of crop plants: design, construction and expression of an artificial plant storage protein gene. In: Bhatnagar D and Cleveland T.E. (eds.) Molecular Approaches to Improving Food Quality and Safety, (pp.1-36). AVI, New York.

Prakash C.S. and Egnin M. (1997) Engineered sweetpotato (*Ipomoea batatas*) plants with a synthetic storage protein gene show high protein and essential amino acid levels. Concurrent session 335. In: Dean J.F.D. (ed.) Abstract, 5th International Congress of Plant Molecular Biology. Singapore.

ANTICARCINOGENIC PROPERTIES OF PLANT PROTEASE INHIBITORS FROM THE BOWMAN-BIRK CLASS

Alfonso Clemente[1], Donald MacKenzie[2], David Jeenes[2], Jenny Gee[2], Ian Johnson[2] and Claire Domoney[1]
[1]John Innes Centre, Norwich NR4 7UH, UK (e-mail: alfonso.clemente@bbsrc.ac.uk)
[2]Institute of Food Research, Norwich NR4 7UA, UK

Keywords Bowman-Birk inhibitor, cell proliferation, inhibitory domains, *Pisum sativum*

1. INTRODUCTION

Protease inhibitors (PI) have long been considered to be antinutritional compounds in legume crops due to their ability to decrease the protein digestibility and adsorption of dietary proteins. In recent years, renewed interest in PI has followed the recognition that certain plant PI are effective at preventing or suppressing carcinogenic processes in a wide variety of *in vitro* and *in vivo* animal model systems (Kennedy, 1998). These reports suggest a positive contribution of PI to the nutritional value of dietary proteins from vegetable sources. The selection of plant PI for plant breeding programs aimed at human health depends on elucidation of the molecular basis for variation in their biological activity. Distinct PI have been identified in legumes with many seed inhibitors inhibiting trypsin and chymotrypsin at distinct sites (Figure 1). The significance of the two active sites is likely to be different in terms of their exploitation. The trypsin inhibitory site has been implicated in the negative effect on bioavailability of dietary proteins in food and feed (Liener, 1994) and in the protection of plants against insects (Volpicella et al., 2000) and fungi (Giudici et al., 2000). The chymotrypsin inhibitory site has been implicated in the anticarcinogenic effect of the soy-derived Bowman-Birk inhibitor (BBI) (Kennedy et. al., 1993). We are investigating the properties of a range of PI homologous to BBI that include sequence variants as well as variants that arise from post-translational processing. We have expressed two pea variant PI and their corresponding C-terminally processed forms as recombinant proteins and are assessing the antiproliferative properties of the variants on colon cancer cells using an *in vitro* cell assay system designed for high-throughput and rapid screening.

I. K. Vasil (ed.), Plant Biotechnology 2002 and Beyond, 429-431.

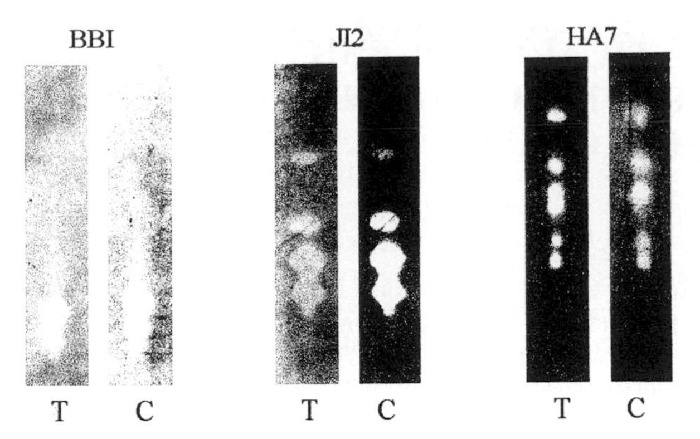

Figure 1. Seed trypsin (T) and chymotrypsin (C) inhibitor isoforms from two pea lines (JI2 and HA7) compared with BBI from soybean, separated by electrophoresis on native gels.

2. ANTIPROLIFERATIVE EFFECTS OF PROTEASE INHIBITORS IN HUMAN COLON CANCER CELLS

2.1. *In vitro* Cell Assay

An *in vitro* system is being used to evaluate changes in the proliferation of human colon cancer cells when different compounds are added. Experimental cells are seeded in 96-well microplates, to which PI at different concentrations (chymotrypsin inhibitor units) are added. A statistically significant and dose-dependent decrease in growth of HT29 human colon adenocarcinoma cells was observed after treatment with BBI (Figure 2).

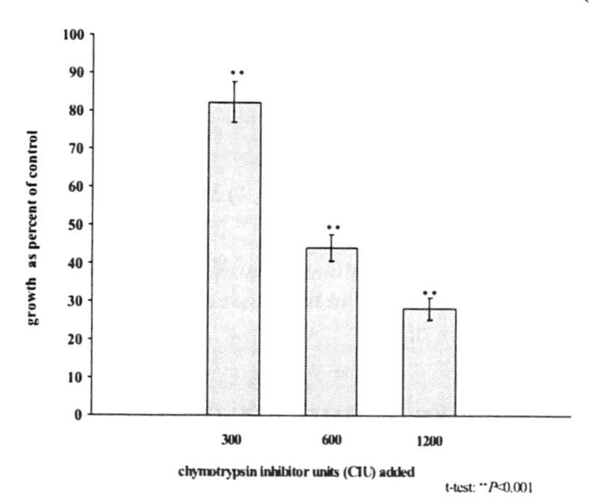

Figure 2. Effect of BBI on in vitro growth of HT29 human carcinoma cells

2.2. Over-expression of Pea Protease Inhibitors

Two pea PI proteins corresponding to *TI1B* and *TI2B* genes, encoding proteins that differ most significantly in amino acids at both sites of enzyme inhibition (Page et al., 2002), have been expressed as mature C-terminally processed and unprocessed active proteins, using a system capable of efficient re-folding of extensively disulphide-bonded proteins. A protein fusion expression system where the target protein is linked to part of a secreted glucoamylase protein (GAM), with a processing site at the fusion junction, has been used (Archer et al., 1994; MacKenzie et al., 1998). Elucidation of the behaviour of the variants will provide fundamental knowledge of variation in the interaction between the relevant protease(s) and plant PI and will establish criteria for the development of highly effective therapeutic PI.

3. ACKNOWLEGMENTS

AC is a recipient of an EU fellowship.

4. REFERENCES

Archer D.B., D.J. Jeenes and D.A. MacKenzie. 1994. Strategies for improving heterologous protein production from filamentous fungi. Anton. van Leeuwen. 65: 245-250.

Giudici T., M.C. Regente and L. de la Canal. 2000. A potent antifungal protein from *Helianthus annus* flowers is a trypsin inhibitor. Plant Physiol. Biochem. 38: 881-888.

Kennedy A.R. 1998. The Bowman-Birk inhibitor from soybeans as an anticarcinogenic agent. Am. J. Clin. Nutr. 78: 167-209.

Kennedy A.R., B.F. Szuhaj, P.M. Newberne and P.C. Billings. 1993. Preparation and production of a cancer chemopreventive agent, Bowman-Birk inhibitor concentrate. Nutr. Cancer 19: 281-302.

Liener I.E. 1994. Implications of antinutritional components in soybean foods. CRC Crit. Rev. Food Sci. Nutr. 34: 31-67.

MacKenzie D.A., J.A.E. Kraunsoe, J.A. Chesshyre, G. Lowe, T. Komiyama, R.S. Fuller and D.B. Archer. 1998. Aberrant processing of wild-type and mutant bovine pancreatic trypsin inhibitor secreted by *Aspergillus niger*. J. Biotech 63: 137-146.

Page D., G. Aubert, G. Duc, T. Welham and C. Domoney. 2002. Combinatorial variation in coding and promoter sequence of genes at the *Tri* locus in *Pisum sativum* accounts for variation in seed trypsin inhibitor activity. Mol. Gen. Genet. (In press).

Volpicella M., A. Schipper, M.A. Jongsma, N. Spoto, R. Gallerani and L.R. Ceci. 2000. Characterization of recombinant mustard trypsin inhibitors 2 (MTI2) expressed in *Pichia pastoris*. FEBS Lett. 468: 137-141.

A GENOMIC APPROACH TO ELUCIDATE GENE FUNCTION DURING WOOD FORMATION

Göran Sandberg
Umeå Plant Science Center Department of Forest Genetics and Plant Physiology, The Swedish University of Agricultural Sciences S 901 83 Umea, Sweden (e-mail: Goran.Sandberg@gnefys.slu.se)

1. INTRODUCTION

Trees are among the most long-lived organisms on earth. They have therefore developed unique traits to cope with problems related to living in harsh climates for very extended periods of time. In boreal ecosystems, trees experience huge annual fluctuations in climate and have been obliged to develop a number of protective mechanisms, operating in a seasonal pattern (Olssen et al., 1997). The most obvious manifestation of these mechanisms is the shedding of leaves in the autumn, but more important is the protection of their most important asset, the stem cells of their meristems (where the great majority of cell divisions take place) by inducing them to enter a dormant state. These seasonal processes also involve bud set, bud break and seasonal alterations between vegetative growth and flowering. These processes that are unique to trees, together with the very extended phase of juvenile growth that occurs in trees cannot be studied in *Arabidopsis*. In addition to processes that occur only in trees, there are also phenomena that could, in principle, be studied in *Arabidopsis* but trees constitute a better experimental system. For example, trees utilise the stem cells of the vascular meristem (the cambium) to produce wood in a strictly organised developmental gradient, and the large physical size of the cambial zone makes it one of the best developmental gradients in any type of organism to study experimentally (Uggla et al.,1996). An important and unique aspect of tree development is the late juvenility-to-maturity transition, which can take up to 20 years in *Populus*. During the juvenile phase the tree is not competent to induce flowering. The reason for this is completely unknown but is presumably due to a complex interaction between plant growth regulators, for example gibberellins, floral repressors and perhaps also epigenetic factors such as alterations in methylation status and/or chromatin structure. Thus, there are many compelling scientific arguments for considering trees important models for developmental biology.

I. K. Vasil (ed.), Plant Biotechnology 2002 and Beyond, 433-438.

2. THE USE OF MODEL SYSTEMS

More than ten years ago we decided to focus the majority of our research on the two model systems *Populus* and *Arabidopsis*. The value of *Arabidopsis* as a test bed for studying "tree" genes was not obvious; it was argued that *Arabidopsis* was too different from a tree to serve as a good model system but we have proved that for studying many biological processes this is not the case. The major advantage with *Arabidopsis*, besides its rapid life cycle and the ease with which molecular and traditional genetics can be performed, is the impressive infrastructure that has been developed for its analysis. However, since trees are, as discussed above, different from *Arabidopsis* in many fundamental aspects, there has also been a need to develop a tree model system. The only real candidate for this is *Populus*. As a result of our efforts an extensive gene catalogue of *Populus* genes, and microarrays almost as comprehensive as those available from *Arabidopsis* have been produced. Moreover, the United States Department of Energy has recently decided to produce the complete genomic sequence of *Populus trichocarpa*. Furthermore, as described below Umeå Plant Science Center will be involved in producing a large-scale gene knockout program in *Populus*,. Taken together this means that in a few years all of the three important infrastructure elements that are currently available in *Arabidopsis* will also be available in *Populus*, opening up tremendous possibilities to perform cutting-edge biology in this tree model system that will complement, and extend, research performed in *Arabidopsis*.

3. TREE FUNCTIONAL GENOMICS

Sweden has developed a comprehensive programme for tree functional genomics with excellent technology platforms not only for most genomic technologies but we are also facilities for large-scale production of transgenic trees and phenotypic analysis. In the *Populus* genome programme to date Umeå Plant Science Center has sequenced, together with the groups of Mathias Uhlén and Joakim Lundberg at the Royal Institute of Technology in Stockholm, over 100,000 poplar ESTs from 16 cDNA libraries representing different *Populus* tissues and developmental stages (Sterky et al, 1998; *Populus* DB), and several additional libraries are in preparation. The UPSC bioinformatics group has developed an EST annotation/functional classification procedure that is virtually automatic (Bhalerao et al., 2002). In collaboration with the Royal Institute of Technology we have also produced a *Populus* DNA microarray consisting of over 13,000 clones. This is, to our knowledge, the most complete array for a eukaryotic organism produced within a single academic project. During 2002 we will produce an array containing the Unigene-set from the analysis of the 100,000 ESTs. Techniques enabling large-scale qualitative and quantitative analysis of metabolites also provide essential complements to the information obtained by the analysis of transcript and protein profiles in living organisms (Fiehn, 2001). The main objective is to analyse quantitatively the metabolite composition, and metabolic fluxes, in samples from well-defined tissues and developmental stages. By analysing different tissues we will be able to create complete metabolite libraries for different poplar tissues. We anticipate that they will cover well over 1500 substances. Through the use of chemometrics the differences between samples can be analysed and visualised. These libraries will form an important base from which we can analyse the impact of different developmental stages on metabolic profiles. The metabolic profiling group is also currently developing techniques for the analysis of metabolic fluxes in a global manner. Studies of wood formation will involve the analysis of specific tissues in the cambium region, dissected by cryo-sectioning. Similarly, it is possible to analyse defined leaf tissue or phloem sap to create metabolic fingerprints. Further subdivision into specific cellular compartments is also possible through the use of a subcellular fractionation protocol developed at UPSC.The emerging field of proteomics has now developed into a mature technology based on high-resolution protein separation combined with database peptide matching, is by far the most commonly used method for protein identification. The need for more complete proteome coverage in a high-throughput fashion has led to the development of alternative methodologies that we currently are exploring. These techniques include use of isotopically coded affinity tags (ICAT) to label proteins from different experimental samples, and multidimensional liquid chromatography separation of peptides prior to tandem mass spectrometry analysis, providing quantitative and qualitative analysis of the proteins, including those not compatible with standard 2DE techniques. It is important to note that the complete poplar genomic sequence will be available soon, increasing the applicability of proteomics to studies in poplar.

4. A FUNCTONAL GENOMICS APPROACH TO ELUCIDATE GENE FUNCTION

We are currently using our technology platforms to analyse transcript, protein and metabolite profiles in specific cell types that form the developmental gradients leading to mature wood cells, leaves and flowers. Based on this information, candidate genes are selected for functional analysis in our two model systems. The challenge now is not large-scale EST sequencing, but to explore gene function! We are selecting candidate genes to knock out based on tissue specific transcript profiling The close link between our two models makes it possible to identify target genes in *Populus* and then isolate the corresponding knockout in *Arabidopsis* for genetic and functional analysis. This has so far been done for almost 50 candidate genes in our laboratory. However, a substantial part of the *Populus* ESTs show low, homology to any

Arabidopsis gene. Although obtaining more sequence information could probably identify the *Arabidopsis* orthologues to a majority of these genes, a large proportion will have to be functionally analysed by knocking them out in *Populus*. In collaboration with one of our industrial partners, SweTreeGenomics AB, we have initiated a large-scale knockout programme in *Populus*, which will initially be based on the *Populus* genes that have no apparent orthologues in *Arabidopsis*. We estimate that this initially will amount to ca. 2500 genes. A substantial part of the functional analysis will be done in *Arabidopsis*. This initial *Arabidopsis* screening, covering thousands of genes, is sorting out the most interesting candidates to subsequently knock out and/or overproduce in *Populus* and thus identify, or confirm, their functions in the tree. Transcript profiling together with metabolite profiling will be extremely valuable for the subsequent analysis of the knockout population since it is well known that knockouts do not always display a visually recognisable phenotype. Following the initial data collection the next logical step is to to deduce gene function. We are concentrating our efforts on genes believed to have roles in signal transduction, cell division, cell elongation, cell wall synthesis, and hormonal regulation of tree growth (Moyle et al., 2002). Genes believed to be involved in regulatory networks will be analysed by transforming *Populus* with the candidate genes under an inducible promoter to explore the network and interacting partners by transcript and metabolite profiling. A central part of our intiative is our ambition to use the combination of transcript, protein and metabolite profiling in our efforts to determine gene function. The function of a specific transcription factor is, for example, best analysed by transcript profiling. On the other hand, if the gene product is involved in cell wall synthesis, the metabolite profile will contain valuable information. Finally, most of the in-depth functional analysis will be done in *Arabidopsis* (when there is an *Arabidopsis* orthologue) and will be based on the possibility of studying genetic interactions, something that is still very difficult in the *Populus* system.

4.1. Selection of Target Genes for Analysis of Gene Function

Selection of target genes is currently based on spatial and or temporal correlation of expression with specific developmental events (Hertzberg et al., 2001a,b). The initial selection of interesting genes to knock out is based on a transcript-profiling of wood-related genes. Our database currently contains expression profiles for over 13,000 sequences. About one third of the genes show differential expression over the wood formation gradient. Genes that have a more global expression pattern or have *Arabidopsis* orthologues is sorted out and the remainder is target genes for RNAi in *Populus* In addition, some sequences are pre-selected based on general knowledge about the genes. Using a similar approach, targets to knock out in *Arabidopsis* is identified.

4.2. Phenotype Analysis

Generation of a large number of transgenic trees requires phenotypic screening that is both fast and accurate. We are developing a screening facility that will allow us to undertake large-scale, two-step screens for modifications in both fibre morphology and cell wall chemistry. The initial bulk screen will employ high throughput techniques in order to filter out candidate lines that will be carefully analysed. This initial screen will include whole plant growth analysis and high throughput analysis giving fibre measurements of the length, width and wall thickness. It is likely that most of the resulting wood phenotypes will deviate only slightly from controls, and not be detectable as visible differences in growth or anatomy. Fine tune analysis will therefore rely on NIR and FT-IR spectroscopy combined with principal component analysis (PCA), to detect chemical and ultrastructural phenotypes. Chemical characterisation will be performed by metabolite screening of extractable components, and solid wood analysis will exploit pyrolysis-MS techniques.

5. REFERENCES

Bhalerao, R., Björkbacka, H., Jonsson Birve, S., Lundeberg, J., Gustafsson, P., Jansson, S. Rapid annotation of plant ESTs. Submitted

Fiehn, O. (2001) Combining genomics, metabolome analysis, and biochemical modelling to understand metabolic network. Comp. Func. Genom. 2: 155.

Hertzberg, M., Sievertson, M., Aspeborg, H., Nilsson P., Sandberg, G. and Lundeberg, J. (2001a) cDNA microarray analysis of small plant tissue samples using a cDNA tag target amplification protocol. Plant J. 25:1

Hertzberg, M., Aspeborg, H., Schrader, J., Andersson, A., Blomqvist, K., Bhalerao, R., Rhaman, D., Uhlen, M., Teeri, T., Lundeberg, J., Sundberg, B., Nilsson, P. and Sandberg, G. (2001b) A transcriptional roadmap to wood formation. PNAS 98:14732.

Moyle, R., Schrader, J., Stenberg, A., Olsson, O., Saxena S., Sandberg, G., Bhalerao R.P.(2002) Environmental and auxin regulation of wood formation involves members of the *Aux/IAA* gene family in hybrid aspen. Plant J., in press.

Olsen, J.E., Juntilla, O., Nilsen, J., Eriksson, M.E., Martinussen, I., Sandberg, G., Moritz, T. (1997) Ectopic expression of oat phytochrome A in hybrid aspen changes critical daylength for growth and prevents cold acclimation. Plant J. 12:1339-1350.

Populus, D.B. http://poppel.fysbot.umu.se/

Sterky, F., Regan, S., Karlsson, J., Hertzberg, M., Holmberg, M., Amin, Bhalerao, R., Larsson, M., Rodhe, A., Villaroel, R., Borjean, W., Montagu, M.V., Sandberg, G., Olsson, O., Terri, T., Gustafsson, P., Uhlen, M., Sundberg, B. and Lundeberg, J. (1998) Gene discovery in the wood forming tissues of *Populus:* Analysis of 5692 Expressed Sequence Tags, PNAS 27:13330.

Uggla, C., Moritz, T., Sandberg, G. and Sundberg, B. (1996) Radial concentration gradients of endogenous IAA over the cambial region of *Pinus sylvestris L.* suggests a role in positional signalling during wood development. PNAS 93:9282.

Olsen, J.E., Juntilla, O., Nilsen, J., Eriksson, M.E., Martinussen, I., Sandberg, G., Moritz, T. (1997) Ectopic expression of oat phytochrome A in hybrid aspen changes critical daylength for growth and prevents cold acclimation. Plant J. 12:1339-1350.

ABIOTIC RESISTANCE AND CHAPERONES: POSSIBLE PHYSIOLOGICAL ROLE OF SP1, A STABLE AND STABILIZING PROTEIN FROM *POPULUS*

W. X. Wang, T. Barak, B. Vinocur, O. Shoseyov and A. Altman
Institute of Plant Sciences and Genetics in Agriculture, The Hebrew University of Jerusalem, P.O. Box 12, Rehovot 76-100, Israel (e-mail: altman@agri.huji.ac.il)

Keywords stable proteins, chaperones, abiotic stress, *Populus,* salinity tolerance

1. INTRODUCTION

The molecular basis of abiotic stress tolerance in plants, especially drought, salinity and extreme temperatures, is rather limited. In view, the worldwide devastating problems of salinization and desertification, efforts are continuing to unravel some of the molecular controls and tolerance mechanisms. Discovery of new genes for abiotic stress tolerance, combined with controlled molecular breeding will have an important role in shaping agricultural plants in the post-genomic era (Wang et al., 2001a).

SP1, a member of a novel class of plant stress-associated protein has been recently characterized from aspen (*Populus tremula* L.) (Wang et al., 2002). SP1 is a boiling-soluble, homo-oligomeric protein composed of 12 subunits of 12.4 kDa polypeptide. At transcriptional level, *sp1* was found to express a basal level, which was modified by several types of environmental stimuli, including salt-, temperature-, osmotic- and desiccation-stress. SP1 was found highly potent in protecting the in vitro activity of citrate synthase, horseradish peroxidase and other enzymes from heat-inactivation, demonstrating that SP1 functions are similar to chaperones (Wang et al., 2001b). In this report, we present studies of *sp1* in transgenic aspen plants under the salinity stress, as well as in the salt resistant *Populus euphratica*, which natively inhabits the Ein Avdat National Park located in the Negev desert, Israel. The site is characterized by a very high salt content in soil and water. Preliminary studies done in the area during the summer revealed a high water conductivity and sodium content in the spring water, while soil water conductivity was 21 folds higher, and Na^+ content was 36 fold higher. Comparison of growth performance and changes in SP1 levels of these two *Populus* species upon salt stress is also presented.

I. K. Vasil (ed.), Plant Biotechnology 2002 and Beyond, 439-443.

2.1. Salt-Tolerance Of *Sp1*-Transgenic Plants

Transgenic *Populus tremula* plants expressing different levels of either *sp1* transcript or protein (data not shown) were employed in 150-mM salt stress experiments. Among the 3 tested transgenic lines (L3, M4, H3), line L3 was found to express low level of *sp1* transcript, and M4 accumulates high level of transcript. Both L3 and M4 express normal levels of SP1 protein, similar to that of the non-transformed (NT) plants. The transgenic line H3 expresses however both higher level of *sp1* transcript and SP1 protein. No significant differences in growth were found among all the tested *sp1*-transgenic lines and NT plants under non-stress conditions. However, when plants were subjected to salinity stress, significant differences were found (Table 1). Early symptoms of damage due to salt stress, such as wilting, necrosis, and death of leaves were first observed in NT and in the transgenic lines M4 and L3. Appearances of these symptoms were considerably delayed in line H3, which had a better growth (36.5% increase) compared to that of NT and the other transgenic line. Surprisingly, line M4 showed a significant reduction in growth compared to NT. During the recovery from 3 week salt stress, line H3 exhibited even better growth performance, e.g. a 68% increase in length and a 51% increment in leaf retention. The better growth performance was further demonstrated when expressed as dry weight of vegetative tissues. In contrast, line M4 almost did not recover from the stress.

Table 1. Growth performance of transgenic aspen (P. tremula) plants under salt stress. Six-week potted plants were subjected to salt stress (irrigating plants once a day with 150 mM NaCl and $CaCl_2$ (in a molar ratio of 6 to 1) and 200 ppm fertilizer) for 3 weeks. Plants irrigated with only 200 ppm fertilizer served as controls. Afterwards, all plants were irrigated as for control plants for 3 additional weeks ("recovery"). Plant length, number of leaves was measured at the end of the stress and recovery. The dry weight was taken after the recovery. Values represent the mean weights ± SE of 6 plants. Values within each column determined to be significantly different ($p \leq 0.05$) by the Student's t test are denoted by different superscript letters. Numbers in parentheses are the percentages of transgenic plants compared to NT plants, which was taken as 100.

line	Growth (cm)		Leaf retention		Dry weight (g)
	stress	recovery	stress	recovery	
NT	6.4^{ab}±1.9(100)	8.0^{a}±2.0(100)	17.8^{a}±1.2(100)	17.0^{ab}±1.2(100)	1.7^{ab}±0.3(100)
H3	8.8^{a}±0.7(136)	13.4^{a}±3.2(168)	19.5^{a}±1.0(109)	25.7^{a}±2.1(151)	2.9^{a}±0.7(169)
L3	5.9^{ab}±0.4(92)	7.3^{a}±2.4(92)	16.6^{a}±0.9(93)	16.7^{ab}±1.8(98)	1.6^{ab}±0.3(93)
M4	5.1^{b}±1.0(79)	0^{b}±0.3(0)	12.7^{a}±1.8(71)	1.7^{b}±3.1(10)	0.3^{b}±0.0(19)

2.2. SP1 Expression In *P. Euphratica*

SP1 homologue was found in the salt resistant *P. euphratica,* using anti-SP1 antibodies (Fig. 1). SP1 homologue detected in *P. euphratica,* sharing with *P. tremula* a similar size (12.4 kDa) and several biochemical properties (e.g.

stable upon boiling, resistant to protease digestion). The internal sequences of this homologue showed that they are identical to SP1 sequence (data not shown).

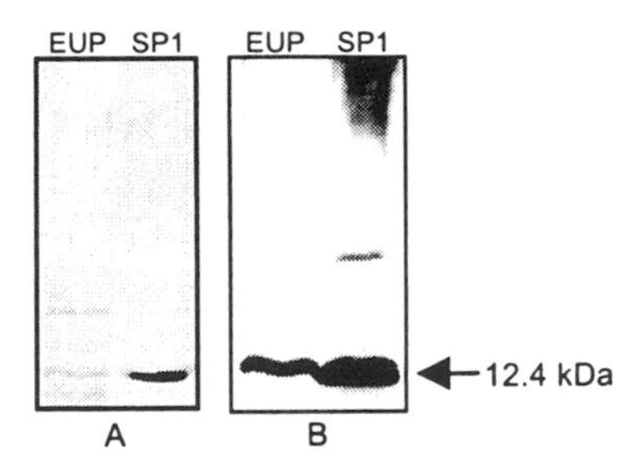

Figure 1. SP1 expression in P. euphratica. Two hundred μg total soluble proteins of P. euphratica (EUP) after boiling and proteolysis was subjected to SDS-PAGE analysis (A) and western blotting (B) using SP1-antibodies. As a positive control, purified SP1 (10 μg) from P. tremula (SP1) was ran in parallel for SDS-PAGE and western blotting.

2.3. Growth Performance And Sp1 Expression In *P. Tremula* And *P. Euphratica* In Response To Salt Stress

The responses to salt stress of in vitro cultured *P. tremula*, *P. euphratica* and *sp1*-transgenic *P. tremula* line H3 were compared. Salt stress treatment was applied by removing plantlets from agar medium and immersing the root system in vials containing 150 mM salt (NaCl:$CaCl_2$ molar ratio at 10:1) and MS liquid medium. Plantlets cultured in MS medium were used as controls. The solutions were replaced every two days and the plant length was measured. Salt stress significantly inhibited growth of all plantlets (Fig. 2). However, different stress symptoms and growth kinetics were observed. Non-transformed *tremula* showed leaf necrosis after 3 days of salt treatment, and died after 4 days. As evident by the growth in the length (TRE-salt), that plantlets stopped growing after only 2 days of salt stress. *Euphratica* was found to respond to the stress similarly, but with at least 2 days delayed in developments of the stress symptoms. Yellowish leaves were appeared after 4 days of salt stress. Although the stress caused cessation of the growth after 4 days treatment (EUP-salt), plants did not die upon longer-term (e.g. 6 days) stress. In contrast, *sp1*-transgenic *tremula* H3 showed a better performance under the same stress treatments. Plantlets did not stop the growing although the growth was significantly slowed down (H3-salt), and the stress symptoms were developed only after 6 days of the stress. Examining the levels of SP1 accumulation during the treatments, it was found that SP1 levels were similar in control plants (data not shown). With salt stress, SP1 levels were varied (inserted figure in Fig. 2). A reduction in the level of SP1 (compared to that of non-treated plants of each species) were observed with non-transformed *tremula* (TRE), up to 26% reduction in SP1 level was found when plants were stressed for 6 days. With *euphratica*, high level of SP1 accumulation (30%) was found at 4 days stress (EUP), which may indicate an adaptation to the stress. Higher SP1 levels were maintained in H3 plants along the whole period of stress.

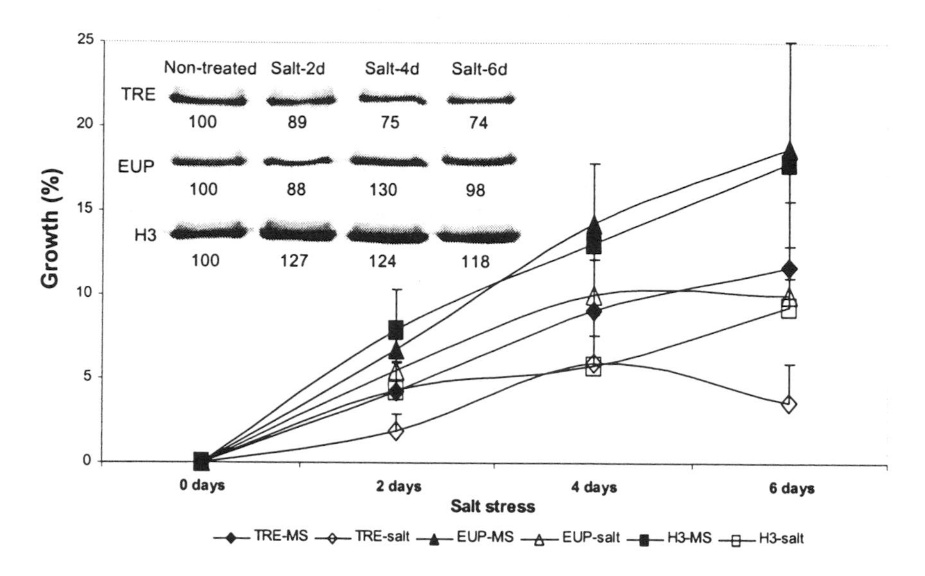

Figure 2. Growth performance and SP1 expression of 3 different poplar plants in response to salt stress. Line drawings: plant growth (growth (%) = ((final length - initial length)/initial length) X 100). Values represent the mean weights ± SE of at least 8 plants Inserted picture: SDS-PAGE analysis and quantification of SP1 expression in the 3 different plants. Same quantity of total soluble protein after boiling and proteolysis was loaded in each lane, and SP1 expression was determined densitometrically. The figures under each band are the percentage of SP1 level relative to non-treated plants (=100%). TRE, tremula; EUP, euphratica; H3, sp1-transgenic tremula.

3. FINAL REMARKS

The transgenic *P. tremula* line H3, which expresses a high level of SP1, evinces a better growth performance and increased tolerance to salinity stress when subjected to NaCl in pot experiments, as judged by stem elongation, leaf retention and dry weight accumulation. The results suggest possible positive correlation between SP1 protein expression and salt stress tolerance. Comparisons between in vitro cultured *P. tremula* and the native salinity-tolerant *P. euphratica* further support our observations that the better performance of plants under salinity is probably associated at least in part with the high levels of SP1 accumulation. This novel homo-oligomeric protein functions in vitro as a chaperone, and may have a similar in vivo function.

4. FINANCIAL SUPPORT

Funding by the European Union (QLK5-2000-01377- ESTABLISH), By Yissum (The Hebrew University of Jerusalem) and by the Chief Scientists, Israel Ministry of Agriculture, is gratefully acknowledged.

5. REFERENCES

Wang, W.X., B. Vinocur, O. Shoseyov and A. Altman. 2001a Biotechnology of plant osmotic stress tolerance: physiological and molecular considerations. Int. Symp. on In Vitro Culture and Horticultural Breeding. S. Sorvari et al. (Eds.), Acta Horticulturae 560. Pp. 285-292.

Wang W.X., D. Pelah, T. Alergand, O. Shoseyov and A. Altman. 2001b Boiling and/or detergent stable, and/or protease resistant, chaperone-like oligomeric proteins, polynucleotides encoding the same, and their uses (Provisional Patent Application No. 60/272,771).

Wang W.X., D. Pelah, T. Alergand, O. Shoseyov and A. Altman. 2002 Characterization of SP1, a Stress-Responsive, Boiling-Soluble, Homo-Oligomeric Protein from Aspen (*Populus tremula* L.). Plant Physiol. (in press).

MODIFICATION OF LIGNIN BIOSYNTHESIS IN FOREST TREES

Vincent L. Chiang
Plant Biotechnology Research Center, School of Forestry and Wood Products, Michigan Technological University, Houghton, MI 49931 USA (email: vchiang@mtu.edu)

1. INTRODUCTION

Many of society's fiber, chemical and energy demands are met through the industrial-scale production of cellulose from wood, during which lignin must be removed at tremendous environmental and process cost. As a result, biotechnology has been used as a promising approach to reduce lignin quantity in trees to improve the efficiency of cellulose production from wood (Bugos et al., 1989; Trotter, 1990; Whetten et al., 1998).

Despite considerable efforts, no successful case of genetic reduction of lignin has been previously reported in tree species. Down-regulating genes encoding caffeate *O*-methyltransferase (COMT) or cinnamyl alcohol dehydrogenase (CAD) have been used as the approaches to lignin reduction but instead have resulted in modified lignin structure (Tsai et al., 1998; Van Doorsselaere et al., 1995; Baucher et al., 1996). 4-Coumarate:CoA ligases (4CLs), a group of enzymes catalyzing the necessary activation of the hydroxylated cinnamic acids to their corresponding thioesters, have been suggested for a role in controlling the biosynthesis of phenylpropanoids, including lignin and flavonoids (Mansell et al., 1972; Knobloch and Hahlbrock, 1975). Thus, manipulating lignin-specific 4CL might provide a means to control lignin content in transgenic trees. However, no lignin-specific 4CL has been identified. Here we present the cloning and characterization of two structurally and functionally distinct 4CL genes, *Pt4CL1* and *Pt4CL2*, from aspen and demonstrate that *Pt4CL1* is devoted to lignin biosynthesis in developing xylem tissue. By antisense down-regulation of the lignin specific *Pt4CL1* expression in aspen, we generated, for the first time, transgenic trees that accumulated structurally normal lignin at substantially reduced levels.

2. EXPERIMENT

Developing secondary xylem tissues were collected from 4-year-old quaking aspen (*Populus tremuloides* Michx.) grown on the campus of Michigan Technological University, and young leaves and internodes from greenhouse-

I. K. Vasil (ed.), Plant Biotechnology 2002 and Beyond, 445-452.

grown quaking aspen. Tissues were immediately frozen and stored in liquid nitrogen until used for protein, RNA, and DNA isolation.

Total RNA was isolated from developing xylem according to Bugos et al.(1995), from which mRNA was isolated using the Poly $(A)^+$ mRNA isolation kit (Tel-test B). A unidirectional λ gt22 expression cDNA library was constructed from xylem mRNA using Superscript λ System (Life Technologies) and Gigapack Packaging Extracts (Stratagene) (Ge and Chiang, 1996). A ^{32}P-labeled parsley 4CL-2 cDNA (a gift from Dr. Carl Douglas) probe was used to screen 5×10^5 plaque-forming units of the aspen xylem cDNA library. Three positive clones were obtained after 3 rounds of plaque purification and end-sequenced using the Δ *Taq* Cycle Sequencing Kit (Amersham). Since end-sequencing revealed that these three clones were analogous to each other, only two of them with longer inserts were completely sequenced for both strands and found to be identical full-length cDNAs and were designated as *Pt4CL1*.

The full-length *Pt4CL1* cDNA was then used for screening an aspen genomic library constructed by cloning *Sau3A* I partial-digested and sucrose gradient-selected genomic DNA fragments into the *Bam*H I site of λDASH II vector (Clontech). Eleven positive clones were obtained and subjected to further purification. Four clones were found to contain the full coding sequence with at least a 2-kb 5' flanking region and one of which, *Pt4CL1g*-4 (≈15 kb), was selected for restriction mapping and partial sequencing to confirm its authenticity. The 5' flanking region was then subcloned into pGEM 7Z for further characterization.

Two sense and one antisense degenerate primers were designed based on the consensus regions of all known plant 4CL amino acid sequences. Two sense primers includeR1S (5'-TTGGATCCGGIA- CIACIGGIYTICCIAARGG) which is located at the first putative AMP-binding domain and H1S (5'-TTGGATCCGTIGCICARCARGTIGAYGG) at 9 amino-acid downstream from the first AMP-binding domain. The antisense primer R2A (5'-ATGTCGACCICKDATRCADATYTCICC) was designed based on the second putative AMP-binding domain, the CGEICIRG motif. Ten micrograms of total RNA each isolated from top (1st to 4th) and lower (6th to 10th) aspen stem internodes, respectively, were reverse transcribed and one-fifth of the reaction mixtures were used as the template for PCR amplification with 2 μM each oligo dT_{20} primer and R1S primer, 200 μM dNTPs, and 2.5 units of *Taq* DNA polymerase (Promega) in a 50 μL reaction. Two microliters of each reverse transcription PCR (RT-PCR) reaction products were subjected to a nested-PCR using H1S and R2A as the primers. Conditions for both runs of PCR were: 94°C/5 min, 30 cycles of 94°C/45 sec, 50°C/1 min, 72°C/1 min 30 sec, and 72°C/5 min. A ≈600 bp fragment amplified from top internode RNA was cloned (Pt4CL2-600) using the TA cloning kit (Invitrogen) and was used to screen the aspen genomic library as described above, and 7 positive clones were identified. One of the clones, Pt4CL2g-11 (≈13 kb), was selected, subcloned and sequenced for its 5' flanking and full coding regions. Based on the genomic sequence, two primers (2A: 5'-TCTGTCTAGATGATGTCGTGGCCACGG and 2B: 5'-TTAGATCTCTAGGACATGGTGGTGGC) were designed around the deduced translation start and stop sites of Pt4CL2g-11, respectively, and used for RT-PCR amplification of top internode RNA as described above. A 1.7-kb cDNA fragment was amplified, cloned, sequenced, and designated as *Pt4CL2*.

PCR was used to introduce a *Bam*HI site at the 5' and 3' ends of *Pt4CL1* cDNA and the amplified product was digested with *Bam*HI and cloned into the same site of pQE-31 containing a histidine tag (6xHis tag, Qiagen). Similarly, a *Bgl*II site was introduced into the 5' end of *Pt4CL2* cDNA and a 400-bp PCR fragment was amplified, sequenced to confirm the fidelity of PCR and used to replace the corresponding 5' end of the original *Pt4CL2* cDNA. The engineered *Pt4CL2* cDNA was cloned into the 6xHis-containing pQE-32 vector (Qiagen) at *Bam*HI and *Kpn*I sites. The expression vector was transferred into *E. coli* strain M15, and the growth and induction of bacterial cells with isopropyl β-D-thiogalactoside were performed according to the manufacturer's protocol (Qiagen). The isolation and purification of the Pt4CL1- and Pt4CL2-6xHis tag fusion proteins were conducted using His•Bind Resin (Novagen) as described by Li et al. (1997).

Plant crude protein extracted from aspen developing xylem tissues (Tsai et al., 1998) and purified Pt4CL1 and Pt4CL2 recombinant proteins were used for enzyme assays as described (Ranjeva et al., 1976). 5-Hydroxyferulic acid was synthesized (Li et al., 1997) and all other substrates were obtained from Sigma. Protein concentrations were determined by the Bradford assay (Bradford, 1976) using BSA as a standard. Southern and Northern blot analyses using *Pt4CL1* or *Pt4CL2* cDNA as a probe were performed as described (Tsai et al., 1998).

3. RESULTS AND DISCUSSION

Two full-length cDNAs, *Pt4CL1* and *Pt4CL2*, encoding 4-coumarate:coenzyme A ligases were cloned and sequenced (GenBank accession No. AF041049 and AF041050, respectively) (Hu et al., 1998). Southern blot analysis of aspen genomic DNA with full-length *Pt4CL1* and *Pt4CL2* cDNAs as probes, respectively, at high stringency revealed that each cDNA probe hybridized to multiple restricted DNA fragments with clearly distinguishable patterns, suggesting that these two cDNAs do not cross-hybridize to each other and that multiple members could be present in both

Pt4CL1 and *Pt4CL2* gene families (data not shown). Northern blot analysis of aspen RNA showed high *Pt4CL1* mRNA levels in secondary developing xylem and lower internodes, moderate level in top internodes, and low level in leaves, as indicated by the intensity of the hybridizing bands of about 1.9 kb that corresponds to the length of the *Pt4CL1* transcript (Fig. 1). Since lignification is known to take place in secondary developing xylem and vascular tissues in leaves and young internodes, these Northern hybridization results are consistent with the interpretation that Pt4CL1 is associated with lignification. In addition, these Pt4CL1 gene expression patterns are in concert with those of other lignin pathway genes encoding C4H, OMT, and F5H (our unpublished results), providing further evidence that Pt4CL1 is involved in lignification. In sharp contrast, no *Pt4CL2* mRNA message could be detected in secondary developing xylem and the message was barely noticeable in lower internodes (Fig. 1). However, a high level of *Pt4CL2* mRNA was observed in top internodes and a weak expression of *Pt4CL2* mRNA was found in leaves (Fig. 1), suggesting that Pt4CL2 could be involved in phenylpropanoid biosynthesis in these green tissues, but the absence of *Pt4CL2* mRNA from secondary developing xylem and lower internodes precludes the involvement of Pt4CL2 as an essential enzyme in lignin biosynthesis in xylem.

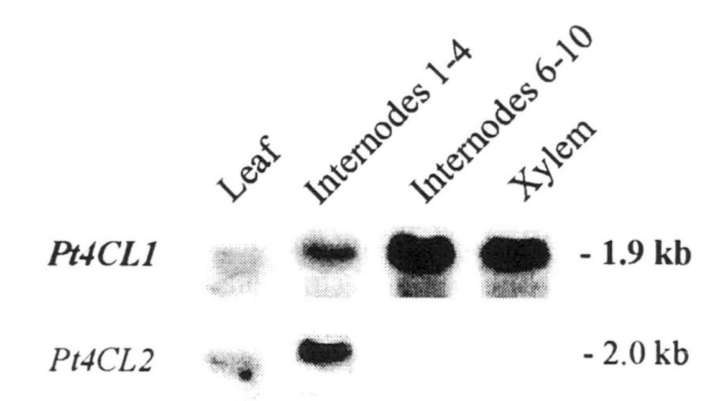

Fig. 1. Northern blot analysis of total RNA (10 µg/lane) from various aspen plant tissues, probed with 32P-labeled Pt4CL1 and Pt4CL2 cDNAs, respectively.

When we repeated the Northern blot analysis of RNAs isolated separately from epidermal layers and developing xylem of aspen stem internodes (6 to 25 from shoot tip), it was found that the expression of *Pt4CL2* mRNA is actually absent from developing xylem tissues but is evident in epidermal layers (Fig. 2), providing clarification that *Pt4CL2* is epidermis-specific in aspen stem. Thus, the specific expression of *Pt4CL2* mRNA in epidermis further supports the biochemical functions of Pt4CL2 protein in the biosynthesis of non-lignin-related phenylpropanoids. When the Northern blot was re-probed with *Pt4CL1* cDNA, the results further validated that *Pt4CL1* mRNA is specifically expressed in developing xylem but not in epidermal layers (Fig. 2).

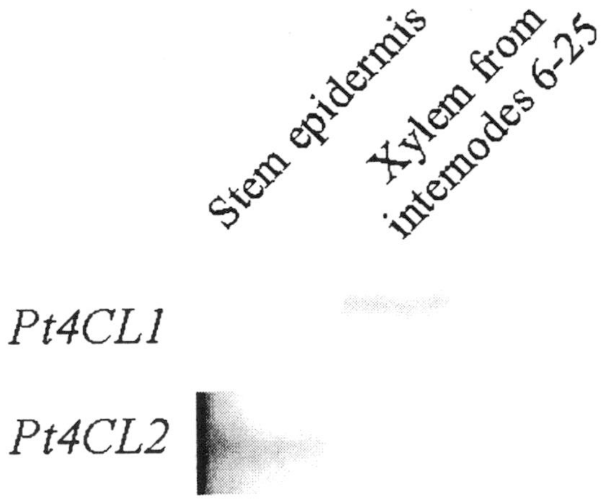

Stem epidermis

Xylem from internodes 6-25

Pt4CL1

Pt4CL2

Fig. 2. Northern blot analysis of total RNA from aspen, probed with 32P-labeled Pt4CL1 and Pt4CL2 cDNAs, respectively.

To compare the biochemical functions of Pt4CL1 and Pt4CL2, *E. coli*-expressed Pt4CL1 and Pt4CL2 proteins were purified to apparent homogeneity with a histidine tag sequence-specific affinity column (Novagen) (Li et al., 1997), and their purity was demonstrated by a single major band of about 60 kDa for Pt4CL1 and 63 kDa for Pt4CL2, as estimated by SDS/PAGE (data not shown). The purified recombinant proteins were characterized for their catalytic activities with various hydroxycinnamic acid derivatives (Fig. 3). The results showed that, in addition to a clear divergence in substrate preference between Pt4CL1 and Pt4CL2, the specific activities of Pt4CL1 with 4-coumaric, caffeic, and ferulic acids were about 6, 10, and 16 times, respectively, higher than those of Pt4CL2, reflecting the markedly different catalytic efficiency between these two aspen 4CLs. The most striking functional difference between Pt4CL1 and Pt4CL2 is that Pt4CL1 can efficiently utilize 5-hydroxyferulic acid, a substrate for the typical syringyl lignin in angiosperms (Kutsuki et al., 1982; Higuchi, 1985), whereas Pt4CL2 is inactive with this compound (Fig. 3). Like Pt4CL1, aspen secondary xylem crude protein extracts also utilized 5-hydroxyferulic acid in addition to 4-coumaric, caffeic, and ferulic acids with relative specific activities that were comparable to the relative specific activities of the purified recombinant Pt4CL1 (data not shown), providing evidence that 4CL activities in aspen xylem extracts are derived mainly from Pt4CL1. These results are consistent with the presence of a high level of Pt4CL1 mRNA in and the absence of Pt4CL2 mRNA (Fig. 1 & 2) from aspen developing xylem.

The preference of Pt4CL1 for lignin-specific substrates, such as ferulic and 5-hydroxyferulic acids (Fig. 3), suggest that Pt4CL1 is involved in lignin biosynthesis. The lignin-related Pt4CL1 protein function is in agreement with the finding that *Pt4CL1* mRNA is specifically expressed in lignifying xylem tissues.

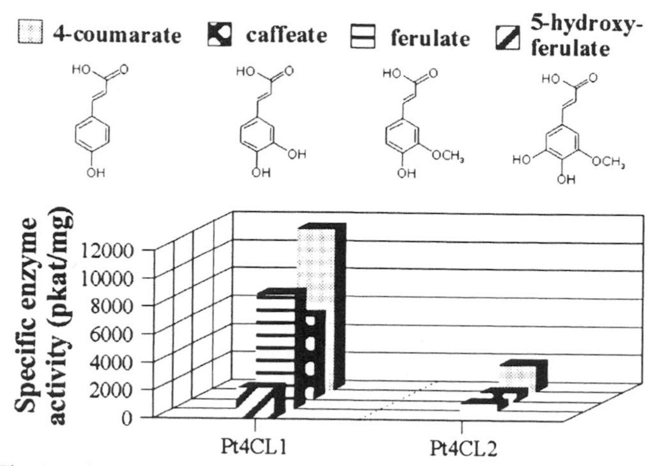

Fig. 3. Substrate specificity of E. coli-expressed and purified Pt4CL1 and Pt4CL2 recombinant proteins.

The fact that the absence of *Pt4CL2* mRNA from lignifying xylem tissues and its presence in young green tissues (Fig. 1), indicates the involvement of Pt4CL2 in converting its hydroxycinnamic acid substrates into early phenylpropanoid pathways other than lignin. In fact, in the context of substrate utilization, Pt4CL1 and Pt4CL2 resemble closely the lignin- and flavonoid-associated isoenzymes, respectively, detected in soybean cell cultures as described by Knoblock and Hahlbrock (1975). Therefore, in aspen Pt4CL1 is a key lignin-pathway enzyme that oversees a continuous metabolic flux for syringyl lignin formation by directing 5-hydroxyferulic acid into the network of lignin biosynthesis. The ability of Pt4CL1 in utilizing multiple substrates *in vitro* implies that Pt4CL1 is potentially capable of regulating the biosynthesis lignin.

To test whether lignin biosynthesis can be controlled by Pt4CL1, transgenic aspen trees with suppressed expression of *Pt4CL1* gene was produced.[18] These trees had a 40 to 45% reduction in lignin quantity, pointing to a role for *Pt4CL1* in regulating lignin biosynthesis. Concomitantly, these lignin-reduced trees had and a 9 to 15% increase in cellulose content, resulting in a cellulose:lignin ratio of 4 compared with 2 in normal aspen. 2-D NMR analysis of MWLs indicated that lignin reduction did not significantly alter its structure. The severe lignin reduction in transgenic trees did not adversely affect growth and development and in fact these transgenic trees had thicker stems, larger leaves, and longer internodes than in control.

4. CONCLUSION

Currently, we are investigating the underlying mechanisms for the enhanced growth in these transgenic trees. The conclusion at this point is that lignin can be genetically reduced without compromising cellulose biosynthesis and the integrity or growth of trees. Most importantly, our results further indicate that in fact the deposition of cellulose and lignin may be regulated in a compensatory fashion in trees such that decreases in one are compensated for by increases in the other to maintain a constant carbon allocation for these two major cell-wall structural components.

5. ACKNOWLEDGMENT

This research is supported in part by grants from the NSF, the USDA-National Research Initiative Competitive Grants Program, and the USDA-McIntir-Stennis Forestry Research Program.

6. REFERENCES

Baucher, M., et al. 1996. Plant Physiology 112: 1479-1490.

Bradford, M. (1976) Anal. Biochem. 72, 248-254.

Bugos, R.C., V.L. Chiang and W.H. Campbell. 1989. Isolation and preliminary characterization of O-methyltransferase from aspen. Proceedings of the Thirty-Fourth Japanese Lignin Symposium at Nagoya, Japan. Nagoya University Press. 25-29.

Bugos, R.C., Chiang, V.L., Zhang, X.-H., Campbell, E.R., Podila, G.K. & Campbell, W.H. (1995) BioTechniques 19, 734-737.

Ge, L. & Chiang, V.L. (1996) Plant Gene Register, 96-075.

Higuchi, T. (1985) in Biosynthesis and Biodegradation of Wood Components, ed. Higuchi, T. (Academic Press, New York), pp. 141-160.

Hu, W.J., Kawaoka, A., Tsai, C.J., Lung, J., Osakabe, K., Ebinuma, H. & Chiang, V.L. (1998) Proc. Natl. Acad. Sci. USA 95, 5407-5412.

Hu, W.J., Harding, S.A., Lung, J., Popko, J.L., Ralph, J., Stokke, D.D., Tsai, C.J., and Chiang, V.L. (1999) Nature Biotechnol. 17, 808-812.

Knobloch, K-H & Hahlbrock, K. (1975) Eur. J. Biochem. 52, 311-320.

Kutsuki, H., Shimada, M. & Higuchi, T. (1982) Phytochem. 21, 267-271.

Li, L., Popko, J.L., Zhang, X.-H., Osakabe, K., Tsai, C.J., Joshi, C.P. & Chiang, V.L. (1997) P roc. Natl. Acad. Sci. USA 94, 5461-5466.

Mansell, R.L., Stöckigt, J. & Zenk, M.H. (1972) Z. Pflanzenphysiol. 68, 286-288.

Ranjeva, R., Boudet, A.M. & Faggion, R. (1976) Biochimie 58, 1255-1262.

Trotter, P.C.. 1990. Tappi 73(4): 198-204.

Tsai, C.-J., Popko, J.L., Mielke, M.R., Hu, W.-J., Podila, G.K., and Chiang, V.L. 1998. Plant Physiology 117(5): 101-112.

Tsai, C.J., Popko, J.L., Mielke, M.R., Hu, W.J., Podila, G.K. & Chiang, V.L. (1998) Plant Physiol. 117:101-112.

Van Doorsselaere, et al. 1995. Plant J. 8: 855-864.

Whetten, R.W., MacKay, J.J., and Sederoff, R.R. 1998. Annu. Rev. Plant Physiol. Plant Mol. Biol. 49: 585-609.

FUNCTIONAL GENOMICS OF WOOD FORMATION IN HYBRID ASPEN

Henrik Aspeborg[1], Kristina Blomqvist[1], Stuart Denman[1], Magnus Hertzberg[2], Anna Ohlsson[1], Torkel Berglund[1], Göran Sandberg[2], Joakim Lundeberg[1], Fredrik Sterky[1], Tuula T. Teeri[1] and Peter Nilsson[1]
[1] Dept. of Biotechnology, KTH – Royal Institute of Technology, AlbaNova University Center, S-106 91 Stockholm, Sweden (e-mail: aspe@biochem.kth.se)
[2] Umeå Plant Science Center, S-901 83 Umeå, Sweden.

Keywords microarrays, *Populus*, xylogenesis

1. INTRODUCTION

In 1997 the SCTFG (Swedish Center for Tree Functional Genomics) a collaboration between Umeå Plant Science Center (Department of Plant Physiology, Umeå University and Department of Forest Genetics and Plant Physiology, SLU Umeå) and the Department of Biotechnology, KTH Stockholm, started a high throughput EST-sequencing project of the hybrid aspen (*Populus tremula* x *tremuloides*) (Sterky et al., 1998). Until spring 2002, the EST-initiative has generated more than 100 000 ESTs from 19 different cDNA libraries. The microarray facility has so far produced one 3000-element array and another 14000-element array which has been used in extensive expression studies. An array including even more elements, representing all 100 000 ESTs, will be produced in the near future. As eight of the cDNA libraries are wood related the SCTFG makes an exceptional platform for the identification of genes and enzymes controlling critical steps in the wood formation process.

2. TRANSCRIPT PROFILING

A first cDNA microarray was designed to include a unigene set consisting of 2995 unique ESTs collected from a broad cambial/developing xylem zone and the arrays were hybridized with targets prepared from different developmental stages of xylogenesis using a novel, in house 3' cDNA tag amplification method (Hertzberg et al., 2001). The analysis revealed that the genes encoding lignin and cellulose biosynthetic enzymes, as well as a number of transcription factors and other potential regulators of xylogenesis, are under strict developmental stage-specific transcriptional regulation

I. K. Vasil (ed.), Plant Biotechnology 2002 and Beyond, 453-454.

(Hertzberg et al., 2001). A number of highly up-regulated genes during secondary cell wall formation were found to encode novel proteins or proteins showing similarity to plant proteins with an unknown function. For a second array, a collection of 33 400 ESTs from seven different cDNA libraries were clustered into almost 14 000 unique clusters. One representative clone from each cluster was included on the microarray. A number of experiments comparing the expression of hybrid aspen cell cultures have been performed, in order to characterize the cultures and their potential use as a system to study carbohydrate biosynthesis and wood formation.

3. FULL-LENGTH SEQUENCING

A bioinformatic analysis of the "unknown" sequences up-regulated during secondary cell wall formation revealed 8 clones which contain modules characteristic of carbohydrate-active enzyme. Closer analysis revealed that each of the 8 clones correspond to different families of glycosyl tranferases. Functional analysis is now underway in order reveal their roles in plant carbohydrate biosynthesis. As a first step towards a functional analysis of the rest of the "unknown" genes, 70 transcripts specifically up-regulated during the secondary cell wall formation, were selected for full-length sequencing. To increase the throughput of the sequencing, a novel, shotgun sequencing approach has been designed.

4. REFERENCES

Hertzberg M, Aspeborg, H, Schrader J, Andersson A, Erlandsson R, Blomqvist K, Bhalerao R, Uhlen M, Teeri TT, Lundeberg J, Sundberg B, Nilsson P, Sandberg G. 2001 A transcriptional roadmap to wood formation. Proc Natl Acad Sci U S A. 4;98:14732-14737.

Hertzberg M, Sievertzon M, Aspeborg H, Nilsson P, Sandberg G, Lundeberg J. 2001 cDNA microarray analysis of small plant tissue samples using a cDNA tag target amplification protocol. Plant J. 5:585-91.

Sterky F, Regan S, Karlsson J, Hertzberg M, Rohde A, Holmberg A, Amini B, Bhalerao R, Larsson M, Villarroel R, Van Montagu M, Sandberg G, Olsson O, Teeri TT, Boerjan W, Gustafsson P, Uhlen M, Sundberg B, Lundeberg J. 1998 Proc Natl Acad Sci U S A, 95:13330

FUNCTIONAL GENOMICS OF WOOD FORMATION

Jae Heung Ko, Sookyung Oh, Sunchung Park, Jaemo Yang, Kyung-Hwan Han*

Department of Forestry, 126 Natural Resources, Michigan State University, East Lansing, MI 48824-1222 USA (e-mail: hanky@msu.edu).

Keywords *Arabidopsis*, secondary growth, wood formation, xylem

1. INTRODUCTION

Wood is of primary importance to humans as timber for construction, and wood-pulp for paper manufacturing. It is also the most environmentally cost-effective renewable source of energy. Resolving the dilemma of preserving forest ecosystems, while meeting the increasing demand of forest utilization, necessitates gaining a fundamental understanding of the biochemical processes involved in tree growth and development.

2. WOOD FORMATION IN *ARABIDOPSIS*

2.1. *Arabidopsis* As A Model For Wood Formation Study

Arabidopsis has been shown to express all of the major components of wood development during its ontogeny and used as a model for the study of wood and fiber production in trees (Lev-Yadun, 1994; Zhao et al., 2000). Recent advances in *Arabidopsis* genomics provide the tools and resources necessary to accelerate this task.

2.2. Gene Expression Profiles in Wood Forming Tissues

Secondary xylem was induced in *Arabidopsis* by either repeatedly removing inflorescence while growing in low density (Lev-Yadun, 1994) or combination of short- and long-day treatment (Fig. 1). We are screening the *Arabidopsis* enhancer trap lines (Campisi et al., 1999). About 3,000 individual lines (~26 % of total 11,370 lines) were screened for reporter gene expression in developing secondary xylem. Sixty-four of the screened lines have shown stem-specific gene expression. We are currently characterizing the selected lines. cDNA-AFLP analysis allowed us to survey about 38,000 transcripts of *Arabidopsis*. A total of 124 transcripts were identified as specific or abundant in secondary xylem tissue, and 45 transcripts in bark

I. K. Vasil (ed.), Plant Biotechnology 2002 and Beyond, 455-456.

tissue. The 8.2k Affymetrix *Arabidopsis* GeneChip® Genome Arrays were used to survey differentially expressed genes during wood formation. Sixty-one transcripts were up-regulated more than ten-fold in the wood-forming stem. The list of the genes whose expression level was dramatically increased includes heat shock transcription factor, molybdopterin synthase, Myb transcription factors, chloride channel protein, and aquaporin.

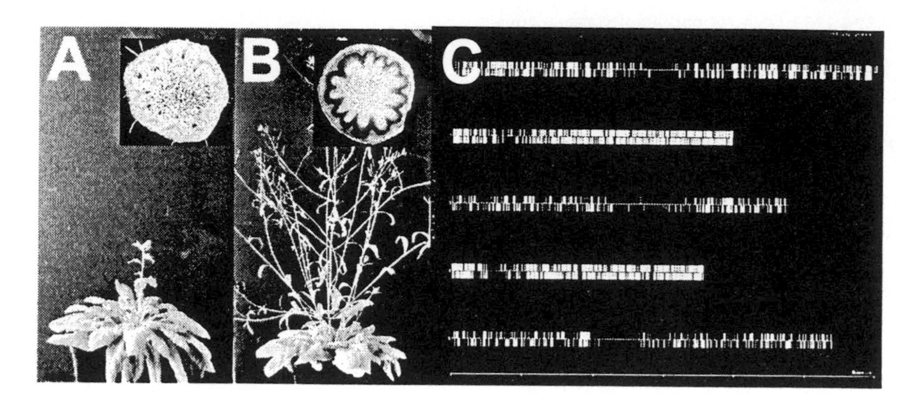

Figure 1. Gene expression profiles in 7 weeks old Arabidopsis *plants: A, control plants; B, wood formation-treated; C, Five chromosomes of* Arabidopsis *showing the gene expression patterns of the 8,200 genes studied. Red color indicates up-regulation, blue for down-regulation, and yellow for no change.*

3. REFERENCES

Campisi, L., Yang, Y. Z., Yi, Y., Heilig, E., Herman, B., Cassista, A. J., Allen, D. W., Xiang, H. J., and Jack, T. (1999). Generation of enhancer trap lines in *Arabidopsis* and characterization of expression patterns in the inflorescence. Plant J. 17:699-707.

Lev-Yadun, S. (1994). Induction of sclereid differentiation in the pith of *Arabidopsis thaliana* (L.) Heynh. J. Exp. Bot. 45:1845-1849.

Zhao, C., Johnson, B.J., Kositsup, B., and Beers, E.P. (2000). Exploiting secondary growth in *Arabidopsis*. Construction of xylem and bark cDNA libraries and cloning of three xylem endopeptidases. Plant Physiol. 123:1185-96.

PHENYLALANINE AMMONIA-LYASE GENE EXPRESSION IN CONDENSED TANNIN-ACCUMULATING AND LIGNIFYING CELLS OF QUAKING ASPEN

Scott A. Harding, Yu-Ying Kao and Chung-Jui Tsai
Plant Biotechnology Research Center, School of Forestry and Wood Products, Michigan Technological University, Houghton, MI 49931 USA (e-mail: chtsai@mtu.edu)

Keywords phenylalanine ammonia-lyase, condensed tannins, lignin, *Populus*

1. INTRODUCTION

Allelochemical pools of condensed tannins and salicylate-derived phenolic glycoside, along with lignins constitute large phenylpropanoid carbon sinks in tissues of quaking aspen (*Populus tremuloides* Michx.). Their synthesis depends on *L*-phenylalanine ammonia-lyase (PAL), which deaminates *L*-phenylalanine to initiate phenylpropanoid metabolism. Despite its apparently generic role in providing *trans*-cinnamic acid for secondary metabolism, PAL often occurs as multiple genes and protein isoforms. In *Populus*, expression of different PAL genes in developing stems and leaves has been most specifically associated with lignification in the past. Here we report identification of two differentially expressed PAL cDNAs from quaking aspen that exhibit distinct metabolic associations in developing tissues. *PtPAL1* was expressed in non-lignifying tissues of shoots and roots, whereas *PtPAL2* was most abundant in heavily lignified structural cells of aspen shoots. We suggest specific metabolic roles for the proteins encoded by these *PAL* genes based on spatiotemporal correlation of their transcript abundance with cellular condensed tannin distribution and lignification. Leaf wounding specifically increased *PtPAL1* expression as well as allocation of carbon to condensed tannins (CT), but did not increase *PtPAL2* expression. Our data provides *in vivo* support for the idea that, although PAL isoforms catalyze only one enzymatic reaction, they are structurally tailored to function in specific metabolic environments. This data may apply to the manipulation of PAL isoforms in species like aspen with allelochemical and lignin sink commitments that can be large enough to affect growth.

I. K. Vasil (ed.), Plant Biotechnology 2002 and Beyond, 457-459.

2. DISTINCT ROLES FOR PAL ISOFORMS

In generating trans-cinnamic acid for phenylpropanoid (secondary) metabolism, PAL diverts carbon from primary metabolic pathways driving cell division and expansion. Control of this gateway can occur by environmental and developmental control of PAL transcription (Shufflebottom et al., 1993), by metabolic feedback inhibition of PAL activity (Bolwell, 1986) and by expression of multiple protein isoforms (Osakabe et al., 1995; Kumar et al., 2001). Thus differential expression of PAL isoforms could provide regulatory flexibility which may be integral to the ability of rapid growing tree species like quaking aspen to coordinate their secondary carbon allocation with carbon fixation and nutrient supply. By correlating the expression pattern of *PAL1* and *PAL2* with the distribution of CTs and lignin, respectively, in developing stems and leaves, we have identified a mechanism by which PAL regulates carbon allocation between two major sinks in aspen. In contrast to previous reports suggesting developmentally distinct roles in lignification for the PAL1-like isoform in primary and secondary tissues of hybrid poplar (Gray-Mitsumune et al., 1999), our data presents aspen PAL1 as an isoform with a relatively minor role in lignification. Together, these reports suggest the possibility of developmental or adaptive flexibility in the role of the PAL1 isoform. Flexibility in phenylpropanoid metabolism is probably conditioned by the formation of channeling complexes that determine branchway activity (Rasmussen and Dixon, 1999). A shift in the metabolic role of PAL1 from lignification to CT metabolism due to developmental or environmental cues could depend on changes in co-expression of PAL1 with other phenylpropanoid genes.

Although *PtPAL1* and *PtPAL2* are differentially localized in aerial tissues, they appear to be co-expressed in many cells of the root tip. Concurrent production of a wide range of products for suberization, signaling and other defensive mechanisms in rapidly developing root tips may require a different level of PAL coordination than in aerial tissues. Our results from shoots and roots may indicate that co-expressed PAL1 and PAL2 isoforms can satisfy root demand because they have complementary rather than redundant roles in secondary metabolism.

3. REFERENCES

Bolwell, G.P., C.L. Cramer, C.J. Lamb, W. Schuch, and R.A. Dixon 1986. L-Phenylalanine ammonia-lyase from Phaseolus vulgaris: Modulation of the levels of active enzyme by trans-cinnamic acid. Planta 169:97-107.

Gray-Mitsumune, M., E.K. Molitor, D. Cukovic, J.E. Carlson, and C.J. Douglas. 1999. Developmentally regulated patterns of expression directed by poplar PAL promoters in transgenic tobacco and poplar. Plant Mol. Biol. 39:657-669.

Kumar, A., B.E. Ellis. 2001. The phenylalanine ammonia-lyase gene family in raspberry. Structure, expression, and evolution. Plant Physiol. 127:230-239.

Osakabe, Y., Y. Ohtsubo, S. Kawai, Y. Katayama, and N. Morohoshi. 1995. Structure and tissue-specific expression of genes for phenylalanine ammonia-lyase from a hybrid aspen, Populus kitakamiensis. Plant Sci. 105: 217-226.

Rasmussen, S., R.A. Dixon. 1999. Transgene-mediated and elicitor-induced perturbation of metabolic channeling at the entry point into the phenylpropanoid pathway. Plant Cell 11:1537-1551.

Shufflebottom, D., K. Edwards, W. Schuch, and M. Bevan. 1993. Transcription of two members of a gene family encoding phenylalanine ammonia-lyase leads to remarkably different cell specificities and induction patterns. Plant J. 3:835-845.

TISSUE CULTURE AND GENETIC TRANSFORMATION OF *CHAMAECYPARIS OBTUSE*

Katsuaki Ishii
Department of Molecular and Cell Biology, Forestry and Forest Products Research Institute, P.O.Box 16, Tsukuba Norinkenkyudanchi-nai, Ibaraki-ken, Japan 305-8687 (e-mail: katsuaki@ffpri.affrc.go.jp)

Keywords *Chamaecyparis obtusa*, tissue culture, transformation, GUS, *bar*

1. INTRODUCTION

Hinoki cypress (*Chamaecyparis obtusa*) is one of the plantation conifer species and the most important timber resource forest tree in Japan. We have reported the *in vitro* propagation through shoot primordia from juvenile seedlings (Ishii, 1986). The regenerated plants were shown to be genetically stable by RAPD, isozyme and flow cytometric analyses (Ishii et al., 2000). We now describe a new liquid culture method for shoot primordium propagation and regeneration of plantlets, and transformation via particle bombardment.

2. MATERIALS AND METHODS

Shoot primordia were induced from buds (3-6 mm long) cultured on Campbell and Durzan's (1975) CD medium containing 10 μM 6-benzylaminopurine (BAP) and 0.027 μM α-naphthaleneacetic acid (NAA) originally from a juvenile seedling (Ishii, 1986). Shoot primordia (c.a. 5 mm diameter) cultured on agar (8 g/l) solidified CD medium were inoculated in liquid CD medium supplemented with different cytokinins (BAP, zeatin, or thidiazuron/TDZ) at different concentrations without NAA and with 20 g/l sucrose. The culture vessel was 25 mm diameter and 200 mm long, and rotated around the axis twice/ minute. Culture room temperature was 25°C and light was 16 hours photoperiod/day of fluorcscent lamps at 100 μmol $m^{-2}s^{-1}$. Dry weight (over one night at 95°C) of shoot primordia cultured in liquid CD medium containing 1 μM zeatin was measured for 2 months. After shoot primordium multiplication in the liquid culture of CD medium containing 1 μM zeatin, they were transplanted to the agar (8 -16 g/l)

I. K. Vasil (ed.), Plant Biotechnology 2002 and Beyond, 461-463.

solidified medium of CD containing 0.027 μM NAA for shoot elongation. Regenerated healthy shoots were subcultured to the rooting medium containing 14.8 μM IBA, 0.54 μM NAA, and 2.66 μM riboflavin solidified with 8 g/l agar for plant regeneration.

Shoot primordia were bombarded with a plasmid containing the kanamycin resistance and β-glucuronidase (GUS) genes driven by the 35S promoter of cauliflower mosaic virus (pBI121, Clontech Laboratories, Inc., Palo Alto, CA). The introduction of a useful herbicide resistance gene (bar) into the shoot primordium with the same promoter (pSLJ2011, Jones et al., 1992) or maize ubiquitin promoter (pAHC20, Christensen and Quail, 1996) was also tried. The plasmid pSLJ2011 is a binary vector plasmid which was cloned into the site of plasmid pRK290 from *E. coli*. The plasmid pAHC20 has Ubi-1 promoter from maize, herbicide phosphinothricin resistance gene (bar), nopaline synthase 3' untranslated sequence and sequence from vector pUC8 of *E. coli*. The Ubi-1 promoter has been shown to be highly active in monocots.

Southern analysis was done using genomic DNA. To the regenerated plantlets from the selection medium, 50 times diluted Basta herbicide (0.37 % of glufosinate in the spray liquid) was sprayed and the response of the transgenic and non-transgenic plants was checked just after regeneration. Three resistant lines were also checked with 50 times diluted Basta spray after one growing season in the containment greenhouse. In a preliminary test, untransformed juvenile Hinoki cypress were killed with 100 times diluted Basta spray treatment.

3. RESULTS AND DISCUSSION

Of the different concentrations of cytokinins, 1 μM of zeatin or 0.1 and 1 μM BAP was good for shoot primordium growth in liquid culture. However, the color of the shoot primordium treated with zeatin was greener and looked fresh. Some callus was observed with the TDZ medium after 1 month. Dry weight increase was 4.6 times after 2 week cultivation. There was a rather stationary phase after week 2 until week 5. Then a second increment of dry weight was observed after 1 to 2 months' culture in the liquid medium. This skewed sigmoid growth curve may indicate that limited culture space for growth after 2 weeks induced temporary retardation of the growth of shoot primordia. So, it is better to subculture at an interval of two weeks for rapid propagation. However, for a moderate and labor-saving method, it may also be decided to adopt to subculture at an interval of two months.

Rooted shoots can be habituated and planted out to the nursery. So far, no abnormality in the morphology or the growth was observed in the regenerated plants. Micropropagated Hinoki cypress has grown steadily. Shoot primordium culture is said to be genetically stable for long-term tissue culture and the regeneration rate was quite high with this system. With the histochemical method, it was apparent that the β-glucuronidase was expressed in the particle bombarded shoot primordium. According to the bar herbicide selection, it appeared that 35S promoter was better than ubiquitin promoter for bar gene expression. The best condition for transformation was with 5 days' preculture using 35S promoter. Nine out of 30 regenerated plantlets showed the positive band (402 bp) of the bar gene. Southern blotting analysis showed the existence of 2 copies of bar genes in two plantlets examined. Surviving plantlets were observed among the regenerated plantlets from the selection medium after Basta spray treatment.

The author thanks Prof. Jonathan Jones of John Innes Centre, Norwich, Dr. Jean Finnegan of CSIRO Division of Plant Industry, Canberra, and Prof. Peter H. Quail of the University of California, Berkeley, for generous permission to use the bar-containing plasmids.

4. REFERENCES

Campbell, R.A. and Durzan, D.J. 1975. Induction of multiple buds and needles in tissue culture of *Picea glauca*. Can. J. Bot. 53:1652-1657.

Christensen, A.H. and Quail, P.H. 1996. Ubiqutin promoter-based vectors for high-level expression of selectable and/or screenable marker genes in monocotyledonous plants. Transgen. Res. 5:213-218.

Ishii, K. 1986. In vitro plantlet formation from adventitious buds on juvenile seedlings of Hinoki cypress (*Chamaecyparis obtusa*). Plant Cell Tissue Org. Cult.7:247-255.

Ishii, K., Yoshioka, H. and Ieiri, R. 2000. Cytogenetic study on in vitro regenerated plantlets of Hinoki cypress (*Chamaecyparis obtusa*). Special issue of the Forest Genetics.: 81-87.

Jones, J., Jones, D.G., Shulumukov, L., Carland, F., English, J, Scofield, S.R., Bishop, G.J. and Harrison, K. 1992. Effective vectors for transformation, expression of heterologous genes, and assaying transposon excision in transgenic plants. Transgen. Res. 1:285-297.

SECONDARY METABOLITES IN THE POST-GENOMIC ERA

Kirsi-Marja Oksman-Caldentey[1], Suvi Häkkinen[1], Alain Goossens[2], Into Laakso[3], Tuulikki Seppänen-Laakso[1], Anna Maria Nuutila[1] and Dirk Inzé[2]

[1]VTT Biotechnology, P.O. Box 1500, FIN-02044 VTT (Espoo), Finland (e-mail: kirsi-marja.oksman@vtt.fi);[2]Department of Plant Systems Biology, VIB, University of Ghent, Ledeganckstraat 35, B-9000 Ghent, Belgium; [3]Division of Pharmacognosy, Department of Pharmacy, P.O. Box 56, 00014 University of Helsinki, Finland

1. INTRODUCTION

Many important pharmaceutical compounds are still isolated from plants. Due to their complex structures the chemical synthesis of these secondary metabolites is usually not applied. Plant cell cultures would offer an attractive alternative production system. However, the yields in cell cultures have only in few cases been commercially feasible. Futhermore, due to the lack of understanding of the biosynthetic pathways, genetic engineering has not had much success until now. It is expected that the application of genomics, proteomics and metabolomics tools will create a paradigm shift in our ability to engineer the often complex biosynthetic pathways of plant secondary metabolites. To illustrate the power of this approach, we present some data on the genome-wide transcript cDNA-AFLP profiling in combination with a GC-MS-SIM alkaloid analysis of methyl jasmonate elicited BY-2 tobacco cells used as a model system.

2. SECONDARY METABOLITES IN PLANTS

Higher plants synthesize a vast amount of low molecular weight organic compounds, the so-called secondary metabolites, many of which have an important function for the survival of the plant in its ecosystem. These compounds play a role in attracting pollinators, protecting against UV-light as well as in various defence related reactions. Besides the importance for the plant itself, secondary metabolites are of interest for people as flavours, fragrances, dyes, pesticides, herbal remedies, and last but not least as pharmaceuticals (Oksman-Caldentey and Hiltunen, 1996). Excellent examples of highly important plant-derived pharmaceuticals are paclitaxel from Pacific yew, vincristine and vinblastine from *Catharanthus roseus* and morphine alkaloids from Opium poppy, just to mention few of them. Often secondary metabolites are produced in low quantities in only certain plant

I. K. Vasil (ed.), Plant Biotechnology 2002 and Beyond, 465-468.

species, and because of their unique and very complex structures their chemical synthesis is in most cases is economically not feasible, Therefore, they are currently isolated from wild or cultivated plant species.

The biosynthetic routes of various plant secondary metabolites are largely unknown, and usually their syntheses are long requiring numerous highly specific enzymatic steps. So far only relatively small number of enzymes involved in the biosynthesis of these plant compounds have been identified and their genes cloned. Notable exceptions are the biosynthetic pathways of the benzophenantridine alkaloid, berberine and some flavonoids which are completely characterized (Verpoorte et al., 2000).

Over the years numerous attempts have been made to produce the specific plant derived compounds in cell and organ cultures. The advantages of applying the cell culture technique are the controlled and reliable production in culture and the fast cultivation time compared to plants. Furthermore, a set of bioreactors especially designed for plant cell cultivation, up to 75,000 liters, has been developed for large- scale production of pharmaceuticals. However, in most cases the contents in cell cultures have not been high enough for commercial production. The strategies used to increase production levels have been purely empirical based on the optimization of the culture conditions and the selection of high-yielding somaclones (Oksman-Caldentey and Arroo, 2000; Oksman-Caldentey and Hiltunen, 1996). In order to understand fully the complexity of secondary metabolism in plants and cell cultures it will be increasingly important to characterize all the genes involved, and their respective functions in a given biosynthetic pathway. We believe that in this post genomic era, a rational approach based on genetic engineering has tremendous potential to improve the production and availability of plant-derived compounds in cell cultures. To demonstrate the potential of this approach we have developed a powerful, novel technology, which allowed us to detect simultaneously the genetic onset of various secondary metabolite pathways and the genetic reprogramming of primary metabolism to sustain secondary metabolism.

3. COMBINING METABOLOMICS AND TRANSCRIPTOMICS: BY-2 TOBACCO CELL CULTURE AS A MODEL SYSTEM

As a model case we performed a genome-wide cDNA-AFLP (cDNA amplified fragment length polymorphism) transcript profiling combined with a targeted metabolome analysis on tobacco BY-2 methyl jasmonate (MeJA) elicited cells. For quantitative chemical analysis GC-MS-SIM technique was used. cDNA-AFLP allows the quantitative detection of the vast majority of transcripts without any prior sequence knowledge. As such it can be used for gene discovery in any organism (Breyne and Zabeau, 2001).

Contrary to whole plants, nicotine was not the predominant alkaloid in BY-2 cell cultures. While the accumulation of nicotine and anatabine started to take place after 12 hours of MeJA elicitation the contents of anabasine and anatalline increased only after 48 hours. Altogether quantitative temporal accumulation patterns of approximately 20,000 transcript tags were determined and analyzed. In total, 591 genes were found to be jasmonate-modulated with different kinetics. Approximately 80% of the genes were up-regulated by methyl jasmonate, whereas only 20% were repressed. More than 50% of the jasmonate modulated genes were early induced, within 1-4 hours after elicitation, which implies that the whole biosynthetic machinery is likely to be fully activated within the first hours of elicitation. Homology searches with the sequences from the unique gene tags (80% of all tags) revealed that 64% of them displayed similarity with genes of known function and 18% with a gene without allocated function. In contrast, no homology to a known sequence was found for 18% of the tags (Goossens et al., 2002; Fig.1).

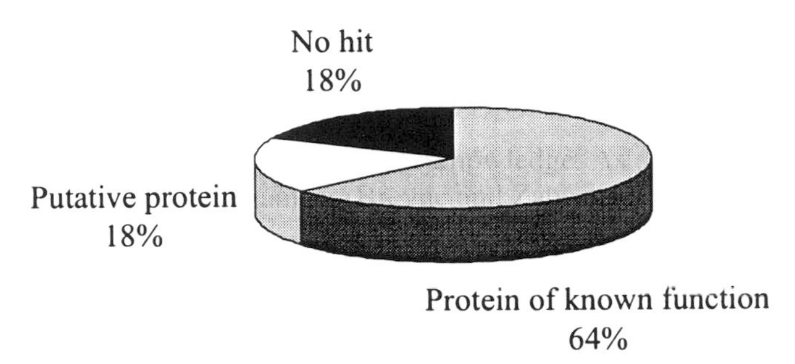

Figure 1. Characterization of MeJA modulated genes.

Identified genes included the genes involved in jasmonate biosynthesis, primary metabolism, stress response, signal transduction and secondary metabolism. Several genes were characterized to be involved not only in nicotine alkaloid pathway but also in phenylpropanoid pathway (Goossens et al., 2002). Functional analysis of the known and novel isolated genes will increase our knowledge of their exact role in regulating the secondary metabolite pathways.

4. CONCLUSIONS

In addition to getting vast information on the links between genomics and metabolomics we were able to collect a lot of knowledge on so far poorly characterized branches of secondary metabolite biosynthetic pathways.

Furthermore, we identified a number of novel genes involved in tobacco secondary metabolism and potentially in plant secondary metabolism in general. It is expected that key regulators controlling the often complex pathways are likely to be discovered using our technology.

5. REFERENCES

Breyne, P. and Zabeau, M. (2001) Genome-wide expression analysis of plant cell cycle modulated genes. Curr. Opin. Plant Biol. 4: 136-142.

Goossens, A., Häkkinen, S., Laakso, I., Seppänen-Laakso, T., Biondi, S., De Sutter, V., Lammertyn, F, Nuutila A.M., Söderlund, H., Zabeau, M., Inzé, D. and Oksman-Caldentey K.M. (2002) From bright yellow to bright orange: linking metabolomics and transcriptomics to unravel plant secondary metabolism. Submitted.

Oksman-Caldentey, K.M. & Hiltunen, R. (1996) Transgenic crops for improved pharmaceutical products. Field Crops. Res. 45: 57-69.

Oksman-Caldentey, K.M. and Arroo, R. (2000) Regulation of tropane alkaloid metabolism in plant and plant cell cultures. In: Metabolic engineering of plant secondary metabolism, Verpoorte, R. and Alfermann, A.W. (Eds), Kluwer Academic Publishers, Dordrecht-Boston-London, pp. 253-281.

Verpoorte, R., van der Heijden, R. and Memelink, J. (2000) General strategies. In: Metabolic engineering of plant secondary metabolism, Verpoorte, R. and Alfermann, A.W. (Eds), Kluwer Academic Publishers, Dordrecht-Boston-London, pp. 31-50.

A FUNCTIONAL GENOMICS STRATEGY TO IDENTIFY GENES THAT REGULATE THE PRODUCTION OF BIOLOGICALLY ACTIVE METABOLITES IN PLANTS

Deane L. Falcone[1,2], Dennis T. Rogers[1], Kil-Young Yun[1], Gabriela Diniello[1], May Fu[1], Irina Artiushin[1], and John M. Littleton[1,3]

[1] Kentucky Tobacco Research and Development Center, University of Kentucky, Lexington, KY 40546, USA (email: dfalcon@uky.edu); [2] Department of Agronomy, University of Kentucky, Lexington, KY 40546, USA; [3] Department of Molecular and Biomedical Pharmacology, University of Kentucky, Lexington, KY 40546, USA

Keywords Callus-based screen; mutagenesis; pharmacology, drug discovery, phytochemistry, genetics

1. INTRODUCTION

Plants produce an extraordinary range of biologically active metabolites and many of these are valuable because of their roles in human health and nutrition. A majority of these natural products are classified as "secondary" compounds, to distinguish them from the essential products of primary metabolism. These are often unique to certain plant families or species. This diversity is one of several factors that has hampered the elucidation of many secondary pathways. Regulatory properties often associated with the biosynthesis of secondary compounds, such as cell-type specific localization and transient expression, also may obscure the true biosynthetic potential of a plant. An additional impediment to studying genes involved in this diverse metabolism is that many plant species have complex genomes and are not amenable to efficient genetic techniques. The strategy described here circumvents these difficulties by applying pharmacological activity screens at the level of undifferentiated plant callus tissue. This strategy allows functional detection of compounds of pharmacological interest and offers a means for applying a functional genomics strategy to mutagenized cell cultures.

2. SCREEN DEVELOPMENT

Screening for natural products in mutagenized callus cultures required the development of several methodologies. These include obtaining large

I. K. Vasil (ed.), Plant Biotechnology 2002 and Beyond, 469-472.

numbers of mutagenized callus lines, methods to nondestructively sample this mutagenized tissue and the development of steps to sensitively screen for activity in the callus extracts. The T-DNA based activation tagging method (Walden et al., 1994) is used as the means to mutagenize the plant cells. The dominant, gain-of-function mutation that generally results from activation tagging makes this mutagenesis technique ideally suited for conducting screens at the callus or individual cell level.

2.1. Plant Cell Mutagenesis, Callus Library Establishment

Mutated cells are prepared by transformation of freshly isolated *N. tabacum* protoplasts by co-cultivation with *Agrobacterium tumefaciens,* harboring the activation T-DNA tagging plasmid, pPCVICEn4HPT (provided by Dr. R Walden) (Fritze and Walden, 1995). After this co-cultivation step, protoplasts are washed and embedded within a low-gelling temperature agarose. Transformed cells develop under antibiotic selection for the T-DNA.

At this stage, the dense population of transformed microcallus tissue is "disembedded", diluted and surface plated. Growth of the diluted microcalli is allowed to proceed until the material is large enough to sample. A small portion of tissue from an individual (0.5 cm dia) callus is removed with a pair of ring forceps and introduced into wells of 48-well microtiter plates. One-half of each sample is used to prepare crude aqueous extracts for displacement assays and the remaining half retained in the plate wells. Sampling and extraction steps are conducted in sterile callus growth media, which allows the retained material to re-grow in the 48-well dishes, and in this way a "master" mutant tissue library is established (Fig. 1). Extracts are prepared by gentle maceration, pelleting the debris by centrifugation and using the supernatant in displacement assays.

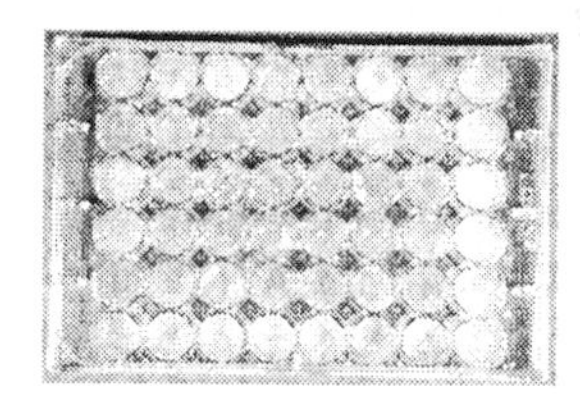

Figure 1. After mutagenesis, microcalli are sampled for assay and arrayed into 48-well microtiter dishes along with callus growth media to permit the sampled tissue to recover and grow.

2.2. High Throughput Screens for Nicotine-like Activity

The development of the biological activity assay was central for this approach to be conducted at the level of plant microcallus tissue. An essential pharmacological property for nicotine-like alkaloids is that they bind to mammalian nicotinic acetylcholine receptors (nicAChRs). To identify cultures that are overproducing such compounds we used a radioligand displacement assay (Gattu et al., 1995). Unlabelled compounds in callus extracts (such as nicotine-like compounds) compete for the binding sites between these and the labeled ligand. As a result, less radiolabel will associate with the membranes.

The extraordinarily high sensitivity provided (we were able to detect 1-10 pmol of nicotine per sample) enabled it to be used to evaluate production of relevant compounds in extracts obtained from small (~50mg) pieces of cultured microcallus tissue. The screen itself is also very rapid because it is based on 96-well plate technology, making it possible to screen up to 500 cultures per week, analyzed in triplicate for nicotine-like alkaloids, and as a control, for polyamines, using displacement of [^{3}H]-spermidine from membrane neuronal receptor preparations.

3. LINES IDENTIFIED AT THE CALLUS LEVEL

Out of 8,500 mutant cultures screened, 2 mutant clones, no's. 1402 and 5094, continue to overproduce nicAChR-displacing activity by at least two orders of magnitude greater than non-mutant cultures. The lines produce polyamines within the normal range, suggesting that alkaloid overproduction is a relatively specific phenomenon. Regenerated plants from these mutant calli also show enhanced epibatidine displacement activity. Line 1402 shows a 3-fold increase in young but not older seedlings, which is most probably due to nicotine by GC analysis. Line 5094, shows a greater degree of displacement activity than line 1402 at both the microcallus stage and in regenerated whole plants. However, all regenerants from this line show reduced growth and a dwarf phenotype. The nicotine content from 5094 is about 5-fold greater than that determined in wild-type plants, although the levels of all other major alkaloids are not correspondingly elevated (Fig. 2). This is curious since variations in alkaloid content in *Nicotiana* species usually correlate with changes in the levels of other alkaloidss, suggesting an alteration more specific to nicotine synthesis rather than a generalized alkaloid pathway upregulation (Roberts et al., 1998). Genomic DNA associated with the inserted T-DNA in both lines has been recovered and several identified open reading frames are being tested for the their involvement in conferring the observed chemical phenotype.

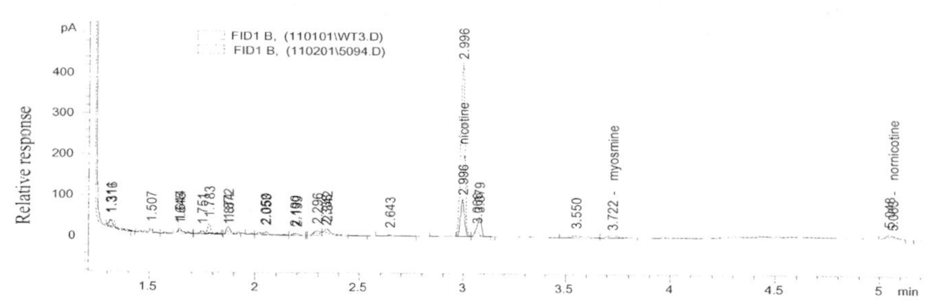

Figure 2. Tobacco line 5094 derived from an individual callus displaying elevated ^{3}H-epibatidine displacement activity in the primary screen exhibits an approximately 5-fold greater amount of nicotine, with no apparent alterations in other major alkaloids, in leaves from the regenerated plants. Dotted line is the chromatographic profile of an alkaloid extract prepared from 5094 plants. Solid line is the wild-type profile.

4. CONCLUSIONS

This combination of activation tagging mutagenesis with high throughput screening for biological activity should enable the isolation of genetic material relevant to synthesis of specific natural products. The advantages of this strategy include that no prior knowledge of the metabolic pathway is required– only a method of screening for the product activity. This general approach may also make accessible the identification of genes from plant species typically recalcitrant to whole plant screening as well as in assisting to establish the roles those genes in plants that do not share homology with known genes.

4. REFERENCES

Fritze, K., and R. Walden. 1995. Gene activation by T-DNA tagging. *Methods Mol Biol* **44**:281-94.

Gattu, M., A.V. Terry, and J. J. Buccafusco. 1995. A rapid microtechnique for the estimation of muscarinic and nicotinic receptor binding parameters using 96 well filtration plates. *J Neurosci Methods* **63**:121-125.

Roberts, M. F., M. Wink, and eds. 1998. *Alkaloids: biochemistry, ecology, and medicinal applications*. New York: Plenum Press, p. 186-189.

Walden, R., K. Fritze, and H. Harling. 1995. Induction of signal transduction pathways through promoter activation. *Methods Cell Biol* **49**:455-69.

Walden, R., K. Fritze, H. Hayashi, E. Miklashevichs, H. Harling, and J. Schell. 1994. Activation Tagging - a Means of Isolating Genes Implicated as Playing a Role in Plant-Growth and Development. *Plant Mol Biol* **26**:1521-1528.

ACKNOWLEDGEMENTS

This work was supported by funds from the Kentucky Tobacco Research Board.

PHYTOREMEDIATION OF TOXIC MERCURY AND ARSENIC POLLUTION

Richard B. Meagher
Department of Genetics, University of Georgia, Athens, Georgia 30602 USA (email: meagher@arches.uga.edu)

Keywords Environment, electrochemical reduction, sequestration, heavy metal, metalloid, hyperaccumulation, elemental and organic pollutants

1. INTRODUCTION

Plants can be used to extract, detoxify, and/or sequester toxic pollutants from soil, water, and air, in a process called phytoremediation. Phytoremediation will likely become an essential tool in cleaning the environment and reducing human and animal exposure to potential carcinogens and other toxins. Alternative physical methods like excavation and reburial of contaminated sediments are too expensive and environmentally destructive. Strategies for phytoremediation need to distinguish between 1) the remediation of elemental pollutants and 2) the remediation of organic pollutants (Meagher, 2000). Elemental pollutants include heavy metals and metalloids (e.g., mercury, lead, cadmium, arsenic) that are immutable. The general goals of elemental phytoremediation are to immobilize, extract, detoxify, volatilize, and/or hyperaccumulate elemental pollutants in above-ground plant tissues for later harvest. Organic pollutants include toxic chemicals such as polyaromatic hydrocarbons (i.e., the benzopyrenes), polychlorinated biphenyls, chlorinated solvents (e.g., trichloroethylene), and nitroaromatics (e.g., the explosive trinitrotoluene). Phytoremediation strategies for organic pollutants are focused on their complete mineralization to harmless products (Doty et al., 2000; Hannink et al., 2001; Meagher, in press).

Plants are particularly and naturally suited to cleaning pollutants from soil and water for several reasons, but three stand out as most significant (Meagher and Rugh, 1996). Firstly, plants can grow 100 million miles of roots per acre in a year. They do this to extract 16 essential nutrients from soil and water and they concomitantly extract pollutants. These same root systems bring elemental pollutants aboveground for later harvest. Secondly, plants are photosynthetic and control 80% of the energy in most ecosystems. They take their fixed carbon energy source, sugars, below ground to feed root growth

I. K. Vasil (ed.), Plant Biotechnology 2002 and Beyond, 473-478.

and use the derived energy to extract nutrients and pollutants. This is a big advantage over microbial remediation systems in which soil must be excavated into fermentation tanks or the organisms must be fed artificially underground. Thirdly, healthy plants growing on a contaminated site also secrete organic acids and proteins that feed a complex rhizosphere of bacteria and fungi. These microorganisms further enhance metabolism of toxic compounds and restoration of a natural ecosystem.

2. PHYTOREMEDIATION OF MERCURY AND ARSENIC

Our long-term goal is use highly productive conservation plant species in the phytoremediation of toxic heavy metals and metalloids. For example, mercury and arsenic pollution are serious world-wide problems affecting the health of hundreds of millions of people. Arsenic is a class A carcinogen and relatively toxic. Arsenic pollution is particularly serious, because of widespread contamination of drinking water. The most serious problems with mercury (Hg) results from the production of methylmercury ($MeHg^+$) by native bacteria in anaerobic sediments at contaminated freshwater and marine wetland sites. $MeHg^+$ is inherently more toxic than metallic Hg(0) or ionic Hg(II), and because it is very efficiently biomagnified up the food-chain, it poses the most immediate threat to animal populations. Our current working hypothesis is that efficient phytoremediation of elemental pollutants like mercury and arsenic can be achieved by controlling their chemical species, electrochemical state, transport, and aboveground binding. Phytoremediation strategies based on this hypothesis will prevent heavy metal toxins from entering the food-chain or water supplies, remove them from polluted sites, and result in an ecologically friendly solution to toxic heavy metal pollution. Combinatorial multi-gene strategies are being tested and giving us tremendous flexibility in designing site-specific mercury-remediating plants.

Examples of our work on mercury and arsenic demonstrate the efficacy of these strategies. First, we successfully engineered *Arabidopsis thaliana* to use the bacterial *merB* gene, methylmercury lyase, to convert methylmercury to less toxic ionic mercury. These plants are highly resistant to methylmercury or phenylmercury at concentrations of 0.1 to 2.5 μM (0.02 to 0.5 ppm). These concentrations are as high as, or higher than those found in the environment and are sufficient to kill control plants in laboratory medium as shown in Figure 2. MerB transgenic plants can store the levels of Hg(II) produced (<0.5 ppm) in their tissues and continue to grow. Second, a highly modified bacterial mercuric ion reductase gene, *merA,* was examined in transgenic tobacco and Arabidopsis. MerA enzyme detoxifies ionic mercury (Hg(II)), electrochemically reducing it to metallic Hg(0). Plants expressing MerA grow on high levels of ionic mercury (25-2500 mM, 5-500 ppm) in media or soil (not shown), levels that kill most native plants (Rugh et al.,

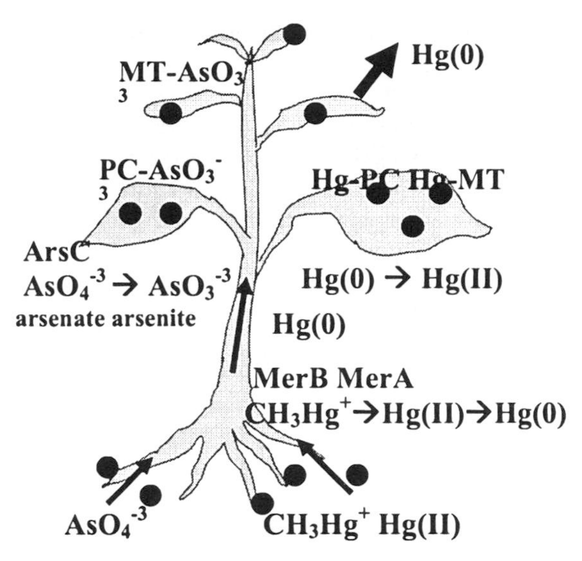

Figure 1. Strategies for the phytoremediation of mercury and arsenic: Solid experimental evidence from model plants expressing several different plant and bacterial transgenes support the following remediation scenarios for mercury (right) and arsenic (left). Right side of diagram. Methyl- (MeHg, CH_3Hg^+) or ionic (Hg(II)) mercury enter the roots of engineered plants. MeHg is converted first to Hg(II) by the enzyme MerB. Hg(II) is electrochemically reduced to Hg(0) by the enzyme MerA. Hg(0) is soluble and transpired aboveground, where it is either a) volatilized from leaves along with waste gases like CO_2 and O_2 or b) reoxidized to Hg(II) by the action of endogenous plant peroxidases and catalases. Hg(II) is highly thiol-reactive and can be hyperaccumulated in thiol-rich complexes. For example, Hg(II) appears to form conjugates with [illegible]-glutamylcysteine peptides like the phytochelatins (PCs), forming HgPC, or coordinates with cysteine rich metallothioneins (MTs), forming HgMT. These strategies are now being applied to field species that can be used in site remediation. Left side of diagram. Arsenic enters the roots primarily as oxidized arsenate (AsO_4^{-3}, oxidation state As(V)). Arsenate is a phosphate analog and is most likely pumped into plant roots by the cryptic activities of phosphate transporters. An endogenous arsenate reductase activity electrochemically reduces most arsenate to arsenite (AsO_3^{-3}, oxidation state As(III)), which is trapped in roots as As(III)-thio-complexes. Blocking this endogenous reductase would enhance arsenic transport aboveground for later harvest. Residual arsenate is transported above-ground and can be reduced by an engineered bacterial arsenate reductase gene, arsC. The arsenite formed aboveground can be trapped safely in thiol-complexes by overexpression of phytochelatin pathway enzymes under control of a light regulated promoter.

1996, 1998a, b). Plants expressing both *merA* and *merB* detoxify methylmercury in two steps to the least toxic metallic mercury, as shown in Figure 3. The *merA/merB* seedlings germinate, and plants grow and set seed at normal rates on even higher levels of methylmercury than are lethal to *merB* plants. Our newest efforts use several additional bacterial and plant transgenes to engineer plants that resist mercury by hyperaccumulation and sequestration. Physiological experiments demonstrate these various plants can either transpire Hg(0) from leaves or trap Hg(II) in aboveground tissues.

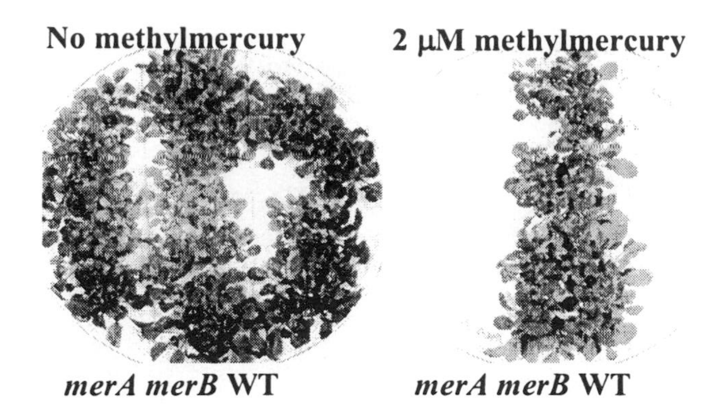

Figure2. Expression of the organomercury lyase gene merB under control of a strong constitutive plant promoter confers high levels of methylmercury resistance. WT (wild-type), transgenic merA and merB seeds were germinated on MS medium with or without methylmercury and grown for three weeks as described previously (Bizily et al., 1999).

Our first experiments on the phytoremediation of arsenic were recently completed with levels of arsenic resistance parallel to those described above for mercury. By controlling the electrochemical state and binding of arsenic we can achieve high levels of resistance and accumulation of arsenic in plant tissues (Li et al., 2002a, b, c; Parkash et al., in preparation). Through several different strategies we can obtain plants, which grow on concentrations five times that which kills control plants.

Arabidopsis and tobacco are powerful, model plants well suited for initial genetic and growth studies. However, transgenic derivatives of native conservation plants are needed for long-term field trials of phytoremediation technologies. For example, *merA* and/or *merB* have been transformed into and partially tested for phytoremediation activity in *Brassica napus (*Canola*), Liriodendron tulipifera* (Yellow poplar), *Populus deltoides (*Cottonwood*)*, and *Oryza sativa var. Japonica (*rice*)* (Heaton et al., in prep.; Heaton et al., 1998; Rugh et al., 1998a, b). More research is needed on the culture, transformation, and regeneration of conservation plants used in the field application of phytoremediation.

We have taken a molecular genetic approach using engineered plants to accelerate the repair of our ailing environment, paralleling that of medical genetics research approach to curing human disease. In this case, enhancing or blocking the action of certain genes in dominant plants in the environment accelerates the detoxification of target compound(s), allowing the ecosystem to repair itself.

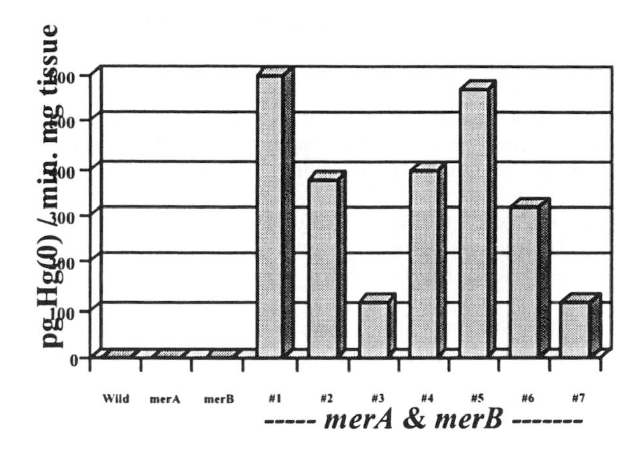

Figure 3. Expression of <u>both</u> the bacterial merA and merB genes are required for efficient conversion of methylmercury to metallic mercury (Hg(0)). Verticle scale is pg of mercury evolved per mg of wet weight tissue for control and transgenic plants. Details are described previously (Bizily et al., 2000).

3. REFERENCES

Bizily, S., Rugh, C. L., and Meagher, R. B. 2000. Phytodetoxification of hazardous organomercurials by genetically engineered plants. Nat Biotechnol. 18:213-217.

Bizily, S., Rugh, C. L., Summers, A. O., and Meagher, R. B. 1999. Phytoremediation of methylmercury pollution: *merB* expression in *Arabidopsis thaliana* confers resistance to organomercurials. Proc Natl Acad Sci USA 96:6808-6813.

Cobbett, C., and Meagher, R. Phytoremediation and the *Arabidopsis* proteome. In Arabidopsis, E. Meyerowitz and C. Somerville, eds. (Cold Spring Harbor, NY: Cold Spring Harbor Laboratory Press). In press.

Doty, S. L., Shang, T. Q., Wilson, A. M., Tangen, J., Westergreen, A. D., Newman, L. A., Strand, S. E., and Gordon, M. P. 2000. Enhanced metabolism of halogenated hydrocarbons in transgenic plants containing mammalian cytochrome P450 2E1. Proc. Natl. Acad. Sci. USA 97:6287-91.

Hannink, N., Rosser, S., French, C., Basran, A., Murray, J., Nicklin, S., and Bruce, N. 2001. Phytodetoxification of TNT by transgenic plants expressing a bacterial nitroreductase. Nat. Biotechnol.

Heaton, A., Rugh, C., and Meagher, R. (in prep.). Phytoremediation of mercury by transgenic rice.

Heaton, A. C. P., Rugh, C. L., Wang, N.-J., and Meagher, R. B. 1998. Phytoremediation of mercury and methylmercury polluted soils using genetically engineered plants. J. Soil Contam. 7:497-509.

Li, Y., Parkash Dhankher, O., and Meagher, R. B. (2002a). Engineering metal hyperaccumulation: I. Overexpression of g-glutamylcysteine synthetase in plants confers high level arsenic and mercury tolerance. (In preparation).

Li, Y., Parkash Dhankher, O., and Meagher, R. B. (2002b). Engineering metal hyperaccumulation: II. Overexpression of glutathione synthetase in plants confers high level arsenic and mercury resistance. (In preparation).

Li, Y., Parkash Dhankher, O., and Meagher, R. B. (2002c). Engineering metal hyperaccumulation: III. Leaf-specific expression of phytochelatin synthase in plants confers high level arsenic and mercury resistance. (In preparation).

Meagher, R. Pink water, green plants, and pink elephants. Nature Biotech. In press.

Meagher, R. B. 2000. Phytoremediation of toxic elemental and organic pollutants. Curr. Opin. Plant Biol. 3:153-62.

Meagher, R. B. 1998. Phytoremediation: An affordable, friendly technology to restore marginal lands in the twenty-first century. In Plants and Population: is there time?, N. Federoff and J. Cohen, eds. (Irvine, CA: National Academy of Sciences).

Meagher, R. B., and Rugh, C. L. 1996. Phytoremediation of heavy metal pollution: Ionic and methyl mercury. In OECD Biotechnology for Water Use and Conservation Workshop (Cocoyoc, Mexico: Organization for Economic Co-Operation and Development), pp. 305-321.

Parkash, O., Li, Y., Agarwal, N., Rosen, B., and Meagher, R. B. (in prep.). The engineered phytoremediation of arsenic. Part I: Controlling electrochemical state and binding in leaf tissues.

Rugh, C. L., Gragson, G. M., and Meagher, R. B. 1998a. Toxic mercury reduction and remediation using transgenic plants with a modified bacterial gene. Hort. Sci. 33:12-15.

Rugh, C. L., Senecoff, J. F., Meagher, R. B., and Merkle, S. A. 1998b. Development of transgenic yellow poplar for mercury phytoremediation. Nat. Biotechnol. 16:925-8.

Rugh, C. L., Wilde, D., Stack, N. M., Thompson, D. M., Summers, A. O., and Meagher, R. B. 1996. Mercuric ion reduction and resistance in transgenic *Arabidopsis thaliana* plants expressing a modified bacterial *merA* gene. Proc. Natl. Acad. Sci. USA 93:3182-3187.

TRANSGENIC PLANTS FOR ENVIRONMENTAL BIOMONITORING: NEW PERSPECTIVES

Olga Kovalchuk[1], Barbara Hohn[2] and Igor Kovalchuk[1]
[1]Department of Biological Sciences, University of Lethbridge, Lethbridge, T1K 3V2, Alberta, Canada (e-mail: olga.kovalchuk@uleth.ca);[2]Friedrich Miescher Institut, 4058 Basel, Switzerland (e-mail: Barbara.Hohn@fmi.ch)

Keywords Environment, contamination, radiation, heavy metals, transgenic plants, mutation, recombination, biomonitoring

1. INTRODUCTION

Modern socio-economic development is often associated with a heavy load of radioactive and chemical contamination of the biosphere. Sources of radiation exposure are almost ubiquitous in our environment due to the growing production of radioactive waste, nuclear tests and accidents in nuclear power plants that lead to the release of radioactive material into the environment. Along with radioactive pollution there exists the problem of chemical contamination. Large territories all over the world are contaminated by wastes of the chemical and petroleum industries. Soils and drinking water supplies are polluted with heavy metals, such as cadmium, lead, copper and others as well as with toxic organic compounds. As a result, living organisms suffer from the cumulative effect of various environmental pollutants - powerful forces that are continuously shaping the genomes of all species

For successful analysis of the genetic effects of exposure to radiation, toxic chemicals and other harmful anthropogenic contaminants, the choice of test system is critical. The ideal biomonitoring system must be eukaryotic, provide a large number of offspring and be easy to handle. Animal based models are difficult to use in environmental studies due to their mobility and ethical issues associated with their testing. Plants provide excellent ethically acceptable alternatives to animal tests in studies of genetoxic effects of various environmental mutagens (Grant, 1994). Being sedentary, plants respond to an exact known level of contaminant over a well-defined territory. Large numbers of progeny and the fast growth rates of many species provide a sufficient number of plants for analysis.

Various classical plant biosensors were used to study the genetic effects of environmental pollutants (Ichikawa, 1992; Grant, 1994; Kovalchuk et al., 2001). Most of the widely used systems to study mutations in plants are based on

I. K. Vasil (ed.), Plant Biotechnology 2002 and Beyond, 479-483.

detection of chromosomal aberrations in *Allium cepa* (Fiskesjo, 1988), *Tradescantia* (Ichikawa, 1992) or *Vicia faba* plants (Kanaya et al., 1994) and micronuclei formation in *Tradescantia* (Rodrigues et al., 1997). These systems were successfully applied to study the genetoxic effects of radiation, toxic organic compounds, heavy metals, various pesticides and other environmental factors.

Some other plant systems were applied to environmental studies. Tobacco plants heterozygous for the Sulfur (Su) nuclear gene which affects chlorophyll content in leaves have also been used for the study of mutagenicity of different chemicals (Friedlender et al., 1996). Mutations in this gene determine a codominant phenotype, exhibiting the genotype of Su/Su (dark green), Su/su (light green) or su/su (albino) plants (Burk and Menser, 1964). These plants were shown to respond to gamma radiation, methyl methanesulfonate (MMS) and N-methyl-N-nitrosourea (MNU) (Friedlender et al., 1996). These plants hold a potential for environmental mutagenesis studies, yet the type of changes which cause the phenotype is unknown at the molecular level (Kovalchuk et al., 2001).

2. TRANSGENIC PLANTS AS BIOSENSORS OF ENVIRONMENTAL CONTAMINATION

Recently we developed new transgenic biomonitoring plants which allow rapid, cheap and precise assays of genotoxicity of radioactively or chemically polluted soil. Both assays are based on the restoration of the transgene activity in *Arabidopsis thaliana* plants transformed with a non-active version β-glucuronidase (*uidA*) marker gene.

2.1. Recombination Assay

For development of "plant recombination monitoring" system model, *Arabidopsis thaliana* plants were transformed with two overlapping, non-functional truncated versions of a chimeric β-glucuronidase (*uidA*) marker gene as a recombination substrate (Swoboda et al., 1994). In cells in which events of homologous recombination (HR) have occurred at this transgenic locus the *uidA* gene was restored. Upon histochemical staining, cells and their progeny expressing β-glucuronidase, could be precisely localized as blue sectors on white plants. Each sector represented a recombination event that restored the disrupted gene, enabling a quantitative assay (Puchta et al., 1995).

Ionizing radiation is a potent mutagen, which causes significant numbers of double-strand breaks in the DNA. The double-strand breaks are in part repaired by homologous recombination (Puchta and Hohn, 1996) leading to the restoration of β-glucuronidase activity in the "plant recombination" system. We conducted a large scale environmental monitoring experiments with the use of transgenic *Arabidopsis thaliana* plants. We performed field and laboratory experiment using soils with various levels of radioactive pollution that were sampled in Chernobyl exclusion zone as well as in different inhabited and contaminated areas (Kovalchuk et al., 1998; Kovalchuk et al., 1999). First we detected a hazardous influence of chronic ionizing radiation on the stability of the plant genome in both sets of experiments (field and laboratory) conducted in the exclusion zone. We observed a dose-dependent increase of homologous recombination in plant populations to 8.4 fold of the control level at pollution levels of up to 300 Ci/km^2 in the open-field and up to 11.0 fold at pollution levels up to 1000 Ci/km^2 in the laboratory experiment. Remarkably, it was possible to correlate recombination frequency not only to the level of the soil contamination but also to the absorbed dose (Kovalchuk et al., 1998). Biomonitoring experiments conducted in the inhabited areas affected by the Chernobyl fallout where the soil contamination ranged between 1 and 40 Ci/km^2 revealed a statistically significant increase of HR (Kovalchuk et al., 1999). Of special importance was the fact that we were able to detect a difference in the frequency of homologous recombination between plants grown on clean soil (0.1) and plants grown on soil with a level of contamination as low as 1.5-3.3 Ci/km^2.

Transgenic recombination plants were also successfully applied to monitor genetic effects of the heavy metals Cd^{2+}, Pb^{3+}, Ni^{2+}, Zn^{2+}, Cu^{2+} and As^{3+} (Kovalchuk et al., 2001). In parallel experiments recombination monitoring plants were used to study genotoxicity of soils sampled in the proximity of an oil refinery that were strongly contaminated with lead and cadmium. We noted a 4-7 fold increase in the frequency of homologous recombination in plants grown in the metal-contaminated soils compared to those grown in clean control soil.

2.2. Point Mutation Assay

Although the recombination system we described before detects the repair of DSBs via homologous recombination, DSBs are not the only type of DNA damage created by various mutagens. It was shown recently that the eukaryotic genome suffers from ten times more nucleotide substitutions than any other types of mutation. Until recently, no system was available to measure somatic point mutation events in plants. In order to analyze the rate of point mutation in higher plants we have created a new reporter system to study the level of point mutation in plants. The system was based on a stop-codon inactivated β-glucuronidase reporter gene introduced into plants. We observed spontaneous restoration of *uidA* activity due to reversion of the stop codons to the original codons - blue sectors indicative of transgene reactivation appeared upon histochemical staining of transgenic plants (Kovalchuk et al., 2000). The newly generated transgenic plants successfully detected the mutagenic effects of DNA damaging factors such as UV-C, X-rays and methyl methanesulfonate (MMS)(Kovalchuk *et al.*, 2000). Additionally we used our "mutation' system to measure possible influences of highly radioactive soils and could confirm its sensitivity.

Recently this novel system was shown to detect the mutagenic influence of the heavy metals Cd^{2+}, Pb^{3+}, Ni^{2+}, Zn^{2+}, Cu^{2+} and As^{3+} . A concentration-dependent increase in the frequencies of T→G and A→G point mutations was observed. Like recombination plants, mutation plants were tested in soils collected at sites in the proximity of an oil refinery station. Soil samples were found to be contaminated to different extents with Pb, Cd, Zn and other chemicals. We observed a 5-10 fold induction of point mutations in plants grown in the contaminated soils compared to those grown in the clean control soil. These results demonstrate the potential applicability of the mutation test plants as biomonitors of chemically and radioactively polluted environments.

3. CONCLUSIONS AND OUTLOOK

We introduced a new transgenic approach, which is fast, sensitive and allows quantitative vizualization of induced recombination and mutation events. It permitted the rapid screening of genotoxicity of the radioactively and chemically contaminated soils. The systems can be broadly used for environmental studies since they provide data after a short period of time (around four weeks) require no sophisticated equipment nor specific knowledge for the detection and scoring of recombination and mutation events. Even large-scale experiments for the evaluation of the genotoxicity of various environmental pollutants may be possible. Moreover, our systems are able to "sense" the presence of very low concentrations of pollutants.

In addition, plants have been shown to be useful as phytoremediation devices, removing pollutants from soil and water, accumulating them in their biomass, detoxifying them in some cases and vaporizing them in other cases (Rugh et al., 1996; Raskin et al., 1997). Our plants can possibly be used as remediation quality control by evaluating the genotoxicity and mutagenicity of the contaminated sols before and after remediation. Thus, transgenic plants are becoming valuable tools for biosensoring, efficient cleanup and post-remediation control of contaminated soils and water.

4. REFERENCES

Burk L.G. and H.A. Menser. 1964. A dominant aurea mutation in tobacco. Tob. Sci. 8: 101-104.

Friedlender M., Lev-Yadun S.;, Baburek I., Angelis K. and A.A. Levy. 1996. Cell divisions in cotyledons after germination: localization, time course and utilization for a mutagenesis assay. Planta 199: 307-313.

Grant W. F. 1994. The present status of higher plant bioassays for detection of environmental mutagens. Mut. Res. 310:175-185.

Hohn B., Kovalchuk I. and O. Kovalchuk. 1999. Transgenic plants sense radioactive contamination. Bioworld 6: 13-15.

Ichikawa S. 1992. *Tradescantia* stamen-hair system as an excellent botanical tester of mutagenicity: its responses to ionizing radiations and chemical mutagens, and some synergistic effects found. Mut. Res. 270: 3-22.

Kanaya N., Gill B., Grover I., Murin A., Osiecka R., Sandhu S. and H. Andersson. 1994. Vicia faba chromosomal aberration assay. Mut. Res. 310: 231-247.

Kovalchuk I., Kovalchuk O., Arkhipov A. and B. Hohn. 1998. Transgenic plants are sensitive bioindicators of nuclear pollution caused by the Chernobyl accident. Nature Biotechnology 16: 1054-1057.

Kovalchuk O., Kovalchuk I., Titov V., Arkhipov A. and B. Hohn. 1999. Radiation hazard caused by the Chernobyl accident in inhabited areas of Ukraine can be monitored by transgenic plants. Mut. Res. 446: 49-55.

Kovalchuk I., Kovalchuk O. and B. Hohn.1999. Transgenic plants as bioindicators of environmental pollution. Review. AgBiotechNet Vol. 1, October, ABN 030.

Kovalchuk I., Kovalchuk O. and B. Hohn. 2000. Genome-wide variation of the somatic mutation frequency in transgenic plants. EMBO J. 19: 4431-4438.

Kovalchuk I., Kovalchuk O. and B. Hohn. 2001. Biomonitoring of genotoxicity of environmental factors with transgenic plants. Trends Plant Sci. 6: 306-310.

Puchta H.;, Swoboda P. and B. Hohn. 1995. Induction of homologous DNA recombination in whole plants. Plant J. 7: 203-210.

Puchta, H. and B. Hohn.1996. From centiMorgans to basepairs: homologous recombination in plants. Trends Plant Sci .1: 340-348.

Raskin I., Smith R.D. and D.E. Salt. 1997. Phytoremediation of metals: using plants to remove pollutants from the environment. Curr. Opin. Biotechnol. 8: 221-226.

Rodrigues G.S., Ma T.H., Pimentel D. and L. Weinstein. 1997 *Tradescantia* bioassays as monitoring systems for environmental mutagenesis: A Review. Crit. Rev. Plant Sci. 16: 325-359.

Rugh C.L., Wilde H.D., Stack N.M., Thompson D.M., Summers A.O. and R.B. Meagher. 1996. Mercuric ion reduction and resistance in transgenic *Arabidopsis thaliana* plants expressing a modified bacterial merA gene. Proc. Natl. Acad. Sci. USA 93: 6667-6671.

DIRECTING METABOLIC FLUX TOWARD ENGINEERED ISOFLAVONE NUTRACEUTICALS IN TRANSGENIC *ARABIDOPSIS*

Chang-Jun Liu and Richard A. Dixon*
Plant Biology Division, The Samuel Roberts Noble Foundation, Ardmore, OK 73402, USA (e-mail: cliu@noble.org and *radixon@noble.org)

Keywords isoflavone, flavonol, metabolic engineering, metabolic flux, *Arabidopsis*

1. INTRODUCTION

Isoflavonoid natural products are primarily limited to legume plants, where they function as antimicrobial phytoalexins or phytoanticipins, and as inducers or suppressors of the nodulation genes in the Rhizobium-legume symbiosis (Dixon, 1999). The simple isoflavones genistein and daidzein are important nutraceutical phytoestrogen molecules found in soybean seeds. They possess perceived chemopreventive activities against hormone-dependent cancers, cardiovascular disease, and post-menopausal ailments (Setchell and Cassidy, 1999) and have the potential ability to improve human memory (File et al., 2001). These properties have led to considerable interest in engineering non-legume plants to contain such phytoestrogens. In addition to isoflavonoids, many other flavonoid-derived compounds also have health-promoting activity. These include flavonols such as quercetin, which occur at significant levels in leaves and fruit of many plant species and have high antioxidant activity.

Isoflavonoids are formed by a branch of the flavonoid biosynthetic pathway, and originate from a central flavanone intermediate that is ubiquitously present in plants. A cytochrome P450 protein, isoflavone synthase (IFS), is the key enzyme for entry into the isoflavonoid pathway (Fig.1).

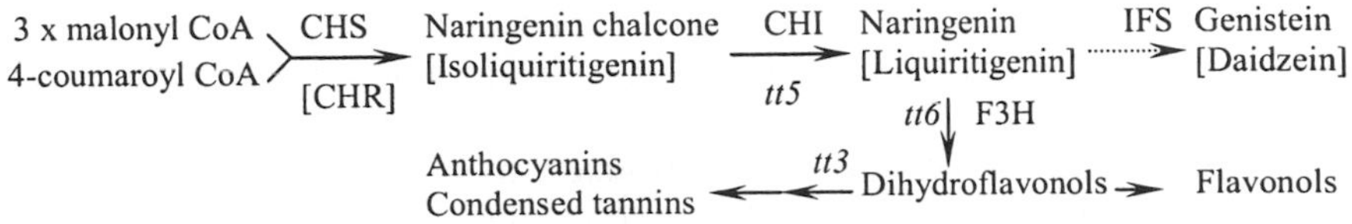

3 x malonyl CoA, 4-coumaroyl CoA → CHS [CHR] → Naringenin chalcone [Isoliquiritigenin] → CHI *tt5* → Naringenin [Liquiritigenin] → IFS → Genistein [Daidzein]

Naringenin [Liquiritigenin] → *tt6* F3H → Dihydroflavonols → Flavonols

Dihydroflavonols → *tt3* → Anthocyanins, Condensed tannins

Figure 1. Scheme of isoflavonoid and flavonoid biosynthesis

I. K. Vasil (ed.), Plant Biotechnology 2002 and Beyond, 485-490.

cDNAs encoding IFS have been cloned from soybean and other species (Steele,1999; Akashi, 1999; Jung, 2000) and IFS has been introduced into *Arabidopsis thaliana*, corn and tobacco. In all cases, only small amounts of genistein were formed (Jung, 2000; Yu, 2001). In an attempt to increase genistein accumulation in transgenic Arabidopsis, we have been focusing on expression of single or multiple flavonoid pathway enzymes, overexpression of a *MYB* transcriptional regulator to increase flux into upstream flavonoid biosynthesis, or use of transparent testa (*tt*) mutants to block the flux into downstream flavonoids.

Our results reveal that increasing upstream flavonoid biosynthesis does not increase the levels of genistein and that complete blockage of downstream flavonoid biosynthesis only has a small effect. We discuss the factors that might limit genistein production in transgenic *Arabidopsis*.

2. RESULTS AND DISCUSSION

2.1. Introduction Of Soybean IFS In *Arabidopsis* Produces Genistein Conjugates And Impairs The Biosynthesis Of Endogenous Flavonols

The soybean IFS (CYP93C1v2) under control of the cauliflower mosaic virus 35S promoter was introduced into the *A. thaliana* ecotype Columbia. HPLC and LC/MS analysis indicated that IFS transgenic plants produced three genistein conjugates: a glucoside of genistein, rhamnose-genistein and hexose -rhamnose-genistein. Among the T1 and T2 generation lines, line 15b produced the highest level of genistein, approximately 7.4 nmol/g fresh weight following acid hydrolysis of conjugates, a level very much lower than the levels of endogenous flavonol conjugates.

To determine whether the low level of isoflavone formation resulted from limiting *IFS* transgene expression as opposed to other factors such as substrate availability *in vivo*, we analyzed several independent homozygous 15b plants for isoflavone and flavonoid levels and the *in vitro* IFS activity (Fig. 2). The levels of genistein in leaf extracts from 15b plants varied from 7.4 to 12 nmol/g FW. There was a log-linear relationship between genistein levels and IFS activity in these various plants (Fig. 2A). However, the two parameters were not directly proportional, indicating the importance of one or more additional factors for genistein accumulation. Interestingly, there was a reverse relationship between levels of genistein and kaempferol in the various lines (Fig. 2B), indicating that introduction of IFS or the resultant accumulation of genistein conjugates inhibited the biosynthesis of flavonoids.

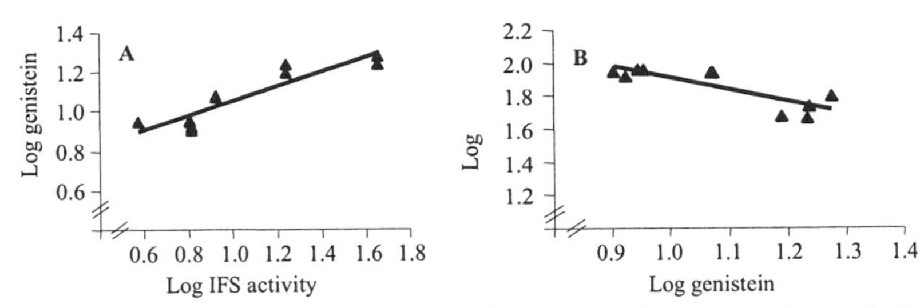

Figure 2.Relationships between IFS activity and production of genistein and kaempferol

2.2. Co-Expression Of Alfalfa CHI With Soybean IFS Does Not Increase The Accumulation Of Isoflavone In *Arabidopsis*

Chalcone isomerase (CHI) precedes IFS in isoflavonoid biosynthesis and catalyzes an intramolecular cyclization of chalcone into flavanone, narigenin. *Arabidopsis* CHI might be involved in an enzyme complex with chalcone synthase, flavonoid 3'-hydroxlase and dihydroflavol 4-reductase, forming a metabolic channel to coordinately control flavonoid biosynthesis (Burbulis and Winkel-Shirley, 1999; Winkel-Shirley, 1999). CHI from legumes is active toward both naringenin chalcone and isoliquiritigenin (2',4,4'-trihydroxychalcone), the latter being the precursor for the 5-deoxy series of isoflavonoids found in many legumes. Alfalfa CHI shares only 43% identity with *Arabidopsis* CHI and lacks the extended N-terminal peptide that contains cysteine residues proposed to be modification sites for acylation and membrane targeting in *Arabidopsis*.

Introduction of alfalfa CHI can complement the *Arabidopsis tt5* mutant (a null AtCHI mutant in the Landsberg erecta ecotype) and led to about a 3-fold increase in flavonol conjugates in the Columbia ecotype, but no naringenin or its conjugates were detected in either the control lines or CHI transgenic lines.

CHI transgenic line 4-11 was crossed with IFS transgenic line 15b. A larger number of F_3 progeny plants were then examined for isoflavone and flavonol levels by HPLC. Homozygous 15b and 4-11 plants (T_4 generation) were included for comparison. The results confirm the reduction in flavonol levels in plants producing genistein (Table 1). Levels of flavonol conjugates in seven progeny were only 58% of the average value for the 4-11 line. However, the data showed no increase in isoflavone levels in plants co-expressing IFS and alfalfa CHI (Table 1), comparied with line 15b.

Table 1. Genistein and flavonol levels in transgenic Arabidopsis expressing soybean IFS and/or alfalfa CHI.[a]

Line	Generation	Alfalfa CHI activity (nmol/min/mg)[b]	Total flavonols (nmol/g FW)[c]	Genistein (nmol/g FW)[c]
Wild-type		0	263.5 ± 63.5	0
CHI 4-11	T_4	49.6 ± 5.2	649.8 ± 224.5	0
IFS 15b	T_4	0	177.2 ± 26.9	9.8 ± 5.5
IFS/CHI	F_3	46.2 ± 9.5	376.7±207.4	10.5 ±4.1

[a]Plants were grown under 12 hr light/12 hr dark, ~150 μE light intensity, for 35 days.
[b]Average of two independent determinations
[c]Mean and s.d of determinations from at least 5 separate plants

2.3. Co-Expression Of IFS And CHI With Activated PAP1-D Leads To Only Small Increases In Genistein Level

Pap1-D is a myb type transcription factor. Overexpression of Pap1-D results in up-regulation of the whole phenylpropanoid pathway and an ~80 fold increase in anthocyanin accumulation (Borevitz et al., 2001). Expression of alfalfa CHI in this background resulted in plants with high levels of both flavonols and anthocyanin. However, F_2 progeny of the cross between IFS transgenic line 6E and *Arabidopsis* line 3-26, which contains alfalfa CHI in the Columbia *pap1-D* background, only accumulate an average of 5.64 nmol/g F.W of genistein. There was only about a 70% increase in genistein levels compared to the parent 6E line (3.3 nmol/g F.W.). At the same time, anthocyanin levels were reduced by as much as 80% in lines expressing IFS that still retained the activation tagged *PAP1* gene.

2.4. Expression Of IFS In The *tt6/Tt3* Mutant

The *tt6/tt3* double mutant in the Arabidopsis Landsburg erecta ecotype lacks functional flavanone-3β-hydroxylase and dihydroflavonol reductase, thus blocking metabolic flux to the biosynthesis of flavonols and anthocyanins (Fig. 3a, c). Expression of IFS in wild type Landsberg only produced approximately 1.88 nmol/g FW of genistein (average from three independent transgenic lines) (Fig. 3b), whereas overexpression of IFS in the tt6/tt3 mutant background produced an average of 5.89 nmol/g FW of genistein from two independent lines (Fig.3d). Although this is a small increase over that observed in the Landsburg wild-type background, the number of plants analyzed was small and the absolute levels of genistein were no greater than observed in other transgenic lines described above.

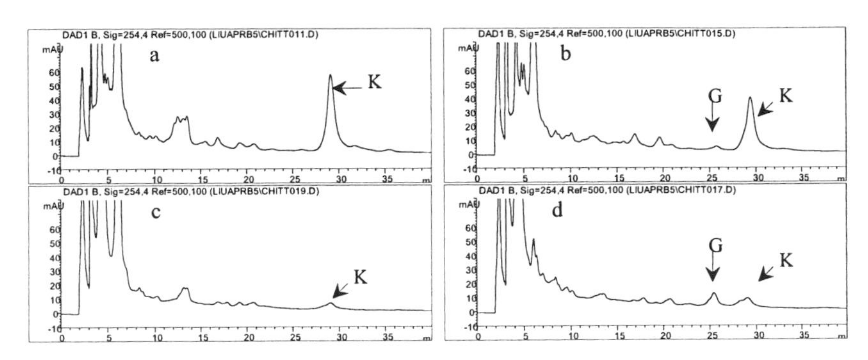

a, Ler wild type; b, IFS in Ler; c, tt6/tt3 mutant; d, IFS in tt6/tt3; K, Kaempferol; G, genistein

Figure 3. HPLC profiling of IFS transgenic lines in Ler wild type and tt6/tt3 mutant

3. CONCLUSIONS

Up-regulation of flux into the early steps of flavonoid biosynthesis does not increase isoflavone accumulation in transgenic Arabidopsis expressing IFS.

Introduction of IFS or the resultant accumulation of genistein conjugates inhibits the biosynthesis of flavonoids.

Blocking the later steps of flavonoid biosynthesis does not appear to result in a significant switch of flux into isoflavone.

The catalytic efficiencies of the introduced P450 protein IFS, and endogenous NADPH: cytochrome P450 reductase, may be other factors impeding accumulation of isoflavone. It may also be necessary to introduce the dehydratase enzyme believed to be involved in the IFS reaction *in vivo* in order to obtain maximum rates of isoflavone formation.

4. REFERENCES

Akashi T., Aoki T., and Ayabe S. 1999. Cloning and functional expression of a cytochrome P450 cDNA encoding 2-hydroxyisoflavanone synthase involved in biosynthesis of the isoflavonoid skeleton in licorice. Plant Physiol. 121: 821-828 .

Borevitz J., Xia Y., Blount J.W., Dixon, R.A. and Lamb C. 2001. Activation tagging identifies a conserved MYB regulator of phenylpropanoid biosynthesis. Plant Cell 12: 2383-2393.

Burbulis I.E., and Winkel-Shirley B. 1999. Interactions among enzymes of the Arabidopsis flavonoid biosynthetic pathway. Proc Natl Acad Sci U S A 96:12929-12934.

Dixon R.A. 1999. Isoflavonoids: biochemistry, molecular biology and biological functions. Comprehensive Natural Products Chemistry, Vol. 1 Sankawa U. (Ed.). Elsevier. Pp. 773-823.

File S.E., Jarrett N., Fluck E., Duffy R., Casey K., and Wiseman H. 2001 Eating soya improves memory. Psychopharmacology. 157: 430-436.

Setchell K.D.R., and Cassidy A. 1999. Dietary isoflavones: Biological effects and relevance to human health. J. Nutr. 129: 758S-767S.

Steele C.L., Gijzen M., Qutob D., and Dixon R.A. 1999. Molecular characterization of the enzyme catalyzing the aryl migration reaction of isoflavonoid biosynthesis in soybean. Arch. Biochem. Biophys. 367: 147-150.

Jung W. et al. 2000. Identification and expression of isoflavone synthase, the key enzyme for biosynthesis of isoflavones in legumes. Nature Biotechnol. 18: 208-212.

Winkel-Shirley B. 1999. Evidence for enzyme complexes in the phenylpropanoid and flavonoid pathways. Physiol. Plant. 107: 142-149.

Yu O. et al. 2000. Production of the isoflavones genistein and daidzein in non-legume dicot and monocot tissues. Plant Physiol. 124:781-794.

IN VITRO PRODUCTION OF SECONDARY METABOLITES BY CULTIVATED PLANT CELLS: THE CRUCIAL ROLE OF THE CELL NUTRITIONAL STATUS

Laurence Lamboursain[1] and Mario Jolicoeur[2]
BIO-P[2], Chemical engineering department, Ecole Polytechnique de Montréal, CP 6079, suc. CV, Montréal, H3C 3A7, Canada ([1]e-mail: Laurence.lamboursain@polymtl.ca; [2]e-mail: Mario.jolicoeur@polymtl.ca)

Keywords Plant cells, secondary metabolites, production, nutritional state, physiology

1. INTRODUCTION

Plants constitute an important source of highly valuable phytochemicals that are widely present in the human pharmacopoeia. But although the demand for these products has increased dramatically in the past few years, the supply of the source plants is sometimes limited. Five decades ago, *in vitro* production of phytochemicals by cultivated plant cells was thought to be an attractive alternative to traditional methods of production. However, very few bioprocesses using plant cell culture achieved commercial profitability. This deceiving result is partly related to the low productivity and poor reproducibility of bioprocesses using plant cells.

In an attempt to have a more global approach of the problem, we made a parallel between whole plants and cultivated plant cells. In whole plants, physiology is strongly dependent upon the nutritional state of the plant (Paul and Stitt, 1993; Yu, 1999; Hell and Hillebrand, 2001; Mcintyre, 2001). In plant cell culture, the composition of the culture medium was also shown to affect cell growth and secondary metabolites (SM) production (Zhong and Zhu, 1995; Sakano et al., 1995; Sato et al., 1996; Yeoman and Yeoman, 1996). However, the data accumulated in the literature are very contradictory and did not allow to draw clear links between physiology and nutrition in plant cell culture.

Our hypothesis is that the same relationship that exists in whole plants between the nutritional status and the physiological state also occurs in cultivated cells. However, since plant cells have the capacity to accumulate large amounts of nutrients in their vacuole, cytoplasm and plastids, this relationship could not be proven if the global nutritional status of the cells

I. K. Vasil (ed.), Plant Biotechnology 2002 and Beyond, 491-495.

(i.e. the extra and intra-cellular concentration in nutrients) is not taken into account. To assess this hypothesis, we compared the effect of N, P and carbohydrate limitations on the physiological state of cultivated cells, and more precisely on their capacity to produce secondary metabolites (SM) after the adding of an elicitor.

2. MATERIAL AND METHODS

2.1. Cell Line

Eschschzoltzia californica (EC) cells were grown in three different culture media. Inoculum 1 was subcultured on B5 medium (Gamborg et al., 1968) containing 30 $g \cdot l^{-1}$ of glucose and supplemented with 0.2 $mg \cdot l^{-1}$ 2,4-Dichlorophenoxy-acetic acid and 0.1 $mg \cdot l^{-1}$ kinetin. Inoculum 2 was subcultured on a B5 modified medium with a reduced content in phosphate (0.8 mM compared to 1.1 mM in plain B5) and Inoculum 3 was grown in a B5 modified medium with a reduced content in nitrate (12.5 mM compared to 25.0 mM in plain B5). All three inocula were subcultured every 10 days at a 1/3 dilution ratio.

2.2. Elicitability Test

3 g of cells (fresh weight) were aseptically transferred in a 75 ml Erlenmeyer flask containing 10 ml of modified MS medium (phosphate free medium without vitamins and hormones), 2 g of wet extractive resin (XAD-7, Sigma) and 2 ml of elicitor solution (crude chitin extract). Flasks were placed on an orbital shaker at 120 rpm under normal laboratory lights for 7 days. The resin and cells were then extracted with acidic methanol and the total alkaloid content of the extract was quantified by fluorescence.

3. ALTERATION OF THE NUTRITIONAL STATUS: EFFECT ON DRY WEIGHT ACCUMULATION

In order to alter the nutritional status of EC cells, single nutrients were added to the three inocula at day 5. At day 10, flasks were harvested and tested for dry weight concentration (DW, figure 1), fresh weight concentration (FW), cell count, elicitability (i.e. capacity of cells to produce alkaloids after elicitation, figure 2) and intracellular composition (table 1). Results show that the three inocula (control flasks) had very different growth and production patterns. For example, the final DW concentration of inoculum 3 was 45 % higher than that of inoculum 1 (8.5 and 12.3 $g\ DW \cdot l^{-1}$ respectively). However, the final cell concentration in inoculum 3 was only 22 % higher than in inoculum 1 (3.5 and 4.3 10^6cells·ml^{-1} respectively). This illustrates that an increase in dry weight is not always attributable to cell division but can also reflect an intracellular accumulation of nutrients (in our case, the accumulation of starch and Pi in inoculum 3). This result shows the inadequacy of following cell growth based on dry weight measurements only.

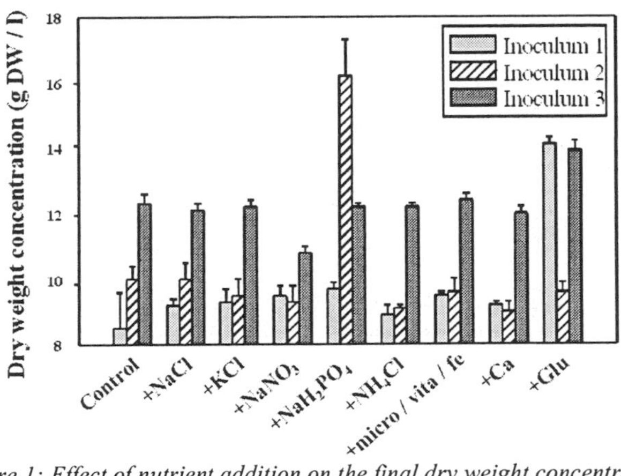

Figure 1: Effect of nutrient addition on the final dry weight concentration

4. EFFECT ON SM PRODUCTION: CONTRADICTORY RESULTS?

EC cells response to the addition of a typical nutrient depends on the nutritional state of the cells at the time of the adding (fig. 2). In fact, the addition or the removal of a nutrient can both stimulate SM production in EC cells depending on the nutritional status of the cells.

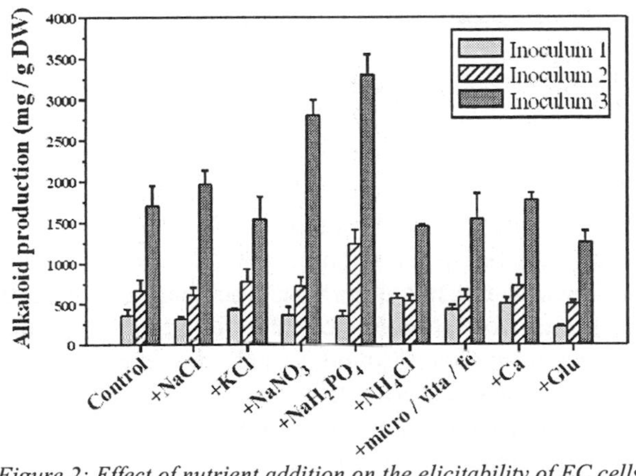

Figure 2: Effect of nutrient addition on the elicitability of EC cells

For example, phosphate limited cells had a higher alkaloid production rate than P-sufficient cells (363.4 mg alkaloid / g for inoculum 1 versus 667.6 mg alkaloid / g DW for inoculum 2 (see control flasks)). However, the addition of Pi at day 5 also increased the alkaloid production rate for inocula 2 and 3 (1233.1 and 3294.9 mg·g DW^{-1} of alkaloid were obtained respectively).

Therefore, Pi limitation and addition seemed both to favor SM production. Although obviously contradictory, those results can in fact be explained after examining the intracellular composition of the biomass (table 1). Those data established that Pi limited cells had a higher carbon content than non limited ones (0.31 and 0.19 g C·g^{-1} DW respectively). Carbon being the major constituent of SM, this could explain why P-limited cells produced more alkaloids than P sufficient cells. In the other hand, when we added Pi to inoculum 2 (which has already accumulated carbon during P limitation), SM production was also increased because the energy level of the C-rich cells was restored, allowing increased metabolic flux through biosynthesis pathways.

Table 1:Intracellular composition of EC cells at day 5

Intracellular concentration	Inoculum 1	Inoculum 2	Inoculum 3
Kjeldahl N (mmole·g^{-1} DW)	3.45	3.30	3.01
Total P (mmole·g^{-1} DW)	0.18	0.16	0.21
Total C (g·g^{-1} DW)	0.19	0.31	0.29

5. CONCLUSIONS

The results presented here showed the inadequacy to use dry weight measurements to follow plant cells growth. When *Eschscholtzia californica* (EC) cells are subjected to P or N limitation, the dry weight concentration (DW) increased even if cell division was stopped. The DW increase was then solely due to the intracellular accumulation of nutrients. We also showed that the accumulation of carbon skeleton by P and N limited cells was the key for an increased SM production rate. The nutritional status of EC cells, which is the intra- and extra-cellular concentrations in nutrients, has a strong influence on the growth and the SM production capacity of the cells. Our results also suggest that the lack of information on the nutritional status of the cells may be responsible for the poor reproducibility reported during plant cell culture, and also for the discrepancies found among previously published results. These findings opens new opportunities toward a better control of secondary metabolites production using plant cell culture, through an increase in both reproducibility and productivity of the process.

6. REFERENCES

Gamborg, O.L.; Miller, R.A. and Ojima, K. 1968. Nutrient requirements of suspension cultures of soybean root cells. Exp. Cell Res. 50:151-156.

Hell, R. and Hillebrand,H. 2001. Plant concepts for mineral acquisition and allocation. Curr. Opin. Biotech. 12:161-168.

Mcintyre, G.I. 2001. Control of plant development by limiting factors: a nutritional perspective. Physiol. Plant. 113:165-175.

Paul, M.J. and Stitt, M. 1993. Effect of nitrogen and phosphorus deficiency on levels of carbohydrates, respiratory enzymes and metabolites in seedlings of tobacco and their response to exogenous sucrose. Plant Cell Environ. 16:1047-1057.

Sakano, K., Matsumoto, M., Yazaki, Y., Kiyota, S. and Okihara, K. 1995. Inorganic phosphate as a negative conditioning factor in plant cell culture. Plant Sci. 107:117-124.

Sato, K., Nakayama, M. and Shigeta, J.I. 1996. Culturing conditions affecting the production of anthocyanin in suspended cell culture of strawberry. Plant Sci. 113:91-98.

Yeoman, M.M. and Yeoman, C.L. 1996. Manipulating secondary metabolism in cultured plant cells. New Phytol. 134:553-567.

Yu, S.M. 1999. Cellular and genetic responses of plants to sugar starvation. Plant Physiol. 121:687-693.

Zhong, J.J and Zhu, Q.X. 1995. Effect of the initial phosphate concentration on cell growth and ginsenoside saponin production by suspended culture of Panax notoginseng. Appl. Biochem. Biotech. 55:241-247.

TRANSGENESIS AND GENOMICS IN MOLECULAR BREEDING OF TEMPERATE PASTURE GRASSES AND LEGUMES

German Spangenberg[1,2], Michael Emmerling[1], Ulrik John[1], Roger Kalla[1], Angela Lidgett[1,2], Eng K. Ong[1], Tim Sawbridge[1] and Tracie Webster[1]

[1]Plant Biotechnology Centre, Agriculture Victoria, La Trobe University, Bundoora, Victoria 3083, Australia;[2]CRC for Molecular Plant Breeding (e-mail: german.spangenberg@nre.vic.gov.au)

1. SUMMARY

Significant advances in the establishment of the methodologies required for the molecular breeding of temperate forage grasses (*Lolium* and *Festuca* species) and legumes (*Trifolium* and *Medicago* species) are reviewed. Examples of current products and approaches for the application of these methodologies to forage grass and legume improvement are outlined. The plethora of new technologies and tools now available for high-throughput gene discovery and genome-wide expression analysis have opened up opportunities for innovative applications in the identification, functional characterisation and use of genes of value in forage production systems and beyond. Selected examples of our current work in pasture plant genomics, xenogenomics, symbiogenomics and microarray-based molecular phenotyping are discussed.

2. PASTURE PLANT TRANSGENESIS

Gene technology and the production of transgenic plants offers the opportunity to generate unique genetic variation, when the required variation is either absent or has very low heritability. In recent years, the first transgenic pasture plants with simple 'engineered' traits have reached the stage of field-evaluation. While gaps in our understanding of the underlying genetics, physiology and biochemistry of many complex plant processes are likely to delay progress in many applications of transgenesis in forage plant improvement, gene technology is a powerful tool for the generation of the required molecular genetic knowledge. Consequently, applications of transgenesis to temperate pasture plant improvement are focussed on the development of transformation events with unique genetic variation and in studies on the molecular genetic dissection of plant biosynthetic pathways

I. K. Vasil (ed.), Plant Biotechnology 2002 and Beyond, 497-502.

and developmental processes of high relevance for forage production (Spangenberg et al., 2001).

Primary target traits for the application of transgenesis to temperate pasture plant improvement are forage quality, disease and pest resistance, tolerance to abiotic stresses, and the manipulation of growth and development. Some representative approaches and selected examples in temperate forage grasses and legumes are discussed (Spangenberg et al., 2001).

Molecular breeding based on transgenesis to overcome limitations in forage quality may be targeted to the individual subcharacters involved: dry matter digestibility, water-soluble carbohydrate content, protein content, secondary metabolites, alkaloids, etc. These molecular breeding approaches may include modification of the lignin profile to enhance dry matter digestibility, genetic manipulation of fructan metabolism to increase non-structural carbohydrate content, genetic manipulation of condensed tannin synthesis to develop 'bloat-safe' forages, and the expression of 'rumen by-pass' proteins to improve the supply of proteins and essential amino acids. Most quality or anti-quality parameters are associated with specific metabolic pathways or the production of specific proteins. This allows target enzymes or suitable foreign proteins to be identified, corresponding genes isolated, and their expression manipulated in transgenic forage plants.

Pathogen and pest infection can considerably lower herbage yield, persistency, nutritive value, and palatability of forage plants. An armory of genes and strategies for engineering disease and pest resistance in transgenic plants has been developed and tested over the last decade, including chitinases, glucanases, plant defensins, phytoalexins, ribosome-inactivating proteins, viral coat proteins, viral replicase, viral movement proteins, *Bt* toxins, proteinase inhibitors, and α-amylase inhibitors. Some of them have been applied to the development of pasture plants, mainly forage legumes, for enhanced disease and pest resistance (Spangenberg et al., 2001).

Plants can be used to express recombinant heterologous proteins. Transgenic plants may be an attractive alternative to microbial systems for the production of certain biomolecules. The perennial growth habit, the biomass production potential, the capacity for biological nitrogen fixation, and the ability to grow in marginal areas exhibited by forage plants, particularly pasture legumes, make them potential suitable candidates for molecular farming. Advances in genetic manipulation technologies that allow high levels of transgene expression and transgene containment may, in the not too distant future, make it possible to exploit some forage plants as bioreactors for the production, among others, of industrial enzymes, pharmaceuticals, vaccines, antibodies and biodegradable plastics. Multidisciplinary efforts will, however, be needed to identify the most feasible targets, to generate transgenic plants with suitable expression levels, and to develop efficient downstream processing technology that could adapt transgenic forage plants for non-forage uses and make them a cost-effective alternative for molecular farming. Significant progress has been achieved in recent years in the production of value-added proteins in transgenic lucerne (Spangenberg et al., 2001).

Small scale planned releases of transgenic plants are required to assess the stability of transgene expression and the novel phenotypes under field conditions and to identify transformation events suitable for transgenic germplasm and cultivar development. Only after the transformation events have been thoroughly evaluated for the stability of the novel phenotype outside of the controlled environment in a glasshouse would it be advisable to continue to integrate these in molecular breeding programs for the development of transgenic cultivars. An illustrative example of design features of such a small scale field trial can be found in a recent field trial of alfalfa mosaic virus (AMV) immune transgenic white clover plants (Kalla et al., 2001).

A range of transformation events in forage legumes and grasses with proof of concept for the technology under containment conditions are being developed. The challenge now is how to best deploy these molecular technologies and tools to evaluate their full potential based on the transgenic transfer of single and multiple valuable genes, to generate novel genetic variability and novel elite transgenic germplasm, and to efficiently incorporate these factors into breeding programs for the development of improved cultivars.

Efficient strategies for the introgression of transgenes into elite parents for the subsequent production of synthetic cultivars have been developed ensuring stable and uniform transgene expression in all plants in the population. One such strategy has been applied to the production of AMV immune transgenic elite white clover plants homozygous for the transgenes. It involves initial top crosses of transformation events chosen after their field evaluation with elite non-transgenic white clover parental lines; selecting for progeny from the harvested seed carrying the transgene and its linked selectable *npt2* marker gene by antibiotic selection or PCR screening followed by diallel crosses between the T_1 progeny. The T_2 offspring plants homozygous for transgenes can be directly identified by high-throughput quantitative PCR transgene detection. The elite white clover plants homozygous for the transgenes are then planted in a selection nursery together with elite non-transgenic parental lines for identification of the new parents of transgenic experimental synthetic cultivars and their subsequent multisite evaluation (Spangenberg et al., 2001).

3. PASTURE PLANT GENOMICS

Forage plant breeding has entered the genome era. The plethora of new technologies and tools now available for high-throughput gene discovery and genome-wide expression analysis have opened up opportunities for innovative applications in the identification, functional characterisation and use of genes of value in forage production systems and beyond. Examples of these opportunities, include 'molecular phenotyping', 'symbio-genomics' and 'xeno-genomics' (Spangenberg et al., 2001).

We have undertaken the discovery of 100,000 ESTs from the key forage crops of temperate grassland agriculture, perennial ryegrass (*L. perenne*) and white clover (*T. repens*) using high-throughput sequencing of randomly selected clones from cDNA libraries representing a range of plant organs, developmental stages, and experimental treatments. The DNA sequences were analysed by BLAST searches, categorised functionally, and subjected to cluster analysis leading to the identification of unigene sets in perennial ryegrass and white clover corresponding to 14,767 and 14,635 genes, respectively (Spangenberg et al., 2001).

We have further developed high density spotted cDNA microarrays with approximately 15,000 unigene sets as a main screening tool for novel ryegrass and clover sequences of unknown function. These EST-based plant microarrays will allow the global analysis of gene expression patterns as a main approach for functional genomics and other applications. Novel applications of EST-based forage plant arrays including 'molecular phenotyping', i.e. the analysis of global or targeted gene expression patterns using complex hybridisation probes from contrasting genotypes or populations and contrasting environments, are now being tested to integrate microarray data with current conventional phenotypic selection approaches used in temperate pasture plant improvement (Spangenberg et al., 2001).

Comparative sequence and microarray data analyses from ryegrass and clover with data from complete genome sequencing projects in *Arabidopsis* and rice as well as from extensive EST discovery programs in the model forage legume *M. truncatula* have been undertaken to provide insight into conserved and divergent aspects of grass and legume genome organization and function.

4. PASTURE PLANT SYMBIOGENOMICS

Pasture legumes and grasses offer unique and exciting opportunities in genome research to study plant-pathogen interactions, legume/nitrogen-fixing bacteria symbiosis, legume/mycrorrhiza associations, and grass/endophyte endosymbiosis, as well as to the application of the knowledge gained from these studies to develop resistance to pathogens and improved beneficial associations in forages.

We have undertaken a gene discovery program in the fungal endophytes of tall fescue and perennial ryegrass, *Neotyphodium coenophialum* and *N. lolii*, respectively. Approximately 8,500 *Neotyphodium* DNA sequences were generated, analysed by BLAST searches, categorised functionally, and subjected to cluster analyses leading to the identification of a 3,806 unigene set in *Neotyphodium*. The program is focused on the discovery of genes involved in host colonization, nutrient supply to the endophytic fungus, and the biosynthesis of active pyrrolopyrazine and pyrrolizidine secondary metabolites (e.g. the insect deterrents peramine and N-formylloline, respectively) and their regulation. It will provide insight into the molecular genetics of the grass endophyte/host interaction as well as into the physiological mechanisms leading to the increased plant vigour and enhanced stress tolerance. These genomic tools and knowledge will underpin the development of technologies to manipulate grass/endophyte associations for enhanced plant performance, improved grass tolerance to biotic and abiotic stresses, and altered grass endophyte host specificity, to the benefit of the grazing and turf industries (Spangenberg et al., 2001).

5. XENOGENOMICS

Genome research with exotic plant species, i.e. 'xeno-genomics', includes gene discovery by high-throughput EST sequencing and large-scale simultaneous gene expression analysis with EST-based microarrays. Xeno-genomics has opened up opportunities for a 'genomic bio-prospecting' of key genes and gene variants from exotic plants. This approach is particularly suited for the discovery of novel genes and the determination of their expression patterns in response to specific abiotic stresses.

We have undertaken a xenogenomic EST discovery focussed on selected Australian native and exotic grasses and legumes that show unique adaptation to extreme environmental stresses. Genes which allow certain plant species to tolerate extreme abiotic stresses including drought, salinity and low fertility soils are being isolated and characterised. The targeted species in the xenogenomic EST discovery program include Australian native grasses, such as the halotolerant blown-grasses (*Agrostis adamsonii* and *A. robusta*) and the aluminium-tolerant weeping grass (*Microlaena stipoides*); as well as exotic species such as antarctic hair-grass (*Deschampsia antarctica*), one of only two vascular plant species native to Antarctica (Spangenberg et al., 2001).

The discovery of novel genes and their functional genomic analysis will facilitate the development of effective molecular breeding approaches to enhance abiotic stress tolerance in forages and other crops.

6. REFERENCES

Spangenberg G., Kalla R., Lidgett A., Sawbridge T., Ong E.K., John U. 2001. Breeding forage plants in the genome era. In: Molecular Breeding of Forage Crops, G Spangenberg (ed.), Kluwer Academic Publishers, Developments in Plant Breeding, Volume 10, p 1-40.

Kalla R., Chu P., Spangenberg G., Kalla. 2001. Molecular breeding of forage legumes for virus resistance. In: Molecular Breeding of Forage Crops, G Spangenberg (ed.), Kluwer Academic Publishers, Developments in Plant Breeding, Volume 10, p 219-238.

ENHANCING TURFGRASS PERFORMANCE WITH BIOTECHNOLOGY

Robert W. Harriman, Eric Nelson and Lisa Lee
Turfgrass Variety Development and Biotechnology Department, The Scotts Company. 14111 Scottslawn Road, Marysville, OH 43041, USA (email: bob.harriman@scotts.com)

Keywords Turfgrass, biotechnology, biolistic transformation, herbicide resistance, quality traits

1. INTRODUCTION

Turfgrasses play an important role in maintaining a healthy environment and enriching our lives. Not only does turf provide an aesthetically pleasing landscape feature or a functional surface for sporting events, turfgrass helps reduce soil erosion and agricultural runoff, and it absorbs carbon dioxide and ozone while releasing life-sustaining oxygen. Turfgrasses trap an estimated 12 million tons of dust each year and an average lawn has the cooling effect of about 10 tons of air conditioning (Beard and Green, 1994).

While advances in breeding and cultural practices are continuously increasing the positive environmental impact of turf, biotechnology has the potential to dramatically enhance our ability to maintain a healthy turfgrass stand with even fewer inputs. Biotechnology can provide breeders with the means to help solve turfgrass management problems that have not, and probably will not, be solved by conventional breeding. Some improved features and benefits potentially available through biotechnology include: Roundup® resistance for weed control with an environmentally benign herbicide. Reduced vertical growth to decrease the frequency of mowing; Broad-spectrum disease tolerance to reduce the need for fungicides; Improved heat or cold hardiness to decrease loss of turf due to winter or summer stress; Increased drought tolerance or water use efficiency to reduce our dependence on potable water sources.

2. BIOTECHNOLOGY PROGRAM AT THE SCOTTS COMPANY

To complement a long history of breeding superior turfgrass varieties and developing advanced turf management tools, The Scotts Company has

I. K. Vasil (ed.), Plant Biotechnology 2002 and Beyond, 503-506.

assembled a biotechnology program to continue our vision of delivering superior turf performance with less cultural maintenance. The traits we focus on include herbicide tolerance, disease resistance and reduced growth. These traits are being introduced into turfgrass species such as creeping bentgrass, Kentucky bluegrass and St. Augustinegrass. Creeping bentgrass is a cool season grass that is used mainly on golf course putting greens, tees and fairways. Kentucky bluegrass is also a cool season grass utilized in both professional and consumer turfgrass stands. St. Augustinegrass is a warm season, predominantly consumer, grass which is grown along the southeast Atlantic seaboard and Gulf Coast, including all of Florida and other tropical environments.

Our transformation technique of choice is the biolistic transformation system. The biolistic DNA delivery system is well documented to transfer and stably integrate foreign DNA into a plant genome (Sanford et al., 1993; Lee, 1996). Embryogenic, regenerable cells are bombarded with DNA-coated gold particles. Transformed cells are identified by culturing on a selective medium containing glyphosate. After 6 weeks on the selective medium, regeneration is performed in the presence of glyphosate. Regenerated plants are transferred to the greenhouse for trait analysis. Herbicide tolerance is confirmed by spraying plants with 128 oz/A of Roundup branded herbicides. Field performance is evaluated for both agronomic characteristics and introduced biotech traits such as herbicide tolerance, disease resistance and/or less mowing. Plants which 'pass the test' are induced to flower for seed production and further agronomic evaluation. The transformed turfgrasses have displayed high levels of herbicide tolerance in both greenhouse and field herbicide spray tests. Field evaluations of other novel traits such as disease resistance and less mowing are being conducted. While we are making solid progress on all of the outlined products, our most advanced biotechnology-enhanced product in relation to commercial release is Roundup resistant creeping bentgrass.

3. ROUNDUP READY CREEPING BENTGRASS

Over 10,000 golf courses in the United States use creeping bentgrass for their greens and/or tees and fairways. Effective control of the grassy weeds, annual bluegrass (*Poa annua*) and roughstalk bluegrass (*Poa trivialis*), is not currently available in creeping bentgrass turf. *Poa annua* may often comprise 50% or more of the turfgrass stand, thus forcing golf course superintendents into co-managing both bentgrass and annual bluegrass. *Poa annua* is more sensitive to abiotic stresses such as temperature extremes and to biotic pressures such as summer patch, anthracnose and annual bluegrass weevil. Failure to properly manage *Poa* may result in massive areas of dead turf, unhappy customers and a golf course superintendent without a job. By introducing the Roundup Ready® gene into creeping bentgrass, a golf course superintendent will be able to eliminate a severe problem and significantly reduce many inputs normally associated with *Poa annua* management by simply spraying the environmentally friendly herbicide Roundup.

Roundup, the brand name for glyphosate, is a broad-spectrum, nonselective, post emergent, systemic herbicide that provides for control of essentially all annual and perennial plants. The mode of action of glyphosate is to inhibit the enzyme EPSP (5-enolpyruvoylshikimate 3-phosphate) synthase and prevent plants from manufacturing three essential aromatic amino acids (phenylalanine, tyrosine, and tryptophan) in the Shikimate metabolic pathway. Animals and humans obtain these amino acids through their diet and do not use this enzyme, thus providing a basis for specific selective toxicity only to plant species. In addition to glyphosate's highly specific mode of action, it does not persist in the environment or bioaccumulate in the food chain. Glyphosate is essentially immobile in almost all types of soils where it is degraded by naturally occurring microbes.

With the use of biolistic transformation, over three hundred independent transgenic events were generated in either our lab or at Rutgers University. The events were screened for agronomic and trait performance. Impressive tolerance to even the highest Roundup rate analyzed (4 gallons/acre) has been demonstrated on turfgrass stands maintained at a fairway cut of a half inch. While resistant to Roundup, Roundup Ready® Creeping Bentgrass can be controlled with other herbicides like Finale or Fusilade. Lead events were introduced into variety development as well as a series of regulatory experiments.

4. REGULATORY ISSUES

Before we can enjoy benefits from a biotech-enhanced product such as Roundup Ready creeping bentgrass, they must be vigorously examined and pass a thorough review by the United States Department of Agriculture and the Environmental Protection Agency. The process can be summarized by addressing five issues (additional information can be obtained at www.usda.aphis.gov).

Issue 1: to determine that the product exhibits no plant pathogenic properties. Issue 2: to show that the product is no more likely to become a weed pest than traditional bred varieties. Issue 3: to prove that the product is unlikely to increase the weediness potential for any other cultivated plant or native wild species with which the product could interbreed. Issue 4: to ensure that the product is unlikely to cause damage to processed agricultural commodities. Issue 5: to show that the product is unlikely to harm organisms beneficial to agriculture. Data to address the topics above is developed over several years and in multiple locations. The analysis determines if the enhanced product is similar (except for the introduced trait) to other genotypes in nature or products currently on the market.

Roundup Ready Creeping Bentgrass was examined in over 50 regulatory studies conducted at 25 locations by scientists from several disciplines and universities. Studies on environmental assessment include biological, reproductive and survival characteristics such as life span, seed productivity, pollen viability and logenvity, self fertility and cross fertility, vegetative capacity, environmental tolerance and herbicide tolerance. No significant difference was noted between conventional cultivars currently on the market and Roundup Ready Creeping Bentgrass. In addition, and as would be expected, the Roundup Ready Creeping Bentgrass plants were equally as susceptible to other broad spectrum and grass herbicides as were conventional, non-transformed cultivars. Furthermore, a rigorous stewardship program will be in place to ensure appropriate management and continuous review of technology for safe and sustained use.

5. SUMMARY

The common turfgrass plant has an anything but common impact on the health of the environment and the health and enjoyment of mankind. While we are enjoying a game of golf or watching our kids playing soccer, the very turf that supports that activity is also filtering air and water while cooling our surroundings. Turfgrass enhanced through biotechnology is not yet commercially available. However, the first products are making their way through the development pipeline and regulatory review. With biotechnology, scientists are developing the tools to assist turfgrass professionals and consumers to maintain a healthier turfgrass stand with fewer inputs for even greater benefits to both the environment and our quality of life.

6. REFERENCES

Beard and Green. 1994. The role of turfgrasses in environmental protection and their benefits to humans. J. Environ. Qual. 23(3):452-460.

Lee, L. 1996. Turfgrass biotechnology. Plant Science 115: 1-8.

Sanford, J.C., F.D. Smith, and J.A. Russell. Optimizing the biolistic process for different biological applications. (1993) Methods in Enzymology, 217: 483-509.

GENETIC MANIPULATION OF COOL-SEASON FORAGE GRASSES AND FIELD EVALUATION OF TRANSGENIC PLANTS

Zengyu Wang
Forage Biotechnology Group, The Samuel Roberts Noble Foundation, 2510 Sam Noble Parkway, Ardmore, OK 73401, USA (e-mail: zywang@noble.org)

Keywords Forage grass, plant regeneration, genetic transformation, transgenic plants, field evaluation

1. INTRODUCTION

Forage grasses provide the backbone of sustainable agriculture and play an important role in soil and water conservation. Ruminant animal productivity is largely dependent on their utilization of forages. Attempts have been made to estimate the value of forages consumed by ruminant livestock. By using estimated feed cost in livestock production, the calculated value of forages far exceeds the cash value of any other crop in the USA (Barnes and Baylor, 1995).

Genetic improvement is one of the most effective ways to increase productivity of forage grasses. Due to the great complexity of forage species and the associated difficulties encountered by traditional breeding methods, the potential of molecular breeding for the development of improved grass cultivars is evident (Spangenberg et al., 1998). Forage grasses are among the most recalcitrant plants to genetically manipulate in vitro (Vasil, 1988; Potrykus, 1991). Significant progress has been made in establishing the methodological basis required for the genetic manipulation of forage grasses in recent years (Wang et al., 2001).

We have been interested in the genetic manipulation of a few important cool-season grass species, including tall fescue (*Festuca arundinacea* Schreb.), Russian wildrye [*Psathyrostachys juncea* (Fisch.) Nevski] and tall wheatgrass [*Thinopyrum ponticum* (Podp.) Liu and Wang]. This report summarizes our work on the development of efficient regeneration and transformation systems for the different species, field evaluation of transgenic plants as well as molecular approaches for forage improvement

I. K. Vasil (ed.), Plant Biotechnology 2002 and Beyond, 507-513.

2. PLANT REGENERATION AND GENETIC TRANSFORMATION

The establishment of morphogenic cell cultures in grasses has proven to be an important prerequisite for gene transfer into these monocotyledonous species (Vasil, 1995). Since the grass species we deal with are largely self-sterile and out-crossing, single genotype-derived embryogenic cell suspension cultures were established and used for plant regeneration and genetic transformation.

A screening for the induction of embryogenic callus, appropriate to initiate morphogenic suspension cultures from single genotypes, was performed for different cultivars and lines of tall fescue, Russian and tall wheatgrass. MS-based media (Murashige and Skoog, 1962), supplemented with the hormone 2,4-D, were used for callus induction and maintenance (Wang et al., 1994, 2002). Embryogenic calli were obtained from most of the cultivars tested. Depending on the cultivar used, 1% to 10% of the seeds or embryos screened produced yellowish friable callus suitable for the initiation of suspension cultures.

Friable embryogenic callus, obtained from mature seeds or embryos of defined single-seed origin, was used to initiate suspension cultures in modified AA medium (Wang et al., 1994, 2002). The suspension cultures were kept at 25°C on a gyratory shaker at 80 rpm and were subcultured biweekly by replacing three-fourths of the culture medium with fresh medium. For all the *Festuca*, *Psathyrostachys* and *Thinopyrum* species considered, embryogenic cell suspensions showing differences in degree of dispersion and growth rate, but mainly consisting of pro-embryogenic cell clusters, could be established after 3 to 6 months. Large numbers of plants were regenerated after plating the suspension cells onto regeneration medium. Frequencies of green plant regeneration between 20% and 90% were observed in the grass species tested. Established embryogenic suspension cultures retained their potential for regeneration of green plants when evaluated over a period of 6 to 13 months.

The establishment of efficient plant regeneration systems provides a basis for the genetic transformation of grasses. Transgenic forage grasses were first obtained by direct gene transfer to protoplasts (Spangenberg et al., 1998; Wang et al., 2001). However, with the development of biolistic techniques, microprojectile bombardment of embryogenic cell cultures has become the prevalent method for producing transgenic grass plants (Wang et al., 2001).

Embryogenic cell clusters from established suspension cultures are ideal targets for biolistic transformation. This is especially true for outcrossing grass species, since the use of single genotype-derived cell suspensions allows the generation of transformants from the same genotype. We have obtained transgenic tall fescue and transgenic Russian wildrye plants by microprojectile bombardment of embryogenic suspension cells. Bombardment parameters with the PDS/1000 biolistic device were partially optimized by transient expression assays of a chimeric β-glucuronidase (GUS) gene construct. For stable transformation, a chimeric hygromycin phosphotransferase gene driven by rice *Act1* regulatory sequences was used. High concentrations of hygromycin (200 mg/l) were applied to the bombarded cells for selection. Resistant calli were recovered after 5 to 7 weeks of hygromycin selection. In vitro plantlets were regenerated from the calli and screened by PCR. The transgenic nature of the regenerated plants was confirmed by Southern and northern hybridization analyses. The frequency of co-transformation, where genes are located on two separate plasmids, was estimated to be 70% to 80% in transgenic tall fescue and Russian wildrye. This frequency allows a fairly good chance for the regeneration of plants with agronomic genes from resistant calli after co-transformation. *Agrobacterium*-mediated transformation is being tested in tall fescue, and the first set of primary transgenic plants have been obtained.

3. FIELD EVALUATION OF TRANSGENIC GRASS PLANTS

Transgenic tall fescue and Russian wildrye plants were transferred to the field to study transmission genetics of the transgenes and to evaluate their agronomic performance.

Transmission of foreign genes to progenies is critical for any potential use of transgenic material in producing novel germplasm or cultivars. However, to date there is no report showing meiotic transmission of transgenes in *Festuca*, although transgenic tall, red and meadow fescues were obtained several years ago (Wang et al., 1992; Spangenberg et al., 1995a,b; Kuai et al., 1999). This is primarily due to the outcrossing nature and vernalization requirement of these species, which make it difficult to obtain progenies under greenhouse conditions. The only report touching transgene inheritance in tall fescue provided confusing information, in which no transgenes were detected in the progenies (Kuai et al., 1999). In our study, fertile transgenic tall fescue plants were obtained after vernalization under field conditions. T1 and T2 progenies were obtained after reciprocal crosses between transgenic and untransformed control plants. PCR and Southern hybridization analyses revealed a 1:1 segregation ratio for both transgenes in the T1 and T2 generations. Southern hybridization patterns were identical for T0, T1, and T2 plants. Our study unequivocally demonstrated for the first time the stable meiotic transmission of transgenes following Mendelian rules in transgenic tall fescue.

Generation of transgenic Russian wildrye has not been reported in the literature. Field-grown transgenic Russian wildrye plants flowered after two winters of vernalization. Fertile plants were obtained and transgenic progenies are being analyzed.

In order to comparatively evaluate agronomical performance of transgenic and non-transgenic tall fescue plants, a two-year study involving primary transgenics (T0), primary regenerants (R0), and seed-grown plants (F0) as well as their progenies (T1, R1 and F1) was carried out. The experimental design was a completely randomized block with three replications. The following agronomical traits were measured on each individual plant: heading date, anthesis date, plant height, growth habit, reproductive tiller number, seed yield and forage yield. Compared with seed-derived plants, the primary transgenics and regenerants were inferior for most of the agronomic characteristics evaluated. However, the agronomic performance of the progenies of the transgenics and regenerants was similar to that of seed-derived plants.

Since wind-pollinated grass species have a high potential to pass their genes to adjacent plants, information regarding gene flow has become extremely important for any future release of value-added transgenic grass cultivars. Transgenic plants provide unique material for studying pollen dispersal and gene introgression into related species. Studies on pollen flow and crossability with related species have been initiated for transgenic tall fescue. The research was aimed at addressing the following questions: (1) What distance can pollen flow and remain viable, and what is the frequency of pollen contamination at different distances from the pollen source? and (2) What is the probability of cross hybridization between tall fescue and other related species under natural conditions? The pollen dispersal experiment used transgenic tall fescue in a central plot, surrounded by exclosures containing recipient non-transgenic plants. The exclosures were 50 m apart and aligned in eight directions, up to a distance of 200 m from the central source plants. The transgene introgression experiment was designed to investigate crossability between transgenic tall fescue and 11 native or introduced *Festuca* and *Lolium* species. A large number of samples is being analyzed for progenies collected from the two experiments. High throughput procedures of molecular analysis have been established to detect the presence or absence of transgenes.

4. APPLICATION OF TRANSFORMATION TECHNOLOGIES FOR FORAGE IMPROVEMENT

Methods required for the genetic manipulation of major forage grasses are now in place. This opens up opportunities to evaluate different genetic engineering strategies for grass improvement. Forage quality improvement, disease and pest resistance, tolerance to biotic and abiotic stresses, manipulation of growth and development, as well as the production of foreign proteins of industrial relevance, represent major targets for molecular breeding of forages.

Currently, we are working on a few applied projects aimed at improving forage quality, drought tolerance, and phosphate uptake of cool season forages. Forage digestibility, especially when plants mature, is a limiting factor for animal production. It is known that lignification of plant cell walls is largely responsible for lowering digestibility of forage tissues. We analyzed lignin deposition and associated changes in anatomy, enzyme activity, gene expression and ruminal degradability in stems of tall fescue at different developmental stages (Chen et al., 2002). Caffeic acid *O*-methyltransferase (COMT) and cinnamyl alcohol dehydrogenase (CAD) are key enzymes involved in lignin biosynthesis. cDNAs of COMT and CAD were cloned from tall fescue. Transgenic tall fescue plants were generated using sense and antisense COMT and CAD gene constructs by microprojectile bombardment. Down-regulation of COMT and CAD resulted in decreased lignin content and increased in vitro digestibility in transgenic plants (Chen et al., in preparation). This is the first case in which it has been shown that transgenic down-regulation of lignin genes led to increased digestibility in a monocot species, although there have been many reports on manipulation of lignin biosynthesis in transgenic dicot species such as Arabidopsis, tobacco, poplar and alfalfa. Besides manipulation of lignin biosynthesis, another strategy to improve forage quality is to delay or inhibit flowering of grasses. The decline of nutritive value in perennial grasses is associated with the onset of stem growth and flowering. Stopping the formation of the less digestible stems or delaying the flowering process is expected to increase forage quality. Large modifications of flowering time in transgenic plants caused by regulating the expression of floral meristem initiation genes have been reported in Arabidopsis (Nilsson and Weigel, 1997). Meristem-identity genes involved in flowering control are being isolated and functionally tested in tall fescue and *Lolium temulentum* (Darnel ryegrass). Drought tolerance is a major target of improvement for cool-season perennial grasses in the southern Great Plains of the USA. Drought stress on perennial forages is a regular feature in this region. A project aimed at isolating genes involved in regulation of stress inducible genes and generating transgenic forage plants was initiated. We have constructed cDNA libraries from drought stressed tall fescue and taken the EST (expressed sequence tags) and microarray approaches to isolate drought related genes. Improving plant phosphate uptake may have an impact on forage production, since phosphate is immobile in soil and very often deficient. A project aimed at improving plant phosphate acquisition by root-specific expression of phytase and/or acid phosphatase genes is in progress. cDNAs specifically expressed in root tissues have been identified in *Medicago trancatula* and expression patterns of the corresponding promoters are being characterized. If it works in the model plant *M. trancatula*, the concept will be applied to other forage species. In addition to the above projects, a collaborative project on large scale sequencing of tall fescue ESTs has been started at the Noble Foundation. The sequencing information obtained from this project will greatly enhance our ability to discover useful agronomic genes in forage grasses.

5. REFERENCES

Barnes, R. F., and J. E. Baylor. 1995. Forages in a changing world. Forages. In: Barnes, R. F., Miller, D. A., and Nelson, C. J. (eds). Iowa State University Press. pp.3-13.

Chen, L., C. Auh, F. Chen, X. F. Cheng, H. Aljoe, R. A. Dixon, and Z. Y. Wang. 2002. Lignin deposition and associated changes in anatomy, enzyme activity, gene expression and ruminal degradability in stems of tall fescue at different developmental stages. J. Agric. Food Chem. (in press).

Kuai, B., S. J. Dalton, A. J. E. Bettany, and P. Morris. 1999. Regeneration of fertile transgenic tall fescue plants with a stable highly expressed foreign gene. Plant Cell Tissue Organ Cult. 58:149-154.

Murashige, T., and F. Skoog. 1962. A revised medium for rapid growth and bioassays with tobacco tissue culture. Physiol. Plant. 15:473-497.

Nilsson, O., and D. Weigel. 1997. Modulating the timing of flowering. Curr. Opin. Biotechnol. 8:195-199.

Potrykus, I. 1991. Gene transfer to plants: assessment of published approaches and results. Annu. Rev. Plant Physiol. Plant Mol. Biol. 42:205-225.

Spangenberg, G., Z. Y. Wang, and I. Potrykus. 1998. Biotechnology in forage and turf grass improvement, Springer, 200 pp.

Spangenberg, G., Z. Y. Wang, M. P. Valles, and I. Potrykus. 1995a. Genetic transformation in *Festuca arundinacea* Schreb.(tall fescue) and *Festuca pratensis* Huds.(meadow fescue). Biotechnology in Agriculture and Forestry. Bajaj, Y. P. S. (ed.). Springer.pPp.183-203.

Spangenberg, G., Z. Y. Wang, X. L. Wu, J. Nagel, V. A. Iglesias, and I. Potrykus. 1995b. Transgenic tall fescue (*Festuca arundinacea*) and red fescue (*F. rubra*) plants from microprojectile bombardment of embryogenic suspension cells. J. Plant Physiol. 145:693-701.

Vasil, I. K. 1988. Progress in the regeneration and genetic manipulation of cereal crops. Biotechnology 6:397-402.

Vasil, I. K. 1995. Cellular and molecular genetic improvement of cereals. Current Issues in Plant Molecular and Cellular Biology. Terzi, M., Cella, R., and Falavigna, A. (eds). Kluwer Academic Publishers. pp.5-18.

Wang, Z., D. Lehmann, J. Bell, and A. Hopkins. 2002. Development of an efficient plant regeneration system for Russian wildrye (*Psathyrostachys juncea*). Plant Cell Rep. 20:797-801.

Wang, Z. Y., A. Hopkins, and R. Mian. 2001. Forage and turf grass biotechnology. Crit. Rev. Plant Sci. 20:573-619.

Wang, Z. Y., G. Legris, J. Nagel, I. Potrykus, and G. Spangenberg. 1994. Cryopreservation of embryogenic cell suspensions in *Festuca* and *Lolium* species. Plant Sci. 103:93-106.

Wang, Z. Y., T. Takamizo, V. A. Iglesias, M. Osusky, J. Nagel, I. Potrykus, and G. Spangenberg. 1992. Transgenic plants of tall fescue (*Festuca arundinacea* Schreb.) obtained by direct gene transfer to protoplasts. Biotechnology 10:691-696.

GENOMIC STRUCTURE OF THE APOMIXIS LOCUS IN *PENNISETUM*

Peggy Ozias-Akins[1], Joann A. Conner[1], Shailendra Goel[1], Zhenbang Chen[1], Yukio Akiyama[1], and Wayne W. Hanna[2]
[1] Department of Horticulture, University of Georgia Tifton Campus, Tifton, GA 31793, USA (e-mail: ozias@tifton.cpes.peachnet.edu); [2] USDA-ARS, Coastal Plain Experiment Station, Tifton. GA 31793, USA (e-mail: whanna@tifton.cpes.peachnet.edu)

Keywords Apospory, bacterial artificial chromosome, fluorescence in situ hybridization, hemizygosity

1. INTRODUCTION

Apomixis is the term used for reproductive processes in flowering plants that result in asexual reproduction through seeds. This final outcome can be achieved through either sporophytic or gametophytic types of development (Nogler, 1984). The sporophytic type is known as adventitious embryony and resembles somatic embryogenesis in that an embryo develops directly from a somatic cell (usually a nucellar cell) within the ovule. Such an embryo typically is nourished by endosperm derived through sexual reproduction, i.e., the meiotically derived embryo sac is fertilized, and adventitious embryos often coexist with zygotic embryos. Gametophytic apomixis retains the alternation of generations (sporophyte-gametophyte-sporophyte) within the apomictic process by replacing meiotically derived embryo sacs with megaspore mother cell- or somatic cell-derived embryo sacs that are unreduced in chromosome number. The egg cell in an unreduced embryo sac is genetically identical to the maternal parent and develops without fertilization into an embryo. The nutritive endosperm develops autonomously (without fertilization) in some apomicts, but in other apomicts, fertilization is required.

In the family Poaceae, the predominant form of apomixis is gametophytic, and apomictic species/cytotypes are relatively abundant in the Panicoid grasses, tribe *Paniceae*. Our work has focused on members of the genus

I. K. Vasil (ed.), Plant Biotechnology 2002 and Beyond, 515-518.

Pennisetum (Brunken, 1977; Stapf and Hubbard, 1934) including one species, *P. ciliare* (L.) Link (syn. *Cenchrus ciliaris* L.) whose taxonomic placement has fluctuated in different treatments. Apomixis in this group of species is aposporous, meaning that a nucellar cell other than the megaspore mother cell develops mitotically into an unreduced embryo sac. These embryo sacs typically have only four nuclei in three to four cells: one egg, one or two synergids, and one or two polar nuclei in the central cell. The distinguishing feature of these embryo sacs when compared with meiotically derived embryo sacs is the absence of antipodals.

2. GENETIC ANALYSIS

The transfer of apomixis from wild relatives to pearl millet (*P. glaucum*) has been attempted with partial success (Dujardin and Hanna, 1989a; Dujardin and Hanna, 1989b). The best donor of the trait has been *P. squamulatum*, an obligate apomict that is hexaploid ($2n=6x=54\text{-}56$), when the recipient has been tetraploid pearl millet ($2n=4x=28$). This cross resulted in both apomictic and sexual offspring that were phenotypically variable (Dujardin and Hanna, 1983). Apomixis was inherited as a simple, dominant trait in this testcross. In the F1 population, molecular markers linked with the trait were isolated by bulking DNA from individuals that fell into discrete phenotypic classes for embryo sac type, i.e., meiotically derived (sexual) or aposporous (Ozias-Akins et al., 1998). All molecular markers were polymerase chain reaction (PCR)-based and were derived from randomly amplified polymorphic DNAs (RAPDs) or restriction fragment length polymorphisms (RFLPs). When single-dose markers were mapped in a population of 397 F1 individuals, all 12 of the PCR-based markers strictly cosegregated with apospory. Such tight linkage indicated repressed recombination in the region of the genome that we designated as the apospory-specific genomic region (ASGR).

3. CHARACTERIZATION OF THE ASGR

Half of the 12 ASGR-linked markers hybridized to low-copy DNA while the remaining markers hybridized to repetitive sequences (Ozias-Akins et al., 1998). Four of the six low-copy DNA markers detected fragments on Southern blots of apomictic individuals, but did not hybridize to any bands in sexual individuals of the population. These DNA sequences were considered to be hemizygous since they apparently were present on a single chromosomal homolog. Hemizygosity has been further demonstrated by fluorescence in situ hybridization (FISH) of bacterial artificial chromosome (BAC) clones on chromosomal spreads. The BAC clones were isolated from two libraries, one from a *P. squamulatum* – pearl millet hybrid and one from *C. ciliaris* (Roche et al., 2002), that provided a combined genome coverage of approximately 8-fold. None of the BAC clones tested thus far have allowed the identification of a chromosomal homolog of the ASGR (Fig. 1a), even though both apomictic species are polyploid. The chromosomal position of the ASGR differs between the two apomictic species. It is near the centromere in *C. ciliaris* but on the distal end of the chromosome in *P. squamulatum*.

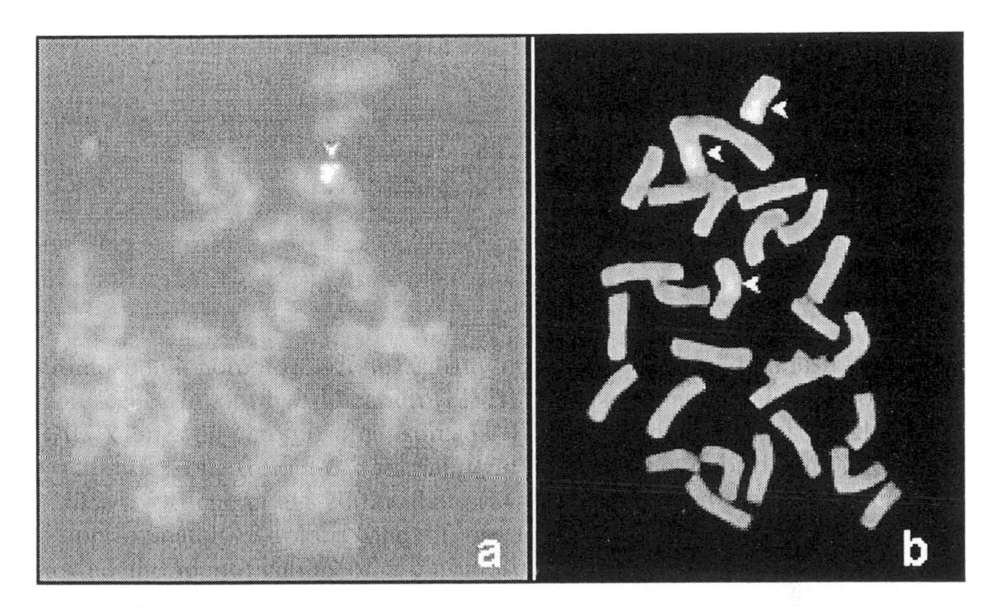

Fig. 1. a) Pennisetum squamulatum chromosomes probed with an ASGR-linked BAC clone. The single hybridizing chromosome is shown with an arrow. b) BC3 chromosomes probed with total genomic DNA from P. squamulatum (3 arrows).

Introgression of apomixis into pearl millet through backcrossing also has been monitored with FISH using both ASGR-linked repetitive sequences and total *P. squamulatum* genomic DNA as dual-labeled probes. A single BC3 line is an obligate apomict, 2n=29, and contains three complete chromosomes that hybridize with *P. squamulatum* genomic DNA (Fig. 1b). Only one of these three chromosomes hybridizes with the ASGR-linked repetitive sequences. It is the transmission of this single chromosome that is required in subsequent backcrosses in order for the apomictic phenotype to be expressed. Some advanced backcross lines contain only one alien chromosome that has been identified by hybridization of ASGR-linked repetitive sequences.

The genetic mechanism underlying apomixis could be as simple as a single regulatory gene or as complex as multiple linked genes whose expression pattern may be controlled by their chromatin context (Roche et al., 2001). The lack of recombination in the ASGR is a significant impediment to positional cloning; therefore, a mutation approach to delineate the essential chromosomal region will be undertaken.

4. REFERENCES

Brunken J.N. 1977. A systematic study of *Pennisetum* sect. *Pennisetum* (Gramineae). Amer. J. Bot. 64:161-176.

Dujardin M., and W.W. Hanna. 1983. Apomictic and sexual pearl millet x *Pennisetum squamulatum* hybrids. J. Hered. 74:277-279.

Dujardin M., and W.W. Hanna. 1989a. Crossability of pearl millet with wild *Pennisetum* species. Crop Sci. 29:77-80.

Dujardin M., and W.W. Hanna. 1989b. Developing apomictic pearl millet - characterization of a BC3 plant. J. Genet. Breed. 43:145-151.

Nogler G.A. 1984. Gametophytic apomixis. Embryology of Angisoperms. B.M. Johri (Ed.) Springer Verlag. pp 475-518.

Ozias-Akins P., D. Roche, and W.W. Hanna. 1998a. Tight clustering and hemizygosity of apomixis-linked molecular markers in *Pennisetum squamulatum* implies genetic control of apospory by a divergent locus which may have no allelic form in sexual genotypes. Proc. Natl. Acad. Sci. USA 95:5127-5132.

Roche D.R., J.A. Conner, M.A. Budiman, D. Frisch, R. Wing, W.W. Hanna, and P. Ozias-Akins. 2002. Construction of BAC libraries from two apomictic grasses to study the microcolinearity of their apospory-specific genomic regions. Theor. Appl. Genet. 104:804-812.

Roche D., W. Hanna, and P. Ozias-Akins. 2001. Is supernumerary chromatin involved in gametophytic apomixis of polyploid plants? Sex. Plant Reprod. 13:343-349.

Stapf O., and C.E. Hubbard. 1934. *Pennisetum*. Flora of Tropical Africa Vol. 6, Part 6. D. Prain (Ed.). Reeve & Co., Ltd. pp. 954-1070.

MOLECULAR IMPROVEMENT OF PERENNIAL RYEGRASS BY STABLE GENETIC TRANSFORMATION

Fredy Altpeter[1*], Jianping Xu[2], Yu-Da Fang[2], Xinrong Ma[2], Joerg Schubert[3], Goetz Hensel[2], Helmut Baeumlein[2], and Vladimir Valkov[2]
[1]University of Florida, Agronomy Department, Laboratory of Molecular Plant Physiology, 2191 McCarty Hall, P.O. Box 110300, Gainesville FL 32611-0300, USA;[2]Institut für Pflanzengenetik und Kulturpfanzenforschung Gatersleben Corrensstrasse 3, 06466 Gatersleben, Germany; [3]Bundesanstalt für Züchtungsforschung, Institut für Resistenzforschung und Pathogendiagnostik, D-06449 Aschersleben, Germany
(*e-mail: faltpeter@mail.ifas.ufl.edu)

1. INTRODUCTION

Perennial ryegrass (*Lolium perenne* L.) is one of the most important turf- and forage grass in the temperate regions (Watschke and Schmidt, 1992). Genetic engineering of grasses will complement traditional breeding in the development of improved cultivars. The integration and expression of selectable marker genes in forage type perennial ryegrass was demonstrated after microprojectile bombardment (Spangenberg et al., 1995; Dalton et al., 1999) or silicon carbide fibre-mediated gene transfer into suspension cells (Dalton et al., 1998) and after direct gene transfer into cell suspension derived protoplast (Wang et al., 1997). The time required from excision of the explants to transfer of the transgenic plants to soil in the earlier protocols was in excess of 10 months. This long tissue culture procedure increases the risk of generating undesirable somaclonal variation and consequently abnormal plants have been reported in perennial ryegrass (Creemers-Molenaar and Loeffen, 1991). We recently presented a more rapid biolistic transformation- and selection protocol for the production of large numbers of fertile transgenic perennial ryegrass plants and demonstrated its applicability to commercially interesting turf type cultivars (Altpeter et al., 2000). Meanwhile an *Agrobacterium*-mediated perennial ryegrass transformation protocol has been developed in our laboratory and is compared with the biolistic gene transfer protocol. Transgenes were introduced into perennial ryegrass plants with the potential to improve ryegrass mosaic virus resistance and to increase tolerance under iron deficiency conditions.

I. K. Vasil (ed.), Plant Biotechnology 2002 and Beyond, 519-524.

2. RESULTS AND DISCUSSION

Both *Agrobacterium*-mediated (AGL1) or biolistic (PDS 1000) transfer of a constitutive *npt II* expression cassette (pCambia vector) in freshly established and vigorously growing embryogenic calli, followed by selection with paromomycin (Altpeter et al., 2000) resulted in transgenic perennial ryegrass plants within 17 to 25 weeks after excision of explants. Between 1.3 and 4.0 % of the bombarded calli and 8 to 16 % of the *Agrobacterium*-inoculated calli regenerated independent transgenic plantlets. The Southern blot analysis confirmed the independent nature of the transgenic plants (data not shown). Most important factors in the development of these highly reproducible perennial ryegrass transformation protocols were genotype dependent tissue culture response, a short tissue culture- and selection period and the use of acetosyringone during *Agrobacterium*-mediated gene transfer. The majority of transgenic lines from both biolistic and *Agrobacterium*-mediated gene transfer had a simple transgene integration pattern with one to four transgene copies and stably expressed the transgene in generative and vegetative progenies (Table 1). The dominant integration pattern usually observed after biolistic gene transfer into grasses and cereals is multiple transgene copy inserts (Hartman et al., 1994; Spangenberg et al., 1995; Altpeter et al., 1996; Stöger et al., 1998). However, Dalton et al. (1999) described that the transgene expressing ryegrass plants carried only one to two transgene copy inserts. They suggested that with a higher copy number transgene expression might be negatively effected as already stated by Matzke and Matzke (1995). In contrast, Stöger et al. (1998) described a higher level of transgene expression in transgenic wheat plants with multicopy inserts. Multicopy inserts after biolisitic gene transfer into perennial ryegrass were commonly inserted at the same locus, whereas the majority of transgenic lines after *Agrobacterium*-mediated gene transfer showed two transgene inserts (Table 1) at independent loci, segregating into single locus events in the following progeny. Consequently a large number of independent single copy events were identified in perennial ryegrass, following *Agrobacterium*-mediated gene transfer and sexual reproduction.

Ryegrass mosaic virus (RgMV) frequently reduces yield and persistence of perennial ryegrass (Clarke and Eagling, 1994). The concept of pathogen-derived resistance (PDR) has been successfully exploited for conferring resistance against viruses in many crop plants (reviewed by Baulcombe, 1999). We introduced an untranslatable RgMV coat protein (RgMV-CP) gene into perennial ryegrass using particle bombardment to explore the potential of RNA-mediated virus resistance. Highly resistant transgenic perennial ryegrass lines against high-dose virion inocula of RgMV-Denmark (data not shown) and -Bulgaria strains were identified, while moderate resistance was observed against RgMV-Czech strain (Table 2) with 100, 95 and 92% sequence identity to the RgMV-CP transgene respectively. Similar

Table 1. Transgene integration pattern, fertility and expression stability of transgenic perennial ryegrass lines after biolistic or Agrobacterium-mediated gene transfer and sexual or one year vegetative reproduction.

Copy No.	**Biolistic** Transgenic lines / Fertile lines / Lines with transgene expression in sexual progenies	**Agrobacterium** Transgenic lines / Fertile lines / Lines with transgene expression in sexual progenies	**Biolistic** Transgenic lines / Lines with transgene expression after 1 year vegetative propagation	**Agrobacterium** Transgenic lines / Lines with transgene expression after 1 year vegetative propagation
1	8 / 6 / 5	5 / 4 / 4	8 / 6	5 / 5
2	8 / 7 / 5	20 / 14 / 14	8 / 7	20 / 16
3	6 / 4 / 3	16 / 11 / 11	6 / 6	16 / 14
4	5 / 3 / 3	4 / 4 / 4	5 / 3	4 / 3
5	2 / 1 / 1	4 / 3 / 3	2 / 1	4 / 4
>5	6 / 4 / 2	0 / 0 / 0	6 / 3	0 / 0
Total	35 / 25 / 19	49 / 36 / 36	35 / 26	49 / 42

observations were described for RNA-mediated virus resistance in the monocotyledonous crops sugarcane (Ingelbrecht et al., 1999) and rice (Pinto et al., 1999). In contrast to the observations in sugarcane (Ingelbrecht et al., 1999) and in accordance to the observations in rice (Pinto et al., 1999), we identified no typical examples of recovery resistance in perennial ryegrass. In previous examples, homology dependence of RNA-mediated virus resistance resulted in the protection against strains that were identical or very similar to the transgene (Mueller et al., 1995). No viral RNA could be detected in the highly resistant transgenic plants and the steady-state RgMV-CP transgene mRNA levels was low or undetectable, although nuclear transcription rates were high as demonstrated by northern blot and nuclei run-off analysis respectively (data not shown). The described characteristics indicate that the observed virus resistance operates by targeted-RNA degradation, resulting in post transcriptional transgene silencing along with inhibition of virus RNA replication. Various models have been proposed to explain PTGS (reviewed by Flavell, 1994). Suppression of PTGS by RgMV differed significantly and consistently in two progenies (28-8, 28-12; Table 3) with identical RgMV-CP transgene integration- and a similar RgMV-CP transgene methylation pattern (data not shown), suggesting the presence of quantitative components controlling the stability of PTGS in perennial ryegrass. This study extends RNA-mediated virus resistance, post transcriptional gene silencing and crop improvement by genetic engineering to perennial ryegrass.

Table 2. RgMV resistance of transgenic perennial ryegrass and the azygous control one, two or nine months after inoculation with RgMV-Bulgaria or -Czech strain.

ELISA means[1] of RgMV content in leaf extracts after inoculation with:						
	RgMV-Bulgaria strain			RgMV-Czech strain		
Line no.	1 mpi	2 mpi	9 mpi	1 mpi	2 mpi	9 mpi
AZ	0.302[b]	3.058[a]	2.245[a]	1.546[a]	3.585[a]	3.611[a]
20pt	0.000[b]	0.069[c]	0.137[c]	0.046[b]	2.329[b]	1.644[dc]
20-8	0.004[b]	0.644[bc]	0.258[c]	0.028[b]	2.483[b]	2.900[ab]
20-12	0.291[b]	0.841[b]	0.270[c]	0.361[b]	3.194[ab]	2.344[bc]
28pt	0.010[b]	0.003[c]	0.000[c]	0.000[b]	0.948[c]	0.762[d]
28-8	0.035[b]	0.120[bc]	0.009[c]	0.416[b]	3.529[a]	2.387[bc]
28-12	1.405[a]	3.454[a]	1.509[b]	1.094[a]	3.860[a]	3.218[ab]

[1)*]means represent ELISA values of virus content at OD 405 nm in ten independent plant extracts. Statistical analysis was performed by SAS. Means were compared by t-test (LSD). Significant difference with $p < 0.05$ are indicated by different letters in the same column. AZ: non-transgenic control segregating from the sexual cross of line20×line28; mpi: months post inoculation. Pt: primary transformants.

Basic cellular processes such as electron transport in photosynthesis and respiration require the precise control of iron homeostasis (Briat et al., 1995). Nicotianamine (NA) is an intermediate in the biosynthetic pathway of the mugineic acid family phytosiderophores (MAs), which are crucial components of the iron acquisition apparatus of graminaceous plants. Recently it was demonstrated that NA chelates both Fe(III) and Fe(II); these complexes are poor Fenton reagents, suggesting a role of NA in protecting cells from oxidative damage (Von Wiren et al., 1999). Thus NA plays a key role in Fe metabolism and homeostasis in all higher plants. Nicotianamine synthase catalyzes the trimerization of S-adenosylmethionine to form one molecule of NA.

NAS-hor1 isolated from barley (Herbik et al., 1999), a plant that is resistant to Fe deficiency, was introduced into perennial ryegrass under control of the constitutive ubiquitin promoter by biolistic gene transfer. Transgene integration and expression of the functional enzyme were confirmed by Sourthern blot, Northern blot and Enzyme activity assay respectively (data not shown). In contrast to wildtype plants, transgenic plants expressed functional NAS activity in leaves independent of iron supply status (data not shown). Wildtype and transgenic perennial ryegrass (*Lolium perenne* L.) plants overexpressing *NAS-hor 1* were grown in hydroponic culture as previously described (Higuchi et al., 1996). Cultures were weekly subcultured to medium with 100 μM iron or without iron and analyzed for chlorophyll content three weeks after culture initiation. Chlorophyll was extracted from randomly sampled 300 mg leaf material according to Cianzio et al., 1979. Chlorophyll content was determined with a spectrophotometer and calculated according to Peterson and Onken 1992. Results from three independent experiments and three replications each are given in figure 1. Three weeks after Fe-deficiency treatment, severe iron chlorosis appeared in the wildtype while several of the transgenic lines showed dark green leaves correlating to the chlorophyll content as shown in figure 1. These results suggest to explore the potential of transgenic plants with ectopic overexpression of nicotianamine synthase under iron deficiency field conditions.

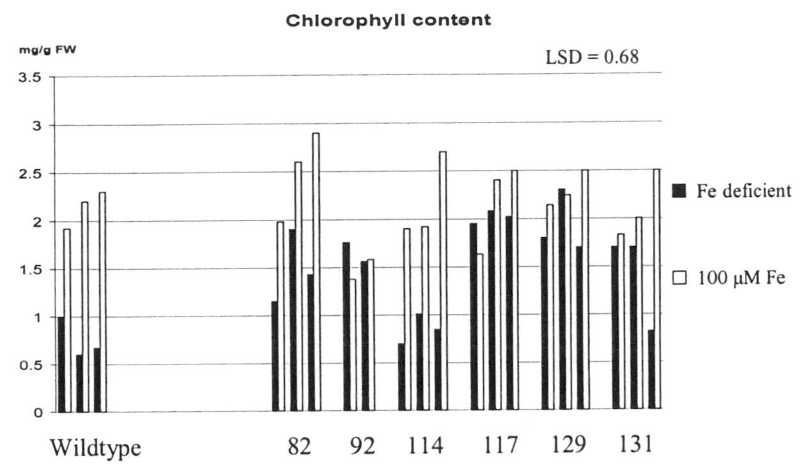

Figure 1. Chlorophyll content of leaf extracts from wildtype and NAS-hor1 expressing perennial ryegrass (82 – 131) after 21 days hydroponic culture with Fe deficiency or 100 μM Fe in two (92) to three (rest) independent experiments.

3. ACKNOWLEDGEMENTS

The authors thank J. Sheen for the sgfp clone, R. Jefferson for pCambia vectors, F. Rabenstein for virus strains and U.K. Posselt for providing seeds of perennial ryegrass cultivars. We thank I. Otto, E. Gruetzemann, M. Nielitz and S. Ahmed for excellent technical assistance. We are grateful to DFG, DAAD and DSE for financial support.

4. REFERENCES

Altpeter, F., V. Vasil , V. Srivastava, and I.K. Vasil. 1996. Nature Biotechnology 14: 1155-1159.

Altpeter, F., J. Xu, and S. Ahmed. 2000. Mol. Breed. 6: 519-528.

Baulcombe, D.C. 1999. Arch. Virol. Suppl. 15: 189-201.

Clarke, R.G. and Eagling, D. 1994. N. Z. J. Agaric. Res. 37: 319-327.

Briat, J.F., Fobis-Loisy, I., Grignon, N., Lobréaux, S., Pascal, N., Savino, G., Thioron, S., Van Wirén, N., Van Wuytswinkel, O. 1995. Biol. Cell 84: 69-81.

Cianzio S., Fehr W. and Anderson I. 1979. Crop Sci. 19: 644-646.

Creemers-Molenaar, J., and J.P.M. Loeffen. 1991. In: A.P.M. Den Nijs, and A. Elgersma, eds.) Fodder Crops Breeding: Achievements, Novel Strategies and Biotechnology, pp 123-128. Proc. 16th Meet. Fodder Crops Section of Eucarpia. 16-22 November 1990. Wageningen, The Netherlands.

Dalton, S.J., A.J.E. Bettany, E. Timms, and P. Morris. 1998. Plant Sci. 132: 31-43.

Dalton, S.J., A.J.E. Bettany, E. Timms, and P. Morris. 1999. Plant Cell Rep. 18: 721-726.

Flavell, R.B. 1994. Proc. Natl. Acad. Sci. USA 91: 3490-3496.

Hartman, C.L., L. Lee, P.R. Day, and E.T. Nilgun. 1994. Bio/Technology 12: 919-923.

Herbik, A., Koch, G., Mock, H.-P., Dushkov, D., Czihal, A., Thielmann, J., Stephan U.W. and Bäumlein H. 1999. European J. Biochem. 265: 231–239.

Higuchi, K., Kanazawa, K., Nishizawa, N.K., Mori, S. 1996. Plant Soil . 178: 171-177.

Ingelbrecht, I.L., Irvine, J.E. and Mirkov, T.E. 1999. Plant Physiol. 119: 1187-1197.

Matzke, M.A. and A.J.M. Matzke. 1995. Plant Physiol. 107: 679-685.

Mueller, E., Gilbert, J., Davenport, G., Brigneti, G. and Baulcombe, D.C. 1995. Plant J. 7: 1001-1013.

Pinto, Y. M., Kok, R.A. and Baulcombe, D.C. 1999. Nature Biotechnol. 17: 702-707.

Peterson, G. and Onken, A. 1992. Crop Sci. 32:964-967.

Shojima, S., Nishizawa, N.K., Mori, S. 1989. Plant Cell Physiol. 30: 673 677.

Spangenberg G., Z.Y. Wang, X.L. Wu, J. Nagel, and I. Potrykus. 1995. Plant Sci. 108: 209-217.

Stöger, E., S. Williams, D. Keen, and P. Christou. 1998. Transgen. Res. 7: 463-471.

Von Wirén, N., Klair, S., Bansal, S., Briat, J.F., Khodr, H., Shioiri, T., Leigh, R.A., Hider, R.C. 1999. Plant Physiol. 119: 1107-1114.

Wang, G.R., H. Binding, and U.K. Posselt. 1997. J. Plant Physiol. 151: 83-90.

Watschke, T.L., and R.E. Schmidt. 1992. In: D.V. Waddington et al. (eds.) Turfgrass Agron. Monogr. 32, pp 129-174. ASA, CSSA and SSSA Madison, Wi, USA.

Zaghmout, O.M.F., and W.A. Torello. 1992. Plant Cell Rep. 11: 142-145.

Sn-TRANSGENIC *LOTUS CORNICULATUS* LINES: A POTENTIAL SOURCE OF DIFFERENTIALLY EXPRESSED GENES INVOLVED IN CONDENSED TANNINS BIOSYNTHESIS

Sergio Arcioni, Francesco Paolocci, Maria Ragano-Caracciolo, Nicola Tosti and Francesco Damiani
Istituto di Ricerche sul Miglioramento Genetico delle Piante Foraggere CNR Via Madonna Alta, 130 06128 Perugia ITALY (e-mail: S.Arcioni@irmgpf.pg.cnr.it)

Keywords Condensed tannins, real-time PCR, gene silencing, genetic transformation

1. INTRODUCTION

Evidence from numerous feeding studies with cows indicates that excessive ruminal protein degradation may be the most limiting nutritional factor in higher-quality temperature legume forages such as alfalfa and clover (Barry and McNabb, 1999). Condensed tannins (CT), found in some minor forage legumes, are known to decrease protein degradation. However, genes involved in CT biosynthesis are still largely unidentified. Since *Lotus corniculatus* is an in vitro highly responsive species and it does accumulate CT in edible tissues, it represents an invaluable legume model species to gain information on regulatory mechanisms leading to CT biosynthesis.

Genetic transformation of *L. corniculatus* S50 genotype with *Sn*, a maize regulatory gene of the anthocyanin pathway (Tonelli et al., 1991), produced transgenic lines with altered flavonoid biosynthesis into leaves. Intriguingly, despite a slight change in the anthocyanin accumulation, *Sn* specifically targeted the leaf condensed tannin (CT) biosynthesis, producing both gain and loss of function CT mutants (Paolocci et al., 1999). Molecular analyses suggested that negative (CT-) mutants are likely due to a homology dependent gene silencing phenomenon which affects, among the others, the expression of *dfr*, the gene encoding the last enzyme shared between the anthocyanin and condensed tannin pathways. Analyses of the leaf transcriptional profiles on CT enhanced (CT+) and CT- mutants by means of cDNA/AFLP, suppressive subtraction hybridisation (SSH) and dissociation subtraction chain (DSC) are in progress. It is expected to clone cDNAs specifically involved in CT biosynthesis to be introduced in high-quality CT-

I. K. Vasil (ed.), Plant Biotechnology 2002 and Beyond, 525-528.

free forage legumes. Long term goal of the present research work is in fact the induction of CT biosynthesis into alfalfa and clover leaves.

2. MATERIALS AND METHODS

L. corniculatus cv. Leo genotype S50 plants were kindly provided by Mark Robbins (IGER, UK). Plant transformation, Southern and northern analyses were carried out as previously described (Damiani et al., 1999). Real-time RT-PCR analyses to assess the leaf *dfr* steady state level was performed by using the SYBR Green 1 master Core (Applied Biosystems) on first strand cDNA of CT-contrasting transgenic lines as well as of Gus-transformed and wild type plants. Elongation factor 1-α (ELF1α) mRNA was used as external standard and reference gene to calibrate *dfr* expression. DFR and ELF1α specific primers were designed by using the Primer Express software (Applied Biosystems). Each primer pair was tested with a logarithmic dilution of a cDNA mix to generate a linear standard curve, which was used to calculate the PCR efficiency. SSH and DSC analyses between CT+ and CT- plants were performed by using the PCR-Select cDNA suppression kit (Clontech) and the method described by Luo et al. (1999), respectively. cDNA/AFLP was carried out basically as described by Vos et al. (1995).

3. RESULTS

Molecular assays showed that CT+ transgenic lines all have a *Sn* single-copy insertion and a low steady state transgene expression level. Conversely, all CT- mutants present a multiple *Sn* insertion, although neither northern nor RT-PCR analyses detected *Sn* specific transcript, at least as mature RNA, in leaves of these mutants. Moreover, *Sn* silencing is paralleled by *dfr* down-regulation in CT – lines, as depicted by quantitative RT-PCR analyses. These data suggest that CT- mutants are likely due to a PTGS (post transcriptional gene silencing) phenomenon which in turn affects the expression of *Sn* and its putative *L. corniculatus* orthologous. This last gene is thought to positively control *dfr* transcription in wild type leaves. The presence of mature *Sn* mRNA, on the contrary, induces a relevant increment of leaf *dfr* steady state mRNA level in CT enhanced lines with respect to wild type plants (Fig.1). Therefore, the level of *dfr* transcript can be likely considered as a marker of leaf tannin biosynthesis in *Lotus.* For a better characterization of *dfr* gene(s) in *Lotus,* Southern hybridisation was carried. This analysis showed that *dfr* is arranged as a small gene family in *L. corniculatus* genome. In addition, the real-time RT-PCR analyses showed that not all *dfr* genes seem to be induced by *Sn.* The dissociation analyses of *dfr* amplicons display in fact a different number of curves between CT+ vs. CT- and wild type plant cDNAs. Specifically, *Sn* seems to trigger the expression of a single *dfr* gene, whereas at least two different *dfr* genes are co-expressed in wild type and CT- leaves. Differences in the promoter region of these *dfr* genes are therefore expected.

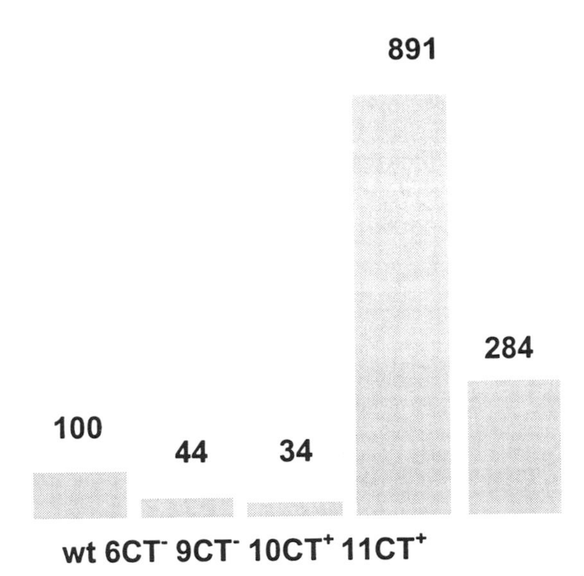

Figure 1. Quantitative expression of dfr *in control (wt) and CT polymorphic (+/-)* Sn *transformed plants*

Upon molecular characterization, the CT+ and CT- mutants have being used as cDNA sources to isolate genes specifically involved in leaf CT biosynthesis following three different strategies: SSH, DSC and cDNA/AFLP. It worths of noting that, differently from SSH and DSC, the cDNA/AFLP protocol allowed the simultaneous comparison of cDNA profiles from two CT+ and CT- mutants with those from mother plants.

Table 1 summarizes the number of clones isolated so far from each techniques. Some of these clones have been sequenced and most of them code for genes of unknown functions.

Table 1. Number of independent clones/bands isolated and studied from each techniques.

Technique	Clones/bands	Sequenced clones/bands	Known sequences
cDNA-AFLP	115	66	35
SSH	347	114	61
DSC	192	7	6

To get rid of clones which are indeed not differentially expressed between CT+ and CT- plants (false positive), reverse northern and/or semi-quantitative RT-PCR analyses are in progress. Those clones proved to be differentially expressed will be used to transform both tannin + and - species in order to assess their involvement into CT biosynthesis.

4. ACKNOWLEDGEMENT

This work has supported by UE project FAIR CT98 4068. However, the content of the paper is the sole responsibility of its authors and it in no way represents the views of the commission or its services.

5. REFERENCES

Barry T.N., and W.C. McNabb. 1999. The implications of condensed tannins in the nutritive value of temperate forages fed to ruminants. British J. Nutr. 81: 263-272.

Damiani F., F. Paolocci, P.D. Cluster, S. Arcioni, G.J. Tanner, R.G. Joseph, Y.G. Li, J. De Majinik, and P.J. Larkin. 1999. The maize transcriptin factor *Sn* alters proanthocyanidin synthesis in transgenic *Lotus corniculatus* plants.

Luo J, Puc J.A., Sloslberg E.D., Yao Y., Bruce J.N., Wright T.C., Becich M.J., and Parsons R. 1999. Differential subtraction chain, a method foridentifying differences in genomic and mRNA. Nucl. Acid Res. 27:

Paolocci F., R. Cappucci, S. Arcioni, and F. Damiani. 1999. Birdsfoot trefoil: a model for studying the synthesis of condensed tannins. Plant Plophenols2: Chemistry, Biology, Pharmacology, Ecology. G.G Gross, R.W. Hemingway and T. Yoshida (eds.). Plenum Publishers, New York. Pp. 343-356.

Tonelli C., G. Consonni, S.F. Dolfini, S.L. Dellaporta, A. Viotti and G. Gavazzi. 1991. Genetic and molecular analysis of *Sn* , a light-inducible tissue-specific regulatory gene in maize. Mol. Gen. Genet. 225: 401-410.

Vos P., Hogers R., Bleeker M., Rejans M., van de Lee T., Hornes M., Frijters A., Pot J., Peleman M., Kuiper M., and Zabeau M. 1995. AFLP: a new technique for DNA fingerprinting. Nucl. Acid Res. 23: 4407-4414.

MICROSPORE EMBRYOGENESIS

Alisher Tashpulatov, Ari Indrianto, Ioulia Barinova, Heidrun Katholnigg, Svetlana Akimcheva, Erwin Heberle-Bors and Alisher Touraev
Institute of Microbiology and Genetics, Vienna Biocenter, Vienna University, Dr. Bohrgasse 9, A-1030, Vienna, Austria (e-mail: Erwin@gem.univie.ac.at)

Keywords Doubled haploids, totipotency, reprogramming, stress, pollen, cell cycle, subtractive suppression hybridization

1. INTRODUCTION

One of the most intriguing questions in developmental biology is the conversion of a somatic cell with restricted developmental options into a totipotent cell, which is able to give rise to an embryo and a reproductively competent organism. In higher plants, totipotency of somatic cells was first experimentally induced by Reinert (1959). The pretreatment of carrot suspension cells by a short pulse of high concentrations of 2,4-dichlorophenoxy-acetic-acid (2,4-D) and subsequent removal of 2,4-D from the medium induced efficient embryogenesis and plant formation. Later, totipotency was shown to be a feature of other cell types in a large number of plant species.

Biochemical, molecular and cell culture studies were carried out to characterize the development of a single cell into an embryo using suspension culture cells. Toonen et al. (1994) used cell tracking to determine individual suspension culture cells of carrot (*Daucus carota* L.) which are competent to develop into somatic embryos. Several cell types were found to give rise to somatic embryos although at different frequencies. The use of computer-controlled cell tracking led to the identification of guard cells as competent cells among sugar beet mesophyll protoplast cultures (Hall et al., 1995). Recently, the somatic embryogenesis receptor kinase (SERK) gene, which is expressed exclusively in cells giving rise to an embryo was isolated from carrot embryogenic suspension cultures (Schmidt et al., 1997). SERK is also expressed in the *Arabidopsis* zygote and was proposed to serve as a *bona fide* marker for the embryogenic state of cultured cells.

I. K. Vasil (ed.), Plant Biotechnology 2002 and Beyond, 529-535.

2. FROM ANTHER TO MICROSPORE CULTURE

Totipotency has been demonstrated also in male reproductive cells (Guha and Maheshwari, 1964; Touraev et al., 1997). Organ cultures of whole anthers at the microspore stage were once believed to contribute to understand pollen development but surprisingly it was discovered that what "hatched" from the anther locules after some time of *in vitro* culture were not mature pollen grains but embryos (Guha and Maheshwari, 1964) bearing the gametic (haploid) number of chromosomes. These embryos developed into haploid plants which were diploidized spontaneously or after treatment with diploidizing agents leading to the formation of homozygous doubled haploids.

For quite some time anther culture remained the technique of choice for doubled haploid production although it had been recognized soon that microspore cultures would allow to analyze microspore embryogenesis properly, and that only this technique would fulfill the dream of the geneticist and breeder to produce a large population of doubled haploids from a cross which would represent the genetic variability produced by male meiosis (Reinert et al., 1975). The breakthrough came when it was recognized that one single factor - stress - was essential to induce embryogenesis in microspores or young pollen grains (Touraev et al., 1997). The critical experiments were those in which immature pollen grains were isolated from well-fed, non-treated plants and were stressed by a starvation treatment in vitro (Kyo and Harada, 1986) as in tobacco or microspores which were exposed to heat-shock as in *Brassica napus* (Custers et al., 1994). Anthers or flower buds were not pretreated in these experiments, thus excluding any effect of the anther wall or the donor plants. These experiments showed clearly that stress is necessary and sufficient to induce embryogenesis in higher plant microspores.

3. MICROSPORE CULTURE OF WHEAT

In recent years the production of embryos from isolated and *in vitro* cultured microspores of several dicots (tobacco, rapeseed) and monocots (barley, wheat) has become very efficient and reproducible (Jähne and Lörz, 1995, Touraev et al., 1997). In many other crop plants, however, the process is still very inefficient and highly genotype-dependent (Ferrie et al., 1995; Touraev et al., 1997). The reasons for recalcitrance are very variable and include factors such contamination of the anthers by bacteria and other microorganisms, size, structure and number of the microspores, or reproductive strategy. In many species only a small fraction of the in vitro cultured microspores undergo embryogenesis. Therefore, the early identification of embryogenic microspores among a population of non-embryogenic ones would facilitate the optimization of doubled haploid formation.

We have developed a very efficient system of embryogenesis in isolated wheat microspore cultures (Touraev et al., 1996a; Indrianto et al., 1999). These protocols were used to follow the full sequence of events which convert an isolated microspore first into an embryogenic microspore and then into an embryo (Indrianto et al., 2001). Isolated wheat microspores were immobilized in low temperature-melting agarose supported by a polypropylene mesh at a low cell density and cultured in a hormone-free, maltose-containing medium in the presence of ovaries serving as a conditioning factor. For the first time it was shown directly that embryos form from individual microspores by direct embryogenesis, without an intervening callus phase. We were able to distinguish three major types of microspores existing in microspore cultures at the start of culture which form a continuum of stages in a process. In this process a normal microspore with a polarized nucleus which is positioned opposite the single germ pore increases in size and the nucleus moves in a central position, connected to the peripheral cytoplasm via numerous cytoplasmic strands which gives these microspores a star-like appearance and is very similar to other mitogenically activated plant cells. This process takes place during stress treatment and at its end the microspore is arrested in development.

In tobacco, male reproductive cells can be reprogrammed during two cell cycles, i. e. in the microspore and bicellular pollen grain stage, and they undergo phase-specific cell cycle arrest during the inductive stress treatment (Zarsky et al., 1992; Touraev et al., 1997). Microspores or young bicellular pollen grains in the G1-phase of the cell cycle undergo DNA-replication during the stress treatment and arrest in G2 while microspores in the G2-phase undergo mitosis and arrest in G1. A cell cycle arrest is also evident in wheat microspores (unpublished results) but not in *Brassica napus* (Custers et al., 1994).

The next step in microspore embryogenesis occurs after relief from the stress. The vacuole disappears and the lumen of the star-like microspores becomes filled with a starch grain-rich cytoplasm. The first division of the stressed microspores or bicellular pollen grains in medium A2 is symmetrical. This type of division is atypical since during normal pollen development the microspore divides by an asymmetric division, while the vegetative cell in the bicellular pollen grain does not divide at all.

Finally, tracking revealed an important developmental mechanism of microspore embryogenesis, i. e. polarity formation and body axis determination in the embryo. Invariably, starch grains became localized in cells in a specific region in all multicellular microspores close to the single germ pore, and the microspore wall always ruptured opposite to the site of starch granule accumulation to release the still globular embryo. Later, in the polarized embryos, the broken microspore wall was often seen attached to the root pole. This latter observation has been made much earlier in three dicot species, i.e. tobacco (Nitsch 1972), *Datura innoxia* and *D. meteloides* (Geier and Kohlenbach, 1973). Apparently, the establishment of polarity precedes rupture of the microspore wall and determines both rupture site and orientation of the body axis of the embryo.

4. GENES INVOLVED IN THE REPROGRAMMING OF MICROSPORES

At present a number of laboratories including our own are active in cloning genes specifically involved in embryogenic induction of microspores. Biochemical and molecular genetic analyses have been performed by many authors to reveal the mechanism of embryogenic induction in microspores (rev. by Cordevener et al., 1995; Touraev et al., 1997). Formation of heat shock proteins, protein phosphorylation and reinitiation of DNA replication take place during the formation of embryogenic microspores/pollen (Cordevener et al.; 1995; Kyo and Harada, 1990; Zarsky et al., 1992, 1995). Several genes have been isolated which have been thought to play an important function in embryogenic tobacco, wheat or *Brassica napus* microspores (Boutilier et al., 1994; Reynolds et al., 1994; Garrido et al., 1992). Unfortunately, no protein or gene has been isolated so far which qualifies as an inducer, or at least specific marker, for embryogenic microspores.

We used suppression subtractive hybridization (SSH, Diatchenko et al., 1996) to isolate cDNA clones differentially expressed in embryogenic tobacco microspores which are formed after carbohydrate and nitrogen starvation during 6 days at 33°C from unicellular microspores. As expected, SSH decreased the concentration of ubiquitously expressed genes and increased the concentration of differential and tissue-specific transcripts. Our results strongly suggest that the concentration of housekeeping genes was considerably decreased in the subtracted pool and was enriched for cDNAs which are comparatively rare and differentially expressed in stressed microspores. To study the expression of the genes in the S_B-pool in a more global way, an expression profile was established by differential screening via Reverse Northern Hybridisation in macroarrays using the subtracted pools as probes.

In total, 435 clones were examined for their expression in non-stressed and stressed microspores, in mature pollen and in microspore-derived embryos at various stages of development. Approximately 20.5% and 0.7% of the clones revealed upregulated and specific, respectively, in stressed embryogenic microspores, compared to non-stressed microspores. After sequencing of this set of transcripts and evaluation of gene redundancy, 27 separate ESTs were selected and subjected to further study. All these ESTs were expressed during microspore embryogenesis while 18 ESTs gave additionally positive hybridization signals in mature pollen. Eight ESTs showed expression in non-stressed and stressed microspores, as well as in microspore-derived embryos, but not in mature pollen. Specific expression in stressed microspores and microspore-derived embryos was shown for one gene.

Sequence analyses and homology searches of the 27 ESTs revealed the involvement of several genes in metabolism, chromosome remodelling, transcription and translation while others had no meaning homologies or showed partial homologies to genes with unknown function.

5. CONCLUSIONS

Tracking of individual microspores revealed that in principle each cultured microspore can be reprogrammed to divide and produce a bipolar embryo. This insight should be of help to further optimize microspore culture for increased doubled haploid production. The ultimate goal is to convert each microspore from a heterozygous F1-plant into a doubled haploid plant such that a population of doubled haploids fully represents the genetic variability of the preceding meiosis. The formation of "star-like", microspores and the efficient transfer of positional information from the anther to the embryo via the microspore are now central questions for further optimization of doubled haploid formation via microspore embryogenesis.

In the future it will be the knowledge of the genes involved in microspore embryogenesis which are expected to give a boost on the further optimization of doubled haploid formation. As cellular reprogramming moves center stage in international research, the knowledge obtained in plants may help to understand the reprogramming of mammalian cells and, vice versa, the knowledge obtained with these cells may support our efforts to better understand and manipulate plant cell totipotency.

6. REFERENCES

Benito Moreno RM, Macke F, Alwen A, Heberle-Bors E (1988) In situ seed production after pollination with in vitro matured, isolated tobacco pollen. Planta 176: 145-148.

Boutilier KA, Gines MJ, Demoor JM, Huang B, Baszczynski CL, Iyer VN, Miki BL (1994) Expression of the *BnmNAP* subfamily of *napin* genes coincides with the induction of *Brassica* microspore embryogenesis. Plant Mol. Biol. 26: 1711-1723.

Cordewener JHG, Custers JBM, Dons HJM, Van Lookeren Campagne MM (1995) Molecular and biochemical events during the induction of microspore embryogenesis. In: Jain SM, Sopory SK, Veilleux RE (eds) *In vitro* haploid production in higher plants. Kluwer Academic Publishers, Dordrecht, pp 107-122.

Custers J.B.M. et al., (1994) Temperature controls both gametophytic and sporophytic development in microspore cultures of *Brassica napus*. Plant Cell Rep. 13: 267-271.

Diatchenko L, Lau Y-FC, Campbell AP, Chenchik A, Moqadam F, Huang B, Lukyanov S, Lukyanov K, Gurskaya N, Sverdlov ED, Siebert DP (1996) Supression subtracted hybridization: A method for generating differentially regulated or tissue-specific cDNA probes and libraries. Proc. Natl. Acad. Sci. USA 93: 6025-6030.

Ferrie AMR, Palmer CE, Keller WA (1995) Haploid embryogenesis. In: Thorpe TA (ed) *In vitro* embryogenesis in plants. Kluwer Academic Publishers, Dordrecht, pp 309-344.

Garrido D, Eller N, Heberle-Bors E, Vicente O (1993) *De novo* transcription of specific messenger RNAs during the induction of tobacco pollen embryogenesis. Sex. Plant Reprod . 6: 40-45.

Garrido D, Vicente O, Heberle-Bors E, Rodriquez-Garcia MI (1995) Cellular changes during the acquisition of embryogenic potential in isolated pollen grains of *Nicotiana tabacum*. Protoplasma 186: 220-230.

Geier T, Kohlenbach HW (1973) Entwicklung von Embryonen und embryogenem Kallus aus Pollenkörnern von *Datura meteloides* und *Datura innoxia*. Protoplasma 78: 381-396.

Guha S, Maheshwari SC (1964) *In vitro* production of embryos from anthers of *Datura*. Nature 204: 497.

Hall RD, Verhoeven HA, Krens FA (1995) Computer assisted identification of protoplasts responsible for rare division events reveals quard-cell totipotency. Plant Physiol. 107: 1379-1386.

Heberle-Bors E (1985) *In vitro* haploid formation from pollen: a critical review. Theor. Appl. Genet. 71: 361-374.

Indrianto A, Heberle-Bors E, Touraev A (1999) Assessment of various stresses and carbohydrates for their effect on the induction of embryogenesis in isolated wheat microspores. Plant Sci. 143: 71-79.

Indrianto A, Barinova A, Touraev A, Heberle-Bors E (2001) Tracking individual wheat microspores in vitro: identification of embryogenic microspores and body axis formation in the embryo. Planta 212: 163-174.

Jähne A, Lörz H (1995) Cereal microspore cultures. Plant Sci. 109: 1-12.

Kyo M, Harada H (1986) Control of the developmental pathway of tobacco pollen *in vitro*. Planta 168: 427-432.

Kyo M, Harada H (1990) Specific phosphoproteins in the initial period of tobacco pollen embryogenesis. Planta 182: 58-63.

Nitsch J-P (1972) Haploid plants from pollen. Z. Pflanzenzücht. 67: 3-18.

Raghavan V (1997) Molecular Embryology of Flowering Plants. Cambridge University Press, Cambridge.
Reinert J (1959) Über die Kontrolle der Morphogenese und die Induktion von Adventivembryonen in Gewebekulturen aus Karotten. Planta 53: 318-333.

Reinert J, Bajaj YPS, Heberle E (1975) Induction of haploid tobacco plants from isolated pollen. Protoplasma 84: 191-196.

Reynolds TL, Crawford RC (1996) Changes in abundance of an abscisic acid-responsive, early cysteine-labeled metallothionein transcript during pollen embryogenesis in bread wheat *(Triticum aestivum)*. Plant Mol. Biol. 32: 823-829.

Schmidt EDL, Guzzo F, Toonen MAJ, De Vries SC (1997) Leucine-rich repeat containing receptor-like kinase marks somatic plant cells competent to form embryos. Development 124: 2049-2062.

Toonen MAJ, Hendriks T, Schmidt EDL, Verhoeven HA, van Kammen AB, de Vries SC (1994) Description of somatic-embryo-forming single cells in carrot suspension cultures employing cell tracking. Planta 194: 565-572.

Touraev A, Indrianto A, Wratschko I, Vicente O, Heberle-Bors E (1996a) Efficient microspore embryogenesis in wheat (*Triticum aestivum.* L) induced by starvation at high temperatures. Sex. Plant Reprod. 9: 209-215.

Touraev A, Pfosser M, Vicente O, Heberle-Bors E (1996b) Stress as the major signal controlling the developmental fate of tobacco microspores: towards a unified model of induction of microspore/pollen embryogenesis. Planta 200: 144-152.

Touraev A, Vicente O, Heberle-Bors E (1997) Initiation of microspore embryogenesis by stress. Trends Plant Sci. 2: 285-303.

Touraev A, Pfosser M, Heberle-Bors E (2001) Microspores- a haploid multipurpose cell. In: Adv. Bot. Res. 35: 53-109.
Zarsky V, Garrido D, Rihova L, Tupy J, Vicente O, Heberle-Bors E (1992) Derepression of the cell cycle by starvation is involved in induction of tobacco pollen embryogenesis. Sex. Plant Reprod. 5: 189-194.

Zarsky V, Garrido D, Eller N, Tupy J, Vicente O, Schöffl F, Heberle-Bors E (1995) The expression of a small heat shock gene is activated during induction of tobacco pollen embryogenesis by starvation. Plant Cell Environ. 18: 139-14.

SOMATIC HYBRIDIZATION IN CITRUS – A RELEVANT TECHNIQUE FOR VARIETY IMPROVEMENT IN THE 21ST CENTURY

Jude W. Grosser
University of Florida,Citrus Research and Education Center (CREC), 700 Experiment Station Road,Lake Alfred, FL 33850, USA (e-mail: jwg@lal.ufl.edu). Article number N-02272 in the Florida Agricultural Experiment Station Journal Series.

Keywords polyploidy, protoplast fusion, scion, rootstock

1. INTRODUCTION

Citrus is one of the few commodities where somatic hybridization is reaching its predicted potential (Grosser and Gmitter, 1990). Somatic hybrid citrus plants have been produced from more than 250 parental combinations, including more than 125 at the CREC (Grosser et al., 2000). Applications of somatic hybridization towards the development of improved seedless scions for the fresh market, and improved rootstocks with potential for tree size control will be discussed. Extensive field research on citrus somatic hybrids combined with emerging molecular analyses of citrus (Nicolosi et al., 2000) has allowed for the development of additional strategies for cultivar improvement. These include targeted cybridization to achieve seedlessness, resynthesizing a better sour orange rootstock at the tetraploid level, and the breeding and selection of new rootstocks at the tetraploid level using somatic hybrids as parents. Ongoing examples of each strategy will be provided.

2. SEEDLESS CITRUS SCION DEVELOPMENT

The international fresh citrus market now demands high-quality seedless fruit, especially when considering new mandarin varieties. New mandarin varieties must also be easy to peel for consumer convenience. Somatic hybridization is now playing a key role in a few strategies to achieve the development of seedless cultivars.

I. K. Vasil (ed.), Plant Biotechnology 2002 and Beyond, 537-540.

2.1. Ploidy Manipulation

Triploid fruits are generally seedless due to the odd number of chromosome sets, and citrus is no exception. Triploids can be produced directly by haploid + diploid somatic hybridization (as demonstrated by P. Ollitrault and co-workers) or by interploid crosses (Grosser et al., 2000). Numerous tetraploid somatic hybrids are now flowering and being used as pollen parents in crosses with selected monoembryonic females to generate seedless triploids (Grosser et al., 2000). At the CREC under the direction of F.G. Gmitter, more than 2000 triploid mandarin hybrids have been recovered from such interploid crosses to date. Overall, this approach is expected to have a major impact on the international fresh citrus market by generating many new competitive seedless mandarin cultivars with a range of maturity dates.

2.2. Cybridization

Seedlessness in diploid citrus generally relates to male and/or female sterility. The seedless Satsuma mandarin is typically male sterile, and its male sterility has been identified to be a cytoplasmic male sterile type (CMS) (Yamamoto et al., 1997). It may be possible to transfer the CMS trait of Satsuma into commercially important seedy diploid cultivars via cybridization. Experiments with a general objective of producing symmetric citrus somatic hybrids have generated cybrid plants on numerous occasions (Moreira et al., 2000). In efforts to transfer the mtDNA from male sterile cultivars to seedy diploid fresh fruit cultivars, we have produced diploid putative cybrid plants of ‘Hirado Buntan Pink’ pummelo, ‘Sunburst’ mandarin, and an unnamed ‘Clementine’ x ‘Murcott’ hybrid from symmetric fusions (W. Guo, D. Prasad, and J. Grosser, unpublished data). If successful, this strategy has potential to improve many high-quality but seedy diploid hybrids.

3. ROOTSTOCK IMPROVEMENT

Diseases such as blight (cause unknown) and quick-decline caused by citrus tristeza virus kill millions of citrus trees annually worldwide. The objective of most rootstock breeding programs is to package all of the necessary biotic and abiotic resistances with wide adaptation and productivity.

3.1. Complementary Hybridization

The primary strategy for rootstock improvement has been to combine complementary diploid rootstocks via protoplast fusion to generate tetraploid somatic hybrid rootstocks. This has been accomplished for numerous combinations, and more than 50 such hybrids have been propagated and entered into commercial field trials. Yield and tree size data from field trials demonstrates that somatic hybrid rootstocks can produce adequate yields of high-quality sweet orange fruit on small trees. If planted at optimum row and tree spacings, somatic hybrids such as sour orange + rangpur and sour orange + Palestine sweet lime can yield approximately 500 field-boxes per acre (field box = 90 pounds of fruit) on trees approximately 3 meters in height. Additional yield data and information on tree survival should result in the release of somatic hybrid rootstocks to the citrus industry.

3.2. Wide-Hybridization

Somatic hybridization has also been used to combine citrus with sexually incompatible genera that exhibit desirable rootstock attributes, including Atalantia, Citropsis, and Severinia (Grosser et al., 1996b). Such hybrids are generally horticulturally inferior, but a few are performing well in field trials including Nova mandarin + *Citropsis gilletiana*, and Succari sweet orange + *Atalantia ceylanica.* If fertile, these hybrids may have value in tetraploid rootstock breeding.

3.3. Building a Better Sour Orange Rootstock

Sour orange was formerly the most important rootstock worldwide due to its wide adaptation, tolerance of citrus blight, and ability to produce good yields of high quality fruit. However, due to its susceptibility to tristeza virus-induced quick-decline disease, it can no longer be used in most situations. A suitable replacement rootstock has yet to be identified. Molecular marker analyses indicates that sour orange is a hybrid of pummelo and mandarin (Nicolosi et al., 2000). We are using somatic hybridization to combine widely adapted mandarins (Amblycarpa and Shekwasha mandarins) with tristeza resistant pummelos or superior pummelo seedlings. To date, 10 putative mandarin + pummelo somatic hybrids have been produced (5 confirmed by RAPD analyses). It should be possible to develop a quick-decline resistant replacement for sour orange that also provides some level of tree size control.

3.4. Rootstock Breeding at the Tetraploid Level

Two somatic hybrids, Nova mandarin + Hirado buntan pummelo (zygotic) and sour orange + rangpur, are performing well in field trials and produce high percentages of zygotic seed. These hybrids can therefore be used as females in crosses at the tetraploid level. Using other high-performance somatic hybrids (i.e. sour orange + Carrizo, Cleopatra + trifoliate orange, sour orange + Palestine sweet lime) as pollen parents offers an opportunity to maximize genetic diversity in tetraploid progeny. Seed from such crosses can be germinated directly in a high pH, calcareous soil/Phytophthora screen. Simultaneously, the selected hybrids are tested for CTV resistance and propagated by topworking and/or rooted cuttings to provide clonal material for further evaluation, which greatly shortens the time required to develop a new rootstock. We began this program at the CREC in 1999, and so far more than 150 genetically diverse "tetrazyg" hybrids have been selected for further evaluation.

4. REFERENCES

Grosser, J.W. and F.G. Gmitter, Jr. 1990. Protoplast fusion and citrus improvement. Plant Breed. Rev. 8:399-374.

Grosser, J.W., Mourao-Fo, F.A.A., Gmitter, F.G. Jr., Louzada, E.S., Jiang, J., Baergen, K., Quiros, A., Cabasson, C. Schell, J.L., and J.L.Chandler. 1996b. Allotetraploid hybrids between *Citrus* and seven related genera produced by somatic hybridization. Theor. Appl. Genet. 92:577-582.

Grosser, J.W., Ollitrault, P., and O. Olivares-Fuster. 2000. Somatic hybridization in Citrus: an effective tool to facilitate variety improvement. In Vitro Cell. Dev. Biol. – Plant. 36:434-449.

Moreira, C.D., Chase, C.D., Gmitter, F.G. Jr., and J.W. Grosser. 2000. Inheritance of organelle genomes in citrus somatic cybrids. Molec. Breed. 6:401-405.

Nicolosi, E., Deng, Z.N., Gentile, A., and S. LaMalfa. 2000. Citrus phyologeny and genetic origin of important species as investigated by molecular markers. Theor. Appl. Genet. 100:1155-1166.

Yamamoto, M., Matsumoto, R., Okudai, N. and Y. Yamada, 1997. Aborted anthers of *Citrus* result from gene-cytoplasmic male sterility. Scientia Hort. 70:9-14.

RECOVERY OF TRIPLOID SEEDLESS MANDARIN HYBRIDS FROM 2N x 2N AND 2N x 4N CROSSES BY EMBRYO RESCUE AND FLOW CYTOMETRY

Luis Navarro, José Juárez, Pablo Aleza and José A. Pina
Department of Plant Protection and Biotechnology, Instituto Valenciano de Investigaciones Agrarias (IVIA), 46113-Moncada, Valencia, Spain (e-mail: lnavarro@ivia.es)

Keywords Citrus, tissue culture, in vitro, tetraploids

1. INTRODUCTION

Production of seedless citrus fruits is required for the fresh market, because consumers do not accept seedy fruits. In mandarins the number of high quality seedless varieties available is very low, and thus, the production of new seedless mandarin varieties has a high priority for many citrus industries worldwide. The recovery of triploid hybrids is the most promising approach to achieve this goal.

Recovery of citrus sexual triploid hybrids (3x = 27) has been reported since the early seventies after 2n x 4n (Esen and Soots, 1972), 4n x 2n (Cameron and Burnett, 1978) and 2n x 2n crosses (Esen and Soots, 1971). In the last case, the triploid embryos are originated by the fertilization of an unreduced diploid female gamete with a normal reduced haploid male gamete. Seeds with triploid embryos are generally underdeveloped or aborted, and there is very difficult to regenerate plants by conventional methods. In addition, analysis of ploidy level of large populations of citrus plants by cytological methods is very difficult. Therefore, the recovery of triploids has not been widely used in citrus breeding.

The development of methodologies for *in vitro* culture of embryos and small seeds and for ploidy analysis by flow cytometry (Ollitrault et al., 1996; Starrantino, 1992), allowed a much more efficient production of citrus triploid sexual hybrids.

In Spain we are carrying a triploid breeding program based on embryo rescue and ploidy analysis by flow cytometry. The objective is to produce new high quality ease peeling and seedless mandarin cultivars.

I. K. Vasil (ed.), Plant Biotechnology 2002 and Beyond, 541-544.

2. EMBRYO CULTURE

Embryos are isolated from aborted or underdeveloped seeds from mature fruits and cultured in the medium of Murashige and Tucker with 5% sucrose, 500 mg/l malt extract and 0.8% Bacto agar. It was found that the germination rate of isolated embryos was 84% in comparison with 42.2% of intact seeds. After germination, the plantlets are recultured in an elongation medium composed of the Murashige and Skoog mineral salts, supplemented with vitamins, 5% sucrose and 0.8% Bacto agar.

3. PLOIDY ANALYSIS

Ploidy analysis of the test tube plants is performed in a single-channel flow cytometer Ploidy Analyser (PA) equipped with a high-pressure mercury lamp HBO 100W. Small pieces of approximately 0.5 cm^2 of leaves of the *in vitro* growing plants are chopped together with a piece of leaf of a control diploid plant in a solution for nuclei isolation, stained with DNA-DAPI and analyzed in the cytometer. Relative peak position, coefficient of variation and ploidy index are easily obtained. The whole process can be done in 10-15 min and the procedure has been proved to be highly reliable.

4. CROSSES

Most citrus genotypes are apomictic and their seeds produce several nucellar asexual embryos in addition to the zygotic embryo, which in many cases do not fully develops. Thus, apomictic genotypes cannot be efficiently used as female parents and in our program only monoembryonic genotypes are used. Pollination was done according to standard procedures in the years 1996 to 2001, and the average fruit set was 55%.

4.1. 2n x 2n Crosses

Female parents mainly included high quality cultivars of clementine (Nules, Fina, Hernandina, Marisol, Bruno, Tomatera, Orogrande) and Fortune mandarin. In these crosses the availability of male parents is not a limitation. Thus, we have used as many as 25 different cultivars. They include the main mandarin cultivars commercially used in Spain and several mandarin and sweet orange genotypes with desirable quality characters. The average number of aborted or underdeveloped seeds per fruit was only 0.9 (Table 1). This number was mainly influenced by the female parent and varied from an average of 0.4 seeds per fruit in clementines to 2.8 in Fortune. Most seeds had embryos with a normal appearance, about 0.1-0.2 mm in size, and were easy to isolate. They had a good germination rate, about 90% of the recovered plants were triploid and they were transplanted to the greenhouse with over 90% survival (Table 1). The growth of the triploid plants was vigorous both in the test tubes, the greenhouses and the field.

4.2. 2n x 4n Crosses

Female parents were mainly clementines. The limitation of this type of crosses is the low number of high quality tetraploid male parents available. Autotetraploid apomictic genotypes are occasionally produced by spontaneous somatic duplication of nucellar cells. Flow cytometry is being used to screen seedling populations to identify autotetraploid genotypes. Willow leaf, Montenegrina, and Nova mandarins and Pineapple sweet orange selected with this procedure have been already used as male parents. The average number of aborted or underdeveloped seeds per fruit was 7.3, but only 67% had embryos (Table 1), 0.1-0.2 mm in size, generally with an abnormal appearance and difficult to isolate. Only about 60% of them produced normal viable plants. Almost all plants recovered were triploid and were transplanted to soil with 80% survival (Table 1). The triploid plants had a slow growth both in the test tubes and in the initial stages after transplanting, but later they grew vigorously in the greenhouse and the field.

Table 1. Recovery of triploid hybrids (3n) by embryo culture in vitro from aborted seeds

Crosses	Nº aborted seeds per fruit	% aborted seeds with embryo	% embryo germination[d]	% triploid plants	% survival in soil	Efficiency Nº 3n plants/ fruit	Efficiency Nº 3n plants/ embryo
2n x 2n[a]	0.9	93.2	89.8	89.3	91.2	0.62	0.73
2n x 4n[b]	7.3	67.5	59	99.3	81.3	1.83	0.4
4n x 2n[c]	11.3	97.0	80	98.3	98.3	8.7	0.79

[a)] Data based on the culture *in vitro* of 4,801 embryos from different crosses.
[b)] Data based on the culture *in vitro* of 4,739 embryos from different crosses.
[c)] Preliminary data based on the culture *in vitro* of 66 embryos from only one cross.
[d)] Germination defined as embryos that produce normal plants *in vitro*.

4.3. 4n x 2n Crosses

The limitation of these crosses is that monoembryonic tetraploid mandarins were not available. Stable non-chimeric tetraploid plants of Nules and Marisol clementines have been recently obtained by culturing *in vitro* small shoot tips with of colchicine and later regenerating whole plants by shoot-tip grafting *in vitro*. The preliminary experiments done with crosses using the tetraploid Nules clementine as female parent are very promising, producing large numbers of seeds with normal embryos that are ease to isolate and have a good behavior *in vitro* and the greenhouse (Table1).

5. DISCUSSION

The three crossing methods represents complementary approaches, but their efficiency is quite different (Table 1). The limitation of 2n x 2n crosses is the small number of aborted seeds produced per fruit, and therefore many pollinations are required to produce a significant number of triploids. This is labor intensive during a two-month period. The limitation of 2n x 4n crosses is the difficulties to regenerate plants from the triploid embryos. This requires higher qualified personnel during at least six months per year. In spite of these limitations, over 3,500 triploid hybrids from 2n x 2n crosses and 2,000 from 2n x 4n crosses have produced and are under evaluation. The 4n x 2n crosses presented the highest efficiency (table 1) once that female tetraploid parents are available. We are now focusing our program to these crosses.

6. REFERENCES

Esen, A., Soost, R.K. 1971. Unexpected triploids in Citrus: their origin, identification and possible use. J. Hered. 62: 329-333.

Essen, A. Soost, R.K. 1972. Tetraploid progenies from 2x X 4x crosses in Citrus and their origin. J. Amer. Soc. Hort. Sci. 97: 410-414

Cameron, J.W., Burnett, R.H. 1978. Use of sexual tetraploids seed parents for production of triploid citrus hybrids. HortSci. 13: 167-169.

Ollitrault P., Dambier D., Jacquemont C., Allent V., Luro F. 1996a. In vitro rescue and selection of spontaneous triploids by flow cytometry for easy peeler citrus breeding. Proc. Int. Soc. Citriculture 1:254-258.

Starrantino, A. 1992. Use of triploids for production of seedless cultivars in citrus improvement programas. Proc. Int. Soc. Citriculture 1:117-121.

AGROBACTERIUM-MEDIATED TRANSFORMATION OF BARLEY POLLEN CULTURES

Jochen Kumlehn[1,2], Liliya Serazetdinova[2], Dirk Becker[2] and Horst Loerz[2]

[1]IPK Gatersleben, Molekulare Zellbiologie, Corrensstr. 3, D-06466 Gatersleben, Germany (e-mail: kumlehn@ipk-gatersleben.de); [2]Institut für Allgemeine Botanik, AMP II, Universität Hamburg, Ohnhorststr. 18, D-22609 Hamburg/Germany (e-mail: loerzamp@botanik.uni-hamburg.de)

Keywords *Agrobacterium*, barley, genetic transformation, *Hordeum vulgare*, microspore culture, pollen culture, pollen embryogenesis

The technology for identification of genes and promoters with putative scientifically and economically relevant function is becoming increasingly powerful. Therefore, reliable and efficient methods to genetically transform important crops are urgently desired for comprehensive functional analyses.

Immature barley pollen at the microspore or early bicellular stage can deviate from the normal process of pollen formation to undergo embryogenesis. Different kinds of stress application, e.g. cold, heat and/or starvation, turned out to be effective in inducing this developmental pathway. Immature pollen shows by far the highest potential for multiple plant regeneration per barley donor plant. From a single spike, up to 250,000 microspores can be isolated, out of which about 25,000 can be induced to undergo androgenic development, eventually resulting in the formation of up to 10,000 plants. About 80 percent of the regenerants have spontaneously doubled their genome during the first sporophytic cell divisions. Androgenic pollen cultures have been used previously for biolistic transformation, but the mechanical impact on the isolated pollen appeared to be too severe, and thus this method turned out to be very inefficient. Based on the huge regeneration potential of pollen cultures, a novel method which is presented here has been developed to genetically transform barley by use of agrobacteria.

Compared with direct gene transfer methods, *Agrobacterium*-mediated transformation has the advantage, that the transfer of the foreign DNA is efficiently guided by an evolutionarily developed system, i.e. the piece of DNA to be tranferred is precisely excised out of the bacterial vector, actively transferred into the plant cell, prevented from being fragmented within the

I. K. Vasil (ed.), Plant Biotechnology 2002 and Beyond, 545-547.

plant cell, transported into the plant nucleus, and eventually integrated into actively transcribed regions of the plant chromosomes by a set of bacterial gene products. In many cases, only a single copy of the transgene is integrated per target cell, which ensures expression stability for subsequent generations. On the other hand, agrobacteria have evolved their system to genetically transform other plants than grasses, which is reflected by the difficulties to employ their potential for these species.

Since embryogenic pollen is commonly cultured in liquid medium in which free diffusion of the media components and complete submergence of the target tissue is always provided, there was the opportunity for a detailed analysis and optimization of the conditions during co-culture of embryogenic pollen and agrobacteria. Several parameters had to be optimized to obtain both successful gene transfer and survival of the pollen culture. Preliminary experiments designed to optimize the pH during co-culture showed that both the cultivated pollen as well as the agrobacteria can dramatically modulate this parameter within a very short period of time. Therefore, some investigations appeared to be necessary to effectively stabilize the pH. A combination of 2-morpholinoethanesulfonic acid and potassiumhydrogenphosphate buffer turned out to be valuable in this respect. Subsequently, it was found that a continuous pH of around 5.9 constitutes the optimum for *Agrobacterium*-mediated transformation, a result which was obtained in suspension cultures of wheat and rice as well as in pollen cultures of barley. In addition, these experiments verified that the growth of the agrobacterial population is strongly pH-dependent. In contrast to weakly buffered cultures in which the pH could increase from below 6 to beyond 7 within two days, agrobacteria multiplied to a much less extent in cultures stabilized at around pH 6 which resulted in a markedly improved survival of the cultured pollen.

In another experimental approach, we have investigated the impact of glutamine on *Agrobacterium*-mediated transformation. It was known that gutamine is essential for inducing androgenic development in pollen cultures of barley, but it can be employed as N- and C-source by the agrobacteria as well. Accordingly, glutamine markedly supported the growth of agrobacteria in co-culture. On the other hand, glutamine did not enhance the transformation activity of the agrobacteria and turned out not to be necessary for androgenic development during 48 hours of co-culture. As a consequence, glutamine was subsequently omitted during co-culture, which resulted in reduced bacterial growth and thereby improved survival of the pollen cultures.

Acetosyringone constitues a phenolic breakdown product of the plant cell wall. It plays a dual role in *Agrobacterium*-mediated transformation: firstly, it serves as chemotacticum for the spatial orientation of the agrobacteria to move towards wounded sites of the plant tissue, and secondly, it acts as an elicitor for the expression of the bacterial virulence genes. In addition, we show that acetosyringone not only enhances the transformation activity of the agrobacteria, but at the same time, it can severely restrict bacterial multiplication. In the system presented here, a concentration of 0.5 mM acetosyringone results in the most efficient transformation of barley pollen cultures.

With regard to the theoretical potential of haploid target cells to directly obtain plants which are homozygous in the transgene due to transformation prior to spontaneous genome doubling, we have carried out experiments with varying pollen preculture time before co-culture. Unfortunately, we found that efficient transformation did require at least 6 days of preculture which coincides with the commencement of bursting of the pollen exines.

Upon extensive revision of the regeneration and selection protocol, transgenic plants were produced by use of the *Agrobacterium* strain LBA4404 along with a co-integrate vector system (kindly provided by Japan Tobacco Inc.) which confers hypervirulence, as well as the strain GV3101 with a non-hypervirulent binary vector system. The two different vector systems gave comparable results with regard to the transformation efficiency. Both phosphinotricin resistance as well as hygromycin resistance turned out to be valuable selectable markers. It was shown that 50 to 75 percent of the T_0 plants carried a single copy of the respective gene transferred. The efficiency of the method is promising: about one transgenic plant can be obtained per barley donor spike. However, we expect that about 20 percent of the transgenic plants are haploids which are generally not fertile. Moreover, the Southern analyses indicate that in rare cases a single transformation event may have lead to more than one transgenic plant.

Compared with immature embryo-based trasformation, the system presented here has several advantages: the spikes are harvested at a markedly earlier developmental stage, i.e. a higher throughput of donor plants can be obtained with a given greenhouse capacity, and as the donor plants can be discarded prior to flowering and seed set, phytopathogens can be controlled more easily. Moreover, the microspores from 30 spikes can be isolated and transferred to culture within about one hour.

Agrobacterium-mediated gene transfer to pollen cultures of barley appears to be a promising way to genetically transform this important crop and genetic model species for cereals. In addition, a valuable experimental system has been established to investigate and optimize the interaction between agrobacteria and target plant cells.

MOLECULAR CHARACTERIZATION OF CITRUS SYMMETRIC AND ASYMMETRIC SOMATIC HYBRIDS BY MEANS OF ISSR-PCR AND PCR-RFLP

M.-T. Scarano[1], L. Abbate[1], S. Ferrante[1], S. Lucretti[2] and N. Tusa[1]
[1]Istituto di Ricerca per la Genetica degli Agrumi-C.N.R., Viale delle Scienze n.11, 90128 Palermo, ITALY (e-mail: tscarano@unipa.it); [2]ENEA CR Casaccia, Sezione Genetica e Genomica Vegetale (026), Via Anguillarese, 301, 00060 S.M. di Galeria (Roma), ITALY

Keywords Citrus, somatic hybrids, ISSR-PCR, PCR-RFLP

1. INTRODUCTION

Molecular marker analysis is the most efficient and reliable procedure for characterizing Citrus somatic hybrids. ISSR-PCR (Inter simple sequence repeat) consists in the amplification of the region between neighboring and inverted microsatellites (SSRs) using primer that are complementary to a single SSR and anchored at both ends with one to three base degenerate oligonucleotide (Zietkiewicz et al., 1994). In presence of citrus asymmetric somatic hybrids, which are diploids and contain the nucleus of the leaf parent and mitochondria of the callus parent, further analysis is necessary to establish cybridization with PCR-RFLP. This method consists of a PCR with mtDNA universal primers followed by digestion with restriction enzymes (Dumolin-Lapegue et al. 1997). Here we report on the application of these two methodologies for the identification of putative Citrus interspecific somatic hybrids and cybrids obtained by symmetric protoplast fusion Somatic hybrids: 'Redblush' grapefruit (*Citrus paradisi* Macfadyen)+'Avana' mandarin (*Citrus deliciosa* Tenore), 'Duncan' grapefruit+'Fortune' mandarin *(Citrus reticulata* Blanco), 'Duncan' grapefruit+'Tardivo di Ciaculli' mandarin (*Citrus deliciosa* Tenore) and cybrids: 'Murcott' tangor (*Citrus sinensis* L. Osb x *Citrus unshiu* Marc.)+'Duncan' grapefruit (*Citrus paradisi* Macfadyen).

2. MATERIALS AND METHODS

Protoplast fusion, regeneration of somatic hybrids, flow cytometry, isozyme analysis and ISSR-PCR experiments have been carried out as previously described (Scarano et al., 2002), and PCR-RFLP experiments as described in Dumolin-Lapegue et al., 1997.

I. K. Vasil (ed.), Plant Biotechnology 2002 and Beyond, 549-550.

3. RESULTS AND DISCUSSION

The following genotypes: 'Redblush' (Rb) grapefruit +'Avana' (Av) mandarin; 'Duncan' (D) grapefruit + 'Tardivo di Ciaculli' (TdC) mandarin; 'Duncan' grapefruit + 'Fortune' (F) mandarin resulted to be all tetraploids and presented all the isozyme (*Pgi* and *Pgm*) and ISSR banding pattern of both parents (Fig.1A). Therefore they were classified as true allotetraploid somatic hybrids. 'Murcott' (M) tangor + 'Duncan' grapefruit genotype was diploid, showed the morphological traits similar to 'Duncan' (leaf parent) and displayed the isozyme (*Pgi* and *Pgm*) and ISSR-PCR banding pattern of 'Duncan' grapefruit (leaf parent) (Fig.1B). Successively, these hybrids were analyzed with PCR-RFLP using universal mtDNA primers (data not shown). All the hybrids presented the banding pattern of the 'Murcott' tangor (callus parent). This new genotype is nevertheless a cybrid (asymmetric hybrid), in which the nucleus is inherited from the leaf parent and mitochondria from callus parent.

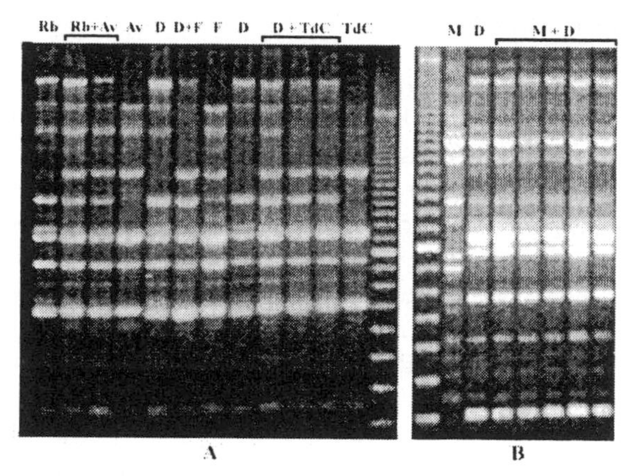

Figure1. ISSR-PCR markers in somatic hybrids of mandarin

4. REFERENCES

Scarano M.-T., L. Abbate, S. Ferrante, S. Lucretti, and N. Tusa. 2002a. ISSR-PCR technique: an useful method for characterizing new allotetraploid somatic hybrids of mandarin. Plant Cell Rep. In press.

Dumolin-Lapegue S., M.-H. Pemonge, and R.J. Petit. 1997 An enlarged set of consensus primers for the study of organelle DNA in plants. Mol. Ecol. 6:393-397.

Zietkiewicz E., A. Rafalski, and D. Labuda. 1994. Genome fingerprinting by simple sequence repeat (SSR)-anchored polymerase chain reaction amplification. Genomics 20:176-183.

A DOUBLED HAPLOID RICE POPULATION AND ITS GENETIC ANALYSIS USING MICRO-SATELLITE MARKERS

Melissa E. Hinga and Yunbi Xu
RiceTec, Incorporated, Alvin, Texas 77512, USA (e-mail: mhinga@ricetec.com; yunbi@ricetec.com)

Keywords anther culture, microsatellite, double haploid

1. INTODUCTION

With the increased study of populations to identify marker-trait associations, the doubled haploid population has become the preference (Lu et al., 1996; Pauls, 1996). Traits important to anther culture have been marked with microsatellites. He et al. (1998) identified markers for four rice anther culture traits on an *indica x japonica* DH population. They found that QTL for green plant yield were not independent of callus induction and green plant differentiation, which confirmed similar findings in maize. We are reporting a DH population derived from a cross between indica and tropical japonica varieties and its genetic analysis.

2. METHODS

A doubled haploid population consisting of 397 plants was created from panicles of an F1 of RT505 (RiceTec Proprietary line) by Jefferson (public variety). The seed from the regenerated plants (DH2) were planted, and phenotypic data for 12 traits were scored and leaf tissue were collected on 205 of the DH lines plus two parents. A second cycle of anther culture was done on 111 of the DH lines, but 28 were lost to contamination. Data was collected on 5 anther culture related traits. The DNA was screened using 135 microsatellite markers that were polymorphic between the two parents. QTL were identified using the program MultiQTL.

3. RESULTS AND DISCUSSION

Forty doubled haploid lines were identical to another line indicating they originated from the same calli and one line was not a doubled haploid. Similar to the results of Afza et al. (2001), we found a very low level of

I. K. Vasil (ed.), Plant Biotechnology 2002 and Beyond, 551-553.

heterogeneity at 0.2% of the 27375 marker-plant combinations. The 164 DH lines were slightly skewed towards one parent (50.6% RT505 *vs.* 49.2% Jefferson) with a Chi-square of 3.869 (Chi-square (1, 5%)=3.841) but much closer to a 1:1 ratio than Lu et al. (1996) observed. Genomic contribution to specific DH lines from one parent ranged from 20 to 80% (Figure 1A). The phenotypic and anther culture traits measured showed transgressive segregation in all but one case. The distributions of anther culture traits were skewed towards a low response (Figure 1B). A total of 36 QTL were identified, with one to four QTL for each of the traits except albino plantlet differentiation (APD). Six QTL were identified for four anther culture traits including two QTL for anther response (AR), two for callus induction (CI), and one for green plantlet differentiation (GPD) and one for green plantlet yield (GPY). Variation explained (VE) by these associated markers ranged from 9.6 to 13.7% (Table 1). Two QTL for CI and one for GPY identified in this study were located at the same regions for the QTL that He et al. (1998) identified. Correlation analysis confirmed the findings of He et al. (1998) that green plant yield is dependent on the callus regeneration or induction rate and the green plant differentiation.

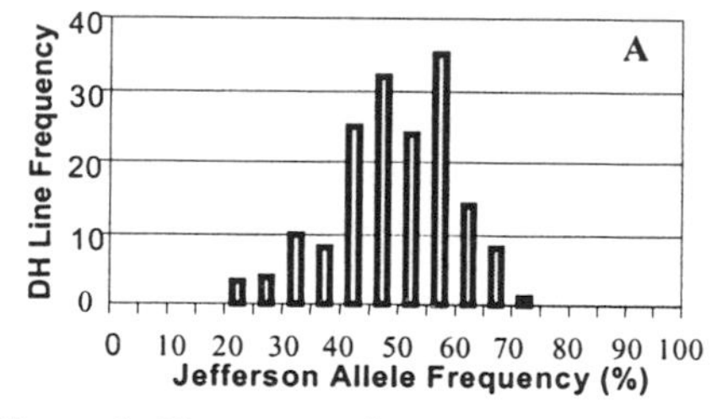

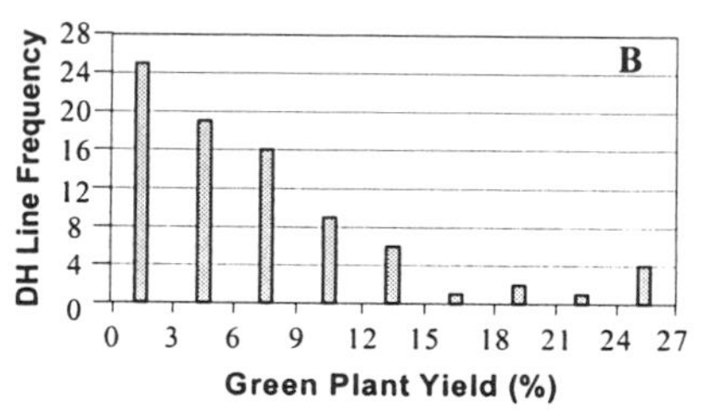

Figure 1. Histograms showing response in percent for (A) parental allele frequency of 164 lines (B) anther culture green plant yield of 83 lines.

Table 1. Microsatellite markers associated with traits related to anther culturability.

Trait	Chr	Linked Marker	LOD	P	m	d	VE(%)
AR	8	RM149-RM281	1.93	0.03520	49.16	-14.08	11.40
AR	10	RM216-RM239	1.84	0.00503	45.98	15.34	11.10
CI	7	RM070-RM248	1.78	0.02010	52.68	-19.34	10.60
CI	8	RM149-RM281	2.65	0.02510	51.75	-21.89	13.70
GPD	11	RM021-RM206	1.91	0.04020	15.86	8.38	12.50
GPY	1	RM009-RM246	1.81	0.03520	7.39	-4.02	9.60

4. REFERENCES

Afza R., J. Xie, M. Shen, F.J. Zapata-Arias, H.K. Fundi, K.-S. Lee, E. Bobadilla-Mucino and A. Kodym. 2001. Detection of androclonal variation in anther-cultured rice lines using RAPDs. In Vitro Cell. Dev. Biol. – Plant 37:644-647.

He, P., L. Shen, C. Lu, Y. Chen and L. Zhu. 1998. Analysis of quantitative trait loci which contribute to anther culturability in rice (*Oryza sativa* L.). Mol. Breed. 4:165-172.

Lu C., L. Shen, Z. Tan, Y. Xu, P. He, Y. Chen and L. Zhu. 1996. Comparative mapping of QTLs for agronomic traits of rice across environments using a doubled haploid population. Theor. Appl. Genet. 93:1211-1217.

Pauls, K.P. 1996. The utility of doubled haploid populations for studying the genetic control of traits determinated by recessive alleles. In vitro Haploid Production in Higher Plants, Vol. 1. S. M. Jain, S.K. Sopory and R.E. Veilleux (eds.). Kluwer Academic Publishers. pp. 125-144.

DEVELOPMENT OF NOVEL WHITE RUST RESISTANT GENETIC STOCKS IN CROP *BRASSICA* BY SOMATIC HYBRIDIZATION

G. Ravi Kumar *, S. R. Bhat, Shyam Prakash and V. L. Chopra
*NRC on Plant Biotechnology, Indian Agricultural Research Institute, New Delhi-110012, INDIA (e-mail: grk_mbio@yahoo.com)

Keywords *Brassica nigra*, *Diplotaxis gomez-campoi*, *Sinapis pubescens*, *Albugo candida* and Protoplast fusion

1. INTRODUCTION

White rust, caused by the fungus *Albugo candida* (Press. Ex. Lev.) Ktz, infects plants of 241 species in 63 genera of *Brassicaceae* leading to substantial yield loss. Considerable efforts have been made to evaluate cultivar resistance to *A. candida* in cruciferous crops and resistance was found in cultivars of radish, *Brassica napus*, *B. juncea* and *B. campestris*, but the variability within the species may not be sufficient to meet the requirements of modern plant breeding and the pathogen may adapt to the resistance. Wild species are often a valuable source of genes important in crop improvement. Various types of sexual barriers can however severely restrict the utilization of germplasm from species distantly related to the crop plants. By somatic hybridization the incompatibility barriers between species can be overcome and a widening of the gene pool of a domesticated crop can be obtained. It involves fusion of protoplasts of hybrid cells into hybrid plants. Unique organelle combinations may also be obtained in the process of protoplast fusion. In the present investigation screening of wild species for white rust resistance was undertaken and the identified resistant species were used for hybridization with *B. nigra*. The regenerated hybrids revealed complete resistance to white rust under artificial inoculation conditions.

2. MATERIALS AND METHODS

Screening of the wild germplasm, sexual and somatic hybrids was undertaken in nethouse conditions. The zoosporangial suspension was sprayed onto the young plants grown in the ambient conditions. The rate of rust development was recorded in terms of percentage of leaf-area covered by rust pustules at

I. K. Vasil (ed.), Plant Biotechnology 2002 and Beyond, 555-558.

weekly intervals, from the time of its appearance to the final intensity and the infection scores were recorded. Hypocotyl protoplasts of *B. nigra* were isolated and fused protoplasts were cultured following the method described by Kirti et al. (1991). Small calli of diameter 3-5 mm were transferred to K3 semisolid medium containing 1.1% sucrose, 0.1 mg/l IAA, 2.0 mg/l Zeatin riboside and 1.0 mg/l BAP for regeneration in *Diplotaxis gomez-campoi* + *Brassica nigra.* 0.1 mg/l IAA, 2.0 mg/l BAP and 2.0 mg/l 2-ip (6-(γ, γ–Dimethylallyl amino) purine) was used for regeneration of *Sinapis pubescens* + *Brassica nigra.* Calli with regenerated shoots were shifted to MS medium with 1% sucrose for shoot elongation. The rooted plants were transferred to small pots and were further analyzed. The flower buds were fixed in Carnoy's fixative. Anthers were squashed in 2% acetocarmine and the coverslips were sealed with wax. International Descriptor of *Brassica* was followed for morphological characterization of hybrids. For RAPD, RFLP and isozyme analysis standard protocols were used.

3. RESULTS

In order to identify potential donor species for white rust resistance, screening of the wild germplasm was undertaken in nethouse conditions. Artificial epiphytotic condition was created to ensure maximum development of disease symptoms to facilitate unambiguous identification of resistant genetic stocks. Two *Brassica nigra* accessions tested were found to be susceptible to white rust. In these susceptible genotypes white to creamy yellow pustules were observed which often coalesced to form patches on leaves, stem and inflorescence. Flower buds were swollen and aborted. *Diplotaxis gomez-campoi* and *Sinapis pubescens* were highly resistant to pathogen showing no symptoms under the standard inoculation conditions.

Protoplasts were isolated from hypocotyl and leaf tissue of *B. nigra* and wild species respectively. The mean frequency of heterokaryons among the cultured protoplasts was 5 % in the combination *S. pubescens* + *B. nigra* (EC No. 426393) and 23 % in *D. gomez-campoi* + *B. nigra* (dwarf). Shoot regeneration of 10.77 % was obtained in the combination *D. gomez-campoi* + *B. nigra* (dwarf) and 0.76 % in the combination *S. pubescens* + *B. nigra.* The number of confirmed hybrids in the fusion experiments are 34 and 1 in the combinations *D. gomez-campoi* + *B. nigra* and *S. pubescens* + *B. nigra* respectively.

All the plants of the somatic hybrid combination *D. gomez-campoi* + *B. nigra* were normal, green and vigorous with typical plant morphology. Most characters were intermediate to both parents. In parents, the pollen fertility was about 100%, whereas, in hybrids it ranged from 35-90%. The single plant obtained in the somatic hybrid combination *S. pubescens* + *B. nigra* exhibited greater resemblance to *B. nigra* than to *S. pubescens*. The hybrid plant was distinctly vigorous from parents but remained throughout in vegetative phase. Meiosis in the intergeneric hybrids was highly organized with many bivalents. Occassionally, a quadrivalent, trivalent and univalent were noticed, but in a very low frequency. Hybridity of the regenerated plants was confirmed by RAPD analysis. The number of bands amplified per primer varied from 4 to 11. The size of the amplified fragments varied from 100 bp to 8000 bp with an average size of 2200 bp. Hybrids showed alteration in pattern in comparison to the parents with unique bands which were not present in either of the parents or absence of bands was also noticed. The inheritance of cytoplasm in the somatic hybrids revealed the presence of mixed cytoplasm with mitochondria from *B. nigra* and chloroplast from *S. pubescens* in the hybrid combination *S. pubescens* + *B. nigra* and mitochondria from one of the parents in the combination *D.gomez-campoi* + *B.nigra* on probing with different chloroplast and mitochondrial genes such as *cox*I, 26S, *nad3*/*rps2* and *psb*D. The hybridity was investigated at biochemical level through isozyme analysis using peroxidase and esterase enzyme activity staining. These isozymes clearly demonstrated hybrid banding patterns among the regenerants. But the absence of the parental bands and/or unique bands were also observed in some regenerants.

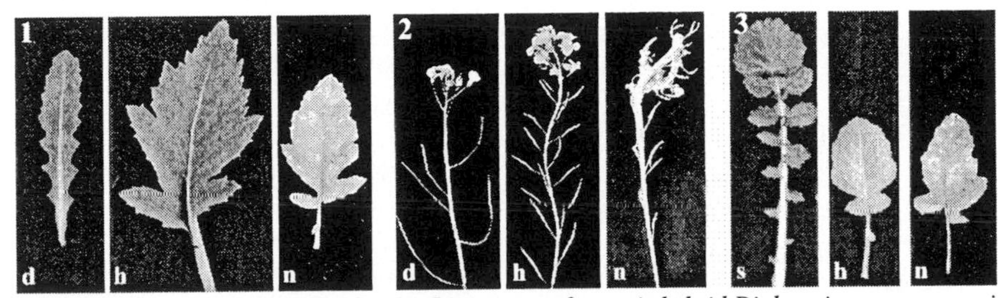

Figure 1 and 2. Response of leaf and inflorescence of somatic hybrid Diplotaxis gomez-campoi + Brassica nigra (h), the parents D. gomez-campoi (d) and B. nigra (n) to white rust. Figure 3. Response of the leaf of somatic hybrid Sinapis pubescens + Brassica nigra (h), the parents S. pubescens (s) and B. nigra (n) to white rust (Albugo candida).

The parents and the somatic hybrids were subjected to white rust resistance screening along with the parents under standard conditions. In all the cases, the parents *B. nigra* showed extreme susceptibility to the fungus by forming white to creamy yellow pustules which often coalesced to patches on leaves, stem and inflorescence leading to swelling, distortion, discoloration and hypertrophy of the affected parts (Figs. 1-3). Unlike *B. nigra*, the resultant hybrids were completely resistant. No disease symptoms were seen as in the wild species *D. gomez-campoi* and *S. pubescens*. This clearly showed that the disease resistance in wild species is heritable and dominant over susceptibility. Therefore, hybridization approach to introgress genes conferring disease resistance appears to be promising. These hybrids can act as bridge species for transferring genes from wild species to crop species.

4. REFERENCES

Kirti, P. B., Prakash, S. and Chopra, V. L. 1991. Interspecific hybridization between *B. juncea* and *B. spinescens* through protoplast fusion. Plant Cell Rep. 9:639-642.

Kumar, G. R. 2002. Development of novel white rust resistant genetic stocks in crop *Brassicas*. Ph. D. thesis, Indian Agricultural Research Institute, New Delhi, INDIA.

THE TRANSATLANTIC DISSENT BETWEEN EUROPE AND THE UNITED STATES ABOUT GMO'S

Klaus Ammann
Botanical Garden, University of Bern, Altenbergrain 21, CH-3013, Bern, Switzerland (e-mail: Klaus.ammann@ips.unibe.ch)

1. DIFFERENCES IN HISTORY AND CULTURE

There is no doubt that in the last few years there has been a growing dissent about various issues between the United States and Europe. Fifty years after the second world war and its very special circumstances – the fact that the USA have sacrificed millions of their own soldiers' lives to free Europe from its worst dictatorship ever - things change gradually, history seems to be forgotten and the rising power of Europe causes inevitably a new independency from the United States. This is all coming at a time when there are several other factors opening divides in other fields of politics and life in general.

It is only fair to point to the huge differences in history between Europe and the USA. Whereas Europe has millions of traces and traditions left from local and national history, this is simply not the case in the relatively young United States. This certainly does not mean that there would be basic differences in the quality of cultural activities, because tradition should not be mixed up with culture.

Just have a look at the differences in agri-culture. Whereas in the United States the transition from old-fashioned agriculture has been thorough and basically carried out decades ago, in Europe there are still lots of small centers of traditional agriculture, with all its intricate marketing network. In the United States, the treeless prairie has certainly favoured industrial agriculture. On the other hand, there are many mountainous regions where small or medium scale agriculture has persisted in Europe.

There are, in connection with agriculture, also lots of other cultural differences to be seen. The variety of food and the enormous diversity of food production in Europe contrasts dramatically with the relatively uniform situation with McDonalds, Kentucky Fried Chicken, etc.in the USA – although one has to admit that in the big centers of the United States there is

I. K. Vasil (ed.), Plant Biotechnology 2002 and Beyond, 559-563.

considerable European (and Extra-European) culture of this sort persisting – and notably gaining ground again – even in rural areas.

This is influencing without any doubt expectations of consumers on both sides of the Atlantic in a different way: While the Americans often prefer fast food, Europeans tend to stick to old habits. Just take the Swiss. They are still fixed stubbornly to the tradition of rushing home at noon to have a rest for two hours including the main meal of the day. Americans often think that this would just be a waste of time and causes unnecessary traffic and thus they prefer a fast lunch.

2. THE CLASH BETWEEN MODERNITY AND POST-MODERNITY

Consider modernity with all its advantages – giving rise to democracy, to enormous progress, to a constantly speeding process in developing new technologies. Those processes in modernity, related to progress, not contested until recently, have been more accentuated in the USA, less conspicuous and less speedy in Europe. In this context science has risen to the mainstream power, and people have been fascinated by all the marvels of new discoveries – just think about the boost space technology has brought. But on the negative side there is the fact that science talks in facts, in a depersonalized "it" language that often does not care about social and cultural contexts. No wonder that in times of uncontested progress people started to notice the dark side of it, since a certain saturation in modern life cannot be denied. Luxury and decadence are clearly connected to this kind of uncritically welcomed technological progress. It was not until people started to fight against environmental problems that the new movement gained ground and today modernity is put into question.

The rise of post-modernity, especially in Europe, has created a new critical view about technological progress, and at the same time, questions, values and norms that hitherto have not been questioned at all. Everything seems to fall apart, anything goes, everything can be seen in relative terms. Science itself is now relative, often connected to unwelcome progress and often seen as a negative cause of environmental problems, problems of modern medicine, etc. There is a strong trend towards alternative views such as organic farming, alternative medicine, etc.

You can follow up the rise of a new "we" language, where the collective thinking, the social and cultural context is often much more important than scientific facts.

Modernity and post-modernity are simultaneously existing and causing lots of dispute among representatives. Disputes that are fruitless, since we should, instead of denouncing each other, try hard to find a common platform among opponents. We urgently need discursive decision making processes to solve those problems.

3. THE TRANSATLANTIC DISPUTE ON GMO'S SEEN IN THE LIGHT OF THE CULTURAL DIFFERENCES

There is no point of cultivating the contrasts and disputing over all the differences, but on the other hand it is also not realistic to force consensus by just ignoring all the differences.

A typical example is the *Precautionary Approach*. It is quite exemplary that the Europeans changed the term swiftly into the 'Precautionary *Principle*', which is by no means justified, since all international legislation is using the Precautionary *Approach*. Europeans tend to raise the stakes by blowing up the term into a principle, and even worse: try to focus it on the negative side of the technology assessment, as if there would be no risk NOT to introduce GMO's anywhere. It also seems to be forgotten, that the Precautionary Approach has been created within the legislative process of the Convention of the Biological Diversity. And here it was without any doubt built on negative facts, on a scientific analysis of negative environmental trends, with hard data and a plethora of uncontested literature on toxicological analysis. But at the same time, nobody was able to predict the future trends, so it was highly justifiable to introduce this kind of a precautionary approach system. But in the subsequent Cartagena protocol legislative process these facts were simply forgotten and the precautionary approach related to the introduction of GMO's was solely built on virtual risks. (I do not consider reductionistic laboratory studies with force-feeding non-target insects with toxic pollen grains of Bt-plants to be clearcut signs of negative trends; they have been contradicted with real-time field studies).

In my opinion the Precautionary Approach should not be abused by opponents of this promising technology to put a halt to the development of transgenic plants by denying field experiments. The Precautionary Approach should be applied as a decision making process with all stakeholders included, a honest decision making process where different kinds of knowledge should be respected, the problem solving process should be conducted in the spirit of a collaborative learning process, which should begin in the light of the symmetry of ignorance, but there should be no room left for denigration of hard scientific facts presented in the process. This kind of decision making process can still lead in certain cases to a unanimously accepted zero solution. In other cases, a free discourse will also reveal elements of a risk balance and also the serious consideration of a non-application of the technological solution.

4. TOWARDS NEW AND UNEXPECTED SOLUTIONS IN MODERN AGRICULTURE

It is of great importance to see the historical context: We are way out of the phase of a primitive agrarian phase of agriculture, as for the moment, we are still working with industrial methods, which brought us the green revolution, often denounced and seldom fairly judged. Now we need a more knowledge based agriculture, which meets the needs of Equitability and Sociability, and – still a dream, evolve into a real Agri-Culture in a new sense, where culture itself becomes an important element.

We should refrain from party thinking and refrain from denigrating each other. There is no use in emphasizing the dark side of genetic engineering and organic farming, because in modern *and* post-modern times there is simply no time left for such sterile disputes. Let us learn from each other and take the best out of all those wonderfully different agricultural strategies. There are possibilities like Precision Biotechnology and a new Organo-Transgenic Farming, which could lead the way. Why not fight the potato blight with the newly discovered resistance genes from well known old potato traits, and why not learn with high tech means to farm with less input, and the dream, to mimic old-fashioned farming with a high genetic diversity with modern means of mutational breeding supported by marker genes and transgenes?

Coming back to the disputes about agriculture between Europeans and Americans this could be the solution, breaking the mould and giving it a new thought, forget about the differences in polls and just simply do a better job. Here are my priorities:

- Create a strong European Agency to deal with safety problems of agriculture and food.
- Initiate research that does not halt at the traditional boundaries which are mainly caused by an unsustainable marketing between the producers of novel and organic food
- Foster transatlantic dialogue, enhance it with decision making processes of the second generation, which contain a certain guarantee to free the participants from prejudice.

In recent months there are signs at the horizon which give reason to a certain optimism within the EU structures, and lately also in Switzerland, my home country, which will adopt a new legislation for transgenic organism *not* including the long debated moratorium.

5. EUROPEANS DIFFER IN VIEWS ABOUT MODERNITY FROM AMERICANS

This is due to the fact that Europeans differ also culturally in many ways: Starting from eating traditions over societal differences up to cultural differences as a whole. Post-modern views prevail in intellectual circles, science is criticized as an instrument of unreflected progress. Back to nature is the slogan and organic farming is becoming a major trend. It is true that the dignity of modernity with all its virtues like liberal democracies, and a comfortable life built on highly refined technology are forgotten and even denigrated in an unfair way. Europeans have gone through so many food, HIV blood and BSE scandals, that with the introduction of GM they think: Now those scientists are coming with another lie. It will take years to build up the trust again, years of hard work and dialogue. Scientists should be involved in debates with a new spirit, believing in good science, but at the same time also obeying the principles of the symmetry of ignorance, respecting fears of the population, which has a finely tuned sensitivity about the good and the bad of biotechnology, knowing that biology has lost its innocence. It is also important to notice that a truly European development of transgenic crops has only just begun and it will still take some years to come up with crops adapted to the needs of Europeans, especially in regions where small-scale agriculture is still prevalent.

SOCIAL ACCEPTANCE OF PLANT BIOTECHNOLOGY

Thomas Jefferson Hoban
N. C. State University, Raleigh, North Carolina 27695-8107, USA (e-mail: tom@sa.ncsu.edu)

Keywords Public perception, consumer acceptance, social impacts

1. INTRODUCTION

We are in the early stages of a revolution in science and technology that will impact almost every aspect of our society and the natural world. Over the last few decades, scientists have made great strides toward a complete unraveling of the blueprint of life contained in DNA. This has led to the development of some relatively simple genetically modified crop plants that have been rapidly adopted by farmers from around the world.

One feature of this revolution is the rapid convergence of a number of areas. Crops (corn and tobacco) are being modified to produce a range of industrial products that may someday replace our dependence on petroleum, thus connecting agriculture directly to other industries. Foods are being designed so that they can serve as therapeutic agents delivering custom formulations to meet an individual's unique genetic profile, thus blurring the line between food and medicine. Scientists need to understand the extent to which society is ready for these and other innovations.

Research has shown that any new technology will take time to achieve broader societal understanding and acceptance. Innovations need to be evaluated in terms of their relative benefits and risks. These new developments will have to be communicated to a variety of stakeholder groups within society. The more dramatic the genetic transformations that are possible, the more important it becomes to have credible policies and programs are in place to facilitate social acceptance and address any unforeseen societal impacts. Scientists must help explain their work to the broader public.

Plant scientists and biotechnologists have been able to provide benefits to farmers, consumers, and the environment. However, these benefits may not be realized in a timely manner if society is unable or unwilling to accept the

I. K. Vasil (ed.), Plant Biotechnology 2002 and Beyond, 565-568.

science and applications as beneficial, safe, and ethical. Such acceptance depends upon how well scientists and others communicate with consumers and opinion leaders about biotechnology. Communication is more effective when we understand the knowledge and attitudes of our target audiences.

2. RESEARCH ON SOCIAL ACCEPTANCE

One purpose of this paper is to explore the extent to which consumers in various parts of the world have similar or different views about biotechnology. It is very costly and complex to conduct global public opinion research. Environics International of Toronto Canada completed interviews with 35,000 consumers from 35 countries. These results allow us to better understand the overall climate for biotechnology acceptance around the world. More information on these results is available from the author or directly from the company.

These results help us assess how the public reaction to biotechnology varies around the world. Respondents to the Environics study were asked the extent to which they agreed or disagreed with the following statement "The benefits of using biotechnology to create genetically modified food crops that do not require chemical pesticides are greater than the risk." This is the most extensive research available with such a global perspective.

Some patterns are clear in public response to this question. Over two-thirds of respondents in the following countries agreed that the benefits of GM crops are greater than the risks: United States, Columbia, Cuba, Dominican Republic, China, India, Indonesia, and Thailand. On the other hand, fewer than 40 percent of consumers in the following countries saw the benefits as greater than the risks: France, Greece, Italy, Spain, and Japan. The overall pattern is that Europe, Japan and South Korea are much more negative than other parts of the world. The US leads the industrialized countries in support for biotechnology. Overall, developing countries tend to be quite supportive of GM crops.

Another purpose of this paper is to evaluate how American consumers' acceptance of biotechnology may have changed over the years. This area has been studied for over ten years (Hoban, 2001.) Selected research results will be summarized here (much more information is available from the author upon e-mail request.) Research results indicate that crop biotechnology has not become a serious concern or even an issue for the vast majority of American consumers.

It is clear from our experience in the US that consumers need to understand and recognize benefits from biotechnology. They want to know why biotechnology is being used. Acceptance of crops with reduced pesticides has remained fairly strong over that time. Between two-thirds and three-quarters of consumers indicated they would intend to consume such products. Willingness to purchase did drop somewhat during 1999 -- but has since rebounded. This corresponded to the point when media coverage of the controversy started. American consumers are even more accepting of crops perceived to be safer (due to reduced pesticides) than they are of crops that are better tasting.

Research has also shown that the controversy over Starlink corn had no significant impact on American consumers' acceptance of crops produced through biotechnology (Hoban, 2001.) Surveys done immediately after the news story broke (along with follow-up research during the following year) reveals that American consumers did not change their attitudes or behavior as a result of the controversy. This was due in part to the relatively high low levels of concern about biotechnology, as well as the specifics of this case.

3. CONCLUSIONS AND IMPLICATIONS

The outlook for North America consumer acceptance remains quite positive despite some notable controversies (e.g., monarch butterflies and Starlink corn.) American consumers will likely remain relatively positive about plant biotechnology. This is due in part to the fact that protest groups have very low credibility in the US -- especially when act like eco-terrorists. It also helps that the organic market provides an outlet for concerned consumers. Plant biotechnology should not become an issue for most consumers (human and animal applications are a different story). Criteria for consumers' food selection remain taste, value, nutrition, and convenience (not seed genetics)

In many ways, the developing countries control the future for plant biotechnology acceptance. As the health and nutritional benefits become more evident and compelling, these nations will rapidly adopt genetically modified seeds. There will be enhanced opportunities for feeding people and finding new market niches in a sustainable way. In fact, most countries are actively pursuing technology development, including Egypt, China, India, and others. Key leaders and scientists are starting to speak for themselves (and hoping to avoid any more pressure from the protest industry or the European Union.) Their support for biotechnology should encourage global public acceptance.

The outlook for Europe is more uncertain, but clearly more negative. Over the past six years, the protest groups have hijacked the public acceptance process in the EU and have effectively held up approvals of any new crops. Decisions are being driven by politics and economics, rather than science and safety. European consumers have been promised that they will not be sold any genetically modified foods. Unfortunately, the current system characterized as "Don't ask, Don't Test, Don't Tell" is misleading to consumers (since the products in the EU generally do contain genetically modified ingredients). In fact, if food prices actually reflected true costs of identity preservation then most consumers may not really want GM-free. Consumer education could be effective if led by brave European leaders and scientists. European resistance to GM crops is hurting farmers and consumers- especially the poor.

Communication with consumers is vital. Research has shown that a number of credible and influential messages can help build public understanding and acceptance of biotechnology. These are as follows:

1. Biotechnology is already providing environmental and humanitarian benefits to society and promises even more in the future.

2. Scientists and governments have determined that the products of biotechnology are as safe or safer than conventional food production systems.

3. Biotechnology is the latest in a series of tools that have already "modified" plants and is more precise than traditional breeding. There are risks with any technology and those from biotechnology are being carefully examined and managed.

4. Biotechnology is vital for future economic prosperity. Those companies and countries that are early developers will be rewarded, while those who are laggards will fall behind.

In addition to enhanced communication, there are other things that plant scientists could do to enhance market acceptance of genetically modified crops. It will be important to speed up development of products with clear consumer benefits. It is also important improve and maintain confidence in science and the government regulatory system. All groups in the food value chain must work together to develop a cost-effective and efficient identity preservation system. This will be particularly important for industrial crops Scientists can also help to address global policy challenges (e.g., the by exposing the costs and nonsense of Europe's proposals for traceability).

4. REFERENCES

Hoban, T.J. 2001. American Consumers' Awareness and Acceptance of Biotechnology. Genetically Modified Foods and Consumers. Ithaca, NY. National Agricultural Biotechnology Council. Pp. 103-114.

GLOBAL EXPERIENCE WITH GENETICALLY MODIFIED CROPS

Randy A. Hautea
ISAAA Global Coordinator and Director, *SEAsia*Center, c/o IRRI, DAPO Box 7777, Metro Manila, Philippines (e-mail: r.hautea@isaaa.org)

1. INTRODUCTION

Modern biotechnology-facilitated crop improvement is undoubtedly one of the most significant technological developments in agriculture. Advancement in gene mapping and gene transfer has allowed the modification of plants that would be difficult or impossible through conventional breeding techniques. Genetically-modified (GM) or transgenic crops currently deployed in the environment exhibit input traits benefiting farmers. Such benefits include increased and protected yields, higher income, and greater flexibility in crop management. Future products are expected to provide broader benefits to consumers that could include enhanced nutritional content.

2. CURRENT STATUS OF GM CROPS

Adoption rates for GM crops are unprecedented and are among the highest for any new technologies by agricultural industry standards (Figs. 1, 2). Transgenic crops are now adopted in 14 countries worldwide, including most recently, India. Global area of GM crops expanded by around thirty-fold from 1996 to 2001 and the number of countries growing these crops has more than doubled (Table 1). From 1.7 million ha in 1996, global GM crop area increased to 52.6 million ha in 2001. Most of the plantings occurred in developing countries, particularly North America, accounting for nearly 75% (USA and Canada combined) of the total area planted in 2001. The rest (25%) is grown in developing countries, with Argentina and China having larger areas planted to GM crops.

Worldwide, extensive commercialization of transgenic crops has been almost exclusively for commodity crops such as maize, soybean, cotton, and canola that have attracted significant interest both from public sector research institutes and private agricultural life science companies. In 2001, of the total global area devoted to the four principal GM crops, 19% was planted to transgenic varieties. Nearly half of the global soybean area is GM soybean (Figure 2). Herbicide tolerant crops represented 77% (40.6 million ha) of the

I. K. Vasil (ed.), Plant Biotechnology 2002 and Beyond, 569-574.

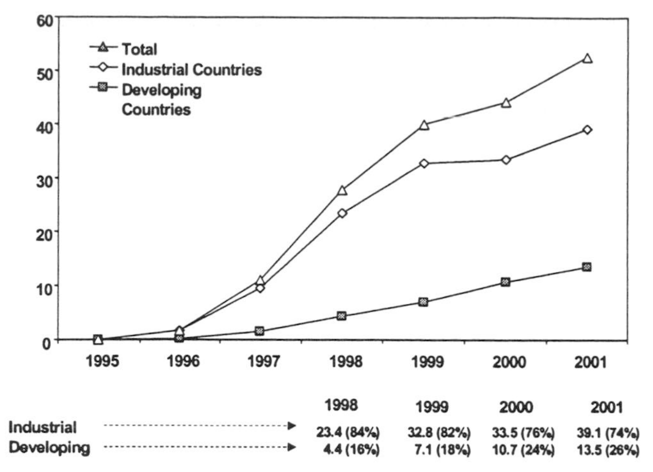

Figure 1. Global area of GM crops (million ha), 1996-2001. Source: C. James (2001) ISAAA Brief 24

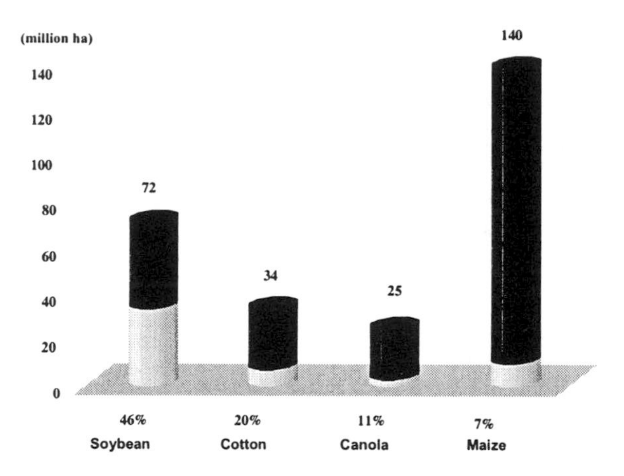

Figure 2. Global adoption of principal GM crops, 2001. Source: C. James (2001) ISAAA Brief 24

global GM crop area in 2001, followed by insect resistant (*Bt*) GM crops planted on 15% (7.8 million ha) of the global GM crop area.

3. GM CROPS BENEFITS

Crop biotechnology can offer farmers significant benefits, as has been documented both in the developed and developing countries. The widespread adoption of major GM crops in the United States and Canada has resulted in significant yield increases, pesticide use reduction, and improved farm

Table 1. Global area of GM crops, by country, 2000 and 2001 (million hectares)

Country	2000		2001		+/-	
	Area	***%***	***Area***	***%***	***Area***	***%***
USA	30.3	68	35.7	68	+5.4	+18
Argentina	10.0	23	11.8	22	+1.8	+18
Canada	3.0	7	3.2	6	+0.2	+6
China	0.5	1	1.5	3	+0.1	+200
South Africa	0.2	<1	0.2	<1	<0.1	+33
Australia	0.2	<1	0.2	<1	<0.1	+37
Mexico	<0.1	<1	<0.1	<1	<0.1	--
Bulgaria	<0.1	<1	<0.1	<1	<0.1	--
Uruguay	<0.1	<1	<0.1	<1	<0.1	--
Romania	<0.1	<1	<0.1	<1	<0.1	--
Spain	<0.1	<1	<0.1	<1	<0.1	--
Indonesia	--	--	<0.1	<1	<0.1	--
Germany	<0.1	<1	<0.1	<1	<0.1	--
France	<0.1	<1	--	--	--	--
Total	**44.2**	**100**	**52.6**	**100**	**+8.4**	**+19**

Source: C. James (2001) ISAAA Brief 24

income. It has been reported that in 2001, US farmers who adopted GM crops increased crop yields by 1.8 million tons, saved US$1.2 billion in production costs, and reduced pesticide use by 21 million kg. Farmers who adopted *Bt* cotton in China, South Africa, and Mexico also experienced increased and protected yields and reduced expenditures on pesticide application. Herbicide tolerant crops also bring considerable benefits to farmers, including reduced herbicide and labor costs, increased flexibility in weed management, and compatibility with conservation tillage regimes.

Biotechnology research is also making rapid progress on developing products with more direct benefits to consumers such as improved food quality and nutritional profile. Crop biotechnology also presents health benefits to farmers, as well as a range of possible environmental benefits.

4. GLOBAL CROP BIOTECHNOLOGY INITIATIVES

Crop biotechnology is gaining acceptance and its enormous potential benefits increasingly being appreciated by farmers adopting the technology. Within Asia, many governments give high priority to crop biotechnology research and development (R&D), with increasing interest on "orphan crops", in the hope of addressing the pressing problems related to improving agricultural productivity, increasing farmers' income, and meeting food security demands. China, in particular, has invested heavily in crop biotechnology research, now ranking second to the USA in terms of public sector investment. It is reported that public research institutes in China have already developed 141 GM crops, 45 of which have been approved for field trials, 65 for environmental release, and 31 for commercialization GM crops already commercialized include cotton, tomato, and sweet pepper.

India has been recently added to the list of GM crops adopters with the approval of the planting of three *Bt* cotton cultivars. Indonesia now extensively grows *Bt* cotton in seven districts of the South Sulawesi province and research on food crops such as maize, peanut, soybean, potato, sweet potato, and rice are ongoing. GM crop researches in Malaysia are conducted primarily in public research institutes and universities and currently limited to laboratory work and glasshouse evaluation. No field trial has been conducted yet, but steps have already been taken to familiarize regulatory bodies with the application and protocols of field testing and biosafety assessment.

The Philippines is already into field testing of *Bt* maize. Current research programs are also focusing on developing transgenic banana with resistance to bunchy top virus, GM papaya resistant to ringspot virus, delayed ripening papaya and mango, and coconut with high lauric content. A directive, which will take effect July 2003, has been recently issued by the government allowing the importation and commercial release of GM products. Thailand focuses on improving crop traits of traditional foods, fruits, and export commodities. Crop biotechnology R&D in Vietnam is largely at the stage of improving technology imported from developed countries. Malaysia has initiated field trial of delayed ripening papaya, the first for a genetically modified food crop developed locally.

Across the African continent, South Africa is the only country to commercialize transgenic crops. *Bt* cotton is grown extensively in the Makhathini Flats of KwaZulu-Natal province. Products resulting from crop biotechnology R&D in Egypt are GM potato resistant to potato tuber moth (PTM), virus resistant squash, and *Bt* maize, all awaiting approval for commercial release. Kenya largely focuses its research on sweet potato. In other parts of Africa, substantial research in crop biotechnology is being carried out, although many countries are still in the early stage of GM crops development.

In Latin America, Argentina is leading the way on the adoption of transgenic crops, limited almost exclusively to commodity crops such as soybeans, maize, and cotton. Much of the investments in crop biotechnology research have come from the private sector with fairly limited activities initiated by the public sector. Mexico is the only other country in Latin America with significant commercial cultivation of GM crops, mostly GM soybean and GM cotton. Commercialization of GM crops remains prohibited in Brazil but it has already produced a number of beneficial GM crops such as herbicide tolerant soybean, GM maize, GM papaya resistant to Brazilian strain of ringspot virus, and common beans resistant to golden mosaic virus.

The European Union still shows strong opposition to genetically modified organisms (GMOs). Environmental release of GMOs is tightly regulated by restrictive guidelines and by labeling and treaceability requirements. Approval process for the commercial release of new GMOs is currently on hold following a *de facto* moratorium on new approvals that took effect in 1998. Despite this, field trials continue to be conducted across Europe, concentrating on commodity crops such maize, oilseed rape, sugarbeet, and potato.

5. PARTNERSHIPS IN CROP BIOTECHNOLOGY PROGRAMS

The development of biotechnology applications is capital intensive, requiring long-term investments. In addition, research costs of developing GM crops are prohibitive, especially as some of the technology used are proprietary and must be licensed from private companies. In developing countries, execution of agricultural biotechnology research is generally highly dependent on the public sector, but funding is often very limited. Over the past decade, growth in spending on agricultural R&D by the public sector continue to decline with parallel shift on spending priorities of international donor agencies away from agriculture. While the private sector contributes immensely to global investments in agricultural research, only a fraction of private sector funds is invested in developing countries. Clearly, the need for stronger public-private sector partnership in agricultural research is evident.

The International Service for the Acquisition of Agri-biotech Applications (ISAAA) facilitates collaborative undertakings between the public and the private sector in crop biotechnology. A number of public-private sector partnerships has been initiated and already showing significant progress in GM crop development. For example, the Papaya Biotechnology Network of Southeast Asia is making strides in developing GM papaya that would benefit farmers from the five member countries of the network—Indonesia, Malaysia, Philippines, Thailand, and Vietnam. ISAAA developed and brokered the project with support from the public and private sectors. Monsanto and scientists from the University of Hawaii are collaborating with the network to develop papaya resistant to papaya ringspot virus (PRSV), while the Syngenta and the University of Nottingham are sharing with the network their technology and knowledge on delayed ripening in papaya. Another project deals on the development and transfer of insect resistant sweet potatoes in Vietnam. ISAAA brokered an agreement with Syngenta, which agreed to donate *Bt* endotoxin genes and to train Vietnamese scientists on how to utilize them.

6. CONCLUDING REMARKS

The deployment of GM crops is changing the global landscape in agricultural production. Future trend in GM crops indicates rapid increase in global adoption largely because of the demonstrable benefits derived from the technology. Documented benefits include increased and protected crop yields, higher income, and greater flexibility in crop management. Results obtained from field trials and farm plantings substantiate crop biotechnology as a practical option to increase food production.

Today, there is an increasing realization that crop biotechnology has the immense potential to contribute to agricultural productivity and sustainability and address the serious problem of meeting global food security demands. An open and broader exchange of information and knowledge on biotechnology will help provide the opportunity to link the needs of societies, particularly in developing countries, with an increasing array of crop biotechnology applications and related innovations that could be vital to their quest for food security.

FDA'S POLICY ON FOOD BIOTECHNOLOGY

Jason Dietz
Food and Drug Administration, HFS-255, 5100 Paint Branch Parkway, College Park, MD 20740 (e-mail: jason.dietz@cfan.fda.gov)

Keywords Biotechnology, food, laws, regulations

The Federal Food, Drug, and Cosmetic Act (FFDCA) provides the United States Food and Drug Administration (FDA) with broad authority to regulate the safety and wholesomeness of food. Developers of new foods have a responsibility to ensure that the foods they offer to consumers are safe and in compliance with all requirements of the FFDCA. FDA stated its policy on foods derived from new plant varieties, including bioengineered plants, in the Federal Register (FR) in 1992 (57 FR 22984, May 29, 1992). FDA recognized in its policy statement that it is prudent practice for developers of new plant varieties to engage in premarket consultations with the agency on safety and regulatory questions, especially with regard to food products developed through new technology. In 1996 (revised October 1997), FDA published guidance on consultation procedures for foods derived from new plant varieties, including bioengineered plants. FDA has completed premarket consultations on over forty bioengineered plants. In 1999, FDA held three public meetings to share its current approach and experience regarding bioengineered foods and to solicit views on whether FDA's policies or procedures should be modified. In 2001, FDA published in the FR (66 FR 4706, January 18, 2001) a proposed rule that would require the submission to the agency of data and information regarding plant-derived bioengineered foods that would be consumed by humans or animals. FDA proposed that this submission be made at least 120 days prior to the commercial distribution of such foods. This proposed action, if finalized, would enhance the Agency's ability to assess whether plant-derived bioengineered foods comply with the safety standards of the FFDCA on an ongoing basis.

I. K. Vasil (ed.), Plant Biotechnology 2002 and Beyond, 575-578.

FDA's Policy on Food Biotechnology

Jason Dietz
Food and Drug Administration
Center For Food Safety and Applied Nutrition
Office of Food Additive Safety

FDA's Policy on Food Biotechnology

- Authority to regulate bioengineered foods
- Consultation procedure
- 1999 Public Meetings
- Proposed pre-market notification rule

Regulatory authority

- Bioengineered foods are regulated by three different agencies:
  - Department of Agriculture (USDA)
  - Environmental Protection Agency (EPA)
  - Food and Drug Administration (FDA)

Regulatory authority

- USDA oversees safety for cultivation.
- EPA oversees the safe use of pesticides.
  - including pesticides produced in bioengineered plants
- FDA ensures products are safe to eat and addresses food labeling issues.

FDA's 1992 policy statement

- FDA stated its policy on foods derived from new plant varieties in 1992. (57 FR 22984, May 29, 1992)
- FDA's authority to regulate foods derived from new plant varieties comes from the Federal Food, Drug, and Cosmetic Act (FFDCA).

FDA's 1992 policy statement

- Food adulteration provision
- Food additive provision
  - New components of food will be regulated as food additives if they are not generally recognized as safe (GRAS).

FDA's 1992 policy statement

- Policy statement also addresses:
  - potential allergenicity of newly introduced proteins
  - nutrient composition and anti-nutrients
  - known toxins and new toxins
  - antibiotic resistance markers
  - decision trees to aid developers

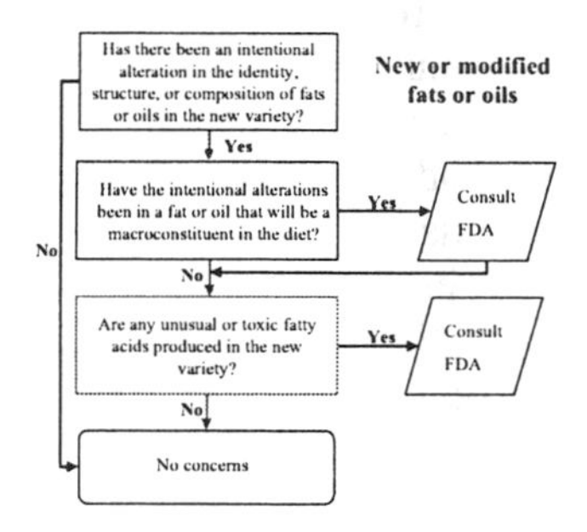

Consultation procedure

- FDA established a consultation procedure to ensure that new products are safe and lawful.
  - FDA believes that all developers of commercially marketed bioengineered foods sold in the U.S. have consulted with FDA prior to marketing their products.

Consultation procedure

- Firms submit a summary of safety and nutritional data of the product.
- When all safety and regulatory issues have been resolved firms receive a letter stating that FDA has no questions at this time.
- Completed consultations are listed on FDA's Internet site.

Consultation procedure

- Most of the introduced traits have been agronomic in nature.
  - pest or herbicide tolerance
- Corn and canola have been the most commonly modified.

Introduced traits

- Introduced traits in crops that have been the subject of completed consultations include:
  - Herbicide tolerance
  - Insect resistance
  - Pollen modification/herbicide tolerance
  - Virus resistance
  - Delayed ripening

New bioengineered varieties

Beet	Flax
Canola	Radicchio
Corn	Squash
Cotton	Tomato
Potato	Papaya
Soybean	Rice

Public meetings in 1999

- FDA held three public meetings in 1999.
  - Communicate policy and solicit opinions on current policy and procedures
  - Received over 35,000 written comments

Public meetings in 1999

- Conclusions from public meetings:
  - No new data to question safety of bioengineered foods currently marketed
  - Divergent views on labeling
  - Concern that current procedures may not be sufficient to deal with future developments

Proposed pre-market notification procedure

- If finalized, would require developers of bioengineered foods to notify FDA 120 d before commercial distribution.
  - Would enhance FDA's ability to assess whether new bioengineered plant varieties comply with FFDCA on an on-going basis

Conclusions

"FDA's scientific review continues to show that all bioengineered foods sold here in the United States today are as safe as their non-bioengineered counterparts. We believe our initiatives will provide the public with continued confidence in the safety of these foods." **Jane E. Henney, M.D.**
FDA Commissioner
May 3, 2000

3. REFERENCES

United States Department of Health and Human Services, "FDA to Strengthen Pre-Market Review of Bioengineered Foods." May 3, 2000.

United States Food and Drug Administration, "Guidance on Consultation Procedures- Foods Derived from New Plant Varieties," October 1997.

United Stated Food and Drug Administration, "Proposed Rule: Premarket Notice Concerning Bioengineered Foods." Federal Register, 66:4706-4738, 2001.

United States Food and Drug Administration, "Statement of Policy: Foods Derived From New Plant Varieties; Notice." Federal Register, 57:22984-23005, 1992.

*all references are available through FDA's Internet site at: http://www.cfsan.fda.gov/~lrd/biotechm.html.

REGULATION AND PUBLIC ACCEPTANCE OF TRANSGENIC CROPS IN CHINA

Zhangliang Chen
National Laboratory of Protein Engineering and Plant Genetic Engineering, Peking University, Beijing, 100871, China (email: zlchen@pku.edu.cn)

Public understanding of importance and potential application of GM technology has been very positive in China. Government has encouraged and strongly supported agrobiotechnology research. Many major crops modified for insects and diseases resistance have been approved to test and release in the fields. Transgenic cotton plants have been approved to commercialization. However, due to the consideration of environmental and food safety, as well as public concerning and international trade issues, China has been very cautious to approve field releases and commercialization of transgenice crops. Transgenic rice, wheat, soybean corn and canola have not been approved for commercialization in China and in field tests or releases. The new regulation on safety agricultural genetically modified organisms (GMO) promulgated by the China State Council on May 23, 2001. Implementation regulation on safety assessment, implementation regulation on labeling and implementation of regulations on the safety of import of agricultural genetically modified organisms will be effected on March 20, 2000. The first category of transgenic plants required to be labeled including:

(1) Soybean seeds, soybean, soybean flour, soybean oil, and soybean meal
(2) Corn seeds, corn, corn oil and corn flour
(3) Rape seeds, rape oil and rape meal
(4) Cotton seeds
(5) Tomato seeds, tomato, tomato sauce

The imported GMOs are divided into 3 groups according to their uses: for research and testing, for production, and for raw materials for processing, with different requirements. Ministry of Agriculture makes the decision of approval or disapproval within 270 days after receiving the application. Soybean is a major imported agricultural good from USA, Argentina, and Brazil, which will be an issue for trade under these new regulations. Detail information will be discussed in the conference.

Abstract only – no manuscript received.

I. K. Vasil (ed.), Plant Biotechnology 2002 and Beyond, 579.

PROGRESS OF TRANSGENIC PLANTS IN CHINA

Zhi-Hong Xu
Institute of Plant Physiology and Ecology, CAS, Shanghai; College of Life Sciences, Peking University, Beijing (e-mail: xuzh@pku.edu.cn)

Great achievements in plant biotechnology have been made in China in the past decade. The projects of plant biotechnology of the National Program of High-Tech Development, initiated in 1986, mainly cover plant transformation and genetic engineering, genetic mapping, marker-assistant breeding, and recently genomics and funtional genomics. The studies on transgenic plants in China focus on insect and disease resistance, stress tolerance, male sterility, nutrition improvement, and bioreactors for producing vaccine or other recombinant proteins by using transgenic plants (Huang et al., 2001; Tein et al., 1998; Xu, 1998).

So far, different Bt toxin and proteinase inhibitor genes from arrowhead and cowpea have been used for increasing insect resistance of different crops (tobacco, rice, maize, cotton, cabbage, cauliflower, poplar tree, etc.; Table 1). Insect-resistant cotton is the first commercialized transgenic product in China, and five other crops (such as chilli pepper, tomato and petunia) have been approved for commercial production. Thirteen cotton boll worm-resistant varieties of cotton have been released, and 1.1 million hectares of transgenic cotton was planted in 2001 (Huang et al., 2001). Transfer of two different insect-resistant genes (Bt and CpTI or arrowhead proteinase inhibitor genes) to cotton has made, the latter with more durable resistance to cotton boll worm than that of the transgenic cotton with Bt gene alone. In China, more than 30% chemical pesticides are applied for cotton. Insect-resistant cotton remarkably decreases the use of pesticides, and benefits the environment due to reduced pesticide pollution. Transgenic wheat with GNA (snowdrop lectin) gene showed reduced leaf loss and aphid number. Transfer of virus coat protein (cp) genes of different viruses increased virus resistance in several transgenic crops (tobacco, tomato, pepper, potato, soybean, rapeseed, rice, wheat). Virus resistant plants have also been obtained by transfer of replicase gene of TMV (tobacco), PVY (potato) and RDV(rice), and by using ribozyme for CaMV(rapeseed) and RSTVd (potato) (Table 2). Xa21 gene has been transferred to different strains of rice for increasing resistance to rice bacterial blight; it can also been used for the production of resistant hybrid rice seeds. Besides Xa21, Xa7 - an another gene resistant to rice bacterial blight - cloned in the Institute of Genetics and Developmental Biology, has been used for increasing disease resistance (Table 3). D-mannitol-1-phosphate dehydrogenase gene (Mt1D) and D-glucitol-6-phosphate dehydrogenase (GutD) cloned from *E.coli*, and BADH gene cloned from *Atriplex hortensis* have been used for increasing salt tolerance in transgenic plants (Table 3). Transgenic technique has also been used to obtain male-sterile plants of wheat, sorghum, soybean and cotton (Table 4).

I. K. Vasil (ed.), Plant Biotechnology 2002 and Beyond, 581-585.

Table 1. Transgenic plants with insect resistance.

Bt toxin	Tobacco, cabbage, poplar, cotton, maize, soybean
Proteinase inhibitor	
CPTI (Cowpea trypsin inhibitor	Tobacco, cotton, rice, poplar
Soybean Kunitz trypsin inhibitor(SKTS)	Tobacco
AHPI(arrowhead ptoteinase inhibitor)	Tobacco, cauliflower, cotton
Bt toxin and CPTI	Cotton
Bt toxin and API (Arrowhead proteinase inhibitor)	Poplar
Pea-Lectin and SKTI	Soybean
PinII	Rice
Snowdrop lectin (GNA)	Tobacco, rice, wheat
Insecticidal peptide from spider, scorpion	Tobacco

Table 2. Transgenic plants with virus resistance.

TMV cp gene	Tobacco, tomato, pepper
CMV cp gene	Tobacco, tomato, pepper
PVX / PVY / PLRY cp gene	Potato
SMV cp gene	Soybean
TuMV cp gene	Rapeseed
WMV cp gene	Watermelon
RYSV N gene	Rice
BYDV cp gene	Wheat
TCS	Tomato

Table 3. Transgenic plants with bacterial disease resistance.

Xa21, Xa7	Bacterial blight	Rice
Antibacterial peptides		
Cecropin B, Shiva-1	Bacterial wilt	Potato, tobacco
T7 Iysozyme		Tobacco
Barley thionin	Wildfire disease	Tobacco
Harpin		Potato
Vst1(grape)	Powdery mildow	wheat

For quality improvement, besides some trials to use transgenic techniques to increase the contents of essential amino acids, and transgenic tomato with longer shelf life, a successful example is rice with changed amylose content. In Prof. Hong's laboratory in Shanghai, Wax genes have been cloned and compared with different rices (Indica, Japonica and glutinous rice), and they found that amylose content of rice endosperm is regulated at the level of Wx transcript processing, specially at the stage of intron 1 excision from the Wx pre-mRNA. Transfer of anti-Wx gene has been used to improve the quality of rice by regulating amylose content (Cheng et al., 2001). Several laboratories in Beijing and Shanghai are working on using transgenic plants as bioreactors to produce different vaccines (Table 6).

Table 4. Transgenic plants with salt resistance.

E. coli:	
D-mannitol-1-phosphate dehydrogenase (MtlD)	Tobacco, rice,
D-glucitol-6-phosphate dehydrogenase (GutD)	strawberry, maize
Atriplex hortensis:	
Betaine adehyde dehydrogenase (BADH)	Rice, strawberry, tobacco, watercress, wheat
Inositol O-methyltransferase (Imt1)	tobacco
Proline gene	rice

Table 5. Transgenic plants with male sterility

Crop Species	Gene transferred for male sterility
wheat	Ta29-anti-actin
Maize	Ta29- Barnase
Rice	Pst-Barnase
Soybean	Anti-RNA of HSP40
Cotton	Ta29-Barnase
Rapeseed	Ta29-Barnase
Tomato	Ta29-Barnase, Ta29-anti-actin
Sesame	Ta29-Barnase

Progress of genomics of *Arabidopsis* and rice has opened a new era for plant science and biotechnology. Chinese scientists have completed the draft sequence of the rice genome (*Oryza sativa* L. spp. Indica) (Yu et al., 2002), are partners in the international Rice Genome Project for Japonica rice, and have just finished the sequencing of chromosome 4 of Japonica rice (Han et al., 2002). Based on these achievements, rice functional genomics project has been initiated, and the projects for other important crops are in consideration. These projects would certainly speed cloning of new important genes and create new ideas and techniques for future development of plant biotechnology. More varieties of transgenic plants would surely be produced and put in field trials in the coming years in China.

Table 6. Transgenic plants used as bioreactor.

hEGF	Tobacco Carrot	Shanghai Inst. of Plant Physiol. CAS
Cholera virus B subunit	Tobacco Carrot (0.1%)	Inst. of Microbiol. CAS
Hepatitis B oral vaccine	Potato (0.05%)	Chinese Acad. of Agri. Sci.
Hepatitis C	Tobacco (5.3% in chloroplasts	A Chinese Acad. of Agri. Sci.

REFERENCES

Chen, W.X., Xiao, G.F., and Zhu, Z. 2002. Obtaining high pest-resistant transgenic upland cotton cultivars carrying cry1Ac3 gene driven by chimeric OM promoter. Acta Bot Sinica 44:963-970.

Cheng, S.J., Ge. H.F., Wang. Z.Y., and Hong, M.M. 2001. Analysis of influence of Wx intron 1 on gene expression in transgenic rice plant. Acta Phytophysiol. Sinica 27:381-386.

Guo, B.H., Zhang, Y.M., Li, H.J., Du, L.Q., Li, Y.X., Zhang, J.S., Chen, S.Y., and Zhu, Z.Q. 2000. Transformation of wheat with a gene encoding for the betaine aldehyde dehydrogenase (BADH). Acta Bot Sinica 42: 279-283.

Han, B,, et al. 2002. Sequence analysis of rice chromosome 4. Nature (in press).

Huang, J.K., Rozelle, S., Pray. C., and Wang, Q.F. 2001. Plant biotechnology in China. Science 295:674-677.

Jia, S.R., and Qu, X.M. (eds) 1996. Antibacterial Peptides Used For Potato Genetic Engineering. Beijing: China Agricultural Science and Technology Publ.

Liu, F.H., Guo, Y., Gu, D.M., Xiao, G., Chen, Z.H., and Chen, S.Y. 1997. Salt tolerance of transgenic plants with BADH cDNA. Acta Genet Sinica 24:54-58.

Tien, B., Xu, Z.H., and Ye, Y. (eds) 1996. Plant Genetic Engineering. Jinan: Shandong Science and Technology Publ.

Wang, G.L., Fang, H.J., Wang, H.X., Li, H.Y., and Wei, Y.T. 2002. Pathogen-resistant transgenic plant of *Brassica pekinensis* by tranfering antibacterial peptide gene and its genetic stability.

Acta Bot Sinica 44:951-955.

Wang, W., Chen, W.X., Zhu, Z., Xu, H.L., Gao, Y.F., Wu, Q., Zhu, Y., and Guo, Z.S. 1999. Studies on highly efficient planting of transgenic cotton. Acta Bot sinica 41:1072-1075.

Xu, Z.H. (ed) 1998. Plant Biotechnology. Shanghai: Shanghai Science and Technology Publ.

Xu, Z.H., and Chen, Z.H. (eds) 1996, Plant Biotechnology for Sustainable Development of Agriculture. Beijing: China Forestry Publ. House.

Yang, G.F., Zhou, P., Zhang, C.F., Zheng, X.Q. 2001. Progress on transgenic plant-derived vaccines. J. Agri. Biotech. 9:301-306.

Yu, J. et al. 2002. A draft sequence of the rice genome (*Oryza sativa* L. spp. *indica*). Science 296:79-92.

TRANSGENIC CROPS IN THE ARGENTINEAN AGRICULTURE

Alejandro Mentaberry
Institute for Genetic Engineering and Molecular Biology, CONICET, and School of Sciences, University of Buenos Aires, Buenos Aires, Argentina (e-mail: amenta@dna.uba.ar)

Keywords Transgenic crops, Argentina

1. REGULATORY FRAMES

In 2001, Argentina ranked second among the countries growing transgenic crops, with a share of about 22% on the total world area. Measured in total hectarage, the introduction of transgenic crops in 2000-2001 grew by 18%, thus maintaining a consistent high adoption rate for five consecutive years (James, 2001). Though complex socio-economical causes could be mentioned to explain this process, it is clear that the existence of an appropriate regulatory frame played a critical role in this achievement. The initial steps to establish such a frame date from 1991, with the creation of the National Advisory Committee on Agricultural Biotechnology (CONABIA). Since then, CONABIA has been acting as a body of consultation and technical support to the Secretary of Agriculture for the experimentation with genetically engineered organisms, establishing the rules and competences to develop field trials, and clearing a considerable number of new transgenic plant varieties for safe release into the local agro-ecosystem. CONABIA is composed by a permanent Secretariat and twenty members representing different technical expertices who have been elected by diverse organizations from the public and private sectors. Its membership acts on an *ad honorem* basis and decisions are taken by consensus.

Argentine regulation is based on the identification of possible risks posed by the biotechnological products themselves, rather than by the procedures by which they were obtained. Each field trial application is reviewed on a case-by-case basis taking into account the biological characteristics of the organism to be released and the features of the local agri-ecosystems. Initial releases are only approved after a careful risk analysis and further evaluation in successive field assays is conducted through an interactive process based on the data obtained in previous stages. Approval of commercial releases include the participation of other regulatory agencies, such those involved in plant protection and seed registration (National Institute for Plant Varieties) and animal and human health (National Service for Animal Health, Ministry of

I. K. Vasil (ed.), Plant Biotechnology 2002 and Beyond, 587-591.

Human Health) and usually involve periods of 3 to 4 years. Finally, marketing of transgenic crop varieties requires a positive report from the Advisory Committee on International Markets.

A total number of 495 field trial permits were granted under this regime during the period 1991-2001 (CONABIA, 2001; Fig. 1). Crops involved in these assays included mainly corn, soybean, cotton, sunflower, wheat, potato, alfalfa and others species in minor percentages. Major traits essayed were tolerance to herbicides, insect resistance and tolerance to virus and fungal diseases. Interestingly, about 10% of these releases corresponded to the introduction of quality traits. Though more of 75% of the applications were made by multinational seed companies, it is worth noticing that 14.5% and 8.3% correspond to national companies and institutions from the public sector, respectively. The participation of several small Argentinean companies, as well as that of public organizations (National Institute for Agricultural Research, National Universities, Research Institutes from CONICET, non-profit organizations) in this process suggests that, despite the current economical crisis, a local agri-biotechnological sector is developing in Argentina.

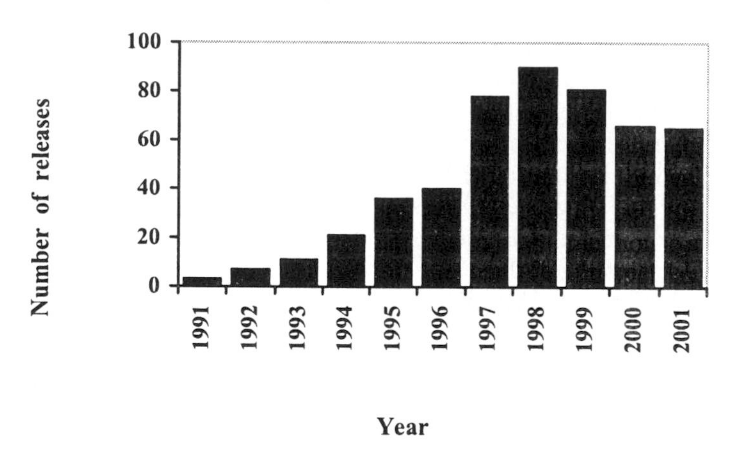

Figure 1. Field trials with transgenic crops in Argentina

2. TRANSGENIC CROPS IN ARGENTINA

Since the introduction of the first transgenic cultivar, the genetically modified varieties have been adopted at an unprecedented high rate by the Argentinean farmers (Fig. 2). It is clear from these data that the adoption of soybean transgenic cultivars constitute a unique case at worldwide scale. In 1996, the year of introduction of the first transgenic cultivar, total soybean area was about 6.0 million ha. This area was increased to 7.8 million ha in 1999; according to a report of the Secretary of Agriculture, 6.1 million of this surface was planted with transgenic varieties. In 2001, the total cultivated area reached

10.2 million ha, 97% of which corresponded to transgenic cultivars. This impressive increase in the cultivated area was made possible by the synergistic

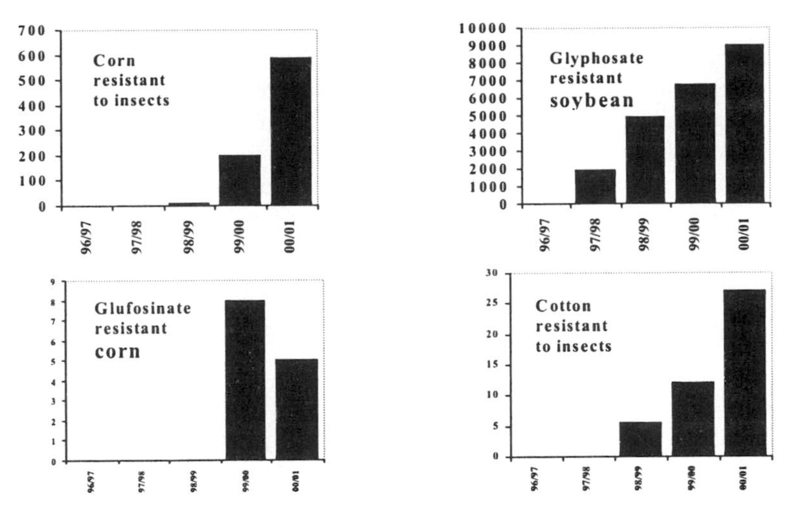

Figure 2. Adoption rates of transgenic crops in Argentina (1996-2001)

complementation of non-tillage farming and transgenic cultivars. The association of these two components is clearly reflected in the rapid evolution of the soybean non-till area, which evolved from 2.86 million ha in 1996 to 5.0 million ha in 2000. The simpler explanation for the rapid adoption of both technologies is the important economical benefits obtained by the growers, which must be attributed not only to lower investments in herbicide, but also to additional savings in terms of fuel and mechanization. Indeed, the application of non-till systems and herbicide resistant varieties and the breeding effort to introduce transgenic resistance into local cultivars adapted to different agricultural regions contributed decisively to the increase of cultivated area. Independent reports by both academic and private sector sources (Gunningham, 2000; Rossi, 2000) concluded that herbicide-resistant soybean resulted in lower production costs as compared with conventional soybean. These studies estimated the national economic benefits for the Argentinean growers between $214 and $356 million in 1999.

Adoption of transgenic corn and cotton varieties expressing resistance to lepidopteran pests followed a similar pattern but at a slower rate. Though in both cases significant economic advantages have been documented in preliminary studies, the lower adoption rate of transgenic varieties in these two crops must is explained by different factors. In the case of corn, the public debate and the delays in establishing a clear regulatory frame in most European countries (which are important exportation markets for Argentina) contributed to postpone the introduction of the new varieties. Despite this factor, it was calculated that 0,56 million ha (17%) from the 3.3 million ha cultivated with corn, were planted with transgenic varieties in 2001. Due to the initial nature of these introductions, information on the large-scale impact of new corn varieties is still scarce. A study by the National Institute for Agronomical Research made in conditions of high insect infestation reported yields 35% higher for transgenic corn in comparison with the non-transgenic controls. Concerning cotton, the area planted with transgenic cultivars comprised about 13% of the total cultivated area in 1999. A report by the National Institute for Agronomical Research made in five different sub-regions of the Chaco ecosystem, established average benefits for the growers of $65 per ha (Elena, 2001). Though the critical economic situation in Northeast Argentina has retarded the adoption of these relatively expensive cultivars, a prospective study made by the private sector estimated a future adoption rate of almost 40% for the period 2004-2005.

3. NON-ECONOMICAL BENEFITS OF TRANSGENIC CROPS

As mentioned before, a remarkable characteristic of soybean production in Argentina is the strong compatibility between herbicide resistant varieties and non-till systems. Two direct consequences of this scheme were the simplification of crop management and a more flexible use of herbicide. These advantages have been highly appraised by the growers and further promoted the adoption of these technologies. Another important consequence of this crop system is the conservation of soil moisture, structure and nutrients, which considerably contribute to control soil erosion and sustainability. Similarly, insect-resistant transgenic varieties offers another example of beneficial environmental effects. As previously shown in studies performed in the United States, preliminary research on transgenic corn and cotton impact in Argentina indicated that substantial decreases in the use of insecticides mitigate environmental contamination and contribute to decrease health risks among rural workers.

4. FUTURE TRENDS

Two different surveys organized by the private sector and the National Institute for Agronomical Research showed that Argentine public attitudes are generally supportive of scientific and technological advancement. Thus, though many of the common issues raised by the global debate on transgenic crops have been discussed in Argentina for almost two years, perception of both growers and consumers seems to be receptive to the introduction of transgenic varieties. Regardless of this relatively favorable context, further adoption of transgenic plant varieties will mainly depend on the attitudes prevailing in the traditional export markets of Argentina.

5. REFERENCES

CONABIA, www.sagpya.mecon.gov.ar

Elena, M.G. Proceedings of the 2001 Beltwide Cotton conference, 2001.

Gunnigham, C. IAE. Universidad Austral Argentina, 2001.

James. C. ISAAA briefs No 23, 2001.

Rossi, R. Matto Grosso Foundation, Cuiaba, Brazil, 2000.

THE STATUS OF PLANT BIOTECHNOLOGY IN SOUTH AFRICA

Jennifer A. Thomson
Department of Molecular and Cell Biology, University of Cape Town, South Africa (e-mail: jat@molbiol.uct.ac.za)

Keywords Genetically modified crops, South Africa

1. INTRODUCTION

In this paper I will cover both the development of genetically modified (GM) crops and the use of molecular genetics in plant breeding. The institutions involved include those of the Agricultural Research Council (ARC), the Council for Scientific and Industrial Research (CSIR), the SA Sugar Experiment Station (SASEX) and a number of universities. Due to space constraints the research will be covered rather superficially but I have included World Wide Web site addresses where possible to allow readers to link at their leisure to more detailed information and descriptions of projects.

2. AGRICULTURAL RESEARCH COUNCIL – INFRUITEC (http://www.arc.agric.za)

This institute is based in the Western Cape province near the home of the stone- and pome-fruit growing industry. It is involved in the development of new fruit varieties, and research to resolve cultivar identification problems. This division is the section most involved in applying biotechnology to assist the breeding program of the Institute. Their objectives include the development of disease resistant fruit and improved shelf life, identification, control and management strategies for fungal and bacterial diseases of deciduous fruit, assisting plant breeders by the identification of molecular markers for disease resistance and other commercially important traits, cloning genetic material for beneficial traits, the construction of vectors for fruit crop transformation and *in vitro* mass propagation of rare material of fruit crops. Their main crops are strawberry, apple, pear and apricot.

I. K. Vasil (ed.), Plant Biotechnology 2002 and Beyond, 593-597.

3. AGRICULTURAL RESEARCH COUNCIL –VEGETABLE AND ORNAMENTAL PLANT INSTITUTE (VOPI) (http://www.arc.agric.za)

This division does research, technology development and technology transfer in support of this agricultural sector from commercial organizations to developing farmers. They support the breeding programs of the major crops, tomato, potato, onion, sweet potato and ornamental flowers. Collaboration with neighbouring countries is developing under the auspices of the UNESCO/BAC BETCEN. The Biotechnology Action Council of UNESCO established a Biotechnology Education and Training Centre for Africa at VOPI in 1995. Its focus is to train African scientists in plant tissue culture and molecular marker technology by the presentation of short courses or extended fellowships.

Important recent achievements include the development and successful field trials of potatoes resistant to potato leaf roll virus. They have recently embarked on a project to develop transgenic sweet potatoes resistant to *Sweet potato feathery mottled virus.*

4. COUNCIL FOR SCIENTIFIC AND INDUSTRIAL RESEARCH – BIOTECHNOLOGY, FOOD SCIENCE AND TECHNOLOGY DIVISION (http://www.csir.co.za)

The focus of research activities primarily involve cereal crops such as sorghum, barley, maize and millet. Specific projects and achievements include:

- Genetic engineering of maize for fungus resistance - A collaborative project with ARC-VOPI is to transform maize to combat cob rot caused by i.a. *Stenocarpella maydis*, one of the most serious fungal pathogens of maize. Genes being used include the polygalacturonic acid inhibitory protein (PGIP).
- Genetic engineering of pearl millet for downy mildew resistance - European Union funding was awarded for this project on *Pennisetum glaucum* which is a valuable staple in the semi-arid areas of Africa. It can be devastated by the fungus *Sclerospora graminicola*.
- Increased methionine and lysine transgenic sorghum.

5. SOUTH AFRICAN SUGAR EXPERIMENT STATION (SASEX) (http://www.sasa.org.za/sasex/about/biotechnology.htm)

SASEX is a cooperative institution that exists to develop and optimise the productivity of sugarcane varieties to provide maximum economic returns, develop environmentally sound agricultural practices, transfer technology to the growing community to promote prosperity and sustain the industry. SASEX is closely involved with the Institute of Plant Biotechnology at the University of Stellenbosch (see below). SASEX undertakes its own research in three main areas:

- Molecular markers: the primary aim of this program is to identify DNA markers which can accurately identify varieties or unambiguously diagnose the presence of important pathogens, and both metabolic and DNA markers which depict important traits in sugarcane.
- Genetic engineering of sugarcane to introduce desirable characters such as resistance to the sugarcane borer, *Eldana saccharina*, sugarcane mosaic virus and *Ustilago scitaminea.*
- Development of techniques and resources to provide specific genetic material and methodologies for use in the above two programs.
- They have developed transgenic sugarcane Nco310 resistant to the herbicide Bastera and conducted successful field trials.

6. FORESTRY AND AGRICULTURAL BIOTECHNOLOGY INSTITUTE (FABI), UNIVERSITY OF PRETORIA (http://www.up.ac.za/academic/fabi/index.html)

Projects include biotic stress tolerance in transgenic citrus and bananas using i.a. cysteine proteinase inhibitors.

7. INSTITUTE FOR WINE BIOTECHNOLOGY, UNIVERSITY OF STELLENBOSCH (http://www.sun.ac.za/wine_biotechnology)

This Institute's research programme focuses on the molecular genetic improvement of wine yeast strains and grapevine cultivars, to increase fermentation and processing efficiency, and to enhance the wholesomeness and sensory quality of wine. Grapevine specific projects include efficient transformation and regeneration of grapevine cultivars, *Vitis* genome sequencing, the construction of genomic and cDNA libraries of grapevine cultivars, improved disease resistance (fungal and viral diseases) and the identification of grape cultivars using genetic marker technology.

Wine yeast strain development includes improved strain management and security, improved fermentation performance, improved efficiency of processing, improved sensorial quality and improved control of microbial spoilage.

8. INSTITUTE FOR PLANT BIOTECHNOLOGY, UNIVERSITY OF STELLENBOSCH (http://www.sun.ac.za/ipb)

This Institute specialises in the characterisation of the primary carbon metabolism of plants, with the ultimate goal of manipulating the relevant metabolic pathways to improve yield and quality, and to introduce novel high price products into plants. This includes genetic manipulation of carbon flow in sugarcane and grapes, characterization of carbon flux in non-photosynthetic plant systems with special reference to sugarcane, fruits and seeds, and isolation and characterization of plant promoters. Much of its research is done in collaboration with SASEX.

9. DEPARTMENT OF GENETICS, UNIVERSITY OF STELLENBOSCH (http://www.sun.ac.za/agric/genet)

Working in close collaboration with the Institute of Wine Biotechnology, research is being carried out on transgenic grapevies resistant to viruses through gene silencing and ribosome inactivating protein. Studies are underway on differentially expressed genes during viral infection using suppression subtractive hybridisation and on grapevine chloroplast transformation.

10. SOUTH AFRICAN NATIONAL BIOINFORMATICS CENTRE, UNIVERSITY OF THE WESTERN CAPE (http://www.sanbi.uwc.ac.za)

This is an internationally renowned Centre and a national asset for bioinformatics in South Africa.

11. DEPARTMENT OF MOLECULAR AND CELL BIOLOGY, UNIVERSITY OF CAPE TOWN (http://www.uct.ac.za/microbiology/mcbdept.htm) (http://www.uct.ac.za/depts/plantstress)

Using a model monocot grass, *Digitaria sanguinalis*, resistance to the African endemic *Maize streak virus* has been achieved using mutants of the replication associated protein (Rep) gene in a 'dominant negative' approach. This approach was based on the hypothesis that as Rep operates as a multimer to initiate viral replication, transgenic plants producing mutant, truncated Reps would render the infecting virus's Reps dysfunctional. Transgenic HiII maize has been produced and testing for resistance is dependent on permission by the SA biosafety committee.

Research is being carried out to develop plants tolerant to drought and other abiotic stresses. The source of the genes is *Xerophyta viscosa*, a monocot 'resurrection plant'. A number of genes have been identified which are induced under conditions of dehydration, cold, heat, high light intensity and salinity. These include genes coding for a membrane protein, a transcription factor, an antioxidant and osmoprotectants. A number have been transferred to *D. sanguinalis*, *Arabidopsis thaliana* and *Nicotiana xanthi*. Initial results indicate significant protection against a variety of abiotic stresses.

Work is underway to produce vaccines against human papilloma virus (HPV), a significant cause of cervical cancer in African women, and the African subtype C of HIV. Genes, including the L1 of HPV have been expressed in tobacco using *Tobacco mosaic virus* as a vector. Virus like particles have been observed in the plants.

12. CONCLUSION

Despite a great deal of research being undertaken in South Africa in plant biotechnology, no locally developed GM crop has yet been commercialised. In May 2002 the Cabinet approved a South African Biotechnology Strategy. This aims to establish two or three Biotechnology Regional Centres within the next few years to develop biotechnology products and enable their commercialisation. Despite the lack of significant venture capital and local biotechnology industries it is hoped that this enterprise will provide the boost required for South Africa to become a player in global agricultural biotechnology.

STATUS OF PLANT BIOTECHNOLOGY IN INDIA

Usha Barwale Zehr
MAHYCO, PO Box 76, Jalna 431023, India (e-mail: uzehr@lsrc.mahyco.com)

1. INTRODUCTION

India today is 1.01 billion people strong, of which 70% of population is rural and derives income from agricultural activities. The food grain production has increased from 50 million tones in 1950s to more than 200 million tones in the year 2000. This is indeed a monumental jump due to continuous utilization of new technologies. In recent years, there is a plateau in the yield and current technologies may not be sufficient to meet the food demand for the coming years. In addition, India also has a large portion of its population undernourished, especially children between the ages of 1 to 4 years.

A recent study "Food Insecurity Atlas of India" (a joint study by M. S. Swaminathan Foundation and World Food Program) indicates that some of the most productive states in India are least sustainable environmentally in the future. Thus, agricultural improvements have to come from technologies which enhance productivity of all resources deployed while conserving our natural resources. Crop productivity in India remains much below world averages and deployment of new technologies holds potential to bridge this productivity gap. The yield gap in rice (as an example) can be due to any combination of the following factors:

1. Pests and diseases: Weeds, brown plant hopper, stem borers, gall midge, tungro virus, bacterial blight, rice blast and sheath blight.

2. Abiotic stresses: Drought, submergence, cold, Fe toxicity, salinity and Al toxicity. For many of these traits conventional resistance sources are not available. Biotechnology provides a unique opportunity to address some of these traits.

Biotechnology activities in India can be broadly viewed in two categories:

1. Transgenic research
2. Other recombinant DNA technologies

I. K. Vasil (ed.), Plant Biotechnology 2002 and Beyond, 599-603.

2. TRANSGENIC RESEARCH

Transgenic research in many crops is being pursued in both public and private sector organizations. Research programs address many of the traits listed for rice above and are at different development stages. The following traits are in the regulatory process for large scale commercial release including insect resistance, virus resistance, male sterility system, nutritional quality, herbicide tolerance and disease resistance. The crops in which these traits are being introduced are cotton, mustard, eggplant, maize, rice, pigeon pea, cabbage, cauliflower and soybean. Second generation transgenics to address broad resistance as well as biosafety issues relating to marker genes are being developed. One such example is the co-transformation with *Agrobacterium tumefaciens* carrying two binary plasmids offers an opportunity for segregating the marker gene out from the desired genotype. This has been implemented in crops such as rice, eggplant, tomato and pigeon pea to name a few.

Crops which have great relevance for Indian agriculture are being actively pursued such as pigeon pea, which is important field crop both from the nutritional value for the Indian population, as well as its good fit in the semi-arid environment. Pod borer is a major pest of pigeon pea causing losses up to 80%. Transgenic pigeon pea carrying cry1Ac gene is being developed to address this pest (*Helicoverpa armigera*). A single technology which controls insects can have multiple benefits for the farmer. The input costs are lower, labor needs are reduced and improved yield potential due to better crop management. There are also positive implications for the environment and health of the farmer.

3. REGULATORY - BIOSAFETY SYSTEMS IN INDIA

India has a well-developed regulatory system in place for evaluation and release of genetically modified crops. The genetically modified crops are governed by Environment (Protection) Act (EPA) 1986, which came into affect from 23rd May 1986. This was followed by formulation of rules and regulation for the manufacture use/import, export and storage of hazardous microorganisms, genetically engineered organisms or cells to be notified, on 5th December, 1989. This constituted the first set of guidelines for working with genetic engineering or GMOs. The guidelines called for constitution of the following:

(i) Recombinant DNA Advisory Committee (RDAC)
(ii) Institutional Biosafety Committee (IBSC)
(iii) Review Committee on Genetic Manipulation (RCGM)
(iv) Genetic Engineering Approval Committee (GEAC)

(v) State Biotechnology Co-ordination Committee (SBCC)
(vi) District Level Committee (DLC)

The Department of Biotechnology (DBT) formulated a set of guidelines "Recombinant DNA Safety Guidelines" in 1990 and subsequently revised the same in 1994 and 1998. The guidelines are exhaustive and deal mainly with transgenic plants, seeds and their evaluation for toxicity and allergenicity. Containment levels for different categories of genetic engineering experiments are also described. Greenhouse design, field evaluation of transgenic crops, import and shipment of genetically modified plants for research use only, are also dealt with in these guidelines. A Monitoring-Cum-Evaluation Committee (MEC) also is proposed for constant monitoring and on-site evaluation by independent experts. An illustration of the procedure needed for import of transgenic Maize seeds.

1. An institution makes an application to its IBSC to import transgenic Maize seeds. Details on the genetic material have to be submitted for review.
2. IBSC examines the proposal if approved; the proposal is forwarded to RCGM via DBT.
3. RCGM reviews the proposal and if approved a letter is sent to institution giving approval for import of seed.
4. Applicant than makes an application to National Bureau of Plant Genetic Resources (NBPGR) for obtaining an import permit.
5. Once the import permit is granted, the same is sent to the party who will be sending the material.
6. Thc seed is received at NBPGR. The seeds on receipt are tested by NBPGR as stipulated and released to the importer on satisfactory results.
    a. The material conforms to the description provided in the permit
    b. The material is free from plant pathogens and pests
    c.A sample of the material is deposited with the NBPGR
7. The applicant receives seeds and notifies its IBSC of the same.
8. A proposal is made to IBSC of follow up action with this material.

The field testing follows the sequence as listed here:

1. Contained field trial: A small area surrounded by trap rows. Usually in one location
2. After review of data; limited field trial permission is granted. Usually more than one location.
3. Followed by Multi-location field trials to determine agronomic performance
4. Farmers field trials.
5. Commercialization

Any of the above steps can be repeated if data collected is insufficient or unsatisfactory. From the initial field trials, biosafety data is also collected such as pollen flow, allergenicity, equivalence, nutritional quality, feed and food safety to list a few. Following these guidelines, Mahyco introduced insect resistant Cotton in 2002. The project started in 1994, with an application for import of 100 gms. of seed for back crossing into Indian elite germplasm. Table 1 gives an overview of the activities:

Table 1: Development of Bt cotton in India

YEAR	ACTIVITY
1994	Formation of IBSC & application for seed import
1995	Permit to import seed (100 gm)
1996	Imported seed, Greenhouse trial
1996	Limited field trial – 1 Location
1996...	Back crossing (ongoing) in green house
1997-1998	Limited field trials-5 Locations
1998	Ruminant (goat) and allergenicity studies
1998-1999	Multi centric replicated trials-15+25 Locations
1999-2000	Multi centric replicated trials-11 Locations
2000-2001	Large scale field trials
2001-2002	Large scale field trials
2002-2003	Commercialization

Since, 1997 Mahyco also conducted the following studies: goat study, aggressiveness, germination, weediness, pollen flow, oil and food/feed properties, soil microflora effect on beneficial and non-beneficial insects, protein expression, fish, poultry, cow and buffalo studies. As can be seen, the regulatory system in India is extensive and thorough and with first successful commercialization of cotton, more products will be available to Indian farmers in coming years.

4. OTHER RECOMBINANT DNA TECHNOLOGIES

Given the diversity of genetic resources and soil microflora that is available in India, many labs are actively pursuing gene discovery programs. Also molecular breeding (marker aided selection or MAS) and genomics strategies are being deployed for genetic improvements. Drought and increased soil salinity are two major factors that limit plant growth and productivity. Using quantitative trait loci (QTL) approach these complex traits are being addressed. Molecular maps are also being generated for crops that are predominantly grown in India such as Chickpea and medicinal plant species. Molecular breeding, either for specific trait or for recovery of desired genotype is being applied but has greater potential for use for making breeding processes more effective.

In summary, transgenics and MAS approaches are being used in Indian agriculture in a complimentary manner. As more high through put DNA marker technologies are available, the costs for the use of some of these technologies will be reduced. With reduced costs, the technologies will be used more and more. Many Indian States have taken up the task of creating "Biotech Parks" to encourage activity in the area of biotechnology. This is creating new infrastructure through out the country and many companies are able to establish collaborations with IT companies for more aggressively exploring bioinformatics and genomics. Greater emphasis is also given to genetic improvements using new tools for India crops by the Government. Given these new developments and the past history of India in terms of extensive agricultural research network, Indian agriculture is well placed to make use of innovations of biotechnology in the coming years.

PLANT BIOTECHNOLOGY IN MEXICO: NEEDS AND CHALLENGES

Rafael F. Rivera-Bustamante
Departamento de Ingeniería Genética, Centro de Investigación y de Estudios Avanzados (Cinvestav), Unidad Irapuato. Irapuato, Gto. México 36500 (email: rrivera@ira.cinvestav.mx)

Keywords Public biotechnology, Biodiversity, Maize controversy

1. INTRODUCTION

According to James (2001), in 2001 there was an estimated and unprecedented global area of 52.6 million hectares for transgenic crops that represented an increase of 19% over the previous year. This increment continued with the tendency shown by transgenic crops during the initial five-year period (1996-2000) where the global area increased from 1.7 million hectares in 1996 to 44.2 million hectares in 2000. This unprecedented adoption rate for a new technology "reflects the significant multiple benefits realized by large and small farmers in industrial and developing countries that have grown transgenic crops commercially" (James, 2001). Beyond these numbers, it is also clear that plant biotechnology faces important challenges and controversial issues as well. Whereas the benefits offered by the technology seemed to be more or less universal (the actual benefit may vary, but the technology is generally applicable to most areas), the challenges and controversies are more specific depending on economic, geographic and/or biological parameters. Mexico represents a very good example of this dilemma. In this presentation, I will address some of the most important problems in Mexican agriculture and how plant biotechnology could help to solve them. Next, I will discuss the important role that public institutions should play to facilitate the adoption and spread of the technology. Finally, I will present the particular situation faced by Mexico in terms of being a Megadiverse country and the Center of Origin and Diversity of some of the most important crops and, at the same time, an important commercial partner of two of the countries with the largest list of released biotechnological products. The confusion around these issues is being reflected in recent modifications to Mexican laws that have instigated uncertainty feelings among Mexican scientists.

I. K. Vasil (ed.), Plant Biotechnology 2002 and Beyond, 605-610.

2. MEXICAN AGRICULTURE AND BIOTECHNOLOGY

A major issue in discussing "Mexican agriculture" is the fact that there is not just ONE. There are many types of agricultures, and farmers, and they both vary in terms of geography (from desertic, arid areas to jungles), economy (from industrialized, large-scale farmers to one-acre subsistence farmers), biology (with more than 500 cultivated species and a large but unknown number of collected and commercialized wild materials) and social organization (communal property to large enterprises). This variability will be an important issue on how the technology is adopted by the farmers.

Some important problems in Mexican agriculture could be addressed by currently available technology. A major issue on the high production costs for many crops in Mexico is the high dependency on agrochemicals to protect the crops from weeds, pests and diseases. Therefore, insect-, herbicide-, and virus-resistance technologies could benefit major crops and seed companies are already producing the required varieties for the market. Cotton is an example on the fast adoption of biotechnological products by Mexican farmers. Although the regulation for commercial releases of transgenic products is still not published, in 2000 more than 25,000 "pre-commercial" hectares were cultivated with Bt cotton, a third of the total production area in Mexico. If the adoption trend continues at the same rate, in less than 10 years almost all cotton grown in Mexico will be transgenic. Similarly, major production constraints in subsistence agriculture are pests and diseases. In this case, however, the challenge is to find a strategy to deliver the technology to the farmers since in most cases the farmers do not use certified seed.

Among the important challenges that remain to be addressed are fungal diseases such the ones caused by *Phytophthora infestans* in potato and *Sclerotium cepivorum* in garlic. Citrus tristeza virus and potato virus Y^{NTN} represent major potential threats for Mexico in addition to the persistent, still important geminivirus problem. Several bacterial diseases are also becoming important, specially in horticultural crops (e.g., *Xanthomonas campestris*). More recently several phytoplasm diseases became important in coconut palm as well as in potatoes. In addition to the typical biotic stress problems, other areas could benefit with current technology. Delayed ripening could be an important contribution for the tropical fruit market, especially with papaya, mangoes, avocados, and bananas.

Finally, probably the most important issue in the next decade will be water stress. If the current water usage continues at the same rate, it is predicted that most Central and Northern Mexico will be considered as arid and semi-arid zone. Strategies that will allow the efficient use of poor soils will be required. An important example of such strategies is the aluminum-tolerant maize plants currently being developed at Cinvestav.

3. MEXICAN PUBLIC BIOTECHNOLOGY

In a country like Mexico with a diverse agriculture, it is obvious that biotech companies will focus primarily on the needs of large-scale farmers, or to highly cultivated varieties. However, it is also clear that the farmers that could benefit the most from this technology are subsistence farmers. This issue is a favorite among biotechnology detractors who insist that Biotechnology will inevitably increase the gap between the rich and the poor. However, an alternative that is often overlooked is the role of public institutions. A strong public biotechnological system could address some of the most important regional/national problems that are below the economic threshold defined by commercial companies as the minimum-size market needed to assure recovery of the investment. Examples such as open-pollinated maize varieties, fresh market potatoes and beans, etc. will probably never be of interest for the companies because of the small market or the difficulties in enforcing intellectual property issues. In these cases, public institutes will have an important niche to develop highly needed biotech products. However, since most of the public institutions usually do not have the capability or expertise to take a product from development to commercialization (including field testing, propagation, registration, regulatory issues, distribution, etc.), complex alliances between several institutions are required. In addition, a strong, funded leadership from the Ministry of Agriculture is also required but difficult to get in place.

In Mexico two public institutions have shown strong leadership on Plant Biotechnology: Center for Research and Advanced Studies (Cinvestav)-Irapuato and UNAM (Biotechnology Institute and other campuses). Other important research groups are located at Autonomous University of Aguascalientes, Center for Scientific Research in Yucatan (CICY), Posgraduate College (Colegio de Posgraduados) and the National Institute for Research in Agriculture, Forestry and Livestock (INIFAP), which can become the "natural" counterpart to provide excellent expertise in field testing, agronomic research and validation of products developed by, or in collaboration with the other Centers and Universities in the country.

An ideal way to promote biotechnology in Mexico will be to organize a "consortium" of biotech companies and Mexican institutions. The industry should be willing to transfer and donate technology to Mexico (through several centers) that could be applied to species and varieties oriented exclusively to subsistence farmers. The national centers (e.g., Cinvestav, UNAM, and other universities) will acquire the compromise of generating a sufficiently large number of transgenic lines that will be selected and evaluated by INIFAP. This strategy follows basically the phases of the Monsanto–Cinvestav potato project. However, an important and still missing link will be to find the proper agency in the Ministry of Agriculture that will propagate and generate a strategy for the distribution of the selected transgenic lines. In the case of the Monsanto-Cinvestav project that step has been difficult to organize since the varieties Rosita and Norteña selected for the small farmers are not of interest to potato seed producers due to the lack of market value, the federal seed producer agency (PRONASE) was closed a few years ago, and neither INIFAP or Cinvestav have the ability to carried out that task.

4. CURRENT STATUS OF MEXICAN REGULATIONS

Since the late 80's when the first application for field-testing transgenic tomatoes was submitted, Mexico has been working on Biosafety issues becoming a pioneer in Latin America in the field. Most of the initial applications were submitted through the Ministry of Agriculture's Plant Health Directorate. From 1988 to 2000, 150 applications for field tests were approved. These controlled field releases included 19 species and 13 phenotypes and were submitted by 24 different applicants. Currently, there are only two products that have been deregulated for production in Mexico. Both are tomato varieties with the Flvr-Savr technology. The products that have been approved by the Health Ministry for human consumption, in addition to the modified-ripening tomatoes, are: Herbicide-resistant (HR) soybean, HR-cotton, Bt-cotton, Bt- and virus-resistant (PVY and PLRV) potatoes, HR-canola, and modified oil-content canola.

In 2000, a new regulatory agency was created with the participation of several ministries (CIBIOGEM: Inter-ministry Commission for Biosafety and Genetically Modified Organisms). The ministries involved are Human Health, Agriculture, Environment, Treasury, Commerce and the National Council for Science and Technology. In theory, this agency was an extraordinary way to share expertise and knowledge about GMOs, and to unify criteria. However, it has been shown that to conciliate opinions is difficult without a strong leadership and a collaborative spirit. In addition, the lack of an independent budget decreases CIBIOGEM's functionality. Hopefully, the recent resign of the scientific advisory committee will need the Federal Government to make decisions and show a stronger commitment.

The important political changes that took place in Mexico in the year 2000 have affected many areas of Mexican life. Science and technology, including Biosafety regulations, are not the exception. A recently approved law submitted through the Environment Commission of the Congress (House of Representatives) has raised severe criticism from the academic community,

This law, poorly written and vague in many areas, opens the possibility of imprisoning scientists, or any person, working with any GMO. To make it more confusing, there are currently at least four new initiatives, from different political parties that attempt to deal with Biosafety issues. Each initiative has been channeled through different Congress Commissions (Agriculture, Health, Environment and Rural Development) and each Commission and the proponent party are lobbying for the required support. It will require a lot of work to compare and conciliate all the initiatives to end up with something that makes sense. It has to be acknowledged, however, that this time, members of the some Commissions are consulting the general public as well as the academic and scientific communities.

5. MEXICO'S DILEMMA ON BIOSAFETY

Mexico is currently in a controversial, unique position in terms of Plant Biotechnology due to economic, geographic, political and scientific reasons. On one hand, Mexico is one of the major commercial partners of the US, the major developer of biotechnological products in the world. On the other hand, Mexico is center of origin or diversity of some of the most important crop species (maize, tomato, cucurbits, etc.). It is obvious that the parameters for product development, field testing, risk assessment and release approval cannot be simply extrapolated.

A central issue in this controversy has been maize. Biotechnology detractors have been using many elements to avoid the release of GM maize in Mexico. The Biosafety committee responded by declaring a *de-facto* moratorium on field tests of GM maize in Mexico. The recent suggestion that transgenes could be detected in local land races of maize in Oaxaca has enhanced the controversy. Although the currently available commercial maize phenotypes (insect and herbicide resistance) probably do no represent possible threats, reports on experimental systems to express pharmaceutical products using maize as a biofactory has encourage several groups to demand a complete ban of GM maize in Mexico. However, a complete ban will imply that many subsistence farmers that could benefit from the technology will not have access to it. In addition, this decision could be used as an important precedent to require the same treatment for other native species. Ironically, Mexico a megadiverse country, with a strong need to improve and modernize its agriculture, could end up with a very restricted use of the technology limited to only species originated elsewhere. Such species, in contrast, are probably not an important part of our culture, therefore their impact will be diminished.

Other people is looking at this situation as an excellent opportunity to study many aspects related to gene flow, environmental impact and social response to GM maize. The only clear thing is that more research and funds are required to obtained useful information that will facilitate the making of decisions. A complete ban will not allow to obtain the information needed, thus no good decision could be taken.

6. REFERENCES

James, C. 2001. Global review of commercialized transgenic crops: 2001. ISAAA Briefs No. 24: Preview. ISAAA Ithaca, NY.

For additional information see Mexican Ministry of Agriculture web Page:
http:// www.sagarpa.gob.mx, and Mexican Ministry of Health web page:
http:// www.ssa.gob.mx

USER-FRIENDLY PROBLEM- FINDING AND SOLVING APPROACHES FOR INTERNATIONAL AGRICULTURAL BIOTECHNOLOGY APPLICATIONS

Kazuo N. Watanabe
Gene Research Center and Institute of Biological Sciences, University of Tsukuba, Tsukuba, Ibaraki, 305-8572, Japan (e-mail: nabechan@gene.tsukuba.ac.jp)

Keywords Sustainable development, technology transfer, risk-benefit analyses

1. INTRODUCTION

Agricultural biotechnology is the one of alternative tools for alleviating the problems in food security and poverty, by improving the food production in quantity and quality and increasing cash-making opportunity in the developing world. Sustainable agriculture is subjected to the appropriate combinatorial use of traditional knowledge and modern sciences, technology and natural resources that include biological resources, soil, water, energy sources and atmosphere (Watanabe and Pehu, 1997). The choices and decisions of such approaches depend on various ways of scientific and individual thinking, creeds, religions, cultures, traditions, ethnic groups and community besides the science and technology associated with the matters. Socio-economic factors are also of importance as well as the policy encumbered with the rural development. Thus, these elements should be carefully considered and case studies should be implemented before disseminating a large-scale agricultural biotechnology application.

2. END-USER BASED PROBLEM FINDINGS AND MULTIDISCIPLINARY APPROACHES

Successful adoption of an agricultural biotechnology in an agriculture-farm community would be totally up to the careful analyses and feedbacks from potential end-users during small scale testing of a technology at a rural community rather than just with a simple problem-solving approach by an immediate dispatch of a technology. The most important aspect in the adoption of agricultural biotechnology in developing countries would be multilateral communications among scientists, policy-makers, regulatory agencies, donors, commercial sector, aid agencies like philanthropic organizations, mediators such as international organizations, and the

I. K. Vasil (ed.), Plant Biotechnology 2002 and Beyond, 611-613.

end-users at a rural community. A strong single scientific discipline and technology and product thereof would be of great help. However, it is also cardinal that scientists should talk each other and inter- and multidisciplinary approaches should be made alleviating the confronting pitfalls and be leading to the peaceful development.

3. PERSPECTIVES, CHALLENGES AND EXPECTATIONS

Very little fund is allocated to public research in the agricultural biotechnology for international development, particularly for the assistance of lesser developed countries, e.g., the fund for the international agricultural research managed by Consultative Group on International Agricultural Research (CGIAR) that consists of sixteen centers, is yearly around US $ 350 million, in contrast, the research expenses spent by the international agricultural private sector would be as much as US$ 2 billion for limited number of specific commodity crop species (UNDP, 2001a).

The Biotechnology development shall have a synergy with the use of plant genetic resources (Watanabe and Pehu, 1997). There are around 80,000 plant species which can be edible in the earth, and only three hundred plant species are actively used in production, and only less than thirty crop species have been heavily invested in plant breeding and commercial production (Raman and Watanabe, 2000). On the other hand, substantial achievement could be made on a small-scale biotechnology applications to landraces of locally important crop species with far smaller investment than what have been made to industrial crop cultivars. Although the social status of these under-utilized crops are regarded somewhat low in terms of modern commercial markets, however, their contributions as crops for food, health and cash-making opportunity, should not be under-estimated to the subsistent farmers and to the production at the marginal lands and/or at a harsh climatic condition (Watanabe and Iwanaga, 1999).

The biotechnologies have various innovative components that shall provide more opportunity for alleviating pitfalls in food security and poverty, however, due to the new history, careful risk assessment of the products derived from the technologies must be conducted, and relevance should be established on risk managements and risk communications (UNDP, 2001b). However, the most important matter would be the public discussion and multilateral communications on the risk-benefit analyses in order to put the technologies and products forward for solving the problems by sharing knowledge, experiences and resources.

4. REFERENCES

Raman K. V., and K. N. Watanabe. 2000. Contribution of Plant Genetic Resources for Plant Breeding. The Proceedings of the Twelfth Toyota Conference: Challenge of Plant and

Agricultural Sciences to the Crisis of Biosphere on the Earth in the 21st Century. Watanabe K. N., and A. Komamine (Eds.). Landes Bioscience, Austin TX, USA. Pp. 159-170.

UNDP 2001a. Chapter 5. Global initiatives to create technologies for human development. Human Development Report 2001: Making New Technologies Work for Human Development. Oxford University Press, Oxford, Pp. 95-117.

UNDP 2001b. Chapter 3. Managing the risks of technological change. Human Development Report 2001: Making New Technologies Work for Human Development. Oxford University Press, Oxford, Pp. 65-78.

Watanabe K. N., and M. Iwanaga. 1999. Plant genetic resources and Its global contribution. Plant Biotechnology 16(1):7-13.

Watanabe K.N., and E. Pehu. 1997. (eds.). Plant Biotechnology and Plant Genetic Resources for Sustainability and Productivity. R. G. Landes Co., Georgetown, Texas, USA, 247p.

AUTHOR INDEX